Annual Conference and Expo of the Association of Avian Veterinarians (AAV 2023)

AA005450

Held at ExoticsCon 2023

Boston, Massachusetts, USA
20 - 24 August 2023

ISBN: 978-1-7138-9471-1

Printed from e-media with permission by:

Curran Associates, Inc.
57 Morehouse Lane
Red Hook, NY 12571

Some format issues inherent in the e-media version may also appear in this print version.

Printed with permission by Curran Associates, Inc. (2025)

For permission requests, please contact Association of Avian Veterinarians
at the address below.

Association of Avian Veterinarians
PO Box 9
Teaneck, NJ 07666
USA

Phone: 720-458-4111
Fax: 720-398-3496

office@aav.org

Additional copies of this publication are available from:

Curran Associates, Inc.
57 Morehouse Lane
Red Hook, NY 12571 USA
Phone: 845-758-0400
Fax: 845-758-2633
Email: curran@proceedings.com
Web: www.proceedings.com

ASSOCIATION OF AVIAN VETERINARIANS

AAV

2023 PROCEEDINGS

AAV.ORG

The 2023 AAV Proceedings contains avian, hybrid and technician manuscripts originally presented at the 2023 ExoticsCon Conference in Boston, Massachusetts, USA, as well as a bonus chapter of avian sessions presented at AAV Avian Medicine Virtual 2023.

EXOTICSCON

Building Exotics Excellence. One City. One Conference.

THE WESTIN BOSTON SEAPORT DISTRICT
SUNDAY, AUGUST 20 - THURSDAY, AUGUST 24

ASSOCIATION OF AVIAN VETERINARIANS

AAV AVIAN MEDICINE VIRTUAL

Introduction

We invite you to enjoy the 2023 AAV Proceedings from ExoticsCon. Within the proceedings, you will find a wealth of information related to the variety of lectures presented at this year's conference.

AAV recognizes and thanks the following committees and staff for their time and hard work in assuring a quality program and publication:

AAV Executive and Conference Committees

Anna Osofsky, DVM, DABVP (Avian Practice), *President, Conference Committee Member*
Stephanie Lamb, DVM, DABVP (Avian Practice), *President-Elect, Conference Committee Member*
Byron De La Navarre, DVM, *Immediate Past President, Conference Committee Chair*
Salina Locke, DVM, *Education Committee Co-Chair*
Julia Shakeri, DVM, DABVP (Avian Practice), *Education Committee Co-Chair*
Susan Tyson, MS, VMD, *AAV Hands-On Workshop Coordinator*
Christina Knight, DVM, *AAV Hands-On Workshop Coordinator*
Len Donato, VMD, DABVP (Avian Practice), *Treasurer*
Jen Riley, DVM, DABVP (Avian Practice), *Advisory Council Representative to the AAV Executive Board*
Robert Groskin, DVM, *Executive Director*

Veterinary Review and Editing Committee

Natalie Antinoff
Gunther Antonissen
Bri Bryan
Sue Chen
John Chitty
Bob Dahlhausen
Byron de la Navarre
Bob Doneley
M. Scott Echols
Jesse Fallon
Jennifer Graham
Cheryl Greenacre
David Sanchez-Migallon Guzman
Michelle Hawkins
Stephanie Lamb
Isabelle Langlois
Salina Locke
Elizabeth Mackey
An Martel
Brynn McCleery
Christina Miller
Mark Mitchell
Deborah Monks
Bianca Murphy
Anna Osofsky
Kristina Palmer
Olivia Petritz
Christal Pollock
Patrick Redig
Jaime Samour
Julia Shakeri
Marcy Souza
Brian Speer
Tom Tully
Paul Welch
Kenneth Welle
Shangzhe Xie

Technical Editor
Charlotte Kirk Baer, MS

Proceedings Layout and Design — BrightLizard, Inc.
Elizabeth Smith

Disclaimer

The material appearing in this publication comes exclusively from the authors and contributors identified in each manuscript. The techniques and procedures discussed reflect the personal knowledge and experience of the authors and contributors, and demonstrate their views of the methods that may be used for these medical procedures. The procedures demonstrated do not incorporate all known techniques, are not exclusive, and other techniques and technology may also be available. Any questions or requests for additional information concerning any of the manuscripts should be addressed directly to the authors.

AAV does not research, review, or otherwise verify any of the information contained in this publication. Opinions expressed in this publication are those of the authors and contributors and not necessarily those of AAV. AAV is not responsible for errors or for opinions expressed in this publication.

AAV expressly disclaims any warranties or guarantees, expressed or implied, and shall not be liable for damages of any kind in connection with the material, information, techniques, or procedures set forth in this publication.

Publication Terms

Conference proceedings manuscripts are accepted as full manuscripts or as scientific abstracts. Scientific abstracts are accepted if the author(s) express(es) intent to submit their work for consideration for future publication in a peer-reviewed scientific journal. Several distinguished veterinary journals, including the *Journal of the American Veterinary Medical Association* and the *American Journal of Veterinary Research* may reject work that has been previously published of 250 words or more and included additional content such as figures, tables, illustrations, and references. Scientific abstracts published in the AAV proceedings are still required to contain data and meaningful information to the reader.

For permission to reproduce or order copies of this publication, contact the AAV association office: www.aav.org, 1802 Antioch Church Road, Watkinsville, GA 30677, USA, email pubs@aav.org.

Visit www.aav.org for information on AAV educational conferences, membership, and other association programs.

Table of Contents

Association of Avian Veterinarians

Section 1: AAV Fundamentals Session

Fundamental Skills in Avian Medicine (Session #3190) 2
Fundamentals of Avian Orthopedics (Session #3124) 6
Nutrition of Psittacine Birds for the Avian Practitioner (Session #3138) 11

Section 2: AAV Plenary Session

Use of Intravenous Lipid Emulsion Therapy for the Treatment of Acute Poisonings in Birds (Session #3436) 21
Agency-Based Parrot-to-Parrot Video Calling as Enrichment for Birds (Session #3447) 23

Section 3: AAV Cutting Edge Session

The Attending Veterinarian's Role in Compliance with Animal Welfare Standards in Birds (Session #3433) 30
Effects of Atorvastatin and Rosuvastatin on Blood Lipids in Quaker Parrots (*Myiopsitta monachus*) (Session #3146) 38
Pharmacokinetics of Single-Dose Oral Atorvastatin and its Metabolites in Orange-Winged Amazon Parrots (*Amazona amazonica*) (Session #3149) 39
Clinicopathologic, Gross Necropsy, and Histopathologic Effects of Repeated Glucocorticoid Administration in Domestic Pigeons (*Columba livia*) (Session #3430) 41
Effects of Capromorelin, Mirtazapine, and Cyproheptadine on Food Intake in Budgerigars (*Melopsittacus undulatus*) (Session #3387) 42
Use of a Continuous Glucose Monitor in a Diabetic Toco Toucan (*Ramphastos toco*) (Session #3426) 43
Radiographic Measurements of Internal Organs in Red Hybrid Hens and Their Evolution During Egg Formation (Session #3121) 44
Use of a Tablet-Based Communication Board and Subsequent Choice and Behavioral Correspondences in a Goffin's Cockatoo (*Cacatua goffiniana*) (Session #3441) 45

Section 4: AAV Wildlife, Anesthesia, and Surgery Session

Juvenile Wild Galah Wasting Syndrome (Session #3269) 52
Retrospective Postmortem Evaluation of Cardiac Valvular Disease in Eagles (Session #3321) 53
Can Packed Cell Volume and Total Solids Guide Wildlife Rehabilitation Decisions for Birds? (Session #3372) 60

Use of Trimethoprim Sulfamethoxazole in the Treatment of Brevetoxicosis in Seabirds and Shorebirds (Session #3123) 61
Pharmacokinetics of Hydromorphone Hydrochloride and its Metabolite in Great Horned Owls (*Bubo virginianus*) (Session #3307) 62
Computed Tomography-Guided Surgical Correction of Air Sac Rupture in a Scarlet Macaw Combined with Pleurodesis-Like Procedure (Session #3467) 63
Hypothermia During General Anesthesia Under Isoflurane, Sevoflurane and Desflurane in Hispaniolan Amazon Parrots (*Amazona ventralis*) (Session #3459) 64
Device Composition Using a 3-Way Stop Cock for Monitoring of Capnography in Small Birds (Session #3209) 65
Comparison of Non-Controlled (Dexmedetomidine-Nalbuphine) and Partially Controlled (Dexmedetomidine-Butorphanol) Sedation Protocols in Chickens (*Gallus gallus domesticus*) (Session #3298) 66

Section 5: AAV Masterclasses Track 1

Avian Zoonoses (Session #3347) 69
Birds' Use of Technology for Enrichment and Socialization (Session #3444) 88
Sedation in Birds (Session #3486) 94

Section 6: AAV Masterclasses Track 2

Avian Wound Management (Session #3220) 100
Vulture Medicine and Conservation (Session #3212) 106
Avian Orthopedic Surgery: From Theory to Practice (Session #3104) 117

Section 7: AAV Masterclasses Track 3

Management of Reproductive and Behavioral Diseases of Captive Psittacine Birds (Session #3186) 123
Veterinary Care of Breeding Collections of Birds of Prey and Psittacine Species: Tips and Tricks to Help Your Client (Session #3108) 133
Cloacal Prolapses in Cockatoos (Session #3136) 139
How to Perform a Salpingohysterectomy (Session #3343) 150
Egg Binding in Small Psittacine Species: Tips, Tricks, Treatment and Outcomes (Session #3455) 154
The Avian Uropygial Gland in Health and Disease (Session #3437) 156

Section 8: AAV Masterclasses Track 4

Avian Bornavirus: A Global Perspective on Pathology, Current Diagnostics, Research and Treatment (Session #3173) 167
Practical Applications of Fluoroscopy in Avian Practice (Session #3217) 181
Recognition, Assessment and Management of Pain Improves Avian Welfare (Session #3166) 185

Section 9: AAV Special Panel on Geriatrics

Avian Geriatric Diseases (Session #3282) 200

Section 10: AAV Potpourri Session

Cryptosporidiosis in Cockatiels and Other Psittacine Birds (Session #3155) 215
Captive Monk Parakeets (*Myiopsitta monachus*) Allowed to Choose Sodium Intake Consume Less Sodium than Their Counterparts (Session #3401) 221
Figure-of-8? More Like, Figure-I-Hate! (Session #3443) 222
Novel 3D Imaging of the Psittacine Beak and a Note on Papillae and Beak Trimming Procedures (Session #3165) 224
Novel Use of Antifungal Nail Lacquer to Treat *Candida* Bill Infections in Ducks (Session #3290) 228
Assessment of Novel Serologic Assays in Avian Species (Session #3227) 233
Effects of a Single Dose of Gabapentin in Adult Domestic Pigeons (*Columba livia domestica*) (Session #3375) 235
Effect of Capromorelin (Entyce®) on Appetite and Weight Gain of Domestic Pigeons (*Columba livia domestica*) (Session #3179) 237
Successful Conservative Management of a Neuro-Compressive Unstable Synsacral Fracture in a Yellow-Headed Amazon Parrot (*Amazona oratri*) (Session #3473) 239
Effects of Benzodiazepines on Food Intake in Budgerigars (*Melopsittacus undulatus*) (Session #3454) 240

Section 11: AAV "I'll Take Topics with the Acronym AI for 500 Please, Alex" Session

Highly Pathogenic Avian Influenza in Raptors (Session #3340) 242
Quarantined Emu (*Dromaius novaehollandiae*) Infected with Highly Pathogenic Avian Influenza H5 Survives, Despite Previous Fulminant Encephalitis Vaccine-Induced Rhabdomyolysis (Session #3383) 244
Artificial Intelligence Assisted Cell Morphology Assessment in Avian Blood Smear (Session #3415) 245
Biosecurity and Quarantine Procedures During a Highly Pathogenic Avian Influenza Outbreak (Session #3341) 246
Avian Influenza: Practical Ways to Protect the Backyard Flock (Session #3134) 247
Measuring Disease Prevalence in Barred Owls (*Strix varia*) Presented to a Louisiana Wildlife Hospital (2014-2021) (Session #3487) 254

Section 12: AAV Potpourri II Session

How to Maximize Foraging Opportunities for Parrots: A Case Study (Session #3275) 256
Desert Rose (*Adenium obesum*) Acute Toxicity and Treatment in 33-Year-Old Blue and Gold Macaw (Session #3380) 258
Patient-Patient Owner Relationships and Their Influence on Budgerigar Husbandry Systems (Session #3327) 259
Clinical Signs Associated with Elevated Copper Levels in Mixed Species Captive Birds: A Case Series (Session #3248) 261
Determining Fecal Microbiome Using Next-Generation Genomic Sequencing in Cockatiels (*Nymphicus hollandicus*) Fed Different Diets (Session #3425) 262
Military Macaw with Chronic Active *Klebsiella* Stomatitis Diagnosed by Next-Generation Sequencing (Session #3468) 264
Fourteen Cases of Naturally Infected Birds with Avian Bornavirus: Usefulness of Histologic Examination of Skin Lesions and Virus Localization of Feather Damaging Behavior for Early Diagnosis (Session #3200) 265
Antemortem Diagnosis and Successful Long-Term Management of Disseminated Xanthogranulomatous Disease in an Eclectus Parrot (*Eclectus roratus*) (Session #3476) 267
Prevalence, Distribution, and Risk Factor Correlates of Adipocytic Tumors in Psittaciformes: 648 Cases (1998-2018) (Session #3361) 268
Radiation Therapy for the Treatment of Squamous Cell Carcinoma in a Barn Owl (*Tyto alba*) (Session #3130) 269
Analyzing Baseline and Post-Restraint Plasma Corticosterone Levels in Hispaniolan Amazon Parrots (*Amazona ventralis*) (Session #3171) 271

Hybrid Sessions

Section 13: Hybrid Session 1

How to Incorporate Exotic Animals into General Practice (Session #3316) 731
How to Convince Your Local Emergency Practice to See Avian and Exotic Animal Patients (Session #3389) 735
How to Support and Train Your Local Emergency Practice to See Avian and Exotic Animal Patients (Session #3391) 738
Compounding Considerations and Making Modified Pharmaceuticals Work for Patients (Session #3411) 742
Recession Proof Your Exotic Animals Practice: For Owners and Associates (Session #3294) 750
The Opioid Epidemic and Why a Human Health Crisis Matters in Veterinary Medicine (Session #3410) 755
The Unkind Truth About Conflict Avoidance and Managing Difficult Conversations (Session #3407) 765
Ten Tips for Supervising and Strengthening Teams When You Are Out of Bandwidth (Session #3446) 773
How to Avoid a Board Complaint and What to Do if You Receive One (Session #3315) 776

Section 14: Hybrid Potpourri Session

Pradofloxacin Pharmacokinetics in Two Exotic Pet Species: Rabbit and Eastern Long-Necked Turtle (Session #3103) 781
VetGyver: How to Apply a Normal Thing to Exotic Pet Equipment (Session #3243) 784
Ecotourism in the Caribbean: The Good, the Bad and the Ugly (Session #3493) 785
Use of Alfaxalone in Exotic Animal Practice (Session #3224) 787
Ooo That Smell! Maggot-Infested Wounds in Wild Mammals (Session #3385) 794
Spatial Analysis of Wildlife Rehabilitation Data for Optimization of Outreach Efforts (Session #3427) 797
Secret Tips for Intubation in Tiny Exotic Pets (Hamsters, Sugar Gliders, Chinchillas, Hedgehogs, Squirrels) (Session #3244) 799
Roles of Veterinarians in Rewilding Projects in Victoria, Australia (Session #3163) 800
How Non-Traditional Companion Animals Impact Climate Change (Session #3292) 807

Section 15: Hybrid Special Session

Use of 3D Printed Surgical Models to Aid Exotic Animal Surgeries (Session #3191) 809

Section 16: Hybrid Marla Lichtenberger Emergency and Critical Care Session

Stayin' Alive 3. You Can't Always Get What You Want: Brass-Tacks versus Gold-Standard Diagnostic and Treatment Options (Session #3268) 812
Containing Contagion and Understanding Zoonotic Disease Potential of Wildlife Species (Session #3404) 820
Wildlife: Coping with Common to Crazy Emergencies (Session #3221) 832
Addressing Dogma in Exotic Animal Practice (Session #3394) 839

Section 17: Hybrid Session 2

No Patient is Too Small to Help: Microsurgery for Exotic Pets (Session #3175) 844
How to Plan and Write a Case Series: Tips from the Editor of the Journal of Exotic Pet Medicine (Session #3501) 848

Section 18: Hybrid Session 3

Electrochemotherapy in Exotic Animals: A Review of the Literature (Session #3135) 853
Introduction to Beekeeping (Session #3368) 867

Technician Sessions

Section 19: Technician Session 1

Gimme Shelter (and UVB and Tube Feeding): Caring for Hospitalized Reptiles (Session #3152) 872
Rabbit Anesthesia 101 for the Veterinary Technician (Session #3234) 884

Section 20: Technician Session 2

Introduction to Zoology for Veterinary Technicians (Session #3510) 895
Getting a Bird's Eye View on Avian Anesthesia and Analgesia (Session #3263) 899

Section 21: Technician Tutor Session

Avian Pain Assessment for Veterinary Technicians (Session #3151) 908
Caudal Intravenous Catheter Placement in Squamate Reptiles (Session #3511) 912
Technique Tutor: Intravenous Catheter Placement in Small Mammals (Session #3335) 918
Technique Tutor: Feeding the Hospitalized Small Mammal (Session #3336) 923

Section 22: Technician Emergencies Session

Exotic Animal Emergencies for Technicians, Part 1: Evaluation of the Emergent Exotic Mammal Patient (Session #3367) 928
Exotic Animal Emergencies for Technicians, Part 2: Evaluation of the Emergent Reptile Patient (Session #3508) 946
Exotic Animal Emergencies for Technicians, Part 3: Evaluation of the Emergent Avian Patient (Session #3509) 956

Author Index

Ahlmann, Annette page – 242, 246
Aihara, Naoyuki page – 265
Ambra, Neta page – 267
Baden, Rachel page – 181
Bailey, Christopher page – 221
Barboza, Trinita page – 38
Barron, Heather Wilson page – 100, 832
Beaufrere, Hugues page – 38, 39, 268, 965
Beekmans, Mandy page – 256
Bellefeuille, Sean page – 809
Berls, Jovanna page – 872, 908
Brandao, Joao page – 66
Bronson, Ellen page – 233
Burnett, Alysha page – 38
Burns, Laura page – 235, 237
Callico, Adriana page – 233
Cattell, Liam page – 245
Chitty, John page – 106
Choo, Katherine page – 797
Clubb, Susan page – 23, 30, 45, 88, 215
Cococcetta, Ciro page – 44
Coutant, Thomas page – 44
Craig, Linden E. page – 53
Cray, Carolyn page – 233
Cremer, Jeannette page – 64
Crocker, Kaitlyn page – 66
Cunha, Jennifer page – 23, 45, 88
Danner, Cécile page – 259
Daugherty, Dawn page – 261
Davis, Kaori page – 245
de la Rosa, Carlos page – 797
Deli, Gianluca page – 807
Diaz Falcón, Noelia page – 64
Di Girolamo, Nicola page – 239, 848
Dimaano, Elyssa page – 245
Divers, Stephen page – 235, 237
Doden, Greta page – 66
Doest, Odette page – 785
Donato, Len page – 750
Doneley, Bob page – 52
Doss, Grayson A. page – 42, 94, 240
Drasch, Katrin page – 259
Duerr, Rebecca page – 100
Echols, M. Scott page – 63, 200, 224, 261
Fallon, Jesse page – 731, 776
Finkelstein, Ariana page – 154
Fithian, Julie page – 66
Flowers, Peter page – 797
Forsythe, Lauren page – 742, 755
Franzen-Klein, Dana page – 242, 246
Gasper, David page – 267
Gonzalez-Jassi, Hugo page – 64
Greenacre, Cheryl page – 69, 150
Gruber, Erika page – 41
Hall, Victoria page – 242, 246
Harris, Laurel page – 154
Harrison, Tara page – 41, 269
Hawkins, Michelle G. page – 62, 185, 965
Heatley, J. Jill page – 221
Hirskyj-Douglas, Ilyena page – 23
Hirskyj-Douglas, Ilyena page – 88
Holliday, Casey page – 224
Hoppes, Sharman page – 167, 200
Huang, Dan page – 245
Hyunh, Minh page – 44
Jacobson, Connor page – 235, 237
Jones, Michael page – 53, 106
Kallenbach, Johannes page – 259
Kamiie, Junichi page – 265
Kanokpornwasin, Rapon page – 799
Keel, Kevin page – 268
Kleinberger, Rebecca page – 23, 45, 88
Knych, Heather page – 39, 62
Korbel, Ruediger page – 259
Kotschwar, Stefanie page – 884, 918, 923
Krumbeck, Janina page – 262
Kunzman, Hannah page – 66
Lamb, Stephanie page – 11
Lattin, Christine page – 271
Leclerc, Antoine page – 233
Lee, Amber page – 123
Lee, Inhyung page – 65
Lee, Jungha page – 65
Lee, Margaret page – 262
Lennox, Angela page – 809
Leonard, Elliott page – 258
Lessner, Emily page – 224
Le, Thanh page – 245

Lex, Zoey page – 60
Lierz, Michael page – 133, 167
Lofgren, Nicolle page – 839
Logsdon, Marcie page – 43
Lubian, Emanuele page – 807
Mackey, Elizabeth page – 867
Macwhirter, Pat page – 800
Makino, Akiko page – 265
Malka, Shachar page – 262
Mans, Christoph page – 42, 94, 240, 267
Martin, Kian page – 258, 264
Martin, Scott page – 258
Mayer, Corinne page – 60
McCleery, Brynn page – 139, 247, 853
McLaughlin, Alicia page – 812
Megan, Lucyanne page – 268
Mikoni, Nicole page – 39
Miller, Christina page – 895, 912, 928, 946, 956
Mitchell, Mark page – 60, 254
Monopoli, Marissa Rae page – 965
Montieth, Stephanie page – 258
Murphy, Bianca page – 245
Music, Meera page – 812
Neuhaus, Laura page – 2
Nixon, J Bruce page – 735, 738
Olsen, Glenn page – 228
Orosz, Susan page – 200
Oster, Seth page – 6
Padlo, Jordan page – 60
Parks, Shelby page – 271
Passarelli, Maria page – 61
Paul-Murphy , Joanne page – 185, 224, 965
Pendl, Helene page – 167
Petritz, Olivia page – 41
Phalen, David page – 52
Phouratsamay, Albert page – 44
Plangsangmas, Tithipong page – 60
Pohly, Andrea page – 457
Ponder, Julia page – 106
Porter, Ian page – 239
Powers, Lauren page – 21, 156
Queiroz-Williams, Patricia page – 64
Rinder, Monika page – 259
Romagnano, April page – 244, 258, 264
Rowe, Madeleine page – 800
Russell, Laken page – 60
Saam, Nicole page – 259
Sanchez-Migallon Guzman, David page – 39, 62, 268, 965
Sands, Nicole page – 262
Schoemaker, Nico page – 256
Schott, Renee page – 222, 773, 794
Settle, Aspen page – 254, 271
Shin, Donhwi page – 65
Smith, Lisa C. K. page – 258
Soga, Reiko page – 265
Son, Won-gyun page – 65
Sosa Higareda, Mariana page – 62
Speer, Brian page – 181, 224
Stark, Ken D. page – 38
St Clair, Laura page – 239
Stern, Laura A. page – 21
Stockman, Jonathan page – 262
Strunk, Anneliese page – 812
Szabo, Zoltan page – 844
Thielen, Lauren page – 63, 239
Thomson, Andrea page – 41
Titel, Caroline page – 42, 240
Tomonaga, Keizo page – 265
Touzet, Chloé page – 44
Trumpp, Kelsey page – 235, 237
Tully, Jr., Thomas page – 64, 271
Turner, Megan page – 735, 738
van Zeeland, Yvonne page – 200, 256
Vecere, Gina page – 262
Vetrano, Liz page – 899
Vinke, Claudia page – 256
Voltura, EV page – 221
Vuong, Kristina page – 53
Whittington, Julia page – 742, 755, 765, 820
Womble, Mandy page – 41
Wood, R. Darren page – 38
Woo, Krystal page – 797
Xie, Shangzhe page – 117, 781
Yeon, Seong Chan page – 65
Yinharnmingmongkol, Chaowaphan page – 784, 799
Yoshino, Souichi page – 265
Zoller, Graham page – 44

Section 1

AAV Fundamentals Session

Cheryl Greenacre, DVM, DABVP (Avian Practice), DABVP (Exotic Companion Mammal Practice)
Session Chair

Fundamental Skills in Avian Medicine

Laura Neuhaus, DVM

Session #3190

Affiliation: From Grafton Animal Hospital, Community Veterinary Partners, 449 Grafton Drive, Yorktown, VA 23693, USA.

Abstract: There are various skills that are needed when first starting to work with avian species. This presentation describes some of the most common techniques. Topics include blood collection, subcutaneous injections, intramuscular injections, intravenous and intraosseous catheters, crop tubing, abdominocentesis, and nasal flush.

Introduction

When first starting to practice avian medicine there are a variety of skills that are necessary. The purpose of this discussion is to further describe some of commonly utilized techniques. For skills not listed, there are a variety of resources available to continue learning. These include but are not limited to continuing education at conferences such as ExoticsCon, the Education Portal and the Golden Egg on the Association of Avian Veterinarians' website (http://aav.org), the Avian Veterinary Clinical Competency Program (https://aav.org/page/avccp), and LafeberVet (http://lafeber.com/vet).

Blood Collection

Blood samples permit the practitioner to perform a wide variety of diagnostic tests. The number and type of diagnostics that can be run, however, can be limited by small patient sizes and blood volume. While blood volume is dependent on the species, a rough estimate is 10% of the body weight.[1] In a healthy bird, a clinician can safely collect 10% of this amount, or 1% of the body weight (for example 1 milliliter in a 100 gram cockatiel).[1] In severely ill birds, smaller sample sizes should be taken to avoid further compromising the patient.[1]

A variety of different sites can be used for blood collection. The right jugular vein is a frequently used site due to the ability to collect a large volume of blood quickly from even small patients.[1,2] To collect a sample, use a small amount of alcohol to expose the overlying aptera (present in most species) then apply pressure to the thoracic inlet to occlude and help stabilize the vessel.[1,2] The disadvantage of this site is that the surrounding large subcutaneous space predisposes the area to hematoma formation.[1]

The basilic vein is another common site for blood collection. For this collection, the patient is placed in dorsal recumbency and the wing is gently extended.[1] Pressure is then applied to the humeral area to occlude the vein prior to collection. Hematomas are common in this area due to movement of the wing.[1]

The medial metatarsal vein is the author's preferred site for blood collection. The patient can be maintained upright during collection, the leg is easily restrained, and hematomas are typically minimal due to protection from the surrounding muscles.[1,2] Due to the smaller size of this vessel, use is typically reserved for medium to large species. When collecting from smaller species, an insulin syringe can help to reduce negative pressure and vessel collapse. To collect a sample, locate the vessel in the region of the tibiotarsus-tarsometatarsal joint and apply pressure proximally over the tibiotarsus to occlude it.

Subcutaneous Injections

The subcutaneous route is one of the most common locations for fluid administration.[3] Fluids are warmed prior to administration and can be given in the intrascapular area, inguinal region, flank, or the axilla with the inguinal region being most common.[2,3] The feathers are parted with a small amount of alcohol and the needle is introduced at a shallow angle with the bevel up.[4] As fluids are injected, the subcutaneous space will expand and the needle can be moved into the middle of the pocket to keep it from inadvertently slipping out.[4] Be careful not to place the needle too deeply, however, as accidental administration into the air sacs can occur with fatal consequences.[4]

Intramuscular Injections

Intramuscular injections are commonly performed for administration of various therapeutic agents. The most common site used is the pectoral musculature.[2,5] The needle is introduced at an acute angle on either side of the keel in the mid to caudal portion of the muscles to avoid the larger vessels located cranially.[2] The plunger is withdrawn prior to administration of the medication to ensure a vessel has not been inadvertently entered.[2] It is important to note that not all medications are suited for intramuscular injection (e.g., Baytril).

Intravenous and Intraosseous Catheters

Intravascular fluids are given to patients that are severely debilitated or undergoing anesthesia. Patients that are more alert may become tangled in fluid lines or remove the catheter which could lead to fatal hemorrhage.[2,3,5] When intravascular fluids are indicated, either an intravenous or intraosseous catheter can be placed. To place a catheter, a local anesthetic, sedation or anesthesia may be needed to reduce patient stress and movement.[5] In the case of an intravenous catheter, placement sites are the same as those for phlebotomy. These catheters can then be secured using tape alone when using the medial metatarsal vein and sutured when using the basilic and jugular vein.[5]

Intraosseous catheters are useful for small or collapsed patients where intravenous access is not possible.[2,5] Intraosseous catheters are placed in the ulna or tibiotarsal bone using either an intraosseous or hypodermic needle.[2,3,5] While hypodermic needles are more readily available, they will frequently become obstructed with bone during insertion and may need to be replaced.[2,3] To place an ulnar catheter, the carpus is flexed and the dorsal condyle of the carpus is identified. The needle is rotated into the distal ulna and directed proximally.[2,5] When placing a tibiotarsal catheter, flex the stifle and insert the needle at or to either side of the insertion of the patellar tendon then rotate and direct the needle distally through the cnemial crest.[2,5] Catheter position for both sites can be checked with radiographs or in the case of ulnar catheters, by visualization of the basilic vein blanching when flushed.[5] Intraosseous catheters that are to be maintained in the awake patient can be secured with tape in addition to a figure-8 style bandage in the case of ulnar catheters.[5] Unlike intravenous catheters, fluids administered through an intraosseous catheter need to be given under pressure either by hand or using a syringe pump.[2]

Ingluvial Gavage

Ingluvial gavage is performed for a variety of reasons including to administer medications, provide nutrition, and collect diagnostic samples. Most commonly either metal gavage needles or red rubber catheters are utilized. Care must be taken when using soft tubes to ensure that the patient does not bite and ingest the material.[2] For this reason, metal tubes are often preferred in psittacine birds and softer red rubber catheters in other species such as Anseriformes and Galliformes.

When placing the tube or gavage needle, the patient is held with the head and neck extended. Lubricate the tube or gavage needle prior to placement to reduce the risk of accidental trauma to the oropharynx and thin walled esophagus.[2,5,6] Next, pass the tube or gavage needle from the left side of the mouth down the right side of the oral cavity. Prior to administration of food, water, or medications, the clinician should palpate for the tip of the tube or gavage needle in the crop to confirm the location.[2,3,6] While administering the syringe contents, the oral cavity should be monitored for regurgitation and the tube removed if seen.[2]

A variety of products are available on the market for nutritional support depending on the species and clinician preference. Up to 50 ml/kg of formula can be given though reduced amounts should be used when first starting gavage feeding.[6] Gavage feeding should be the last treatment performed to reduce the risk of regurgitation secondary to stress and feedings can be administered two to four times per day.[6]

An ingluvial lavage can be performed using the same technique. Rather than food, warm water is infused with smaller amounts than those used for gavage feeding to take into account current crop contents.[2] After infusing water into the crop, gently massage it and aspirate the fluid back into the syringe.[2] Do not forcibly aspirate the fluid back into the syringe as this can cause damage to the crop.[2]

Nasal Flush

Nasal flushes are used for both therapeutic and diagnostic purposes. To perform a nasal flush, position the patient so that the head is directed towards the floor, place a syringe containing warm saline against the nare, and flush the fluid through the nares.[7] The fluid expressed can be collected in a sterile container or on a swab for cytology and culture and sensitivity.

Abdominocentesis

Abdominocentesis is performed in birds with coelomic effusion to help relieve pressure on the air sacs and lungs as well as to collect samples for diagnostics.[8] The skin over the abdomen is disinfected and a 23- to 27-gauge needle attached to a syringe is introduced along the midline.[3] If introduced lateral to midline, the air sacs may be inadvertently entered.[3] A fluid extension set and stopcock may be added to aid in removal of larger volumes of fluid.[8] Be aware that when removing large amounts of fluid, protein shifts and changes in abdominal pressure can cause serious or fatal side effects.[3,8]

References

1. Campbell TW, Grant KR. *Exotic Animal Hematology and Cytology.* 5th Ed. Hoboken, NJ: Wiley-Blackwell; 2022.

2. Longley LA. *Anesthesia of Exotic Pets.* Philadelphia, PA: Elsevier; 2008.

3. Harrison GJ, Lightfoot TL. *Clinical Avian Medicine.* Palm Beach, FL: Spix Publishing, Inc.; 2006.

4. LafeberVet. Subcutaneous fluids in birds. Available at https://lafeber.com/vet/subcutaneous-fluids-in-birds/. Accessed March 20, 2023.

5. Speer BL. *Current Therapy in Avian Medicine and Surgery.* St. Louis, MO: Elsevier, Inc.; 2016.

6. LafeberVet. Tube feeding birds. December 9, 2010. Available at https://lafeber.com/vet/tube-feeding-birds/. Accessed March 20, 2023.

7. LafeberVet. Avian respiratory emergencies: an approach to the dyspneic bird. Available at https://lafeber.com/vet/respiratory-emergencies/. Accessed March 20, 2023.

8. LafeberVet. Abdominocentesis in birds. Available at https://lafeber.com/vet/abdominocentesis-in-birds/. Accessed March 20, 2023.

Fundamentals of Avian Orthopedics

Seth C. Oster, DVM

Session #3124

Affiliation: From the Department of Clinical Sciences, Auburn University, College of Veterinary Medicine, 1220 Wire Road, Auburn, AL 36849, USA and the Southeastern Raptor Center, Auburn University, College of Veterinary Medicine, 1350 Pratt-Carden Dr, Auburn, AL 36849, USA.

Abstract: Fundamentals of Avian Orthopedics is a 1-hour masterclass designed to update new and experienced clinicians with the fundamentals of, triaging, diagnosing and treating orthopedic injuries in avian patients. The course begins with a discussion of initial case presentation and management, including triage and short-term stabilization of injuries. The discussion then focuses on guiding the selection of conservative versus surgical management and exploring those methodologies.

Introduction

Orthopedic injuries affecting birds include fractures, degenerative joint disease, and developmental abnormalities. Avian cortical bone is thinner and more brittle than their mammal counterparts. This is due to the pneumatic medullary cavity and increased mineral content. In this author's experience, fractures are the most common orthopedic injury that presents in practice.

Fractures

Fractures commonly present as an emergency after acute trauma. Clinical signs include wing droops, lameness, instability of long bones, and crepitation on palpation. The principles of managing fractures in avian patients are similar to those of mammalian orthopedics. These include initial stabilization and triage, fracture assessment, and conservative or surgical stabilization.

Initial triage consists of supporting normal respiration and circulation, as many traumatic injuries such as fractures can leave the patient in shock. This can include fluid therapy, supplemental oxygen, and heat based on the patient's initial assessment. The remaining elements of triage include stabilization of the fracture site, usually through external coaptation, and providing analgesic support. In addition, antimicrobial therapy and local cleaning and disinfection may be required if the bones fragments are exposed (compound fracture).

External coaptation of fractures varies based on bones involved. A figure 8 bandage can be used to stabilize fractures involving the radius, ulna, or metacarpal bones. A body bandage can be used to stabilize fractures of the clavicle, coracoid, and scapula. Fractures of the humerus are stabilized with a combination of a figure 8 and a body bandage. Fractures of the tibiotarsus or tarsometatarsus can be stabilized using a roll gauze splint with side bars of ridged material such as aluminum or thermoplastic tape for support.

Analgesic support for orthopedic injuries is important from both a healing and ethical standpoint. Multimodal pain support is required to reach adequate levels of analgesia in an acute fracture. Based on available pharmacokinetic data a combination of meloxicam and buprenorphine SR are this author's preferred combination.

Radiographs are required to evaluate most fractures suitability for surgery and to guide the selection of the appropriate surgical fixation method. Due to the pain and potential for further injury while properly positioning a bird for diagnostic radiographs, sedation or anesthesia is required. Orthogonal views of the affected limbs should be used to assess the type and location of the fracture, and the morphology and the positioning of the fragments.

Positioning is key to diagnostic radiographs. Having a few items can make positioning easier. Foam wedges or pads can be very useful to elevate heads or synsacrum to get proper alignment for lateral views. Foam can also be useful in separating legs in a lateral view. White tape or masking tape can be used to secure limb in position with minimal damage to feathers. Bird Boards have also been used to aid in positioning. These devices are made of acrylic and are more appropriate for larger birds. Most of these devices have straps or ties to secure wings and legs and an acrylic plate to secure the head in place. This author prefers the use of foam and tape.

A properly positioned ventrodorsal view of the coelom should, if possible, include the thoracic inlet cranially and legs caudally. The patient's dorsum should be laid flat on the table with the keel pointed straight towards the ceiling. This should result in the keel being superimposed directly over the thoracic vertebrae in the resulting imaging.[1] Both legs and wings should be extended equally to prevent rotation of the body. Wings should also be extended equally to prevent rotation of the body. When using tape to restrain, wing tape should touch the wing over the distal radius/ulna. Leg tape should contact the leg over the distal tarsometatarsus. The keel should be aligned with the + created with the light of the beam and the beam should be centered over the caudal portion of the keel. The entire coelomic cavity should be fit within a single image in all but the largest bird. Labeling with a metal R or L is of course part of standard procedure.

When positioning for a lateral view of the coelomic cavity, both wings should be fully extended dorsally. It is important to not superimpose the wings directly on top of each other. The down wing is commonly place more cranial to allow easy identification on the resulting image. Tape can again be placed over the distal radius/ulna. In large birds, a foam pad or wedge between the wings can prevent overextending wings. Legs should be extended ventrocaudally. Again, avoid superimposing legs by moving the down leg more cranially. Tape can again be placed around the distal tarsometatarsus. In most birds, the hips are narrower than the shoulders. This will cause most birds to rotate their hips towards the downside, twisting the coelom. Placing a small foam wedge under the hips will help prevent this rotation and make a well aligned image. To confirm correct positioning, the heads of the humerus should superimpose with each other as well as the head of the femur on the final image.

Lateral images of the wings are designed to have the beam base from the ventral aspect of the wing through the dorsal aspect. To position this view, place the patient on the table in dorsal recumbency. Extend the wing that you wish to view until it reaches the natural end of its range of motion. In this position, it is best to use tape passing over the metacarpus to make sure both the elbow and carpus are fully extended. In larger birds, tape can be added over the mid radius and ulna to spread the pressure of the tape out. Center the radiographic tube over the mid radius and ulna if you can fit the entire wing in the field. In larger birds, it may take two or three images to get the entire wing radiographed.

The caudal-cranial view of the wing is another often used position. This positioning starts with the patient in dorsal recumbency on the table. Feet should be towards the restrainer and the head should point away from the restrainer. While keeping the patient's head flat on the table, the restrainer grasps the legs and tail in a glove hand and raises them towards the ceiling. Be aware this technique has an increased risk of regurgitation. The other hand holds the primary feathers of the wing and extends the wing out from the body. The beam should pass from the caudal aspect of the wing through the cranial aspect. Be sure to avoid getting fingers in the view. The beam can be centered over the proximal radius and ulna. In large birds, multiple views may be required to get the entire wing imaged. If the image is properly positioned, the radius and ulna will be super imposed in the resulting image.

Lateral images of the legs can be accomplished by placing the patient in the same position as the lateral whole-body images. In the case of the legs, the down leg should be moved cranially (example, the right lateral view of the legs should have the right leg moved cranially). The toes should be spread out and taped in position to prevent superimposition. The beam should be centered over the proximal tibiotarsus and the entire leg from the femur to the toes should be included in the image if possible. Large birds or long-legged birds may require multiple images.

A ventrodorsal view of the synsacrum is achieved by placing the patient in dorsal recumbency. The legs are fully extended and taped in place similar to the positioning to the ventrodorsal images of the whole body. Digits should be spread out and taped in place to prevent superimposition. The beam should be centered over the proximal tibiotarsus and the entire leg from the femur to the toes should be included. In larger birds or long-legged birds, multiple images maybe required.

The H-view (oblique view, perhaps) is a recent addition to the toolkit of avian practitioners. It is used to examine the coracoid, clavicle, and scapula without these bones being super imposed upon one another. This image is acquired by placing the patient in the same position that would be used for a standard whole body ventrodorsal view. It is important that the patient's head and feet are aligned with the length of the table. This technique requires a machine that will allow for the head of the radiographic unit to rotate to a 30-45° angle. The exact angle depends on the species of bird being examined. This practitioner routinely starts at 30° for most patients and increases the angle if there is still superimposition of the bones of interest. Rotate the head of the unit to the desired angle and then move the unit caudally in relation to the patient. The beam should be aimed at the shoulders of the patient passing through it in a caudoventral-craniodorsal angle. As this beam is at an angle to the patient, traditional measurement for the determination of technique can be difficult. Technique set based on ventrodorsal measurements of the patient tend to result in overexposed images. This practitioner uses a ventro-dorsal measurement and decreases the technique by 1 centimeter in most cases.

Surgical correction of fractures in birds can be done with a variety of methods including external fixators (EXT), intramedullary pins (IM), EXF-IM tie-ins, ring fixators, and bone plates. Each of these methods has their own benefits and contraindications. External fixators are relatively easy to place as pins can be placed through incision or percutaneous. Reduction of the fracture can be achieved either close or open depending on the surgeon's experience and overall displacement of the distal segment. This type of fixation is ideal for controlling rotation and compressional forces. It also provides adequate stabilization against bending forces. It is also possible to combine this method with an intramedullary pin. By tying the two fixation methods together, this provided increased stabilization.

Intramedullary pins are another common method of fracture repair in birds. The placement of this fixation is more difficult, as the surgeon must plan where the pin will exit the bone. In some bones, such as the ulna or humerus, the anatomy of the bones allows the pin to exit without interfering with a joint surface. In other bones, such as the metacarpus or tarsometatarsus, it is not possible to place an IM pin without damaging the joint. Place of the pin is commonly done by entering the fracture site through an incision and retrograding the pin into place. This type of fixation method is superior at stabilization against bending forces but has minimal affect to rotational or compressional forces.

In both surgical and conservative management of fracture, appropriate follow up care is key to recover. Patients should have limited activity and use of the affected limb until evidence of bony healing is present on radiographs. Avian bone healing is often quicker than mammalian bone healing. This author often observes bridging bony callus formation on radiographs 4 weeks after the injury. Ideally, return to normal activity is delayed until there is evidence of such healing. Once stabilized, pain from a fracture is significantly reduced. In most cases, this author continues multimodal analgesic support for 1 week after a surgical procedure. After that, the analgesic plan is decreased to an NSAID alone, usually for another week. Injuries to the pelvic limb can cause offloading of weight of the injuried limb and potentially cause pododermatitis on the contralateral limb. Both feet should be monitored for development of pododermatitis and changes in substrate, perching materials, or use of protective bandaging should be used if it is noted.

Regarding injuries of the wing, it is important to prevent contraction of the patagial membrane or loss of range of motion of the joints of the wing. This author's preferred protocol is passive range of motion (PROM) therapy performed with either sedation or anesthesia depending on the patient's temperament. Warm packing the affected joints or tissue for 5-10 minutes prior to start of a PROM session can be of value as well. While sedated or anesthetized, the bandage is removed and measurements of ROM of both elbow and carpus are taken prior to start of the session. Then the wing is gently extended and flexed, with particular attention paid to achieving full extensions. This may require the practitioner to apply gentle pressure during this process. It is important for the practitioner to support any areas that still are unstable during this process. This process is continued for 5 minutes or until full ROM is achieved. The bandaging can then be replaced and the patient returned to the enclosure. Patients with extensive contracture may require additional analgesia after PROM, in which case this author will usually resume use of along acting opioid for 24 hours. These physiotherapy sessions are performed up to 3 times a week based on patient needs and tolerance of the procedure.

Degenerative Joint Disease

Degenerative joint disease is commonly seen in older psittacine patients but can be seen in all species and younger birds when there is previous trauma or injury to joints. Clinical signs depending on the limb affected, but include lameness, loss of flight, loss of range of motion of joint, falling or difficulty ambulating, and lethargy. In some cases, this author's patients have demonstrated feather destructive behavior around the affected joints. Diagnosis is made by the characteristic radiographic lesions of sclerosis of subchondral bone, deceased joint space, and osteophyte formation.

Treatment includes analgesic support and husbandry changes. The most common analgesic used with degenerative joint disease is meloxicam due to the ability to administer orally at home and the relative safety of prolonged use. In addition, joint supplements such omega-3 fatty acids and glucosamine have been reported to improve clinical signs. Husbandry changes will include wider perches to allow easier stability and reducing access to vertical height to minimize the chance of falls.

Developmental Abnormality

Perosis, commonly called slipped tendon, is the most common developmental abnormality seen in young growing birds. This is commonly seen in poultry, waterfowl, cranes, ratites, and other long-legged birds. The cause of this deformity is due to cartilage deformity in the hock joint leading to medial luxation of the gastrocnemius tendon. This results in swelling of the hock joint, lateral deviation of the foot of the affected leg, and lameness in the affected leg. The cause is associated with diets high in protein, manganese deficiency, choline deficiency, and biotin deficiency. Treatment of this condition primarily involves surgical fixation of the tendon in the normal anatomical position. This type of intervention is best applied as early as possible to minimize any deformation of the tibiotarsus or tarsometatarsus. In addition, supplementing manganese, choline, and biotin along with reducing the protein content of the diet is recommended.

References

1. Darrow B, Bennett RA. Avian Orthopedics. In: Bennet RA, Pye GW, eds. *Surgery of Exotic Animals.* Hoboken, NJ: Wiley-Blackwell; 2022:112-153.

2. Calvo Carrasco D. Fracture management in avian species. *Vet Clin North Am Exot Anim Pract.* 2019;22: 223-238.

3. Hawkins H, Guzman D, Beaufrere H, et al. Birds. In: Carpenter JW, ed. *Exotic Animal Formulary.* 5th Ed. St. Louis, MO: Elsevier; 2018:167-375.

4. Guzman DSM, Ceulemans SM, Beaufrère H, et al. Evaluation of the thermal antinociceptive effects of a sustained-release buprenorphine formulation after intramuscular administration to American kestrels (*Falco sparverius*). *J Avian Med Surg.* 2018;32:1-7.

5. Harcourt-Brown NH. Orthopedic conditions that affect the avian pelvic limb. *Vet Clin North Am Exot Anim Pract.* 2002;5:49-81.

6. Hoybers Y, Bosmans T, Risselada M, et al. General anesthesia for the surgical repair of a tarsometatarsal fracture in a Harris's hawk (*Parabuteo unicinctus*). *Vlaams Diergeneeskd Tijdschr* 2008;77:309-3014.

7. Longley L. Avian anaesthesia. In: *Anaesthesia of Exotic Pets.* Elsevier, 2008;129–170.

8. Taiyari H, Abu J. External skeletal fixator to stabilize the orthopedic conditions in avian species: a systematic review of case reports and case series. *Pertanika J Sci Technol.* 2022;30:1815-1839.

9. Hatt JM, Christen C, Sandmeier P. Clinical application of an external fixator in the repair of bone fractures in 28 birds. *Vet Rec.* 2007;160:188-194.

10. Yoon H-Y, Fox DB, Jeong S-W. Tibiotarsal and ulnar fracture repair in a great horned owl (*Bubo virginianus*). *J Vet Clin.* 2008;25:218-220.

11. Naldo J, Saggese M. Radiography. In: Samour J, ed. *Avian Medicine.* 3rd Ed. St. Louis, MO: Elsevier; 2016:130-153.

12. Krautwald-Junghanns M, Pees M. Imaging techniques. In: Tully TN, Dorrestein GM, Jones AK, Cooper JE, eds. *Handbook of Avian Medicine.* St. Louis, MO: Elsevier; 2009:85-100.

13. Krautwald-Junghanns M. In: Coles B, ed. *Essentials of Avian Medicine and Surgery.* 2nd Ed. Ames, IA: Blackwell;1997:56-102

14. Silverman S, Tell L. *Radiology of Birds: An Atlas of Normal Anatomy and Positioning.* St. Louis, MO: Elsevier Saunders; 2010.

Nutrition of Psittacine Birds for the Avian Practitioner

Stephanie Lamb, DVM, DABVP (Avian Practice)

Session #3138

Affiliation: From Arizona Exotic Animal Hospital, 744 N Center St #101, Mesa, AZ 85201, USA.

Abstract: Optimal nutrition is critical to maintain health and well-being. Currently dietary recommendations for psittacine birds are derived from research in other species, several studies on specific nutrients in individual psittacine species and first-hand experience in aviculture. This lecture is geared towards providing the veterinary practitioner with the basics for understanding psittacine nutrition using available scientific literature.

Introduction

To determine if a diet is healthy and appropriate for a psittacine bird one must look at the nutrients that are required by the individual. Macronutrients, including proteins, fats and carbohydrates, must be evaluated. Micronutrients including vitamins and minerals must also be critiqued. Not all micronutrients have been studied in psittacine birds; however, there is information available on certain vitamins and minerals that can help guide our understanding of nutrition in psittacine birds and allow us to make appropriate recommendations for certain species. All nutrients must be provided within certain ranges or levels in the diet to provide adequate nutrition.

Water

The most important nutrient that is often forgotten is water. It is required for many functions including maintenance of appropriate cellular structure, transport nutrients throughout the vascular system, provide cushioning to joints, maintain body temperatures, and encourage normal passage of excretory products from the body. Generally, younger animals, larger birds, and those living in warmer environments have higher water needs. Studies in parrots show that they require 2.4% of their body weight per day as water when they are under thermoneutral conditions. If they are in hot environments though this can increase by 12 times to account for evaporative losses of water.[1,2]

Energy

Energy requirements of the individual bird can be determined by doing simple mathematical calculations. The basal metabolic rate (BMR), also known as the resting energy requirement (RER), is calculated according to the following equation.[1]

$$\text{BMR/RER} = 73.6 \times \text{BW (kg)}^{0.73} \text{ OR } 0.697 \times \text{BW (g)}^{0.705}$$

Once the BMR is established, this value is then multiplied by a factor based on the bird's energy needs for its activity level, health status, or age to calculate the metabolic energy requirement (Table 1).[1]

Table 1. Metabolic energy requirements of birds at different health status, activity level, and life stage.

Health status, activity level, life stage	Multiplication factor
Healthy juvenile	BMR × 2-3
Sick juvenile	BMR × 3-4
Perching, resting	BMR × 1
Perching, alert	BMR × 1.9-2.1
Preening	BMR × 1.6-2.3
Singing, grooming, bathing	BMR × 2.9
Mild stress	BMR × 1.2
Moderate stress	BMR × 1.5
Major stress, sepsis	BMR × 1
Gliding flight	BMR × 2
Sustained flight	BMR × 11
Aerial flight	BMR × 2.7-5.7

An example of this can be illustrated by calculating the energy requirements of an adult, female African grey parrot that is not laying eggs but is overweight. The bird presents with a weight of 450 grams and a body condition score of 7/9. The bird's BMR is calculated as $78 \times (0.45)^{0.75} = 42.8$ kcal. Since the parrot is overweight the perching resting rate (BMR × 1) or mild stress rate (BMR × 1.2) can be used to determine metabolic energy requirement. For this patient that would be a metabolic energy requirement that equals 42.8-51.4 kcal of food per day. The caloric content of the diet offered is then calculated to determine the volume of food that is needed to meet caloric needs.

Macronutrients

Carbohydrates

Carbohydrates are important as energy sources, function for storage within cells and provide cellular structure. They are also known as sugars and are made up of monomers, known as monosaccharides, that combine to make larger polymers, known as disaccharides and polysaccharides. Examples of monosaccharides are glucose, galactose and fructose. When these combine to make larger molecules the different linkages between the monomers create the overall larger structural form. The two main types of linkages are alpha and beta linkages and different enzymes are required to break these links down to release individual monomer units. Beta linkages are often broken down by microbes that can be present in the gastrointestinal tract, particularly in species that have a cecum or ceca. Psittacine birds have either absent or vestigial ceca which means they do little to no hind gut fermentation of carbohydrates with beta linkages. Absorption of carbohydrates occurs across intestinal microvilli and some frugivorous and nectivorous species have exceptionally long intestinal microvilli to aid in the absorption of free sugars found in their food.[2]

Protein

A wide range of dietary protein levels, from 6-31%, has been described for psittacine birds.[1-6] This is a wide range and considering the needs of individual species of birds, studies have shown variability. When sourced from high-quality ingredients, 11% dietary protein was adequate for maintenance in adult cockatiels.[1] Free-ranging cockatiels will select seeds containing 8.8-14% protein.[7] Interestingly, a nectarivore like the lorikeet can be maintained on 3% dietary protein, but the source of protein must be a good-quality source.[1] Experiments in African grey parrots indicate dietary protein must be 10-15% to maintain plasma protein levels.[2]

Studies in cockatiel chicks have offered variable levels of protein with different outcomes. Chicks were severely stunted when given only 5% dietary protein. When fed 10-15% dietary protein, chicks demonstrated slow growth and were slightly stunted. Optimal growth was associated with 20% protein with 0.8% lysine; however, when dietary protein levels reached 25%, the chicks continued to grow well but exhibited aggression. Even higher levels of 35% protein caused growth to slow and continued aggression.[8,9] It has been theorized that larger psittacine birds may require higher protein levels during growth. In one study evaluating the crop contents of wild scarlet macaw chicks being parent fed, 18-26% protein was present.[4] In studies where Amazon parrots were given free choice food selection, they chose diets that contained about 15.5% protein; however, it was not determined if this was a healthy level.[5]

Protein malnutrition in the bird can manifest as weight loss, immunodeficiency, and a loss of cell integrity and feather structure. Essential amino acids found in keratin include histidine, isoleucine, lysine, methionine, threonine, tryptophan and valine. A deficiency in any one of these may result in poor feather condition. Methionine deficiency during growth or molting results in dark, horizontal stress marks on the feathers. In addition, this amino acid deficiency can lead to a loss of structural proteins in feathers that help make the blue coloration observed. With a loss of this structural protein those feathers that are green lose their blue structural component and appear yellow.[10] A deficiency of this particular protein is also one of the components that can lead to fatty liver syndrome.[11]

Excess dietary protein has been studied. Many have had concerns that too much protein could lead to gout. A relationship between excess dietary protein and gout has been proposed in psittacine birds; however, research does not support this assumption. When African grey parrots were offered 30% dietary protein, this level was not associated with an increase in uric acid levels.[3] In cockatiels fed 11%, 20%, 35% and even 70% dietary protein, uric acid levels did increase as protein levels increased but gout was not induced. This is likely because of the increased breakdown of protein into the nitrogenous waste product, uric acid but it does not necessarily correlate with kidney damage. These studies also showed an increase in metabolic enzymes, ALT and AST, as protein levels increased but this is likely because of the increased demand for processing the protein that was entering the body. One negative finding in these studies was that the incidence of lipogranulomas within the liver increased as protein levels increased.[7]

Lipids

Published ranges for dietary fat levels in the diets of psittacine birds are between 4-20% showing there is quite a bit of variability; however, for maintenance purposes many sources recommend the levels be lower around 4-5%.[1,4,6,9] Studies show that wild living birds and birds fed free choice select much higher levels of dietary fat. Studies with scarlet macaws show that parents will feed their young food items containing 21-31% fat.[4] When fed free choice, captive Amazon parrots will select food items that contain 20.6% fat.[5] The benefits or harms of these levels of fats were not evaluated in these observational studies and further investigation is needed.

There are essential fatty acids that need to be in bird diets. Although birds possess desaturase enzymes that can form double bonds at the delta 5, 6 and 9 positions of fatty acid chains, they do require linoleic acid (n-6) and linolenic acid (n-3) in the diet.[2,6,12,13] There are some studies in psittacine birds that support the benefits of omega-3 fatty acid supplementation.[14-16] For instance, a custom-formulated diet containing 3% fish oil led to a reduction in triglyceride and cholesterol levels.[14] In addition, a postmortem study comparing the incidence of atherosclerosis with linolenic acid levels in breast and adipose tissue found that birds without atherosclerosis had higher tissue levels of linolenic acid.[15]

Micronutrients

Vitamins

Many sources recommend a level of vitamin A between 3,000-8,000 IU/kg of feed.[13,17] Adult cockatiels given 2,000 IU vitamin A per kg feed did not have any issues but when given 10,000 IU/kg the birds displayed hyper-excitability, modified vocalizations and an impaired antibody response. Higher vitamin A levels (100,000 IU/kg) also led to more vocalizations and impaired antibody response as well as deteriorating feather quality and pancreatitis.[17] Lorikeets fed 20,000 to 45,000 IU/kg of vitamin A in the diet developed liver damage, skin and feather changes, and splenic hemosiderosis.[18] Hypovitaminosis A has been well described in psittacine birds with the development of squamous metaplasia occurring in epithelium of the respiratory, renal and reproductive systems when the nutrient is chronically under supplemented.[10] A study in cockatiels suggested that vitamin A can be stored in the liver for up to 2 years, therefore adverse effects are more profound in young, growing birds of those with severe nutritional deficiencies.[17]

Recommended dietary vitamin D levels in bird diets range from 500 to 2,000 IU/kg of the diet.[10,13] Budgerigars have specifically been suggested to receive 10 IU per day according to one source.[3] Vitamin D deficiencies will lead to thin-shelled eggs, poor reproductive success and rickets. Excesses can be problematic as well with anorexia, lameness, lethargy, diarrhea, polyuria and metastatic mineralization seen.[10] A study looking at various levels of calcium and vitamin D in budgerigars diets found that metastatic mineralization was more dependent upon calcium levels and not effected by vitamin D levels. They showed that vitamin D could range from 500 to 2,000 IU/kg of the diet and may be able to go up to 3,300 IU/kg of the diet without adverse effects.[19] Growing macaws are thought to be more susceptible to hypervitaminosis D and historically there have been clinical reports of nutritionally imbalanced batches of commercial hand-feeding formulas containing excess levels of vitamin D and causing pathology in birds. In one report, a juvenile cockatoo and a blue and gold macaw chick fed vitamin D at 26,790 IU/kg feed developed hypercalcemia, hyperphosphatemia, hyperuricemia and renomegaly.[20]

Vitamin E is poorly studied in psittacine birds; however, suggested dietary levels range between 11-50 ppm in the diet.[6,13] Deficiencies in vitamin E can lead to encephalomalacia, exudative diathesis, muscular dystrophy, and poor reproduction.[6,9,10] Vitamin E levels that exceed requirement levels 100 fold can manifest as decreased growth, anemia, decreased bone mineralization, as well as reduced storage of vitamin A within the liver.

Vitamin K is also poorly studied in psittacine birds. Approximately 1 ppm has been suggested in the diet.[13] Deficiencies can lead to clotting disorders and skeletal abnormalities.[6] Excesses of this nutrient are said to need to be 1000 times the requirement to manifest as problems, such as anemia and interference with vitamins B and C.

Several of the B vitamins, including B_1, B_3, B_6, B_7, B_9 and choline, have not been well studied in psittacine birds. For these nutrients, we extrapolate information from other birds like chickens to supply us with information on what is believed to be appropriate. Vitamin B_1 (thiamin) is thought to be needed at 2-4 ppm in diet and deficiencies could result in polyneuritis, polioencephalomalacia, paresis, paralysis, opisthotonos, and convulsions.[9,10,13] Vitamin B_3 (niacin) is suggested to be at 39-50 ppm in the diet and deficiencies can lead to scaley dermatitis and hock disorders in chickens.[9,10,13] Vitamin B_6 (pyridoxine) is required at 4-6 ppm in the diet of chickens and if supplied lower than this it could cause a reduced appetite, convulsions, dermatitis and anemia.[9,10,13] Excesses have been reported to cause problems in other birds including gyrfalcons having green urates, regurgitation, hyporexia and death within 24-36 hours and pigeons having liver degeneration and death with 24-48 hours of getting excessive levels.[21] Vitamin B_7 (biotin) is thought to be needed at 0.25 ppm in the diet and if it is at lower levels problems like perosis, fatty liver and kidney syndrome, dermatitis, swollen eyelids, and mandibular swellings can occur in those species it has been studied in.[10,13] Vitamin B_9 (folic acid) is suggested to be at 1-1.5 ppm in the diet and deficiencies can result in anemia, slow growth, poor feathering, and reduced egg production and hatchability.[9,10,13] Choline (sometimes referred to as vitamin B_4) is thought to be required at 1500 ppm of the diet and deficiencies could play a role in liver disease, atherosclerosis, neurologic disorders and in cockatiels unpigmented feathers.[10,13] All of these nutrients do need further investigation in psittacine birds.

There are studies on vitamin B_2 (riboflavin) in cockatiels. If birds lack this nutrient in the diet they can develop achromatosis of the feathers or a failure of development of the normal pigment in the feathers. Cockatiels will also have a reduced number of eggs in the clutch and reduced hatchability. [6] This nutrient is supposed to be at 4-6 ppm in the diet.[9,13]

Vitamin B_5 (pantothenic acid) also has some limited studies in cockatiels. When this nutrient was not available, cockatiel chicks would die 7-10 days post hatching. When fed low levels of pantothenic acid, chicks would die within 3-3.5 weeks and display an absence of body feathers but wing and tail feathers were present.[6] This nutrient is suggested to be at 11-20 ppm in the diet.[9,13]

Vitamin B_{12} (cyanocobalamin) is suggested to be at 0.01 ppm in the diet.[13] There has been documentation of a macaw developing anemia when it was not receiving adequate levels of this nutrient.[22]

Vitamin C is not required in the diet of psittacine birds because they are able to synthesize this nutrient on their own.[6,9] It is produced within the kidneys of psittacine birds and therefore, may be required in birds suffering from kidney disease (Orosz, personal communication). Addition of this nutrient to the diet must be done with caution in those species who have a propensity towards the development of iron storage disease as vitamin C will enhance the absorption of iron.[1]

Minerals

Several studies have evaluated the calcium needs of different psittacine birds. Recommended adult maintenance levels of this nutrient range from 0.3-1.1%.[4,6,9,13] In cockatiels, 0.35% dietary calcium was found to be adequate for maintaining appropriate shell thickness and egg production.[6] A study evaluating various levels of calcium and vitamin D in budgerigar diets found that metastatic mineralization was more dependent upon calcium levels and not affected by vitamin D. Although both nestling and adult budgerigars did well with 0.3% dietary calcium, concentrations of 0.7% and 1.5% calcium were toxic. The most severe metastatic mineralization lesions were seen in chicks that died spontaneously.[19]

Maintenance dietary phosphorous levels range between 0.3-0.4% of the diet.[13] The budgerigars in the aforementioned study were fed 0.7% dietary phosphorous. This level was well tolerated with 0.3% calcium.[19] Deficiencies in phosphorous can result in rickets, weakness, thin eggshells, and a depressed appetite.[10]

Recommended dietary magnesium levels range between 500-667 ppm in the diet.[9,13] Low levels of this nutrient can result in poor egg production or result in neuromuscular disorders.[10] It has been suggested that seizures occur in African grey parrots with hypomagnesemia; however, a case report investigating this problem looked at plasma levels of magnesium and not dietary nutrient levels.[23]

There are several minerals that have been poorly studied in psittacine birds and recommendations for these nutrients are based on data for other granivores. Potassium, sodium, chloride, manganese, zinc, copper, selenium, iron, and iodine recommendations are in need of more studies in psittacine birds. Potassium is suggested to be at 0.3-0.4% of the diet and deficiencies could result in retarded growth, reduced egg production with thin shells and muscle weakness.[9,10,13] Sodium is recommended at 0.1-0.15% of the diet and low levels could result in decreased egg production, poor growth, softened bones, decreased cardiac output, decreased plasma volume and decreased gonadal inactivity.[6,9,10,13] Chloride is thought to need to be at 0.1-0.15% of the diet. Deficiencies could lead to poor growth, dehydration, hemoconcentration and mortality.[6,9,10,13] Manganese is recommended to be at 60-75 ppm in the diet and deficiencies in this could lead to decreased egg production and shell strength and in chickens it is one of the nutrient deficiencies that leads to perosis.[6,9,13] Zinc is suggested to be at 40-50 ppm in the diet and deficiencies could result in retarded growth, reduced egg production and hatchability, frayed feathers and poor wound healing.[6,9,10,13] Copper is suggested to be in the diet at a level of 6-8 ppm and deficiencies can result in anemia, cardiac hypertrophy, bone deformities, and dissecting aneurysms of the aorta in waterfowl.[6,9,13] Excesses of copper could result in kidney or neurologic disorders. Selenium is believed to be required at 0.1 ppm and deficiencies can result in poor growth and chick mortality, white muscle disease, liver necrosis and pancreatic fibrosis.[6,9,10,13,24] Excesses of this nutrient can cause impaired hatching success.[10]

Recommended dietary iron levels are extrapolated from what is known in other species and ranges from 60-80 ppm.[9,13] This particular nutrient bears important consideration for those individuals or species that are prone to iron storage disease. Granivorous psittacine birds are not generally susceptible to iron storage disease, but disease has been documented in lorikeets, blue and gold macaws and African grey parrots.[1,25,26,27] Based on what is known for other species it is recommended that those individuals who are prone to this disease do not get more than 50 ppm of iron in the diet.[1] Deficiencies in iron can result in anemia.[10]

Suggested dietary iodine levels range between 0.3-0.4 ppm; and deficiencies can result in low thyroid levels, goiters, and poor growth.[9,10,13] Budgerigars on iodine-deficient diets have been found to develop goiters.[28]

Dietary Recommendations

Although many nutrients have not been specifically studied in psittacine birds, based on our current level of understanding, the general recommendations extrapolated from other granivorous species seem to meet the needs of psittacine birds. The formulated diets currently on the market for psittacine birds do meet the requirements set forth for other granivores. Therefore, formulated diets for psittacine birds are generally recommended as the base of the psittacine diet. The ratio of formulated diets in relation to other food items is debated with suggestions ranging from 60-80% of the diet. The remainder of the diet should consist of fresh items, like vegetables, fruits, grains, beans, nuts and seeds.[1] Some aviculturists are proponents of all-natural diets that do not contain formulated foods but balancing these diets can be tricky and difficult for the average owner. In addition, birds can preferentially select items out of a mix, which can create nutritional imbalances.[5]

The eclectus parrot is thought to need a higher level of vitamin A in the diet and some anecdotal reports suggest this species may be more sensitive to high levels of pellets or formulated foods in the diet. Color variety cockatiels, budgerigars, lovebirds, and parrotlets may also be sensitive to too high proportions of pellets in the diet as well and have been seen anecdotally to develop kidney disease.[20] For them it is recommended that they receive no more than 50-60% of their diet as pellets. These observations are from aviculture and in-depth studies are needed to further specify the legitimacy of these concerns.

An all-seed diet has been historically used for captive psittacine birds due to observations of what they consume in the wild; however, these diets are not currently recommended as they can contain excessive fat, low levels of calcium, phosphorus, sodium, zinc, iron, lysine, and vitamin A.[5] Problems noted in previous sections regarding these nutrients have been seen in psittacine birds kept on all seed diets.

Nutritional Strategies

There are certain differences among species that must be considered because a variety of nutritional strategies are employed by psittacine birds. Generally, most psittacine birds are classified as florivores. These are animals that consume a plant-based diet. Among these, there are granivores that consume grains, frugivores which consume fruit, and nectarivores, which consume nectars. Some birds are classified as exhibiting a mix of these nutritional strategies. There are also some omnivorous psittacine birds.[2] Many macaws, queen of Bavaria conures and Palm cockatoos are believed to require higher levels of dietary fat and fiber.[29] Lories and lorikeets are specialists that consume nectar and pollen along with some seeds, fruits, buds and insects.[1] When compared to other psittacine birds, lorikeets have lower protein needs and possibility lower vitamin A requirements as well.[1,18] Aviculturists have said African grey parrots need a higher protein and fat diet, although good scientific references for this do not exist. African grey parrots in the wild have been found to forage from 14 different tree species and will consume different parts of these trees. They have been seen to eat the seeds and arils of four of these tree species, the fruits of seven to nine of these trees, the flowers of two of these trees, and the buds of one of these trees.[30] They also have been observed to eat the vegetative parts of an aquatic plant of the *Lemma* species.[31,32] Some of the birds classified as omnivores include the Eastern rosella, Major Mitchell cockatoo, and the sulphur-crested cockatoo. These individuals still consume seeds, fruits, and flowers like other birds but they also will consume insects.[2] It is important to note that although these may be the items birds are consuming in the wild, a wild life and a captive life are vastly different in terms of energy expenditure and risks to the birds so caution must be exercised when looking to the wild diet to dictate the captive diet.

Additional psittacine species have been studied in the wild to observe what food items they are selecting. Hyacinths consume acuri and bocaiuva palm nuts. *Calyptorhyncus* genus cockatoos eat seeds from the cones of sheoak and bulloak genera coniferous trees.[29] Austral parakeets have been observed to eat pollen, seeds, honeydew, lichens, and parasitic mistletoe.[29] Cockatiels have been documented to eat 17 species of native grasses and herbaceous dwelling plants in their native environment along with imported cereal grains.[29]

Nutritional Additives

Understanding of the importance of nutritional additives and supplements, such as prebiotics and probiotics, is poor in psittacine nutrition at the time of this writing; however, it is important to at least acknowledge that there are prebiotics, probiotics, and dietary textures that contribute to an individual's nutrient basis and health status. Knowledge about the gut microbiome is a hot topic in human medicine and surely its importance will likely be further elucidated in time with birds as well.

Conclusions

Further work is needed to investigate various aspects of nutrition for different species of psittacine birds. With time, more knowledge of individual nutrients in different species of psittacine birds will be gained, which will help practitioners.

References

1. Koutsos E, Gelis S, Echols MS. Advances in nutrition and nutritional therapy. In: Speer BL, ed. *Current Therapy in Avian Medicine and Surgery.* St. Louis, MO: Elsevier; 2016:142-176.

2. Koutsos EA, Matson KD, Klasing KC. Nutrition of birds in the order Psittaciformes: A review. *J Avian Med Surg*. 2001;15(4):257-275.

3. Harper EJ, Skinner ND. Clinical nutrition of small psittacines and passerine. *Sem Exot Pet Med.* 1998;7: 116-127.

4. Brightsmith DJ, Mcdonald D, Matsafuji D, Bailey CA. Nutritional content of the diets of free-living scarlet macaw chicks in southeastern Peru. *J Avian Med Surg*. 2010;24:9-23.

5. Brightsmith DJ. Nutritional levels of diets fed to captive Amazon parrots: does mixing seed, produce, and pellets provide a healthy diet? *J Avian Med Surg*. 2012;26:149-160.

6. Roudybush TE. Nutrition. In: Altmann RB, Clubb SL, Dorrestein GM, Quesenberry K, eds. *Avian Medicine and Surgery.* Philadelphia, PA: WB Saunders; 1997:27-53.

7. Koutsos EA, Smith J, Woods LW, Klasing KC. Adult cockatiels (*Nymphicus hollandicus*) metabolically adapt to high protein diets. *J Nutr*. 2001;131:2014-2020.

8. Roudybush TS. Food and water interrelations and the protein requirements for growth of an altricial bird, the cockatiel (*Nymphicus hollandicus*). *J Nutr*. 1986;116:552-559.

9. Brue RN. Nutrition. In: Ritchie BW, Harrison GJ, Harrison LR, eds. *Avian Medicine: Principles and Application.* Lake Worth, FL: Wingers Publishing Inc; 1994:63-95.

10. Macwhirter P. Malnutrition. In: Ritchie BW, Harrison GJ, Harrison LR, eds. *Avian Medicine: Principles and Application.* Lake Worth, FL: Wingers Publishing Inc; 1994:842-861.

11. Lumeij JT. Hepatology. In: Ritchie BW, Harrison GJ, Harrison LR, eds. *Avian Medicine: Principles and Application.* Lake Worth, FL: Wingers Publishing Inc; 1994:522-537.

12. Strunk A. Psittacine nutrition: focus on lipid metabolism and obesity. *Proc Annu Conf Am Board Vet Pract.* 2012.

13. Kollias GV, Kollias HW. Feeding passerine and psittacine birds. In: Hand MS, Thatcher CD, Remillard RL, Roudebush P, eds. *Small Animal Clinical Nutrition,* 4th Ed. Marceline, MO: Walsworth Publishing Co; 2000:979-991.

14. Heinze CR, Hawkins MG, Gillies LA, et al. Effect of dietary omega-3 fatty acids on red blood cell lipid composition and plasma metabolites in the cockatiel, *Nymphicus hollandicus*. *J Anim Sci*. 2012;90: 3068-3079.

15. Bavelaar FJ, Beynen AC. Severity of atherosclerosis in parrots in relation to the intake of alpha-linolenic acid. *Avian Dis*. 2003;47:566-577.

16. Petzinger C, Hetley JJ, Cornejo J, et al. Dietary modification of omega-3 fatty acids for birds with atherosclerosis. *J Am Vet Med Assoc*. 2010;236:523-528.

17. Koutsos EA, Tell LA, Woods LW, Klasing KC. Adult cockatiels (*Nymphicus hollandicus*) at maintenance are more sensitive to diets containing excess vitamin A than to vitamin A deficient diets. *J Nutr*. 2003;133: 1898-1902.

18. Park F. Vitamin A toxicosis in a lorikeet flock. *Vet Clin North Am Exot Anim Pract*. 2006;9:495-502.

19. Phalen DN. Renal failure associated with metastatic mineralization of the kidneys in birds. *Proc Annu Conf Assoc Avian Vet*. 2009.

20. Echols MS. Avian renal disease: part I and II. *Proc Annu Conf Am Board Vet Pract*. 2012.

21. Samour J, Perlman J, Kinne J, et al. Vitamin B_6 (pyridoxine hydrochloride) toxicosis in falcons. *J Zoo Wildl Med*. 2016;47:601-608.

22. Tully TN, Beaufrere H, Sirninger JR, et al. Suspected diet-related vitamin B_{12} anemia in a hyacinth macaw (*Andorhynchus hyacinthinus*) *Proc Annu Conf Assoc Avian Vet*. 2010:317-320.

23. Kirchgessner MS, Tully TN, Nevarez J, et al. Magnesium therapy in a hypocalcemia African grey parrot (*Psittacus erithacus*). *J Avian Med Surg*. 2012;26:17-21.

24. Schmidt RE, Reavill DR, Phalen DN. *Pathology of Pet and Aviary Birds*, 2nd Ed. Ames, IA: Wiley Blackwell; 2016.

25. Rasidi EK, Xie S, Hsu C. Iron storage disease in a blue and gold macaw (*Ara ararauna*). *Proc ExoticsCon*. 2019;384.

26. Lamb S, Reavill D, Biswell E. Pathology in practice: hepatic iron overload in a Congo African grey parrot (*Psittacus erithacus*). *J Am Vet Med Assoc*. 2022;259:S2. https://doi.org/10.2460/javma.21.04.0209.

27. O'Connor MR, Garner MM. Iron storage disease in African grey parrots (*Psittacus erithacus*) exposed to a carnivorous diet. *J Zoo Wildl Med*. 2018;49:172-177.

28. Loukopoulos P, Bautista AC, Puschner B, et al. An outbreak of thyroid hyperplasia (goiter) with high mortality in budgerigars (*Melopsittacus undulatus*). *J Vet Diagn Invest*. 2015;27:18-24.

29. Toft CA, Wright TF. *Parrots of the Wild*. Oakland, CA: University of California Press; 2015.

30. Tamungang SA, Ajayi SS. Diversity of food of the grey parrot (*Psittacus erithacus*) in Korup National Park, Cameroon. *Bull African Bird Club*. 2003;10:33-36.

31. Fotso R. Survey status of the distribution and utilization of the grey parrot in Cameroon. Secretariat CITES; 1998.

32. May DL. The behavior of African grey parrots in the rainforest of the Central African Republic. *PsittaScene*. 1996;8:8-9.

Section 2

AAV Plenary Session

Anna Osofsky, DVM, DABVP (Avian Practice), ***AAV President***
Session Chair

Use of Intravenous Lipid Emulsion Therapy for the Treatment of Acute Poisonings in Birds

Lauren V. Powers, DVM, DABVP (Avian Practice), DABVP (Exotic Companion Mammal Practice), and
Laura A. Stern, DVM, DABVT

Session #3436

Affiliation: From Avian and Exotic Pet Service, Carolina Veterinary Specialists, 12117 Statesville Road, Huntersville, NC 28078, USA (Powers), The ASPCA Animal Poison Control Center, 1806A Fox Drive, Champaign, IL 61820, USA (Stern).

Note: Abstract format selected by author(s). See Front Matter for description of conference proceedings manuscripts submitted as scientific abstracts.

Abstract: A 4-year-old male white bellied caique (*Pionites leucogaster*) was presented for cannabis (water pipe resin) ingestion that was observed 2 hours prior. On presentation the bird was stuporous and hyperesthetic. The bird began having seizures responsive to intramuscular midazolam approximately 7 hours after admission. The bird's condition progressively worsened over the following 3 hours although bradycardia was not observed. An intravenous catheter was placed and intravenous lipid emulsion (ILE; Intralipid 20%, Baxter Healthcare, Deerfield, IL, USA) was administered at 16 ml/kg/h with an estimated total volume of 62 ml/kg. A rapid favorable response to treatment was observed. Monitoring for lipemia was not performed due to patient size. The bird was discharged the following day and remains well at the time of this writing.

ILE has been used to treat lipophilic toxicoses in a variety of animal species including dogs and cats,[2-5,14] horses,[6] inland bearded dragons,[7] rabbits,[8] guinea pigs,[9] and several avian species.[10-12] ILE has previously been used as a successful therapy for tetrahydrocannabinol (THC) toxicosis in dogs and humans.[1,2,14] From 2018-2022, the ASPCA Animal Poison Control center received 38 calls regarding birds (excluding poultry and waterfowl) exposed to cannabis, representing 1.8% of total bird cases.[15]

ILE has occasionally been associated with significant adverse effects,[3-5] including a dog with THC toxicosis that developed acute respiratory distress syndrome (ARDS).[2] The comparatively high dosage of ILE used in this caique (62 ml/kg vs. 9.5-20.1 ml/kg suggested for dogs[16-19] and 16.5-30 ml/kg used in other avian species[10,11]) was well-tolerated in this bird.

References

1. Aksel G, Güneysel Ö, Taşyürek T, et al. Intravenous lipid emulsion therapy for acute synthetic cannabinoid intoxication: Clinical experience in four cases. *Case Rep Emerg Med.* 2015:1-5.

2. Botha H, Jennings SH, Press SA, et al. Suspected acute respiratory distress syndrome associated with the use of intravenous lipid emulsion therapy in a dog: a case report. *Front Vet Sci.* 2019;6:1-6.

3. Romans CW, Day TK, Smith JJ. Oxidative red blood cell damage associated with propofol and intravenous lipid emulsion therapy in a dog treated for 5-fluorouracil toxicosis. *J Vet Emerg Crit Care.* 2020;30(4): 481-486.

4. Seitz MA, Burkitt-Creedon JM. Persistent gross lipemia and suspected corneal lipidosis following intravenous lipid therapy in a cat with permethrin toxicosis. *J Vet Emerg Crit Care.* 2016;26(6):804-808.

5. Epstein SE, Hopper K, Farrell KS. Manual plasma exchange to treat an accidental overdose of intravenous lipid emulsion in a dog with baclofen toxicosis. *J Am Vet Med Assoc.* 2022;260(6):650-656.

6. Vieitez V, Gómez de Segura IÁ, Martin-Cuervo M, et al. Successful use of lipid emulsion to resuscitate a foal after intravenous lidocaine induced cardiovascular collapse. *Equine Vet J.* 2017;49(6):767-769.

7. DeMel D, Gleeson M, Schachterle K, et al. Successful treatment of ivermectin overdose in a bearded dragon (*Pogona vitticeps*) using gastric lavage and intravenous lipid emulsion. *J Vet Emerg Crit Care.* 2022;32(5):680-684.

8. Branco SEMT, Mattoso CRS, Botelho AFM, et al. Intravenous lipid emulsion treatment in rabbits with ivermectin toxicosis. *J Vet Emerg Crit Care.* 2021;31(3):340-350.

9. Ebel JJ, Gannon K, Kraenzlin M. The use of intralipid emulsion therapy to treat status epilepticus in a guinea pig (*Cavia porcellus*) secondary to severe ivermectin toxicity. *J Vet Emerg Crit Care.* 2023;33(1):107-111.

10. Schmidt LK, Keller KA, Tonozzi C, et al. Intralipid emulsion therapy for the treatment of suspected toxicity in 2 avian species. *J Avian Med Surg.* 2022;36(4):394-399.

11. Lubian E, Capitelli R, Nappi S, et al. Use of intralipid emulsion therapy to treat suspected oleander toxicosis in a domestic goose (*Anser anser domesticus*). *J Exot Pet Med.* 2021;39:78-80.

12. Barron HW, Kehoe S, Bast R, et al. Use of intravenous lipid emulsion therapy as a novel treatment for brevetoxicosis in double-crested cormorants (*Phalacrocorax auritus*). *Proc Annu Conf Assoc Avian Vet.* 2019:151.

13. Amissah RQ, Vogt NA, Chen C, et al. Prevalence and characteristics of cannabis-induced toxicoses in pets: Results from a survey of veterinarians in North America. *PLoS One.* 2022;17(4):e0261909.

14. Williams K, Wells RJ, McLean MK. Suspected synthetic cannabinoid toxicosis in a dog. *J Vet Emerg Crit Care.* 2015;25(6):739-744.

15. American Society for the Prevention of Cruelty to Animals. Unpublished data. AnTox™. ASPCA Animal Poison Control Center; 2023.

16. Epstein SE, Hollingsworth SR. Ivermectin-induced blindness treated with intravenous lipid therapy in a dog. *J Vet Emerg Crit Care.* 2013;23(1):58-62.

17. Herring JM, McMichael MA, Corsi R, et al. Intravenous lipid emulsion therapy in three cases of canine naproxen overdose. *J Vet Emerg Crit Care.* 2015;25(5):672-678.

18. Royle K, Bandt C. Intravenous lipid emulsion to treat suspected cocaine toxicosis in a dog. *Can Vet J.* 2020;61(1):49-52.

19. Wright HM, Chen AV, Talcott PA, et al. Intravenous fat emulsion as treatment for ivermectin toxicosis in three dogs homozygous for the ABCB1-1Δ gene mutation. *J Vet Emerg Crit Care.* 2011;21(6):666-672.

Agency-Based Parrot-to-Parrot Video Calling as Enrichment for Birds

Rebecca Kleinberger, PhD,
Ilyena Hirskyj-Douglas, PhD,
Jennifer Cunha, JD,
Susan Clubb, DVM

Session #3447

Affiliation: From Northeastern University, 360 Huntington Ave, Boston, MA 02115, USA (Kleinberger), Massachusetts Institute of Technology, 77 Massachusetts Ave, Cambridge, MA 02139, USA (Kleinberger), University of Glasgow, Glasgow G12 8QQ, UK (Hirskyj-Douglas), Parrot Kindergarten, Inc., 601 Heritage Dr, Ste 424, Jupiter, FL 33458, USA (Cunha), Rainforest Clinic Veterinary Clinic for Birds and Exotics, 3319 E Road, Loxahatchee, FL 33470, USA (Clubb).

Abstract: Over 20 million parrots are kept as pets in the United States, often lacking appropriate stimuli to meet their high social, cognitive, and emotional needs. An approach was developed based on strong ethics and animal agency to allow parrots to engage in video-calling other parrots. A pilot experiment was conducted with feedback from bird experts and a 3-month study with 18 pet birds was run to evaluate the potential, value and usability of a parrot-to-parrot video-calling system. The system was evaluated with regard to perception, agency, engagement, and overall perceived benefits. Results show that every bird used the system (147 corroborated bird-triggered calls), that most of them exhibited high motivation and intentionality, and that all caretakers reported perceived benefits, including learning to forage or even fly by watching others.

Introduction

Parrots are a highly social species of birds that often suffer from lack of companionship and species-specific socialization in captivity.[1] This can result in stereotypical behaviors such as self-mutilation, feather-destructive behavior, and aggression, which can be indicative of poor psychological well-being.[2] Various approaches have been recommended to alleviate these issues, including environmental enrichment, positive reinforcement-based behavior modification, and the use of restraints or collars.[3] While enrichment and training are important for parrots, none of these approaches replace the importance of socialization, particularly with other parrots. In addition, co-housing parrots with other birds or arranging in-person meetups theoretically may improve bird well-being,[4] but there are many drawbacks, including the practical difficulties of arranging in-person social gatherings for birds, and the high rate of fatal transmissible diseases such as avian ganglioneuritis, which inflicts 43% of the parrot population and calls into question the ethics of parrots' proximal contact and socialization. In recent decades, the human-computer interaction (HCI) community has worked on providing tools, approaches, and applications to help support the well-being and socialization of captive animals. This paper focuses on the development of agency-based parrot-to-parrot calling devices to help meet parrots' social needs. The aim is not to assess parrot's intelligence but to provide them with at-home enrichment.

The contributions of this project are 1) an ethical framework and novel protocol for bird-triggered video calls, 2) an 18-bird longitudinal study over 3 months assessing the potential of tablet-based video calling for parrots, 3) analyses and insights into how birds use their new agency and engagement in the video-calling process, 4) interviews and feedback from the human caretakers and 5) new perspectives and a design proposal for future parrot-centered video-calling systems. The Ethics Committee approved this project for Animal (number EA 01/22) and Human Use as Subjects (number 300210172) at Glasgow University.

Background

Assessing the use of tablets for birds requires determining if parrots have the biologic and physiologic ability to perceive and understand screen-based stimuli and the feasibility of using off the shelf devices. Although parrot vision differs from humans as well as within parrot species, decades of research suggests that they can make sense of screen-based stimuli, with early previous work pigeons pecking at targets.[5] Although perceiving single dots on screens is different from complex scenes, some birds also appear to make sense of moving images despite reduced color spectrum and flicker.[6] Standard validation tests have been used to compare birds' attention to the object of interest on the screen with objects in real life.[7] In this study, sense-making is assessed by measuring parrots' attention to the other bird appearing on screen, moving off screen, or moving within the screen in the initial exposure to video calls in Phase 1.

Research in the HCI and animal-computer interaction (ACI) communities have focused on developing tools to provide enrichment and socialization to captive animals, including pets left home alone for long periods of time. Such research and technology development includes investigating how animals can use technology to interact with each other as well as humans.[8] Interactivity and agency are important considerations to ensure ethical and meaningful animal experience, and both animal consent and capacity are necessary for animals to express agency. To measure animal intention, an economics-based approach has been used to assess how highly an animal values a proposed option, often implemented in the context of operant conditioning (e.g., how much will an animal pay with behavior for a certain outcome).[9] The current study implements corroboration tasks to increase the pay and measure the value of consent from participating birds by requiring them not only to ring a bell but also to touch the photo of a bird to call.

Interactive systems that give animals agency in shaping their surroundings have shown potential benefits in reducing stereotypies and improving behavior.[10] Researchers have developed choice-based technology for environmental control for animals, such as cockatoos[11] and macaws[12] to choose food, toys, music, and even call their caretakers.[13] The perception of self-agency can be beneficial to animals, and there is potential for interactive tools that maintain animal control and agency. The concept of competency is central to understanding agency. Here, the authors assessed agency quantitatively by counting the number of outgoing calls made independently by birds during a specific phase. Engagement can also be assessed quantitatively through time-on-task, and in this research, the time-on-call was used as a method for measuring engagement. Measured positive engagement provides further validation of the birds' agency and intent around making the video calls.

Materials and Methods

Following the expert survey and pilot, a longitudinal at-home study on 18 pet parrots was conducted over a 3-month period. The study was conducted in two phases. The first phase, called "Meet-and-Greet", focused on establishing the association between 1) touching a bell and 2) touching the image of another bird on the screen to trigger a call. The second phase called "Open Calls" focused on measuring the birds' social interaction through pairwise video calls. The participants were grouped according to their time zone, availability, species, and bird size, and each bird was individually supported. The study involved home-based parrots, over 1 year old, with no known behavioral issues, and that were comfortable with looking at screens and touching objects.

Data was gathered from video submissions and diary inputs. Parrots' interaction times with the bell and tablet, as well their engagement times as caller and callee birds via the recorded video, were measured. Although human participants were instructed to stop calls when birds exhibited disengagement behaviors, the end call time was confirmed by an expert parrot behaviorist to verify if the birds showed earlier signs of disengagement that may have gone unnoticed. Eight such cases were adjusted in the analysis, and caretakers were notified to improve their understanding of disengagement behaviors.

To assess whether parrot participants perceived the other bird on the screen and potentially made sense of the video calls, meet-and-greet sessions (Phase 1) were examined since they represented the first introduction to video calling. The examination focused on changes in behavior between instances when the other bird was present versus absent from the screen, looking for signs of visual and auditory perception and interactive behaviors. Specifically, the distance the bird chose to stand from the screen, presence of timely responses to the other bird either appearing on or leaving the screen, clear indications that the bird was following the other bird on the screen, and timely responses to auditory stimuli were noted.

More details about the methods, ethics, setup, training protocol for both humans and parrots, and analysis methods can be found in Kleinberger, et al.[13]

Results

Based on the principles of animal behavior and cognition and ACI/HCI, the data analysis method aimed to evaluate parrots' calls based on four variables: perception, agency, engagement and perceived benefits.

Two authors independently labeled each video, with an inter-rater reliability of 89%. Regarding sense-making assessment, 50% of the birds stood close to the screen. Fourteen out of 18 (78%) birds exhibited clear timely responses to the other bird appearing on screen, six of which came closer, while three reacted with head bobbing behaviors. Five out of 18 (28%) parrots reacted when the other bird moved off-screen. Five out of 18 (28%) birds appeared to follow the other birds on the screen.

To evaluate whether birds made use of their newfound agency during Phase 2, for each bird and each session, measurements were made of how many times the bird triggered an outgoing call (up to two total outgoing calls per bird per session), including the distribution and variability between and within groups. Birds' individual social scores were measured as the ratio between the number of calls triggered and the total potential number of calls over the eight sessions. Pearson's correlation coefficient was used to test the linear relationships between the number of calls the parrots made and the number of calls they received. All the birds participated in video calls and were observed to make outgoing calls. One key finding was the strong correlation between the number of calls made by a bird and the number of calls it received. There are various possible explanations for this connection, such as increased familiarity with other birds or the voice of the other caretaker, which may lead to rapport and more attention from caretakers. Another interesting finding was which birds initiated the calls.

Aside from the number of outgoing calls triggered, the study also examined quantitatively the engagement of the birds through the calls, a potential marker of their level of interest, through how long they were engaged in the calls they themselves triggered versus those they received using the corrected duration of calls. Correlation between use of their agency (individual social scores) and their engagement during calls was determined. A significant correlation between high social scores and low disengagement from calls was observed.

Various factors may influence engagement, including size, age, household composition, and human daily availability; however, no significant correlation was found between bird size and engagement. Additional longitudinal validation of the study and co-specific grouping are needed to establish the effects of specific factors. Regardless of individual factors, the more a bird uses its agency, the more likely it is to stay engaged during the calls it triggers.

In both the pilot and full study, the human's role in the process was deemed crucial; from training the birds in the calling system to providing continued presence and encouragement during the calls. The video call system aimed to enhance the birds in their regular life context, where humans already play a significant part. It required careful monitoring and balance as caretakers' enthusiasm and desire for the birds to engage could cause stress for the parrots. This emphasizes the importance of caretaker training and adherence to protocols in any future systems centered around birds' agency.

Bird Caregiver Interview and Survey

The study found that all caretakers believed the experience was beneficial for their birds and for themselves. The caretakers reported various new observed behaviors supporting a beneficial outcome to the birds, including more confidence, calmer behavior, new foraging behaviors and new flying behaviors. One caregiver said, "She came alive during the calls." Another reported, "He would clearly ring the bell, and very decisively select a friend" and another said, "After the calls with [another participant bird], she started flying more. I think she realized that flying is an okay behavior. [...] I'm very happy about this!"

No negative behavior was reported by human participants. In addition, caretakers reported interactive-mirroring behaviors between the birds during calls, including both birds playing with toys, foraging, head-bobbing, dancing, waving, saying hello, vocalizing, learning new sounds and singing together. Those observations were further supported by analysis of the video recordings.

Human participants also reported having learned more about their birds and feeling better able to recognize behaviors. The sessions also provided bonding time between humans and birds, mediated by technology. The improved understanding of the caretakers for their animals may provide a healthy ground for interspecies bonding.

Discussion

Results of the perception analysis suggest that birds exhibited a diverse range of sense-making responses triggered by visual and auditory stimuli, including reacting to various sound sources from the correspondent's side, the presence of another bird on the screen, following the bird on-screen or reacting to their correspondent disappearing.

A significant correlation between the number of outgoing calls and the number of received calls was observed, which led to the development of a social score for each bird and group. This connection may be explained by increased familiarity with the other bird leading to rapport and increased special attention from caretakers. It was also found that the role of the human in the loop was instrumental, from initial training in the calling system to their continued presence and encouragement during the calls. Caretaker enthusiasm and desire for their bird to engage could be stressful for a parrot, emphasizing the importance of caretaker training and protocol adherence.

Screen-based parrot-to-parrot video-calling could be used for social enrichment, but further longitudinal validation and co-specific grouping are needed to establish the effects of specific factors. In addition, groups of two birds may allow for more and easier enrichment opportunities through socialization, as coordinating schedules was easier. Overall, the study highlights the potential of screen-based video-calling for parrots and the importance of considering their agency and needs when developing such systems.

The study required a large time commitment from caretakers and synchronizing schedules, which may make it difficult to sustain the calls long-term without support and structure provided by the research team. The limited range of the touchscreen device-housed cameras for video calls made the viewing experience inconsistent in some settings. Future work could investigate whether the birds would react similarly to pre-recorded videos and if ergonomic improvement on the tablet app display could better support their experience. While exposing birds to live, interactive stimuli could be argued as more natural and appropriate for their species identity, video calls offer more flexibility and ease of usage. There are ethical concerns about deceiving animals in playback experiments and birds' cognitive processing was not evaluated in this study. Further research could investigate whether grouping birds with others of the same species affects interaction levels, whether different populations of parrot species find video calls more or less meaningful, and optimal call duration and frequency.

Conclusions

Following a pilot experiment and expert survey, a 3-month study was conducted with 18 pet birds to evaluate the potential, value and usability of a parrot-to-parrot video-calling system. The system was assessed in terms of perception, agency, motivation, engagement strength and overall perceived benefits. With 147 corroborated bird-triggered calls, results show that every bird used the system, that most of them exhibited high engagement, and that all caretakers reported perceived benefits. The role of the caretaker appeared central in the birds' interaction with technology. The experience appeared to not only provide benefits to the parrots through bird-specific interactions and behavior learning, but also to the human participants by learning more about their animal companions. Leveraging internet calls ultimately allowed the birds to bond socially with other birds while also intimately reconnecting with their human companion.

References

1. Bradshaw GA, Yenkosky JP, McCarthy E. Avian affective dysregulation: psychiatric models and treatment for parrots in captivity. *Proc Annu Conf Assoc Avian Vet.* 2009.

2. Ebisawa K, Nakayama S, Pai C, et al. Prevalence and risk factors for feather-damaging behavior in psittacine birds: Analysis of a Japanese nationwide survey. *PLoS ONE.* 2021;16(7):e0254610.

3. Duncan B. Avian intelligence, parrot behavior and toys. *AFA Watchbird.* 1997;24(4):13-16.

4. German Federal Ministry of Food, Agriculture and Consumers. 1995. German Minimum Requirements for the Keeping of Parrots. Available at: http://www.thebirdschool.com/caring-for-parrots/parrot-cages-parrot-cages/german-minimum-requirements-for-the-keeping-of-parrots/. Accessed September 10, 2022.

5. Gibson BM, Wasserman EA, Frei L, Miller K. Recent advances in operant conditioning technology: a versatile and affordable computerized touchscreen system. *Behav Res Methods Instrum Comput.* 2004;36(2): 355-362.

6. Pepperberg IM, Wilkes SR. Lack of referential vocal learning from LCD video by grey parrots (*Psittacus erithacus*). *Interact Stud.* 2004;5(1):75-97.

7. Lea SEG, Dittrich WH. What do birds see in moving video images. In: Fagot J, ed. *Picture Perception in Animals.* New York, NY: Taylor & Francis Routledge; 2000:143-180.

8. Hirskyj-Douglas I, Lucero A. On the internet, nobody knows you're a dog...unless you're another dog. *Proc Conf Human Factors Comput Syst.* 2019;1-12.

9. Duncan IJH. Measuring preferences and the strength of preferences. *Poult Sci.* 1992;71(4):658-663.

10. Grandin T, Curtis SE, Widowski TM, Thurmon JC. Electro-immobilization versus mechanical restraint in an avoid-avoid choice test for ewes. *J Anim Sci.* 1986;62(6):1469-1480.

11. Cunha J, Rhoads C. Use of a tablet-based communication board and subsequent choice and behavioral correspondences in a Goffin's cockatoo (*Cacatua goffiana*). *Proc Conf Anim-Comput Interact.* 2020;1-9.

12. Kleinberger R, Harrington AHK, Yu L, et al. Interspecies interactions mediated by technology: an avian case study at the zoo. *Proc Conf Human Factors Comput Syst.* 2020;1-12.

13. Kleinberger R, Cunha J, Vemuri M, Hirskyj-Douglas I. Birds of a feather video-flock together: agency-based parrot-to-parrot video-calling system for interspecies ethical enrichment. *Proc Conf Human Factors Comput Syst.* doi.org/10.1145/3544548.3581166.

Section 3

AAV Cutting Edge Session

Jennifer Graham, DVM, DABVP (Avian Practice), DABVP (Exotic Companion Mammal Practice), Dipl ACZM
Session Chair

The Attending Veterinarian's Role in Compliance with Animal Welfare Standards in Birds

Susan Clubb, DVM

Session #3433

Affiliation: From Rainforest Clinic for Birds and Exotics, and Hurricane Aviaries, Inc., 3319 E Road, Loxahatchee, FL 33470, USA.

Abstract: On February 21, 2023 the U.S. Department of Agriculture (USDA) Animal and Plant Health Inspection Service (APHIS), published the final rule, Standards for Birds Not Bred for Research Under the Animal Welfare Act (9 CFR Parts 1, 2, and 3 Docket No. APHIS 2020-0068). The rule is effective March 23, 2023. Current Animal Welfare Act (AWA) licensees and registrants must comply with this rule by August 21, 2023. New AWA licensees and registrants, must comply with this rule by February 21, 2024.

Introduction

In 2002, Congress amended the definition of animal in the AWA by limiting the exclusion of birds to only those bred for use in research. In 2020, in response to lawsuits from animal welfare organizations, the US Court of Appeals for the District of Columbia ruled that USDA must publish a proposed rule by Feb 22, 2022, and a final rule no more than a year later. The final rule was 237 pages. The most important points for attending veterinarians are summarized here.

At least five requirements for attending veterinarians were removed between publication of the proposed rule and the final rule. An attending veterinarian must be a graduate of an accredited school of veterinary medicine, but a state license is not required, and the attending veterinarian could even reside in a different state. Ideally the attending veterinarian would be experienced in avian medicine. Board certification is not required. The attending veterinarian must make at least the first inspection. There is no required frequency of subsequent inspections, but annually is recommended. If the attending veterinarian lives far from the facility, s/he can work in conjunction with a surrogate local veterinarian who has less experience. The attending veterinarian can work with the aviculturist by telemedicine between visits to the facility. Regulations, forms and more can be found at www.aphis.usda.gov/animalwelfare.

Part 1 - Definitions

Bird - Any member of the class Aves, excluding eggs.

Carrier - Operator of any airline, railroad, motor carrier, shipping line or other enterprise engaged in transporting animals for hire. (Except anyone transporting a migratory bird, MBTA).

Exhibitor - Any person public or private, exhibiting any animals including carnivals, circuses, animal acts, free flight shows, zoos, educational exhibits whether operated for profit or not.

Exemptions - Retail pet stores, pigeon races, household pets, county fairs or livestock shows, falconry, bird fancier shows, exhibitions intended to advance agriculture.

Farm animals - Poultry (chickens, turkeys, swans, partridges, guinea fowl, peafowl, ducks, geese, pigeons and doves, grouse, pheasants and quail.

Intermediate handler - Any person who receives custody of birds in connection with transportation in commerce.

Pet animal - Includes canaries, cockatiels, lovebirds and budgerigar parakeets. This term excludes exotic or wild animals. See complete list on the USDA website.

Retail pet store - A place of business of residence at which the seller, buyer and animals available for sale are physically present so buyer may observe.

Weaned - A bird that has become accustomed to take food and has done so, without supplemental feeding … for a period of 5 days.

Also Exempt – Any person who sells 200 or fewer birds 250 grams or less, or sells eight or fewer birds more than 250 grams determined by average adult weight of the species.

Any person maintaining four or fewer raptors for exhibition, and holds a valid U.S. Fish and Wildlife Service (USFWS) permit.

License types

Class A- Breeder - Sells birds bred at the licensed facility for pets or exhibition.
Class B - Dealer - Buys and sells birds not bred by the licensee.
Class C - Exhibitor - Any person public or private, exhibiting any animals including carnivals, circuses, animal acts, free flight shows, zoos, educational exhibits whether operated for profit or not.

Registration types

Carrier (Exp - airlines)
Intermediate carrier
Research facility

Application can be made on the USDA APHIS Animal Care website.

Part 2 - Regulations

2.40. Attending veterinarian and adequate veterinary care

The role of an attending veterinarian is much more comprehensive than providing veterinary medical care to sick animals. They also provide important guidance to help licensees and registrants keep their animals healthy. In general, an attending veterinarian's duties include:

- Providing each individual animal veterinary care that ensures its health and well-being.
- Working with the facility to develop a program that covers all aspects of animal health: communication with the veterinarian, disease monitoring and prevention, husbandry, nutrition, provision of emergency care, employee knowledge, and humane euthanasia.

- A written program of veterinary care is required for all facilities with a part-time attending veterinarian or consultant arrangement. The attending veterinarian is not required to sign the program of veterinary care.

- Conducting regular visits to the premises. There is no minimum frequency of visits for birds. APHIS recommends annually, but this is at the discretion of the attending veterinarian. Visits should occur as often as necessary to ensure that the attending veterinarian is aware of the conditions of the animals and facilities; and has the information needed to ensure the provision of adequate veterinary care and to oversee the adequacy of animal care and use. Additional site visits may be necessary for disease outbreaks; ill or injured animals; training of caretakers, or for the proper introduction of new animals.

- Animals in need of care may be taken to the clinic or practice of the attending veterinarian. These visits do not replace the need for regularly scheduled visits to the premises of the licensee or registrant.

- The attending veterinarian is also responsible for overseeing and approving additional practices for certain covered species and at research facilities.

2.35. Recordkeeping requirements

New information collection requirements created by the regulation include bird identification records, environmental enrichment plan records, cleaning and sanitation records, consignment documents and certifications for shipments of birds. Individual animal identification is not specifically required for birds. The birds should be on their acquisition records. If the birds are transported they must be identified at the time of delivery by a label (attached to the primary enclosure with the number of birds, species, distinctive physical features and identifying marks), or by either a leg or wing band applied to each bird, or by microchip

2.50. Time and method of identification

When one or more birds are confined in a primary enclosure, the birds shall be identified by:

1) A label attached to the primary enclosure which shall bear a description of the birds included in the primary enclosure: number of birds, species, any distinctive feathers or identifying marks.

2) A leg or wing band.

3) A transponder placed in a standardized location (facility must have a compatible reader).

2.75. Breeders, exhibitors and dealers

They will be required to maintain records of all birds. See regulations for specific details.

Part 3 - Standards

New facility standards are extensive. While the attending veterinarian should be familiar with those standards, they are not expected to evaluate compliance with those standards; however, some of the facility standards require approval or consultation by the attending veterinarian.

Space requirements

Both part-time and full-time attending *veterinarians* must consult with the facility to ensure that the space in all enclosures housing birds is adequate and allows for normal postural and social adjustments. Inadequate space may be indicated by evidence of malnutrition, poor condition, debility, stress or abnormal behavior patterns.

The normal postural and social adjustments of a bird may be restricted: When the attending *veterinarian* determines that making species-typical postural or social adjustments, such as dust-bathing, foraging, or running, would be detrimental to the bird's good health and well-being. The attending veterinarian must document the reason and recommended duration for the restriction and make such records available for review by an APHIS inspector.

§ 3.154 Environment enhancement to promote psychological well-being

Dealers, exhibitors, and research facilities must develop, document, and follow a species appropriate plan for environment enhancement adequate to promote the psychological well-being of birds. The plan must be approved by the *attending veterinarian* and must be in accordance with the regulations in this subpart and with currently accepted professional standards as cited in appropriate professional journals or reference guides. This plan must be made available to APHIS upon request, and, in the case of research facilities, to officials of any pertinent funding agency. The plan, at a minimum, must address each of the following:

(1) Social grouping. The environment enhancement plan must include specific provisions to address the social needs of species of birds known to exist in social groups in nature. Such specific provisions must be in accordance with currently accepted professional standards as cited in appropriate professional journals or reference guides. The plan may provide for the following exceptions:

(a) If a bird exhibits vicious or overly aggressive behavior, or is debilitated as a result of age or other conditions (e.g., arthritis), it can be housed separately;

(b) Birds that have or are suspected of having a contagious disease must be isolated from healthy animals in the colony as directed by the *attending veterinarian*. When an entire group or room of birds is known to have been or believed to be exposed to an infectious agent, the group may be kept intact during the process of diagnosis, treatment, and control.

(c) Birds may not be housed with other species of birds or animals unless they are compatible, do not prevent access to food, water, or shelter by individual animals, and are not known to be hazardous to the health and well-being of each other. Compatibility of birds must be determined in accordance with generally accepted professional practices and actual observations as directed by the *attending veterinarian*, to ensure that the birds are in fact compatible. Individually housed social species of birds must be able to see and hear birds of their own or compatible species unless the attending veterinarian determines that it would endanger their health, safety, or well-being. If individually housed social species of birds are unable to see and hear birds of their own or compatible species then special attention regarding enhancement to their environment must be provided.

(2) Environmental enrichment. The physical environment in the primary enclosures must be enriched by materials or activities that would provide the birds with the means to express non injurious species-typical activities. Species differences should be considered when determining the type or methods of enrichment. Examples of environmental enrichments include providing perches, swings, mirrors, and other increased cage complexities; providing objects to manipulate; varied food items; using foraging or task-oriented feeding methods; and providing interaction with the care giver or other familiar and knowledgeable person consistent with personnel safety precautions.

(3) Special considerations. Certain birds must be provided special attention regarding enhancement of their environment, based on the needs of the individual species and/or individual bird and in accordance with the instructions of the *attending veterinarian*. Birds requiring special attention are the following:
- Nestlings, chicks, or fledglings;
- Those that show signs of being in psychological distress through behavior or appearance;
- Those used in research for which the Institutional Animal Care and Use Committee (IACUC)-approved protocol requires restricted activity; and
- Individually housed social species of birds that are unable to see and hear birds of their own or compatible species.
- The *attending veterinarian* may exempt an individual bird from participation in the environment enhancement plan because of its health or condition, or in consideration of its well-being. The basis of the exemption must be recorded by the attending veterinarian for each exempted bird. Unless the basis for the exemption is a permanent condition, the exemption must be reviewed at least every 30 days by the *attending veterinarian*.

(4) Restraint devices. Birds must not be permitted to be maintained in restraint devices unless required for health reasons as determined by the *attending veterinarian* or by a research proposal approved by the IACUC at research facilities. Any restraining actions must be for the shortest period possible.

§ 3.155 Feeding.

(a) The diet for birds must be appropriate for the species, size, age, and condition of the bird. The food must be wholesome, palatable to the birds, and free of contamination. It must be of sufficient quantity and nutritive value to maintain a healthy condition and weight range of the bird and to meet its normal daily nutritional requirements. Birds must be fed at least once a day except as directed by the *attending veterinarian*, normal fasts, or other professionally accepted practices. If birds are maintained in group housing, measures appropriate for the species must be taken to ensure that all the birds receive a sufficient quantity of food.

§ 3.156 Watering.

Potable water must be provided in sufficient quantity to every bird housed at the facility, unless restricted by the *attending veterinarian*

§ 3.160 Compatibility and separation.

(a) Socially dependent birds, such as clutch-mates, must be housed in social groups, except where the *attending veterinarian* exempts an individual bird because of its health or condition, or in consideration of its well-being, or for specific management needs, or where such social grouping is not in accordance with a research proposal and the proposal has been approved by the research facility IACUC.

Birds that have or are suspected of having a contagious disease or communicable condition must be separated from healthy animals that are susceptible to the disease as directed by the *attending veterinarian.*

Transportation and Shipping

Carriers and intermediate handlers may accept un-weaned birds for transport as long as the *attending veterinarian* approves the un-weaned birds can be transported safely, provides written instructions for the conditions of transport, and is signed within 10 days of shipment. The attending veterinarian must approve that the un-weaned birds can be transported safely and provide written instructions of transport, and must sign those instructions within 10 days of the shipment. The veterinarian is not required to physically see the birds, this can be done virtually.

§ 3.164 Food and water requirements.

(a) All weaned birds must be offered food and potable water within 4 hours before being transported in commerce, unless the *attending veterinarian* approves a delay or a delay is in accordance with professionally accepted standards.

(b) Dealers, exhibitors, research facilities, and operators of auction sales must provide potable water to all weaned birds transported in their own primary conveyance at least every 12 hours after such transportation is initiated, except for birds which, according to professionally accepted standards or under the direction of the *attending veterinarian*, require watering or feeding more or less frequently.

(c) Carriers and intermediate handlers must provide potable water to all live, weaned birds at least every 12 hours after accepting them for transportation in commerce, except for birds which, according to professionally accepted standards or under the direction of the *attending veterinarian,* require watering or feeding more or less frequently.

(d) A sufficient quantity of food and water or other source of hydration must accompany the bird to provide food and water for such bird during period of transport, except as directed by *veterinary* treatment and other professionally accepted standards.

§ 3.165 Care in transit.

(a) Surface transportation (ground and water). During surface transportation, any person subject to the Animal Welfare regulations in this part and parts 1 and 2 of this subchapter transporting birds in commerce must ensure that the operator of the conveyance, or a person accompanying the operator, visually observes the birds as frequently as circumstances may allow, but not less than once every 4 hours, to ensure that the birds are receiving sufficient air for normal breathing, that climatic and environmental conditions are being maintained in accordance with the requirements in § 3.168, and that all other applicable standards are met. The regulated person must ensure that the operator or person accompanying the operator determines whether any of the birds are in physical distress and obtains *any veterinary* care needed for the birds as soon as possible.

(b) Air transportation. When transported by air, live birds must be visually observed by the carrier as frequently as circumstances may allow, but not less than once every 4 hours, if the animal cargo space is accessible during flight. If the animal cargo space is not accessible during flight, the carrier must visually observe the live birds whenever they are loaded and unloaded and whenever the bird cargo space is otherwise accessible to ensure that they are receiving sufficient air for normal breathing, that climatic and environmental conditions are being maintained in accordance with the requirements in § 3.168, and that all other applicable standards are met. The carrier must determine whether any such live birds are in physical distress and arrange for any needed veterinary care as soon as possible.

(c) Prohibition on the transport of ill, injured, or distressed birds. Any person subject to the Animal Welfare regulations in this part and parts 1 and 2 of this subchapter may not transport in commerce birds that are ill, injured, or in physical distress, except to receive *veterinary* care for the condition.

Determining if a license is needed by a breeder

The following groups and species are considered to be small pet bird species (250 grams or less as determined by average adult weight of the species). A breeder must be licensed if they sell more than 200 small birds.

- Finches and finch-like birds from the family Fringillidae*
- Cockatiels
- Parrotlets
- Lovebirds
- Caiques
- Senegals and other *Poicephalus* parrots
- Mini-macaws, 250 grams or less by average adult weight of species
- Conures
- Pionus
- Lories and lorikeets
- Parakeets and other parakeet-like birds, such as but not limited to ringnecks, monks, lineolated, kakarikis, and rosellas
- Doves and pigeons in the order Columbiformes 250 grams or less by average adult weight of species
- Poultry, for use as pets, that are 250 grams or less by average adult weight of species

The following groups and species are considered to be large pet bird species (more than 250 grams as determined by average adult weight of the species). A breeder must be licensed if they sell more than eight large birds.

- Cockatoos
- African greys
- Eclectus
- Macaws
- Amazons
- Toucans
- Hornbills
- Turacos

- Kookaburras
- Doves and pigeons in the order Columbiformes more than 250 grams by average adult weight of species
- Poultry, for use as pets, more than 250 grams by average adult weight of species

References

1. 9 CFR Parts 1, 2, and 3 (Docket No. APHIS 2020-0068).

Effects of Atorvastatin and Rosuvastatin on Blood Lipids in Quaker Parrots (*Myiopsitta monachus*)

Hugues Beaufrère, DVM, PhD, Dipl ACZM, DABVP (Avian Practice), DipECZM (Avian),
Trinita Barboza, DVM, DVSc,
Alysha Burnett,
Ken D. Stark, PhD,
R. Darren Wood, DVM, DVSc, DACVP

Session #3146

Affiliation: From the Department of Medicine and Epidemiology, School of Veterinary Medicine, University of California–Davis, Davis, CA 95616, USA (Beaufrère), the Department of Clinical Sciences, Cummings School of Veterinary Medicine, Tufts University, North Grafton, MA 01536, USA (Barboza), the Department of Pathobiology, Ontario Veterinary College, University of Guelph, Guelph, ON N1G 2W1, Canada (Wood), the Department of Kinesiology and Health Studies, University of Waterloo, Waterloo, ON N2L 3G1, Canada (Stark).

Note: Abstract format selected by author(s). See Front Matter for description of conference proceedings manuscripts submitted as scientific abstracts.

Abstract: Statin drugs are the most effective class of hypolipidemic and anti-atherosclerotic drugs, with atorvastatin and rosuvastatin being the most effective. While the use of statins would be a tremendous asset in the treatment of dyslipidemia and lipid-accumulation disorders in birds, there are limited data available in psittacine species.

Two consecutive randomized cross-over trials on quaker parrots were performed to study the effect of these two statins. Ten birds were used in an initial balanced cross-over experiment with five oral treatments for 2 weeks each (control, atorvastatin 10 mg/kg q12h and q24h, rosuvastatin 10 mg/kg q12h and q24h). Plasma lipidomics and lipoprotein profiling were performed after each treatment. Twelve birds were used in a second experiment consisting of two parallel cross-over studies, each on six birds either fed their regular diet or a 0.3% cholesterol diet and receiving two oral treatments for 2 weeks each (control, atorvastatin 20 mg/kg q12h). Plasma lipidomics, lipoprotein profiles, and β-Hydroxy-β-methylglutaryl-CoA reductase activity (HMG-CoA) were subsequently measured. Results were analyzed with serial linear mixed models and trends were assessed graphically.

No statistically significant effect of any statin treatment was detected on plasma lipids, lipoproteins, creatine phosphokinase, and HMG-CoA. In the first trial, all the rosuvastatin treatments led to some non-significant decreases in several triacylglycerol species while in the second trial, this was observed only in birds on atorvastatin 20 mg/kg q12h being fed their regular diet. These results suggest that quaker parrots may require much higher doses of statin drugs to show significant and clinically useful lipid-lowering effects.

Acknowledgments: Funding for this study was provided by the Morris Animal Foundation (grant D19ZO-301) and the OVC Pet Trust Foundation. This publication has not been reviewed or endorsed by the Morris Animal Foundation and the views expressed do not necessarily reflect the views of the Foundation, its officers, directors, affiliates, or agents. We also thank Hagen Inc. for providing and manufacturing the experimental diets. The authors thank the Analytical Facility of Bioactive Molecules, the Hospital for Sick Children, Toronto, Canada for performing metabolomics analysis.

Pharmacokinetics of Single-Dose Oral Atorvastatin and its Metabolites in Orange-Winged Amazon Parrots (*Amazona amazonica*)

Nicole Mikoni, DVM,
David Sanchez-Migallon Guzman, LV, MS, DipECZM (Avian), DipECZM (Small Mammal), Dipl ACZM,
Heather Knych, DVM, PhD, DACVCP,
Hugues Beaufrère, Dr med vet, PhD, DABVP (Avian Practice), DipECZM (Avian), Dipl ACZM

Session #3149

Affiliation: From the Veterinary Medical Teaching Hospital (Mikoni), Department of Medicine and Epidemiology (Guzman, Beaufrère), K.L. Maddy Equine Analytical Pharmacology Laboratory (Knych), School of Veterinary Medicine, University of California Davis, Davis, CA, 95616, USA.

Note: Abstract format selected by author(s). See Front Matter for description of conference proceedings manuscripts submitted as scientific abstracts.

Abstract: Dyslipidemias are common in captive psittacine species and have been associated with lipid-accumulation disorders such as atherosclerosis and hepatic lipidosis.[1-5] Amazon parrots are especially prone to atherosclerosis.[2,3] Use of statins is considered a mainstay of human hyperlipidemia and atherosclerosis treatment.[6] Previous studies evaluating atorvastatin (10 mg/kg q24h for 30 days) in hypercholesterolemic Hispaniolan Amazon parrots did not find significant hypolipidemic effects at studied dosages;[7] however, evaluation of statin drug pharmacokinetics in psittacine species is limited and could help determine dosages required for hypolipidemic effect. This study reports the pharmacokinetics of a compounded oral atorvastatin (10 mg/ml) administered via gavage to orange-winged Amazon parrots. A pilot study using escalating doses from 20-80 mg/kg was performed on six birds. A follow-up study using eight birds (four males/four females, varying ages) was then conducted using a single dose of 20 mg/kg. For both studies, plasma concentrations of atorvastatin and the two main active metabolites (para- and ortho-hydroxyatorvastatin) were measured in each bird by LC-MS/MS at 10 different time points over 30 hours. Pharmacokinetic analysis was performed using non-compartmental analysis. Mean atorvastatin half-life, T_{max}, and C_{max} were 5.96 hours, 1.6 hours, and 82.6 ng/ml, respectively. For para-hydroxyatorvastatin, the half-life, T_{max}, and C_{max} were 6.46 hours, 5 hours, and 34.1 ng/ml, respectively, and 5.58 hours, 3.38 hours, and 7.35 ng/ml for ortho-hydroxyatorvastatin. These results provide pharmacokinetic data on atorvastatin and its main metabolites in orange-winged Amazon parrots that can be used to inform therapeutic recommendations for clinical patients and plan future pharmacodynamic studies.

Acknowledgments: This study was supported by the UC Davis Center for Companion Animal Health and the Richard M. Schubot Parrot Wellness and Welfare Program.

References

1. Beaufrère H, Reavill D, Heatley J, et al. Lipid-related lesions in quaker parrots (*Myiopsitta monachus*). *Vet Pathol.* 2019;56(2):282-288.

2. Beaufrère H, Ammersbach M, Reavill D, et al. Prevalence of and risk factors associated with atherosclerosis in psittacine birds. *J Am Vet Med Assoc.* 2013;242(12):1696-1704.

3. Beaufrère H, Cray C, Ammersbach M, et al. Association of plasma lipid levels with atherosclerosis prevalence in Psittaciformes. *J Avian Med Surg*. 2014;28(3):225-231.

4. Martín-Castillo A, Castells MT, Adánez G, et al. Effect of atorvastatin and diet on non-alcoholic fatty liver disease activity score in hyperlipidemic chickens. *Biomed Pharmacother.* 2010;64(4):275-281.

5. Adánez G, Castells MT, García Pérez B, et al. Effects of atorvastatin on progression-regression of renal injury in hyperlipidemic chickens. *Histol Histopathol.* 2008;23(9):1131-1142.

6. McKenney JM, Roth EM. Statins. In: Ballantyne CM, ed. *Clinical Lipidology: A Companion to Braunwald's Heart Disease E-Book.* Elsevier; 2015:227-256.

7. Robertson JA, Guzman DSM, Graham JL, et al. Evaluation of orally administered atorvastatin on plasma lipid and biochemistry profiles in hypercholesterolemic Hispaniolan Amazon parrots (*Amazona ventralis*). *J Avian Med Surg*. 2020;34(1):32-40.

Clinicopathologic, Gross Necropsy, and Histopathologic Effects of Repeated Glucocorticoid Administration in Domestic Pigeons (*Columba livia*)

Olivia A. Petritz, DVM, Dipl ACZM,
Mandy Womble, DVM, PhD,
Erika Gruber, DVM, DACVP (Clinical),
Andrea E. Thomson, BS,
Tara M. Harrison, DVM, MPVM, Dipl ACZM, Dipl ACVPM,
DipECZM (Zoo Health Management), CVA

Session #3430

Affiliation: From the Department of Clinical Sciences (Petritz, Thomson, Harrison), Population Health and Pathobiology (Womble, Gruber), North Carolina State University, College of Veterinary Medicine, Raleigh, NC 27607, USA.

Note: Abstract format selected by author(s). See Front Matter for description of conference proceedings manuscripts submitted as scientific abstracts.

Abstract: Steroids are commonly used in chemotherapy protocols yet the safety of these medications in any avian species, a reportedly steroid-sensitive taxon, has not been evaluated. Previous publications have focused on the avian pituitary-adrenocortical axis and its response to steroid administration, rather than the safety or side effects of long-term oral steroid administration.[1,2] Eleven adult unknown sex Carneau pigeons (*Columba livia*) aged 7-9 years of age were randomly assigned to a control (n = 5) or treatment group (n = 6). All treatment birds received prednisolone 1 mg/kg PO q24h for 29 days, while the control group received a similar volume of saline PO q24h. Complete blood counts and body weights were performed once weekly. Plasma biochemistries, *Aspergillosis* antibody, and protein electrophoresis were performed prior to the start of the study and at the end of the treatment course on all birds. At the end of the treatment period, the birds were humanely euthanized, and complete necropsies including histopathology were performed. No adverse clinical signs were noted in any bird throughout the study period. There was no statistically significant difference in the white blood cell count, absolute lymphocyte numbers, or *Aspergillosis* antibodies between treatment and control groups at day 0 compared to day 29. Post-mortem examination of all birds in both groups was largely unremarkable. There was no evidence of infectious disease on gross or histologic postmortem examination. Overall, no adverse clinical, clinicopathologic, or histopathologic changes were noted in the treated birds compared to the control birds following repeated administration of oral prednisolone.

References

1. Westerhof I, Van den Brom WE, Mol JA, et al. Sensitivity of the hypothalamic-pituitary-adrenal system of pigeons (*Columba livia domestica*) to suppression by dexamethasone, cortisol, and prednisolone. *Avian Dis.* 1994;38:435-445.

2. Westerhof I, Pellicaan CHP. Effects of different application routes of glucocorticoids on the pituitary-adrenocortical axis in pigeons (*Columba livia domestica*). *J Avian Med Surg.* 1995;9:175-181.

Effects of Capromorelin, Mirtazapine, and Cyproheptadine on Food Intake in Budgerigars (*Melopsittacus undulatus*)

Christoph Mans, Dr med vet, Dipl ACZM,
Caroline E. Titel, BS,
Grayson A. Doss, DVM, Dipl ACZM

Session #3387

Affiliation: From the School of Veterinary Medicine, University of Wisconsin-Madison, 2015 Linden Drive, Madison, WI 53706, USA.

Note: Abstract format selected by author(s). See Front Matter for description of conference proceedings manuscripts submitted as scientific abstracts.

Abstract: Hyporexia or anorexia frequently occur in sick and/or hospitalized psittacine birds. Reduced caloric intake in these species can lead to a variety of detrimental effects and prolong hospitalization. Several medications have been used to stimulate appetite in dogs and cats, but only benzodiazepines have been evaluated in psittacine species.[1,2] In a series of randomized, blinded, placebo-controlled complete crossover studies, the effects of mirtazapine (1 and 5 mg/kg), capromorelin (10 and 40 mg/kg), and cyproheptadine (0.5 and 2.5 mg/kg) on food intake were evaluated in healthy budgerigars (*Melopsittacus undulatus*, n = 12). After administration of a single dose of each drug via crop gavage, food intake was measured in hourly intervals for the first 8 hours and then one measurement for the 8- to 12-hour period. The birds were monitored by video recording for potential adverse effects. No clinically relevant or statistically significant effects on food intake were found following the administration of mirtazapine and cyproheptadine, regardless of the dose administered. Following administration of capromorelin at either dose, food intake was significantly increased for the first 8 hours compared to the control treatment. A dose-dependent difference in capromorelin efficacy was not found. Regurgitation was observed in 42% and 92% of birds following administration of capromorelin at 10 and 40 mg/kg, respectively, while gavage of the control treatment resulted in regurgitation in 25% of birds. Therefore, capromorelin at 10 mg/kg should be considered for future studies, and lower doses may be effective in increasing food intake with potentially reduced adverse effects.

References

1. Martel A, Doss G, Berg C, Mans C. Effects of midazolam on food intake in budgerigars. *J Avian Med Surg.* 2022;36:53-57.

2. Scagnelli A, Titel C, Doss G, Mans C. Effects of midazolam and lorazepam on food intake in budgerigars. *J Exot Pet Med.* 2022;41:42-45.

Use of a Continuous Glucose Monitor in a Diabetic Toco Toucan (*Ramphastos toco*)

Marcie Logsdon, DVM

Session #3426

Affiliation: From the Exotics and Wildlife Department, Washington State University Veterinary Teaching Hospital, 100 Ott Rd, Pullman, WA 99164, USA.

Note: Abstract format selected by author(s). See Front Matter for description of conference proceedings manuscripts submitted as scientific abstracts.

Abstract: An 11-year-old male toco toucan (*Ramphastos toco*) previously diagnosed with diabetes mellitus[1] presented for acute necrosis of the tongue. On examination the patient was severely emaciated at 512 grams with a body condition score (BCS) of 1/5. Discussion with the owner revealed that the toucan's overall condition was declining despite glipizide administration and dietary management. Hospitalization and insulin trial were initiated. Initial blood glucose (BG) reading was 1790 mg/dl[2] (Demand xPAND Plus Chemistry Analyzer, Siemens Medical Solutions, Malvern, PA, USA). Due to cost and sample volume requirements of the analyzer, attempts were made to use a glucometer for monitoring. This was unsatisfactory as the glucometer (not validated for birds) maxed out at 750 mg/dl[3] and provided inconsistent results if dilutions were attempted. It also required frequent restraint and blood draws for sampling. In pursuit of a less invasive monitoring system continuous glucose monitor (CGM)[4] (FreeStyle Libre 2, Abbott, Abbott Park, IL, USA) was placed on the lateral aspect of the left distal leg in the flexor muscles caudal to the tibiotarsus. The monitor was well tolerated by the patient and took continuous readings, recording every 5 minutes, for 4 days at which point the patient was euthanized due to overall decline. At four timepoints blood was drawn and compared with the clinical pathology lab. The CGM values were comparable, measuring 6-28 mg/dl higher with greatest deviation happening at levels ≥ 500 mg/dl. Given the physiologic lag between blood and interstitial glucose levels, this was felt to be adequate alignment for aiding in clinical decisions.

References

1. Rae M. Endocrine disease in pet birds. *Semin Avian Exot Pet Med.* 1995;4:32-38.

2. International Species Information System. ISIS Physiological Reference Intervals for Captive Wildlife: A CD-ROM Resource. Ramphastos_toco_Males_only__All_ages_combined_Conventional_American_units__2013_CD.html.

3. Zoeits (2017). AlphaTrak 2 Product Insert. Available at: https://www.zoetisus.com/content/_assets/docs/vmips/package-inserts/alphatrak_2.pdf. Accessed 4/24/23.

4. Wiedmeyer CE, Johnson PJ, Cohn LA, Meadows RL. Evaluation of a continuous glucose monitoring system for use in dogs, cats, and horses. *J Am Vet Med Assoc.* 2003;223(7):987-992.

Radiographic Measurements of Internal Organs in Red Hybrid Hens and Their Evolution During Egg Formation

Thomas Coutant, Dr med vet, IPSAV,
Ciro Cococcetta, Dr med vet,
Albert Phouratsamay, Dr med vet,
Chloé Touzet, Dr med vet,
Graham Zoller, Dr med vet, IPSAV, DipECZM (Avian),
Minh Huynh, Dr med vet, DipECZM (Avian), Dipl ACZM

Session #3121

Affiliation: From Service des Nouveaux Animaux de Compagnie (Coutant, Cococcetta, Phouratsamay, Zoller,* Huynh); Service d'imagerie médicale (Touzet), Centre Hospitalier Vétérinaire Fregis, 43 Avenue Aristide Briand, 94110 Arcueil, France.

*Present address: Clinique Vétérinaire Only Vet, 7 rue Jean Zay, 69800 Saint-Priest.

Note: Abstract format selected by author(s). See Front Matter for description of conference proceedings manuscripts submitted as scientific abstracts.

Abstract: There is no study that describes radiographic measurements of internal organs in backyard hens. Moreover, the effect of egg laying on these measurements has not been studied in avian species even though it could affect radiographic measurements, depending on the stage of egg formation. This issue is of particular concern in laying hens since they can lay on a daily basis. Thirteen laying adult hens and three non-laying juvenile Rhode Island Red hybrid hens were used to provide preliminary radiographic reference values for internal organ measurement in clinically healthy chickens as previously described in psittacine birds.[1] Whole body radiographs of five adult hens were taken every 2 hours during a 24-hour period to study the evolution of these measurements throughout egg formation. The only reliable organ measurements were heart and liver width on ventrodorsal views, which were significantly greater in adult compared to juvenile hens ($P = 0.024$). Among the different possible ratios, only heart width:total was significantly different between age classes, being greater in juvenile hens ($P = 0.024$). Hepatic silhouette width and ratios with coelomic width were found to increase over time ($P < 0.02$) but did not follow egg formation since liver width does not decrease immediately after the oviposition. Eggshell thickness was found to be a promising parameter to evaluate the stage of egg formation. This study demonstrates that egg formation should be considered when trying to establish reference ranges for internal organ radiographic measurements in laying hens in future studies.

References

1. Geerinckx L, Van der Vekens E, Saunders JH, et al. Literature review of radiographic measurements of internal organs in Psittaciformes. *J Exot Pet Med.* 2019;28(1):9.

Use of a Tablet-Based Communication Board and Subsequent Choice and Behavioral Correspondences in a Goffin's Cockatoo (*Cacatua goffiniana*)

Jennifer Cunha, JD,
Rebecca Kleinberger, PhD,
Susan Clubb, DVM

Session #3441

Affiliation: From Parrot Kindergarten, Inc., 601 Heritage Dr, Ste 424, Jupiter, FL 33458, USA (Cunha), Northeastern University, 360 Huntington Ave, Boston, MA 02115, USA (Kleinberger), Rainforest Clinic for Birds and Exotics, 3319 E Road, Loxahatchee, FL 33470, USA (Clubb).

Abstract: An enculturated individual is one who learns the content of a culture and assimilates it into their practices. An enculturated Goffin's cockatoo (*Cacatua goffiniana*) was taught to use a commercially available, android-based communication board. Through associative conditioning, the subject learned to press pictures representing items in categories for foods, beverages, activities, objects, and interactions. Over the course of 22 days, communication opportunities were tracked. Each of cockatoo's responses to the question "What do you want?" was corroborated by follow-up questions and body language verification (voluntary transfer to hand and other locations, acceptance or rejection of request execution). In all 33 reviewed communication opportunities, the researcher responded to the request with a corroboration task, and 26 of the opportunities involved two corroboration tasks. The interactions comprised 81 request and follow-up question tasks, and the cockatoo's response-positive corroboration consistency rate was 95%. Inter-observer agreement was independently collected on 30% of the reviewed communication opportunities with a score of 97.5% fidelity. In addition, the cockatoo used the communication board for the following requests: 42% - interactive activity; 33% - food/beverage; 18% - extra-familial interaction; 6% - object. These results show positive interactions with a communication board and warrant further replication with additional birds.

Introduction

Many animals have demonstrated the ability to use a variety of symbols to communicate. Since 1953 research has tracked a variety of species' abilities to use symbols to represent choice requests. Rats press levers to request food[1]or to remove aversive stimuli.[2] Horses have also demonstrated the ability to request the application or removal of blankets, and their choices were corroborated by the environmental temperature.[3]

Computer interfaces have also been used by researchers to communicate with non-human animals. Keyboard interfaces have been developed for dolphins to request objects, events, and locations.[4] A bonobo utilizes a touch-screen computer for lexigram use to communicate requests and ideas.[5]

Birds have demonstrated an impressive level of cognitive processing in wild and experimental settings. Corvids communicate information across generations[6] and have solved complex problems, outperforming 7- to 10-year-old children.[7]

Goffin's cockatoos evidence a limited form of the inference-through-exclusion.[8]They have also used and manufactured tools,[9] and can outperform 7-year-old children in puzzle tasks.[10] An African grey parrot in the Pepperberg laboratory, learned and spoke in context more than 100 words,[11] identified shapes and numbers by name,[12] understood the concept of zero,[13] and the ordinality of numbers.[14]

Choice and control over environment is a primary reinforcer among animals. Brent and Weaver (1996) discovered that greater degrees of environment control led to beneficial effects on the behavior of singly-housed animals.[15] Control is a vital part of psychological wellbeing, and some researchers urge that animals should even be allowed to choose their preferred enrichment objects and activities.[16]

Some parrots exhibit challenging behaviors. Nearly 40% of cockatoos and African grey parrots engage in feather destructive behavior, a form of self-injury.[17] Many also display aggressive behavior and excessive vocalization. Although the origin of these behaviors is complex, some behaviors may result from captive environments that lack choice and control. Some researchers have found that captive parrots that are free-roaming for 4-10 hours per day display lower levels of feather destructive behavior.

Augmentative Alternative Communication and Choice-Making

Augmentative alternative communication (AAC) is a graphic-symbol based communication method that has been developed for use by some special-needs individuals.[18] One type is comprised of a touchscreen computer board that allows the user to touch a picture to communicate requests and participate socially with others. Some special needs individuals exhibit challenging behaviors including self-injury, aggression toward others, damage to the physical environment, and tantrums.[19] AAC devices aid in social interaction, cognitive development, increase independence and control over the individual's environment through choice making, and decrease challenging behavior.[19]

In a meta-analysis of twenty-four single subject case studies, AAC training intervention was found to decrease these behaviors as individuals developed the ability to communicate requests using pictures as an alternative to the challenging behavior, and their quality of life was enhanced through the facilitation of communication.[19]

With millions of animals in home-settings, facilitating an accessible means of communication via commercially available AAC apps on a tablet can open up pathways to choice requests and enhance quality of life.

This study analyzed whether a Goffin's cockatoo can be trained to use an AAC device through associative conditioning and observed behavioral corroborations and communication consistency as evidence of intention matching the symbols requested.

Methods

This study involved an enculturated female Goffin's cockatoo that participated in previous research on symbol-representation. The bird was housed with other cockatoos and had constant access to food and water, regular mental enrichment training, and access to engagement with natural foraging materials. It was trained on word and picture symbols based on associative conditioning. It was also trained to use the symbols for choice tasks that are generalized to other objects and concepts, including toys, food and beverage items, and activities. The bird has seemingly learned to use the symbols on the CommBoard app on a tablet. Historically, the bird scored 70% accuracy on a match-to-sample paradigm using color cards and its use of the symbols, and 70% on sound-text correspondence tasks. This project followed applicable international and national guidelines for the care and use of animals. The training method involved voluntary training sessions three-to-five times per week for 20-45 minutes per session, and the use of positive reinforcement.

In this project, a Samsung Galaxy Tab A tablet with a protective case and the CommBoards AAC Speech Assist for Android, purchased in the Playstore, were used. The picture stimuli were derived from in-app pictures, Google images, and photos taken by the researcher. The bird was trained using varied reinforcers such as sunflower seeds and pine nuts. The researcher taught the bird to ring a toy bell to request access to a speech board. The training procedure included frustration avoidance techniques and pexigram training using a higher value (pine nut) and a lower value (sunflower seed) reinforcer, and if it got two answers wrong in a row, it was invited to engage in a simpler target touch task for a treat, to maintain momentum and lower frustration in learning.

Results

Inter-observer agreement (IOA) data were collected during observations of each of the communication opportunities. For 30% (10) of the 33 included communication opportunities, each researcher scored the recorded videos independently. IOA was calculated for each intervention step through point-by-point agreement by dividing the total agreements by the total agreements plus disagreements and multiplying by 100%. For the steps collected with event recording, point-by-point agreement was calculated by comparing each recorded response. IOA averaged 97.5% between the two researchers.

The 33 included communication opportunities comprised 81 cockatoo requests and researcher follow-up questions on the tablet. In all interactions, the cockatoo ultimately selected a pexigram that it corroborated. The cockatoo had a 94% corroboration rate on the initial request-corroboration task questions. Corroboration of the tasks comprised the following: follow-up questions (e.g., repeated question or yes-no question) - 85%; body language corroboration - 97%; two forms of corroboration - follow-up questions and body language corroboration - were observed in 79% of the communication opportunities.

The cockatoo's requests comprised: interactive activities (e.g., tracing, play outside) - 44%; food or beverage (e.g., treat, tea) - 36%; interactions with other humans - 12%; and object requests - 8%.

The bird's incorrect answers comprised a submenu item that was directly under the menu selection and may have been an inadvertent press and another request that was declined and not corroborated. In addition, the two excluded interactions involved corroboration attempts which were declined by the cockatoo: the cockatoo pushed away the offered hand for voluntary transport based on the requests in the two excluded sessions.

Some additional observations were made. Although the cockatoo was not specifically trained in-depth on the use of the want pexigram, in one session, it spontaneously began pressing "want - treat," which mirrored the question format ("What do you want?") The cockatoo also was not trained on the "to drink" menu, and spontaneously touched the picture of a glass with the label "To Drink," requested tea by picking a picture of a teacup with a label "Tea," and corroborated with a "yes" response. The researcher equidistantly offered the bird a cup of water and a cup of tea, and it drank five gulps of warm tea, rejecting the water.

Finally, most communication opportunities concluded with a yes-no corroboration question (i.e., "Do you want a treat? Yes or no.") In some later communication opportunities, the cockatoo selected a request picture and then immediately touched "yes" (e.g., treat-yes).

Discussion

Goffin's cockatoos have demonstrated remarkable skill on cognitive tasks, including puzzle tests [12], inference through exclusion [26], and tool manufacturing [2].

The goal of this pilot project was to see whether an enculturated Goffin's cockatoo could learn to touch pictures on a communication board to make requests, and whether after selecting the picture, the request selection would be corroborated by consistency in follow-up question responses and body language (accept/reject of selection, voluntary transfer on to hand to execute request). The two corroboration types were selected because of the bird's training in the use of "yes" and "no" symbols, its prior testing results indicating basic knowledge of yes and no symbols, and its history of body language (accepting/refusing objects, dropping food items, or refusing to engage/walking away from offered activities).

The subject cockatoo in these communication opportunities demonstrated high levels of consistency (95%) between its selection requests and its corroboration tasks. The IOA evidenced a high level of fidelity (97.5%) between the two researchers' observations on video-recorded requests and corroboration tasks.

The cockatoo most often requested play and learning activities and favorite food items, but also engaged in unexpected generalization of requests to non-associated pictures. This study demonstrates the potential for using AAC devices and communication training as a behavioral intervention for challenging behaviors in animals. The study is significant because it demonstrates the potential of using tablet devices and AAC apps to train nonhuman animals in communication.

Conclusions

This study investigated whether a Goffin's cockatoo's communication board selections after associative training would be positively corroborated with follow-up questions and body language. Specifically, the research comprised two questions: 1) can an enculturated Goffin's cockatoo be trained through associative conditioning and positive reinforcement training to select pictures on a commercially available tablet communication board? and 2) are the picture requests positively corroborated by follow-up questions and body language?

Following associative conditioning, the cockatoo was presented with the board by making requests across 33 analyzed sessions. The bird demonstrated a 95% corroboration rate with high IOA fidelity on instances of clear selection, indicating that it was able to use the tablet communication board app to communicate requests.

References

1. Davey GC, Cleland, GG. Food anticipation and lever-directed activities in rats. *Learn Motiv.* 1984;15(1):12-36. doi.org/10.1016/0023-9690(84)90014-6.

2. Barker DJ, Sanabria F, Lasswell A, et al. Brief light as a practical aversive stimulus for the albino rat. *Behav Brain Res.* 2010;214(2):402-408. doi.org/10.1016/j.bbr.2010.06.020.

3. Mejdell C, Buvik T, Jørgensen G, Bøe K. Horses can learn to use symbols to communicate their preferences. *Appl Anim Behav Sci.* 2016;184.

4. Herzing D. Interfaces and keyboards for human-dolphin communication: what have we learned? *Anim Behav Cogn.* 2016;3:243-254. doi:10.12966/abc.04.11.2016.

5. Rafaelle P. Speaking bonobo. Smithsonian Magazine. 2006. Available at: www.smithsonianmag.com/science-nature/speaking-bonobo-134931541. Accessed June 1, 2023.

6. Marzluff JM, Walls J, Cornell HN, et al. Lasting recognition of threatening people by wild American crows. *Anim Behav.* 2010;79(3):699-707. doi.org/10.1016/j.anbehav.2009.12.022.

7. Logan CJ, Jelbert SA, Breen AJ, et al. Modifications to the Aesop's Fable Paradigm Change New Caledonian Crow Performances. *PLoS ONE.* 2014;9(7):e103049. doi.org/10.1371/journal.pone.0103049.

8. O'Hara M, Auersperg AMI, Bugnyar T, Huber L. Inference by exclusion in Goffin cockatoos (*Cacatua goffini*). *PLoS ONE.* 2015;(10)8:e0134894. doi:10.1371/journal.pone.0134894.

9. Auersperg AM, Szabo B, von Bayern AM, Kacelnik A. Spontaneous innovation in tool manufacture and use in a Goffin's cockatoo. *Curr Biol.* 2012;22(21):R903-R904. doi:10.1016/j.cub.2012.09.002

10. Habl C, Auersperg AMI. The keybox: Shape-frame fitting during tool use in Goffin's cockatoos (*Cacatua goffiniana*). *PLoS ONE.* 2017;12(11):e0186859. doi.org/10.1371/journal.pone.0186859.

11. Pepperberg IM. Functional vocalizations by an African grey parrot (*Psittacus erithacus*). *Zeitschrift Für Tierpsychologie.* 1981;55(2):139-160.

12. Pepperberg IM. Acquisition of the same/different concept by an African grey parrot (*Psittacus erithacus*): learning with respect to categories of color, shape, and material. *Anim Learn Behav.* 1987;15(4):423-432. doi.org/10.3758/BF03205051.

13. Pepperberg IM. Grey parrot (*Psittacus erithacus*) numerical abilities: addition and further experiments on a zero-like concept. *J Comp Psychol.* 2006;120(1):1-11. doi.org/10.1037/0735-7036.120.1.1.

14. Pepperberg IM, Carey S. Grey parrot number acquisition: the inference of cardinal value from ordinal position on the numeral list. *Cognition.* 2012;125(2):219-232. doi.org/10.1016/j.cognition.2012.07.003

15. Brent L, Weaver O. The physiological and behavioral effects of radio music on singly housed baboons. *J Med Primatol.* 1996;25(5):370-374. doi.org/10.1111/j.16000684.1996.tb00031.x.

16. Washburn DA. The four Cs of psychological wellbeing: lessons from three decades of computer-based environmental enrichment. *Anim Behav Cogn.* 2015;2(3):218-232. doi.org/10.12966/abc.08.02.2015.

17. Jayson S, Williams D, Wood J. Prevalence and risk factors of feather plucking in African grey parrots (*Psittacus erithacus erithacus* and *Psittacus erithacus timneh*) and cockatoo (*Cacatua* spp.). *J Exot Pet Med.* 2014;23:250-257. doi:10.1053/j.jepm.2014.06.012.

18. Drager K, Light J, McNaughton D. Effects of AAC interventions on communication and language for young children with complex communication needs. *J Pediatr Rehabil Med.* 2010;3(4):303-310. doi.org/10.3233/PRM-2010-0141.

19. Ganz JB, Earles-Vollrath TL, Heath AK, et al. A meta-analysis of single case research studies on aided augmentative and alternative communication systems with individuals with autism spectrum disorders. *J Autism Dev Disord.* 2012;42(1):60-74. doi.org/10.1007/s10803-011-1212-2.

Section 4

AAV Wildlife, Anesthesia, and Surgery Session

Seth Oster, DVM;
Sayrah Gilbert, DVM
Session Chairs

Juvenile Wild Galah Wasting Syndrome

Bob Doneley, BVSc, FANZCVS (Avian Medicine), and
David Phalen, DVM, PhD

Session #3269

Affiliation: From the School of Veterinary Science, The University of Queensland, The UQ Gatton Campus, Gatton, Queensland 4343 Australia (Doneley), Sydney School of Veterinary Science, University of Sydney, 425 Werombi Rd, Camden 2570 Australia (Phalen).

Note: Abstract format selected by author(s). See Front Matter for description of conference proceedings manuscripts submitted as scientific abstracts.

Abstract: Over the last 20 years, the authors have observed a wasting syndrome in wild juvenile galahs (*Eolophus roseicapilla*) and little corellas (*Cacatua sanguinea*) in eastern Australia. The syndrome is characterized by mucoid, watery, or tarry diarrhea and emaciation. *Macrorhabdus ornithogaster* (MO) and/or a *Spironucleus* sp. may be present in the feces. Few birds respond to treatment and most die. To determine if there was a correlation between MO or *Spironucleus* infection and the wasting syndrome, 21 wild galahs and corellas were allocated into two groups: those with possible wasting syndrome (n = 14; group 1) or those with other causes for presentation (e.g., road trauma; n = 7; group 2). Wet preparations of feces were examined at 400x magnification. All birds were euthanized, necropsied, and samples were collected for PCR and histopathology. The mean weight of group 1 birds was significantly less than those in group 2. MO and *Spironucleus* were identified in a significantly greater proportion of the birds in group 1 than in group 2. PCR proved to be more sensitive than direct fecal examination for the detection of infection with MO and *Spironucleus* (Table 1). A chronic lymphoplasmacytic enteritis with blunting and fusing of the duodenal villi was found in group 1 birds, but not in group 2 birds. In conclusion, a correlation between wasting disease and infection with either or both MO and *Spironucleus* was found. Therefore, treatment for both is indicated. Duodenal lesions in group 1 birds were most consistent with those caused by a recent enteric viral infection, suggesting the involvement of a third pathogen contributing to this syndrome. The intestinal lesions may be sufficiently severe to cause a maldigestion or malabsorption syndrome and treatment of this may be required to improve treatment outcomes.

Table 1. Weight and faecal examination results of galahs and corellas.

	Group 1		Group 2	
Mean weight (galahs)	205 g (n = 8)		305 g (n = 5)	
Mean weight (corellas)	238 g (n = 6)		350 g (n = 2)	
	Number	%	Number	%
Direct fecal examination: *Macrorhabdus*	3/12	25.0	0/7	0.0
PCR: *Macrorhabdus*	11/14	78.6	4/7	57.1
Direct fecal examination: *Spironucleus*	1/11	0.09	0/7	0.0
PCR: *Spironucleus*	7/12	58.3	2/6	33.0

Retrospective Postmortem Evaluation of Cardiac Valvular Disease in Eagles

Kristina S. Vuong, Dr med vet,
Michael Jones, Dr med vet, DABVP (Avian Practice),
Linden E. Craig, Dr med vet, PhD, DACVP

Session #3321

Affiliation: From the Department of Small Animal Clinical Sciences (Vuong, Jones), and the Department of Biomedical and Diagnostic Sciences, University of Tennessee, College of Veterinary Medicine, 2407 River Drive, Knoxville, TN 37996, USA (Craig).

Abstract: Cardiovascular disease in birds of prey is underreported compared to their companion avian counterparts.[1,2] Literature regarding valvular lesions in birds of prey is scarce and include singular reports of left atrioventricular valvular endocarditis in an adult, free-ranging, male bald eagle (*Haliaeetus leucocephalus*)[3] and aortic valvular endocarditis in an adult, free-ranging, female red-tailed hawk (*Buteo jamaicensis*).[4] The purpose of this presentation is to describe the prevalence, signalment, gross necropsy findings, and histologic lesions of valvular lesions in eagles. In this retrospective study, necropsy reports of 24 free-ranging and captive eagles were evaluated over a 15-year period (July 3, 2006 - February 28, 2021). Five bald eagles (*Haliaeetus leucocephalus*) and one golden eagle (*Aquila chysaetos*) met the inclusion criteria demonstrating gross and histologic cardiac valvular lesions (25% prevalence). All six birds were captive adults (> 1 year of age) with the majority having a known age equal to or greater than 25 years (5/6; 83.3%). Four of the affected birds were female and two were male. The aortic and left atrioventricular valves were equally affected. Myxomatous valvular degeneration was observed in four eagles and endocarditis including *Staphylococcus aureus* vegetative endocarditis was observed in two birds. Acute or chronic cerebral infarcts were present in all six of the birds, which contributed to their death. Valvular cardiac disease should be considered as a differential diagnosis in eagles exhibiting respiratory distress, neurologic signs, syncope, and/or sudden death. This presentation contributes to the limited literature regarding cardiac valvular disease and secondary cerebral infarcts in these avian species.

Introduction

Limited data is available about cardiac diseases in birds of prey despite their prevalence as avian patients.[1,2] Cardiomyopathies, endocarditis, atherosclerosis, and pathogenic arrhythmias (e.g., supraventricular premature complexes, second degree atrioventricular blocks) have been described in raptor species.[1,3,4] A retrospective study of postmortem examinations conducted on 75 captive birds of prey found that 24% of the birds had cardiac lesions, including myofiber degeneration, myocarditis, endocarditis, pericarditis, atherosclerosis, and thrombi.[5] These lesions can have infectious (e.g., West Nile virus, bornavirus, sarcocystis), nutritional, toxic, inflammatory, or idiopathic origins.[6] Information pertaining to cardiac valvular diseases in birds of prey are scarce, and secondary cerebral infarcts have not been reported in these species.

The most common valvular diseases described in other veterinary species, such as companion birds and dogs, include vegetative endocarditis and myxomatous degeneration of the left atrioventricular valve.[7-9] Myxomatous valvular degeneration is the non-inflammatory deposition of myxomatous connective tissue (e.g., glycosaminoglycans) within a disorganized extracellular matrix.[6, 10-13] Right-sided congestive heart failure is more common in avian species.[6,9,14] During the initial stages of right-sided congestive heart failure, the right ventricle undergoes concentric hypertrophy in response to pressure overload.[14] Dilation of the right ventricle occurs in response to volume overload, leading to regurgitation of blood through the spiral right atrioventricular valve in birds.[7,14] Birds are more prone to pulmonary hypertension due to the limited ability of the pulmonary capillaries to dilate within non-expansile lungs.[14] Right-sided congestive heart failure often progresses to biventricular heart failure.[6] Clinical signs of cardiac valvular disease are typically subclinical but can progress to dyspnea, syncope, and arrhythmias.[1] The prognosis becomes guarded if secondary congestive heart failure develops.[7]

Bald eagles (*Haliaeetus leucocephalus*) are often admitted to rehabilitation centers and veterinary teaching hospitals in North America.[15] The objective of this retrospective study was to evaluate the prevalence, gross necropsy findings, and histopathologic lesions of cardiac valvular disease in eagles.

Materials and Methods

Necropsy reports completed by the pathology service at the University of Tennessee, College of Veterinary Medicine (Knoxville, TN, USA) were searched to identify eagles with cardiac valvular lesions. Necropsies were evaluated over a 15-year period (July 3, 2006 - February 28, 2021). The pathology archival database was surveyed for the term "valve" within the necropsy findings of eagles during this period. Only cases with gross and/or microscopic valve lesions were included in this retrospective study. The database included necropsies conducted on free ranging eagles brought to the facility by good Samaritans and captive, non-releasable eagles housed at the American Eagle Foundation (Pigeon Forge, TN, USA). Information regarding signalment, diet, housing, and previous medical information was recorded when available. Captive eagles that resided at the American Eagle Foundation were housed in approximately 50-meter × 100-meter outdoor enclosures. The diet consisted of commercially sourced quail, rats, mice, and trout, and the birds also had continuous access to water. The outdoor enclosures were large enough for birds to fly short distances. No extra exercise regimens were enforced on the eagles. Annual physical examinations with complete blood counts, plasma biochemistry panels, West Nile virus vaccinations (1 ml intramuscularly; Vetera West Nile Equine Vaccine, Boehringer Ingelheim Vetmedica Inc, Saint Joseph, MO, USA), and ivermectin injections (0.4 mg/kg intramuscularly; Durvet, Blue Springs, MO, USA) were completed by the Avian and Zoological Medicine service at the University of Tennessee.

Results

A total of 24 eagles had necropsies completed between July 3, 2006 - February 28, 2021. Six of the eagles (25%, 95% confidence interval: 8.9-58.9% of the eagles) were reported to have cardiac valvular lesions. Five bald eagles and one golden eagle (*Aquila chrysaetos*) met the inclusion criteria. All six birds were captive adults (> 1 year of age), with the majority having a known age of equal to or greater than 25 years (83%; 5/6). Four of the affected birds were female and two were male. The main clinical signs noted in the study birds prior to death were respiratory (dyspnea, inspiratory and expiratory wheezes), neurologic (seizures, torticollis, nystagmus, head tremors), or cardiovascular in origin (syncope). Three cases were found dead in their enclosures, two were euthanized due to a poor prognosis, and one died despite aggressive supportive care with intraosseous fluids, oxygen support, and antibiotic treatment (ceftiofur 10 mg/kg IM; Excede, Zoetis, Parsippany, NJ, USA).

The most common heart valves to have pathologic changes were the left atrioventricular valve (cases 1, 2, 4, 5) and aortic valve (cases 1, 2, 3, 6). The valvular lesions were characterized as either myxomatous valvular degeneration (cases 1, 3, 4, 6) or endocarditis (cases 2, 5). Thickened myxomatous valves with chondroid metaplasia were present in the left atrioventricular valves of cases 1 and 4 and the aortic valves of cases 1, 3, and 6. Valvular mineralization was present in cases 3, 4, and 6. The cause of the valvular changes was unknown in most cases. Heterophilic endocarditis affecting the left atrioventricular and aortic valves was present in case 2, which also was diagnosed with a pulmonary adenocarcinoma that had metastasized to the kidneys, liver, and pleura. Left atrioventricular vegetative endocarditis secondary to a *Staphylococcus aureus* infection was present in case 5. Acute or chronic cerebral infarcts were present in all cases. The valvular lesions and cerebral infarcts were diagnosed as the cause of death in 83.3% (5/6) of the cases. Pulmonary adenocarcinoma with multisystemic metastasis was the main reason case 2 was euthanized.

Discussion

Myxomatous valvular degeneration (MVD) was the most prevalent cardiac valvular lesion identified in this retrospective study of eagles. Valves grossly appear smooth and thickened due to the non-inflammatory deposition of myxomatous connective tissue.[6,10-13] Myxomatous valvular degeneration is rarely reported in avian species but there are descriptions of MVD affecting the left atrioventricular valve in an Indian ringneck parakeet (*Psittacula krameria*),[9] ostrich (*Struthio camelus*),[16] and Amazon parrot (*Amazona aestiva*),[17] as well as the right atrioventricular valve in an African black-footed penguin (*Spheniscus demersus*).[18] Conversely, MVD is the most prevalent cardiovascular disease in dogs, often progressing to left atrioventricular valve regurgitation and left-sided congestive heart failure.[8,11,12,19] The left atrioventricular valve is the most commonly affected valve in the dog (62%), with 32.5% of cases involving both the left and right atrioventricular valves.[8,11,13] The pathophysiology of MVD in dogs is unclear but there is a predisposition in small to medium breeds, older animals (> 7 years), and males.[8,11-13] Myxomatous valvular degeneration is rarely reported in other common animal species treated by veterinarians.[12,19] There are histopathologic similarities between the structure of the left atrioventricular valve in humans and dogs.[12] On scanning electron microscopy, there is an uneven distribution of endothelial damage, especially on the edge of the valvular leaflets in dogs with MVD.[19] Myxomatous valvular degeneration in humans affects the left atrioventricular valve and aortic valve, similar to the results we found in the six eagles.[20,21] The cause in humans is also unclear but is correlated to age (>65 years), genetics, and is more common in females.[12,20] The prevalence of MVD in our study (16.6%, 95% confidence interval: 0.1-31.5; 4/24) was higher than that of small breed dogs (14.4%)[22] and humans (2.5%).[20] No conclusions can be made given the limited sample size from this study.

Endocarditis was present in two eagles in this study. Vegetative left atrioventricular valve endocarditis with multisystemic necrotizing vasculitis was present in case 5. *Staphylococcus aureus* cultured from blood and the vegetative valve was thought to have been spread hematogenously from a previous traumatic skin wound. Bacteria, including *Staphylococcus* species, *Streptococcus* species, *Pasteurella multocida*, *E. coli*, and *Pseudomonas* species, are commonly cultured in avian vegetative endocarditis with the most common isolate being *Staphylococcus aureus*.[2] These bacteria originate from acute traumatic wounds and focal infections such as salpingitis, pododermatitis, or gut translocation.[1,3,7] The lesions in this case were consistent with endocarditis lesions in other species.[10] The most commonly affected valves in birds are the left atrioventricular and aortic valves.[23] Clinical signs associated with vegetative endocarditis include lethargy, decreased appetite, and sudden death.[3] Bacterial thrombi from vegetative lesions can embolize to multiple organs including the lung, liver, spleen, kidney, pancreas, and brain. Vegetative endocarditis has been described most frequently in poultry,[24] with additional reports of non-hemolytic, coagulase positive *Staphylococcus* species on the aortic and left atrioventricular valve in an emu (*Dromaius novaehollandiae*),[23] *Streptococcus* species on both atrioventricular valves in a Waldrapp ibis (*Geronticus eremita*),[25] *Staphylococcus aureus* on the left atrioventricular valves of mallard ducks (*Anas platyrhynchos*),[26] *Enterobacter cloacae* on the left atrioventricular valve of a blue and gold macaw (*Ara ararauna*),[27] and *Staphylococcus aureus* on the left atrioventricular valve of a cockatoo (*Cacatua alba*).[28] In birds of prey, left atrioventricular valvular endocarditis caused from suspected *Pasteurella multocida* with disseminated thromboembolic lesions was described in an adult free-ranging male bald eagle.[3] Suspected staphylococcal aortic valvular endocarditis with multisystemic thrombi was reported in a free-ranging female red-tailed hawk (*Buteo jamaicensis*) with an acute left proximal radial fracture.[4] Diagnosis of the underlying cause of vegetative endocarditis is made through blood culture and echocardiogram but prognosis is poor despite long-term antibiotic therapy.[29]

The acute cerebral infarctions in cases 1, 3, 4, and 6 were attributed to a turbulence-induced thromboembolus produced from the degenerate left atrioventricular and/or aortic valves. Case 3 had an acute cerebellar infarct in addition to a chronic cerebral infarct that likely resulted from endocarditis and left atrioventricular valve regurgitation, which were diagnosed through echocardiographic imaging 5 months prior to necropsy. The relationship between valvular disease and cardioembolic cerebral infarctions in birds is not well established.[14] There is a high risk of a cerebrovascular event in humans with valvular heart disease, especially if the person has concurrent atrial fibrillation, myocardial infarction, prosthetic valves, and dilated cardiomyopathy.[12,19,30-32] Mitral valve prolapse is seen with left atrioventricular MVD in humans and surgical reconstruction increases the risk of thromboemboli.[19,33] There are no reports that MVD increases the risk of a thromboembolism in animal species (dogs) commonly treated by veterinarians despite observing platelets adhered to the exposed subendothelial matrix under scanning electron microscopy.[11,12,19] The differences observed between the risk of cardioembolic cerebral infarctions in dogs and humans remains unknown but are thought to involve differences in platelet function.[19] Avian thrombocytes have longer clotting times and have less tight aggregates that may result from differences in the GPIb*alpha* receptor.[34] Vascular encephalopathies in dogs have been reported to be associated with atherosclerosis secondary to hypothyroidism, parasitic emboli from *Dirofilaria immitis*, bacterial vegetative endocarditis, sepsis, neoplasia, coagulopathies, and trauma.[35,36] Thrombi move through the middle cerebral artery in both dogs and humans and result in weakness, syncope, altered mentation, and seizures, as observed in the eagles.[31,35]

Thromboemboli are rarely reported in birds. Fat and egg yolk emboli in psittacine birds and have been associated with hepatic lipidosis and atherosclerosis.[37] Avian thrombocytes have different aggregation properties and are less able to form shear-resistant thrombi.[14] Atherosclerosis is a progressive, degenerative disease that leads to the dystrophic calcification, chondroid metaplasia, necrosis, collagen, lipid, foam cell and fibrin deposition, luminal stenosis, and decreased elasticity of arteries;[2,10,38] however, there was no evidence of atherosclerosis in the six eagles.

The left atrioventricular and aortic endocarditis found in case 2 were unexpected findings. One cause of illness in this bird was pulmonary adenocarcinoma with widespread metastasis to the kidneys, liver, and pleura. The cerebral infarct resulted in the neurologic signs, which contributed to the decision to euthanize this animal and was more chronic than the identified left atrioventricular and aortic endocarditis. Primary respiratory tract neoplasms (e.g., pulmonary adenocarcinomas) are typically malignant and rare, affecting approximately 0.1-0.2% of birds.[39-42] Pulmonary carcinoma causing progressive weakness and paraparesis has been reported in a great horned owl (*Bubo virginianus*),[43] Moluccan cockatoo (*Cacatua moluccensis*),[39] hyacinth macaw (*Anodorhynchus hyacinthinus*),[41] African Grey parrot (*Psittacus erithacus*),[44] and blue and gold macaw (*Ara ararauna*).[45] These neoplasms often metastasize to the air sacs, pneumatized humerus, vertebrae, and spinal cord.[39,41,43-45] Multiple thromboembolic events secondary to a nonbacterial thrombotic endocarditis was observed in a human with concurrent pulmonary adenocarcinoma.[46] The nonbacterial endocarditis was attributed to a paraneoplastic hypercoagulation in the human patient.[46] We speculate that the same pathogenesis was involved in the eagle identified as case 2 in this study.

The cause of the valvular lesions in this retrospective study was undetermined, but age is a suspected factor in the MVD cases. Valvular disease was not encountered on any wild eagle necropsies. Captive birds of prey are susceptible to cardiac disease due to their higher fat diet, less active lifestyle, and ability to reach geriatric ages.[1] Five of the six eagles were > 25 years old. Females may have a higher prevalence of cardiac disease given the mobilization of cholesterol, low density lipoproteins, and triglycerides during ovulation (egg laying).[1,47] Limitations of this study include a potential geographic bias as well as similar lifestyles of the captive eagles at this one facility.

Acknowledgments: We thank the American Eagle Foundation for their contributions to the health and wellbeing of these non-releasable eagles.

References

1. Oster SC, Pariaut R. Cardiac disease of raptors. *J Avian Med Surg.* 2021;35(4):382-389.

2. Jones MP. Vascular diseases in birds of prey. *J Exot Pet Med.* 2013;22(1):348-357.

3. Jessup DA. Valvular endocarditis and bacteremia in a bald eagle. *Mod Vet Pract.* 1980;61(1):49-51.

4. Lemon MJ, Pack L, Forzán MJ. Valvular endocarditis and septic thrombosis associated with a radial fracture in a red-tailed hawk (*Buteo jamaicensis*). *Can Vet J.* 2012;53(1):79-82.

5. Cooper JE, Pomerance A. Cardiac lesions in birds of prey. *J Comp Pathol.*1982;92(1):161-168.

6. Knafo SE, Rapoport G, Williams J, et al. Cardiomyopathy and right-sided congestive heart failure in red-tailed hawk (*Buteo jamaicensis*). *J Avian Med Surg.* 2011;25(1):32-39.

7. Strunk A, Wilson GH. Avian cardiology. *Vet Clin North Am Exot Anim Pract.* 2003;6(1):1-28.

8. Borgarelli M, Buchanan JW. Historical review, epidemiology and natural history of degenerative mitral valve disease. *J Vet Cardiol.* 2012;14(1):93-101.

9. Oglesbee BL, Lehmkuhl L. Congestive heart failure associated with myxomatous degeneration of the left atrioventricular valve in a parakeet. *J Am Vet Med Assoc.* 2001;218(3):376-380.

10. Schmidt RE, Reavill DR, Phalen DN. Cardiovascular system. In: Schmidt RE, Reavill DR, Phalen DN, eds. *Pathology of Pet and Aviary Birds.* 2nd Ed. Hoboken, NJ: John Wiley & Sons, Inc; 2015:1-19.

11. Keene BW, Atkins CE, Bonagura JD, et al. ACVIM consensus guidelines for the diagnosis and treatment of myxomatous mitral valve disease in dogs. *J Vet Intern Med.* 2019;33(1):1127-1140.

12. Aupperle H, Disatian S. Pathology, protein expression and signaling in myxomatous mitral valve degeneration: comparison of dogs and humans. *J Vet Cardiol.* 2012;14(1):59-71.

13. Fox PR. Pathology of myxomatous mitral valve disease in the dog. *J Vet Cardiol.* 2012;14(1):103-126.

14. Fitzgerald BC, Beaufrère H. Cardiology. In: Speer BL, ed. *Current Therapy in Avian Medicine and Surgery.* St. Louis, MO: Elsevier; 2016:252-328.

15. Deem S, Terrell S, Forrester D. A retrospective study of morbidity and mortality of raptors in Florida: 1988-1994. *J Zoo Wildl Med.* 1998;29(1):160-164.

16. Kubba MAG, Al-Azreg SA. Endocardiosis and congestive heart failure in a captive ostrich (*Struthio camelus*). *Open Vet J.* 2013;3(2):121-125.

17. Kagohara A, dos Santos Filho M, Marinho JPM, et al. Myxomatous degeneration of the left atrioventricular valve in a true parrot (*Amazona aestiva*): a case report. *Braz J Vet Med.* 2020;42(1):e104720-e104720.

18. McNaughton A, Frasa S Jr, Mishra N, et al. Valvular dysplasia and congestive heart failure in a juvenile African penguin (*Spheniscus demersus*). *J Zoo Wildl Med.* 2014;45:987-990.

19. Corcoran BM, Black A, Anderson H, et al. Identification of surface morphologic changes in the mitral valve leaflets and chordae tendineae of dogs with myxomatous degeneration. *Am J Vet Res.* 2004;65(2):198-206.

20. Lung B, Vahanian A. Epidemiology of valvular heart disease in the adult. *Nat Rev Cardiol.* 2011;8(1): 162-172.

21. Shipton B, Wahba H. Valvular heart disease: review and update. *Am Fam Physician.* 2001;63(11):2201-2208.

22. Serfass P, Chetboul V, Sampedrano CC, et al. Retrospective study of 942 small-sized dogs: prevalence of left apical systolic heart murmur and left-sided heart failure, critical effects of breed and sex. *J Vet Cardiol.* 2006;8(1):11-18.

23. Randolph JF, Moise S, Graham DL, et al. Bacterial endocarditis and thromboembolism of a pelvic limb in an emu. *J Am Vet Med Assoc.* 1984;185(11):1409-1410.

24. Chadfield MS, Christensen JP, Christensen H, et al. Characterization of streptococci and enterococci associated with septicemia in broiler parents with a high prevalence of endocarditis. *Avian Pathol.* 2004;33(6): 610-617.

25. Greenwood AG, Marshall J, Tinsley EGF. Vegetative endocarditis in a Waldrapp ibis. *Avian Pathol.* 1996;25(1):387-391.

26. Wobeser G, Kost W. Starvation, Staphylococcosis, and vitamin A deficiency among mallards overwintering in Saskatchewan. *J Wildl Dis.* 1992;28(2):215-222.

27. Isaza R, Buergelt C, Kollias GV. Bacteremia and vegetative endocarditis associated with a heart murmur in a blue-and-gold macaw. *Avian Dis.* 1992;36(1):1112-1116.

28. Steinagel A, Quesenberry K, Donovan T. Vegetative endocarditis due to Staphylococcus aureus in an umbrella cockatoo (*Cacatua alba*). *J Avian Med Surg.* 2019;33(4):419-426.

29. MacDonald K. Infective endocarditis in dogs: diagnosis and therapy. *Vet Clin North Am Small Anim Pract.* 2010;40(1):665-684.

30. Fleming HA, Bailey SM. Mitral valve disease, systemic embolism and anticoagulants. *Postgrad Med J.* 1971;47(1):599-604.

31. Arbox A, Alió J. Acute cardioembolic cerebral infarction: answers to clinical questions. *Curr Cardiol Rev.* 2012;8(1):54-67.

32. Petty GW, Khandheria BK, Whisnant JP, et al. Predictors of cerebrovascular events and death among patients with valvular heart disease. *Stroke.* 2000;31(1):2628-2635.

33. Cohn LH, Couper GS, Aranki SF, et al. Long-term results of mitral valve reconstruction for regurgitation of the myxomatous mitral valve. *J Thorac Cardiovasc Surg.* 1994;107(1):143-151.

34. Ribeiro AM, Zepeda-Mendoza L, Bertelsen MF, et al. A refined model of the genomic basis for phenotypic variation in vertebrate hemostasis. *BMC Evol Biol.* 2015;15(1):1-8.

35. Cook LB, Coates JR, Dewey CW, et al. Vascular encephalopathy associated with bacterial endocarditis in four dogs. *J Am Anim Hosp Assoc.* 2005;41(1):252-258.

36. Garosi L, McConnell JF, Platt SR, et al. Results of diagnostic investigations and long-term outcome of 33 dogs with brain infarction (2000-2004). *J Vet Intern Med.* 2005;19(1):725-731.

37. Zwart P, Samour J. The avian respiratory system and its non-infectious disorders: a review. *J Exot Pet Med.* 2021;37(1):39-50.

38. Lujan-Vega C, Keel K, Barker CM, et al. Evaluation of atherosclerotic lesions and risk factors of atherosclerosis in raptors in northern California. *J Avian Med Surg.* 2021;35(3):295-304.

39. Jones MP, Orosz SE, Richman LK, et al. Pulmonary carcinoma with metastasis in a Moluccan cockatoo (*Cacatua moluccensis*). *J Avian Med Surg.* 2001;15(2):107-113.

40. Reavill DR, Tumors of pet birds. *Vet Clin North Am Exot Anim Pract.* 2004;7(1):537-560.

41. Baumgartner WA, Guzman DSM, Hollibush S, et al. Bronchogenic adenocarcinoma in a Hyacinth macaw (*Anodorhynchus hyacinthinus*). *J Avian Med Surg.* 2008;22(3):218-225.

42. Garner MM. Overview of tumors: section II. In: Harrison GJ, Lightfoot TL, eds. *Clinical Avian Medicine.* Palm Beach, FL: Spix Publishing; 2006:566-571.

43. Rettenmund C, Sladky KK, Rodriguez D, et al. Pulmonary carcinoma in a great horned owl (*Bubo virginianus*). *J Zoo Wildl Med.* 2010;41(1):77-82.

44. André JP, Delverdier M. Primary bronchial carcinoma with osseous metastasis in an African grey parrot (*Psittacus erithacus*). *J Avian Med Surg.* 1999;13(3):180-186.

45. Fredholm DV, Carpenter JW, Shumacher LL, et al. Pulmonary adenocarcinoma with osseous metastasis and secondary paresis in a blue and gold macaw (*Ara ararauna*). *J Zoo Wildl Med.* 2012;43(4):909-913.

46. Kaufmann CC, Wessely E, Huber K. Non-bacterial thrombotic endocarditis in the context of pulmonary adenocarcinoma: a case report. *Eur Heart J.* 2020;4(1):1-5.

47. Beaufrère H, Ammersbach M, Reavill DR, et al. Prevalence of and risk factors associated with atherosclerosis in psittacine birds. *J Am Vet Med Assoc.* 2013;242(12):1696-1703.

Can Packed Cell Volume and Total Solids Guide Wildlife Rehabilitation Decisions for Birds?

Zoey Lex, BS,
Laken Russell, DVM,
Corinne Mayer, DVM,
Jordan Padlo, CVT,
Tithipong Plangsangmas, DVM,
Mark Mitchell, DVM, MS, PhD, DipECZM (Herpetology)

Session #3372

Affiliation: From the Department of Veterinary Clinical Sciences, School of Veterinary Medicine, Louisiana State University, Skip Bertman Drive, Baton Rouge, LA 70803, USA.

Note: Abstract format selected by author(s). See Front Matter for description of conference proceedingsmanuscripts submitted as scientific abstracts.

Abstract: Wildlife are being presented to veterinarians because of negative encounters with humans. Many of these cases are critical and require medical and surgical management; however, because of limited resources, it is important to develop triage protocols to ensure resources are applied to cases with the best chance for success. The Louisiana State University Wildlife Hospital of Louisiana (LSU-WHL) currently receives over 1,000 wildlife cases a year, primarily avian species (68.8%, 2021). As a donation-based institution, it is imperative that resources are used wisely. Because of this, there is a need to understand how economical diagnostics could better evaluate prognosis in these species. Packed cell volumes (PCV) and total solids (TS) are examples of inexpensive physiologic data that could assist veterinarians with triage. The purpose of this study was to evaluate if PCV and TS measurements obtained on intake could be used as significant prognostic indicators for avian cases presented to the LSU-WHL. The hypotheses for this study were that PCV and TS would be predictive for case outcome, with cut-off values being useful in deciding the prognosis for a case. A cross-sectional study was performed using LSU-WHL records from 2020-2022. Avian cases that had a PCV and TS recorded on intake before treatment and a final disposition were included. Birds with PCV < 20% were always found to have a negative outcome (died or euthanized); TS <1.8 g/dl were also found to be a negative predictor of outcome. These results suggest that these inexpensive tests are useful to triage avian cases.

Use of Trimethoprim Sulfamethoxazole in the Treatment of Brevetoxicosis in Seabirds and Shorebirds

Maria E. Passarelli, VMD

Session #3123

Affiliation: From Save Our Seabirds, 1708 Ken Thompson Pkwy, Sarasota, FL 34236, USA.

Note: Abstract format selected by author(s). See Front Matter for description of conference proceedings manuscripts submitted as scientific abstracts.

Abstract: The toxic dinoflagellate *Karenia brevis* blooms frequently in the warm waters of the Gulf of Mexico. These harmful algal blooms can lead to significant morbidity and mortality in marine wildlife populations of the Florida Gulf Coast. In the rehabilitation setting, treatment has traditionally focused on the primary effects of the toxin;[1,2] however, brevetoxins can also significantly alter the immune response of the patient,[3,4] which may lead to secondary complications such as infection.[5] We examined the success of a broad-spectrum antibiotic treatment protocol in comparison to a supportive care-only protocol in the treatment of brevetoxicosis in 86 seabird and shorebird patients. The species most commonly affected were double-crested cormorants (*Nannopterum auritum*), laughing gulls (*Leucophaeus atricilla*), and sanderlings (*Calidris alba*), although other species were also affected. The supportive care-only group received fluid therapy and deworming as needed with fenbendazole 20 mg/kg q24h for 3 days. The treatment group was administered trimethoprim sulfamethoxazole 30 mg/kg q12h for 7 days, fenbendazole 20 mg/kg q24h for 3 days, and fluid therapy. In the supportive care-only group, 39 out of 56 died (69.6%) and 17 out of 56 survived to release (30.4%). In the treatment group, 8 out of 30 died (26.7%) and 22 out of 30 survived to release (73.3%). This study demonstrates that the addition of a broad-spectrum antibiotic to a red tide treatment protocol may help to counteract secondary infection due to brevetoxin-induced immunocompromise in avian species.

References

1. Fauquier DA, Flewelling LJ, Maucher JM, et al. Brevetoxicosis in seabirds naturally exposed to *Karenia brevis* blooms along the central west coast of Florida. *J Wildl Dis.* 2013;49:246-260.

2. Perrault JR, Barron HW, Malinowski CR, et al. Use of intravenous lipid emulsion therapy as a novel treatment for brevetoxicosis in sea turtles. *Sci Rep.* 2021;11:1-12.

3. Gebhard E, Levin M, Bogomolni A, De Guise S. Immunomodulatory effects of brevetoxin (PbTx-3) upon in vitro exposure in bottlenose dolphins (*Tursiops truncatus*). *Harmful Algae.* 2015;44:54-62.

4. Walsh CJ, Butawan M, Yordy J, et al. Sublethal red tide toxin exposure in free-ranging manatees (*Trichechus manatus*) affects the immune system through reduced lymphocyte proliferation responses, inflammation, and oxidative stress. *Aquat Toxicol.* 2015;161:73-84.

5. Niedringhaus KD, Shender LA, DiNuovo A, et al. Mortality in common (*Sterna hirundo*) and sandwich (*Thalasseus sandvicensis*) terns associated with Bisgaard taxon 40 infection on Marco Island, Florida, USA. *J Comp Pathol.* 2021;184:12-18.

Pharmacokinetics of Hydromorphone Hydrochloride and its Metabolite in Great Horned Owls (*Bubo virginianus*)

Mariana Sosa-Higareda, MVZ,
David Sanchez-Migallon Guzman, LV, MS, DipECZM (Avian), DipECZM (Small Mammal), Dipl ACZM,
Heather Knych, DVM,
Michelle G. Hawkins, VMD, DABVP (Avian Practice)

Session #3307

AAV House Officer Manuscript Competition Winner

Affiliation: From William T. Pritchard Veterinary Medical Teaching Hospital, School of Veterinary Medicine, University of California-Davis, Davis, CA 95616, USA (Sosa-Higareda), Department of Medicine and Epidemiology, School of Veterinary Medicine, University of California-Davis, Davis, CA 95616, USA (Sanchez-Migallon Guzman, Hawkins), Maddy Equine Analytical Pharmacology Laboratory, School of Veterinary Medicine, University of California-Davis, Davis, CA 95616, USA (Knych).

Note: Abstract format selected by author(s). See Front Matter for description of conference proceedings manuscripts submitted as scientific abstracts.

Abstract: Hydromorphone hydrochloride is a semisynthetic full μ-opioid more potent than morphine for acute pain.[1] A recent study demonstrated that hydromorphone has thermal antinociceptive effects in great horned owls following IM administration of 0.3 and 0.6 mg/kg for 3 and 1.5 hours respectively.[2] The objective of this study was to determine the pharmacokinetic parameters of hydromorphone and its metabolite, hydromorphone-3-glucuronide (H3G) after a single IV and IM dose in great horned owls (*Bubo virginianus*). Six healthy adult great horned owls were administered a single dose of hydromorphone (0.6 mg/kg) either IM or IV with a 6-week washout period between experiments. Blood samples were collected at 5 minutes and 0.5, 1.5, 2, 3, 6, 9, and 12 hours after drug administration. Plasma hydromorphone and H3G concentrations were determined with liquid chromatography–tandem mass spectrometry, and pharmacokinetic parameters were calculated with a non-compartmental model. Hydromorphone had high IM bioavailability 130%, with rapid elimination after administration and a large volume of distribution. Mean C_{max} was 225.46 ± 0.21 ng /ml at 13 minutes after IM injection. Mean $t_{1/2}$ were 1.62 ± 0.36 and 1.35 ± 0.59 hours after IM and IV administration, respectively. The metabolite H3G was readily measured after IM and IV administration. Mean C_{max} was 41.14 ± 12.0 ng /ml and $t_{1/2}$ 4.65 ± 3.37 hours for the IM route. Hydromorphone was well tolerated in all birds. This study is the first to document the presence of the metabolite H3G in avian species, suggesting a hydromorphone metabolism similar to that of mammals.

References

1. Murray A, Hagen NA. Hydromorphone. *J Pain Manag*. 2005;29(5):57-66. doi:10.1016/j.jpainsymman.2005.01.007

2. Monopoli MR, Sanchez-Migallon Guzman D, Paul-Murphy Joanne, et al. Evaluation of the thermal antinociceptive effects of hydromorphone hydrochloride in great horned owls (*Bubo virginianus*). Proc *ExoticsCon*; 2022:61.

Computed Tomography-Guided Surgical Correction of Air Sac Rupture in a Scarlet Macaw Combined with Pleurodesis-Like Procedure

Lauren Thielen, DVM, DABVP (Avian Practice), and
Scott Echols, DVM, DABVP (Avian Practice)

Session #3467

Affiliation: From Cornell University Hospital for Animals, Ithaca, NY 14853, USA (Thielen), Texas Avian & Exotic Hospital, Grapevine, TX 76051, USA (Thielen), Mobile Avian Surgical Services, Oakley, CA 94561, USA (Echols), Parrish Creek Veterinary Hospital & Diagnostic Center, Centerville, UT 84014, USA (Echols).

Note: Abstract format selected by author(s). See Front Matter for description of conference proceedings manuscripts submitted as scientific abstracts.

Abstract: An 11-year-old male scarlet macaw (*Ara macao*) presented with a 2-year progressive history of chronic air-sac rupture. A large accumulation of subcutaneous air, distending a space of 15 × 6 centimeters, along the ventrocaudal left side of the neck extending onto the dorsal head, was appreciated as a ruptured left cervical air sac. A computed tomography (CT) scan was performed to locate a rent location for surgical planning. CT findings diagnosed the open communication between the cervical and interclavicular air sacs via two individual rents. Upon surgical investigation, the largest rent (7 millimeters) was identified by visual inspection; however, the smaller rent was present deep to the initial rent and located based on CT findings. Closure of the small rent (2 millimeters) was made possible by performing surgery through the larger rent, followed by closure of the large rent. The day after surgery, although dramatically reduced, a mild to moderate amount of subcutaneous air was present. A pleurodesis-like procedure was performed by injecting doxycycline hyclate into the air sac-subcutaneous space, to generate a reaction to the skin and the air sac tissues and keep it collapsed.[1] An intravenous catheter was used to deflate the air and then inject the doxycycline, and then maintained as an air sac cannula for 1 week to prevent subcutaneous air recurrence. The parrot is currently symptom-free 1 year post operation. Despite correction of air sac rupture being described in previous literature, this is the first report of a combined CT-guided surgical correction and pleurodesis-like procedure.[2]

References

1. Antinoff N. Attempted pleurodesis for an air sac rupture in an Amazon parrot. *Proc Annu Conf Assoc Avian Vet.* 2008.

2. Browning GR, Eshar D, Tucker-Mohl K, Berke K. Diagnosis and surgical repair of a chronic ruptured cervical air sac in a double yellow-headed Amazon parrot (*Amazona ochrocephala oratrix*). *J Exot Pet Med.* 2019;29:45-50.

Hypothermia During General Anesthesia Under Isoflurane, Sevoflurane and Desflurane in Hispaniolan Amazon Parrots (*Amazona ventralis*)

Hugo Gonzalez-Jassi, MVZ,
Thomas N. Tully Jr., DVM, MS, DABVP (Avian Practice), DipECZM (Avian),
Jeannette Cremer, Dr med vet, DACVAA, CVA,
Noelia Diaz Falcón, DVM, MS,
Chin-Chi Liu, MS, MApStat, PhD,
Patricia Queiroz-Williams, DVM, MS

Session #3459

Affiliation: From the Department of Veterinary Clinical Sciences, College of Veterinary Medicine, Oklahoma State University, 2065 W. Farm Rd., Stillwater, OK 74078, USA (Gonzalez-Jassi), Department of Veterinary Clinical Sciences, School of Veterinary Medicine, Louisiana State University, Skip Bertman Dr., Baton Rouge, LA 70803, USA (Tully Jr, Cremer, Diaz Falcón, Chin-Chi Liu, Queiroz-Williams).

Note: Abstract format selected by author(s). See Front Matter for description of conference proceedings manuscripts submitted as scientific abstracts.

Abstract: Hypothermia caused by general anesthesia using isoflurane (iso), sevoflurane (sevo), and desflurane (des) was evaluated in Hispaniolan Amazon parrots (HAP). Twelve adult parrots were premedicated with butorphanol (0.5 mg/kg IM) and randomly assigned to three groups: iso, sevo, and des. Anesthesia was induced via face mask (4% iso, 6% sevo and 12% des) followed by orotracheal intubation and maintenance with 1.4-2% iso, 2.4-3% sevo, and 8.5-9.2% des. Each parrot received each inhalant on different occasions (7-day washout period). Hypothermia was defined as body temperature (BT) below 100 °F.[1] External heat support was provided when BT dropped to 99.5 °F. Time for heat loss from 102 °F to 99.5 °F (T1), time to first increase in BT from 99.5 °F (T2) and time from initial heat support to 102 °F (T3) were statistically analyzed. Heart rate, respiratory rate, oxygen saturation, end-tidal carbon dioxide, and non-invasive blood pressure were also analyzed. All three inhalants rapidly caused hypothermia (T1 Mean ± SD in mins iso:13.92 ± 8.58, sevo:11.42 ± 4.14 and des:12.58 ± 4.62). There were no statistically significant differences among inhalants over T1 and T3, but T2 had a significant difference ($P = 0.0412$) between des and sevo. Sevoflurane caused transitory apnea and respiratory depression. When hypothermic, 10 animals had a systolic arterial pressure below 100 mmHg (iso:4, sevo:3 and des:3). Two types of bradyarrhythmia were identified with all inhalants. Hypothermic HAP under general anesthesia using iso, sevo and des, had similar complications. When heat support is provided, the inhalant used does not play a major role in the time to correct this complication.

References

1. Rembert MS, Smith JA, Hosgood G, et al. Comparison of traditional thermal support devices with the forced-air warmer system in anesthetized Hispaniolan Amazon parrots (*Amazona ventralis*). *J Avian Med Surg*. 2001;15(3):187-193.

Device Composition Using a 3-Way Stop Cock for Monitoring of Capnography in Small Birds

Jungha Lee, DVM,
Donhwi Shin, DVM,
Seongchan Yeon, DVM, PhD,
Inhyung Lee, DVM, PhD,
Won-gyun Son, DVM, PhD

Session #3209

Affiliation: From the Department of Veterinary Clinical Sciences, College of Veterinary Medicine,Seoul National University, Seoul 08826, Korea.

Note: Abstract format selected by author(s). See Front Matter for description of conference proceedings manuscripts submitted as scientific abstracts.

Abstract: In small birds, perioperative monitoring of vital signs is challenging due to their size. End-tidal carbon dioxide ($EtCO_2$) monitoring is especially difficult due to relatively small tidal volume and large dead space of airway devices. In this study, without specialized equipment minimizing dead space, a low-cost device composition using a 3-way stop cock for intraoperative capnography monitoring was evaluated in small birds (eight species) weighing 70-260 grams. Medical records were reviewed for nine birds receiving general anesthesia for bone or beak fracture repair. The birds were premedicated with none, sedatives (medetomidine, midazolam, alfaxalone) and/or analgesics (hydromorphone, ketamine, tramadol) intramuscularly (IM). After mask induction, intubation was performed using an intravenous catheter (14- to 22-gauge), and was maintained with isoflurane (end-tidal concentration 2.0-2.5%) for 25-100 minutes. Monitoring of heart rate, oxygen saturation, temperature, $EtCO_2$ was attempted, but temperature and $EtCO_2$ monitoring was only possible due to the small size of the patients. To minimize dead space, an $EtCO_2$ sampling line was connected to the intubated catheter on the opposite side via 3-way stop cock for effective sampling of exhalation gas. The other way that a 3-way stop cock was connected to the breathing tube was through an extension tube. Extubation was performed after surgery and the patients recovered uneventfully except for one case; hypoperfusion and hypovolemic shock was estimated to be the reason of death in a cachexic patient. Complications associated with intubation and ventilation were not observed during peri-anesthetic period.

Comparison of Non-Controlled (Dexmedetomidine-Nalbuphine) and Partially Controlled (Dexmedetomidine-Butorphanol) Sedation Protocols in Chickens (*Gallus gallus domesticus*)

Greta Doden, DVM,
Kaitlyn Crocker, DVM,
Hannah Kunzman, DVM,
Julie Fithian, DVM,
João Brandão, LMV, MS, DipECZM (Avian), Dipl ACZM

Session #3298

Affiliation: From the Department of Veterinary Clinical Sciences, College of Veterinary Medicine, Oklahoma State University, 2065 W. Farm Rd., Stillwater, OK 74078, USA.

Note: Abstract format selected by author(s). See Front Matter for description of conference proceedings manuscripts submitted as scientific abstracts.

Abstract: Typical avian sedation protocols include controlled drugs, creating regulatory challenges for veterinarians and field researchers.[1] Recent avian studies have investigated the use of dexmedetomidine, a non-controlled α2-agonist.[2,3] Butorphanol is a commonly used κ-agonist/μ-antagonist opioid and a Schedule IV controlled substance. Nalbuphine, also a κ-agonist/μ-antagonist opioid, is currently not controlled. Few studies have assessed nalbuphine in avian species.[4-7] An effective, non-controlled sedation protocol would be particularly useful in field situations. The purpose of this study was to determine the effective sedation dosage and compare the sedation quality between non-controlled nalbuphine-dexmedetomidine (ND) and partially controlled butorphanol-dexmedetomidine (BD) protocols. Effective dose 99% of each protocol was estimated using an Up-and-Down method in 20 healthy, adult Buff Orpington hens (*Gallus gallus domesticus*). Ten hens were subsequently included in a blinded, randomized, complete crossover design. Each animal received 2 mg/kg butorphanol and 50 μg/kg dexmedetomidine or 12.5 mg/kg nalbuphine and 20 μg/kg dexmedetomidine intramuscularly. Cardiorespiratory parameters and sedation score were evaluated by a blinded observer every 10 minutes until recovery. A lower score indicated effective sedation. Both protocols induced sedation. Heart and respiratory rates did not differ between treatments, but heart rates were significantly different from baseline. BD induced lower scores for recumbency at 10, 40, and 60 minutes; muscle tone at 30, 40, and 90 minutes; and radiographic positioning at up to 30 minutes. Although BD provided lower scores at most time points, the ND combination also induced sedation and may be a useful alternative to controlled substances for light to moderate sedation in avian patients.

References

1. González MS, Adami C. Psittacine sedation and anesthesia. *Vet Clin North Am Exot Anim Pract.* 2022;25: 113-114.

2. Mumm LE, Mans C. Evaluation of dexmedetomidine-midazolam sedation in budgerigars (*Melopsittacus undulatus*). *J Am Vet Med Assoc.* 2022;260(10):1194-1199.

3. Sha J, Kongara K, Singh P, et al. Pharmacokinetics and pharmacodynamics of butorphanol and dexmedetomidine after intranasal administration in broiler chickens (*Gallus gallus domesticus*). *Vet Sci.* 2022;9(212):1-13.

4. Guzman DSM, KuKanich B, Keuler NS, et al. Antinociceptive effects of nalbuphine hydrochloride in Hispaniolan Amazon parrots (*Amazona ventralis*). *Am J Vet Res.* 2011;72:736-740.

5. Keller DL, Guzman DS-M, Klauer JM, et al. Pharmacokinetics of nalbuphine hydrochloride after intravenous and intramuscular administration to Hispaniolan Amazon parrots (*Amazona ventralis*). *Am J Vet Res.* 2011;72(6):741-745.

6. Guzman DSM, Braun JM, Steagall PVM, et al. Antinociceptive effects of long-acting nalbuphine decanoate after intramuscular administration to Hispaniolan Amazon parrots (*Amazona ventralis*). *Am J Vet Res.* 2013;74(2):196-200.

7. Guzman DS-M, KuKanich B, Heath TD, et al. Pharmacokinetics of long-acting nalbuphine decanoate after intramuscular administration to Hispaniolan Amazon parrots (*Amazona ventralis*). *Am J Vet Res.* 2013;74(2):191-195.

Section 5

AAV Masterclasses Track 1

Chistal Pollock, DVM, DABVP (Avian Practice);
Michael Bignall, DVM
Session Chairs

Avian Zoonoses

Cheryl B. Greenacre, DVM, DABVP (Avian Practice), DABVP (Exotic Companion Mammal Practice)

Session #3347

Affiliation: From College of Veterinary Medicine, University of Tennessee, Knoxville, TN 37996, USA (Emeritus). Current address: Office of Research, University Research Animal Resources, University of Georgia, 501 D.W. Brooks Drive, Room 3504, Athens, GA 30602, USA.

Abstract: There are many diseases that can be spread from birds to humans, the most well-known being chlamydiosis. There are other less notable zoonotic diseases from birds including those that can spread directly (salmonellosis, campylobacteriosis, yersiniosis, influenza A, Newcastle disease, by mosquito (viral encephalitides) or indirectly (histoplasmosis and cryptococcosis). An overview of typical clinical signs, diagnostic and treatment options for birds are provided for common avian zoonotic pathogens.

Introduction

Diseases that can be spread from birds to humans are collectively called avian zoonoses. These diseases can be categorized by causative agent, in this case, bacterial and chlamydial, viral, fungal, and parasitic. Although the reported incidence of disease from avian zoonotic pathogens is low, and most if not all can be prevented by proper sanitation and hygiene practices and/or proper use of personal protective equipment (PPE), it is best to be familiar with the clinical signs, and diagnostic and therapeutic options available to manage these diseases. There are many resources available for further information on avian zoonoses including books, published articles, websites, and free informational pamphlets from governmental websites.[1-21]

Avian Chlamydiosis

Taxonomy and nomenclature

The disease psittacosis goes by the names parrot fever, ornithosis, chlamydiosis, and chlamydophilosis.[1-17,19-21] The term psittacosis refers to the disease in people originating from a parrot (a psittacine bird), whereas the term ornithosis refers to the disease in people originating from any species of bird. The term chlamydiosis is a more generic term and refers to infection in any animal or person with any organism in the *Chlamydia* or *Chlamydophila* (previous name) genus. The term avian chlamydiosis is used to specify an infection in a bird with any *Chlamydia* sp., including the recently identified *Chlamydia gallinacea* (turkeys and chickens) and *Chlamydia avium* (pigeons). Many changes in the nomenclature of organisms in the Chlamydiaciae family have occurred over the years. For this discussion, the organism will be called by its current name, *Chlamydia psittaci*. Note that *Chlamydia psittaci* should not be confused with a related organism in people, *Chlamydia trachomatis*, a sexually transmitted disease of people, or another related organism, *Chlamydia pneumoniae*, a common respiratory pathogen of people.

History

In the late 1800s psittacosis was recognized as a clinical entity and was called pseudotyphus. In 1892, 51 cases and 16 deaths occurred in Paris, France. In 1929-1930 a pandemic occurred involving 12 countries and approximately 800 cases. This outbreak stimulated research into the disease and it was found to be associated with parrots, particularly those recently imported from South America. Immediately prior to the pandemic a large shipment of birds from South America was distributed worldwide for sale during the Christmas season. The infective stage, known as elementary bodies, were first described independently by three different researchers, Leventhal, Cole and Lillie in 1930, which lead to the alternate name of LCL bodies. In 1945 the term chlamydia, from the Greek word chalmus for cloak, was introduced. As late as 1965 the organism was thought to be a virus. It wasn't until the organism was observed with electron microscopy to have bacterial RNA, ribosomes and cell wall structures that it was proven to be an obligate intracellular bacterial organism (a bacteria that, like viruses and the Rickettsia, is incapable of extracellular living). The *Chlamydia* sp. were lumped in taxonomically with the Rickettsia until 1966 when Page validated the placement of the organism into its own family, Chlamydiaceae.

Epidemiology

The organism has been found in 130 species of birds worldwide and a variety of mammals, including humans. There are numerous potential avian species that may act as a source of infection for people. The most common source of infection (70% of all cases in the 1980's) is exposure to a recently acquired psittacine bird, most often a budgerigar, cockatiel or lovebird. Birds smuggled across the United States borders as well as those raised in captivity can potentially be infected with avian chlamydiosis. Other birds can be a potential source of infection, such as domestic or wild pigeons, passerines (soft-billed birds) or poultry.

The general prevalence of chlamydial infection in captive birds is thought to be less than 5%, and in wild birds less than 1%, but may increase dramatically in birds stressed by shipping, crowding, chilling and breeding. If an outbreak of psittacosis occurs in a facility with multiple birds or a facility that buys and sells birds, then the potential for rapid spread to other birds and people is high. Common high-traffic situations are pet stores, bird breeding facilities, and flea markets. Often it is difficult to later identify the seller of birds at a flea market.

People can become infected with psittacosis even after brief, passing exposure to infected birds or environments and may not present with an obvious bird-related exposure history. In the 1980s, 43% of the psittacosis cases reported in humans involved bird fanciers or pet bird owners. Others at vocational risk include pigeon fanciers. Others at occupational risk include pet store employees, veterinarians, veterinary technicians, laboratory workers, workers in avian quarantine stations, farmers, wildlife rehabilitators, zoo workers and employees of poultry slaughtering and processing plants (usually turkey).[20,21] Occasionally exposure to wild pigeon roosts is a source of infection to the general public.

The distribution of this disease in birds and people is worldwide. From 1988 to 1998 the Centers for Disease Control and Prevention (CDC) received reports of 813 human cases in the United States, but since psittacosis is only a notifiable disease (physicians are not required to report it) and diagnosis is sometimes missed, the actual number of cases is probably much higher. During the 1980s, 70% of the reported cases were associated with exposure to infected pet birds. The prevalence of disease in birds is considered to be low (< 5%), but can reach 100% during an outbreak. In birds, the morbidity during an aviary outbreak is generally up to 50-80% and mortality is up to 10-30%. An excellent source of yearly documented cases in humans is the CDC's Morbidity and Mortality Weekly Report. Through the CDC there is a graph in PowerPoint form of reported human psittacosis cases in the United States between 1971 and 2001 at this website: http://www.cdc.gov/epo/dphsi/annsum/1999/slides/psitgraph.ppt.

Pathogenesis

The *C. psittaci* organism is transmitted by either inhalation or ingestion of the spore-like elementary body phase of the organism. Shedding in birds can be activated by stress, such as shipping, crowding, chilling and breeding. Person to person transmission has been suggested, but never proven. Those individuals that are immunosuppressed are more susceptible to the disease and its effects.

The organism *C. psittaci* is relatively resistant, surviving in the soil for 3 months or within a bird dropping for up to 1 month. *C. psittaci* is a dimorphic organism, meaning it exists in two different forms called phases, the infectious phase and the replicating phase. In general, the *C. psittaci* organism enters the host via inhalation or ingestion and replicates in the host's cells. The infectious form is small (0.2-0.6 microns) and called an elementary body. After the elementary body is inhaled or ingested it is endocytosed by a host cell where it transforms into a reticulate body. The reticulate body is larger (1.5 microns) than the elementary body and is the replicating, or metabolically active, form of the organism inside the host's cell. The reticulate body uses nutrients from the host cell and undergoes multiple rounds of binary division before releasing multiple elementary bodies from the cell when it ruptures. Two or more days pass between the time the host cell is infected and the elementary bodies are released. These infectious elementary bodies then infect other of the host's cells or are released into the environment via feces, nasal secretions, sputum, blood or infected tissues. The elementary body is metabolically inactive and resistant to environmental forces, so it can survive long enough to be inhaled or ingested by another host.

The *C. psittaci* organism is a documented hazard among laboratory personnel and appropriate precautions should be taken when working with this organism or birds naturally or experimentally infected with this organism. *C. psittaci* may be present in tissues, feces, nasal secretions and blood of birds and present in tissues, sputum and blood of humans. The organism easily aerosolizes making it easy to inhale the infective phase of the organism. Persons performing a necropsy of any bird suspected of having avian chlamydiosis should wear gloves and wet the carcass with soapy water to decrease aerosolization or organisms, as well as practice Animal Biosafety Level 2 practices, containment equipment and facilities and respiratory protection.

Clinical signs

Parrots

There are many species of parrots and each species, even each individual, may respond differently to being infected with *C. psittaci*. For example, some birds present quite ill whereas others exhibit very subtle signs of disease. Generally, the parrot with psittacosis presents with depression, lethargy, anorexia, dyspnea, nasal or ocular discharge, conjunctivitis, and biliverdinuria. Rarely, birds present comatose, which has been observed in sensitive species such as macaws. Commonly both the spleen and liver are enlarged.

Pigeons and passerine birds

Pigeons and passerine birds seem to exhibit little if any clinical signs of disease while infected with the *C. psittaci* (or *C. avium*) organism and therefore are sometimes referred to as asymptomatic carriers of the disease.

Turkeys and chickens

Turkeys and chickens seem to exhibit little if any clinical signs of disease while infected with the *Chlamydia* sp. and therefore are sometimes referred to as asymptomatic carriers of the disease.

Diagnosis

Suggestive

Consistent clinical signs and radiographs showing splenomegaly, ± hepatomegaly are suggestive. A CBC showing a heterophilic, monocytic leukocytosis and a mild non-regenerative anemia are also suggestive. A plasma electrophoresis may be suggestive of either acute or chronic disease.

More specific

There are many tests available for use in birds including tests to detect antibodies in the serum (elementary body assay [EBA] and immunofluoescent antibody [IFA]) and test to detect antigen in the feces or blood (enzyme linked immunofluorescent antibody assay [ELISA] and polymerase chain reaction [PCR]). It is best to perform a panel of three tests including PCR of blood, PCR of feces and IFA of serum. Also, a fluorescent antibody (FA) test can be performed on tissue such as liver tissue from a biopsy or necropsy. For legal purposes, cell culture from the feces is the best test, but the organism does not consistently grow, and shedding of the organism in the feces is intermittent, and there is also risk to laboratory personnel when the organism is grown in the laboratory. There is one commercial laboratory, Texas Medical Diagnostic Laboratory, offering culture.[1]

The following definitions have been accepted by the American Veterinary Medical Association (AVMA) and the Association of Avian Veterinarians and can be found in the Compendium of Measures to Control *Chlamydia psittaci* Infection Among Humans (Psittacosis) and Pet Birds (Avian Chlamydiosis), 2017 by the National Association of State Public Health Veterinarians (NASPHV).[1]

A confirmed case of avian chlamydial infection is defined on the basis of at least one of the following:

- Isolation of *C. psittaci* from a clinical specimen.
- Identification of chlamydial DNA by use of in situ hybridization to detect chlamydial DNA followed by specific *C. psittaci* DNA detection using PCR-based testing of in situ hybridization-positive tissues or secondary *C. psittaci*-specific DNA probes in combination with characteristic pathology. The commercial antibodies used for immunohistochemistry staining cross-react with non-chlamydial epitopes and are not diagnostic.
- A fourfold or greater change in serologic titer in two specimens from the bird obtained at least 2 weeks apart and assayed simultaneously at the same laboratory.
- Identification of suggestive intracellular bacteria within diseased cells in smears or tissues (e.g., liver, conjunctival, spleen, respiratory secretions) stained with Gimenez or Macchiavello stain in combination with detecting *C. psittaci* DNA in the same tissue sample using a *C. psittaci*-specific PCR-based detection assay.

A suspected case of avian chlamydial infection is defined as a compatible illness and at least one of the following:

- Identification of chlamydial nucleic acid by PCR-based testing in conjunctival, choanal, or cloacal swabs; blood; or feces.

- Chlamydiaceae antigen (fluorescent antibody) in feces, a cloacal swab specimen, or respiratory tract or ocular exudates. The commercial antibodies used for fluorescent antibody staining cross-react with non-chlamydial epitopes and are not diagnostic.
- Is epidemiologically linked to a confirmed case in a human or bird.

There may be cause for concern and further investigation if one of the following occurs:

- Identification of chlamydial nucleic acid by PCR-based testing in conjunctival, choanal, or cloacal swabs; blood; or feces in a healthy-appearing bird, or
- Compatible illness with positive results from a nonstandardized test or a new investigational test.

Treatment

Treatment of birds should be supervised by a licensed veterinarian. The treatment of choice in birds is oral doxycycline, since it is absorbed better and eliminated slower than other tetracyclines. Care should be taken to observe for signs of toxicosis while treating such as lethargy, anorexia, biliverdinuria. If it occurs, the doxycycline should be discontinued and the bird offered supportive care until recovered, and then a lower dose can be attempted. Doxycycline monohydrate or calcium syrup formulations are administered at a dose of 40-50 mg/kg/day for most species, although a lower dose of 25 mg/kg/day is used in African grey parrots, Goffin's cockatoos, blue and gold macaws and green-winged macaws.[13,14]

A medicated feed can be administered to budgies made of one part steel cut oat groats and three parts small-hulled millet (measured by volume) and 5-6 milliliters of sunflower oil per kilogram of feed, then mixing to coat seeds, then adding and mixing in 300 milligrams of doxycycline hyclate (from capsules) per kilogram of oat/seed/oil mix. Make fresh daily.

It has been shown through pharmacologic studies that alternatively medicated water can be offered to cockatiels at 200-400 milligrams of doxcycline hyclate per liter of drinking water, or can be offered to Goffin's cockatoos at 400-600 milligrams per liter of drinking water, or can be offered to African grey parrots at 800 mg per liter of drinking water.[13,14]

The intramuscular formulation of doxycycline called Vibravenos (Pfizer, London) is available in Europe and used at a dose of 75-100 mg/kg body weight every 7 days for four or more doses. Realize the intravenous (IV) formulation of doxycycline hyclate is not acceptable for intramuscular use, but for IV use only.

The injectable, long-acting oxytetracycline (LA-200, Pfizer, Exton, PN) has been administered in some species (Goffin's cockatoos, blue-fronted and orange-winged Amazon parrots and blue and gold macaws) subcutaneously at a dose of 75 mg/kg body weight every 3 days, but is only used short term in birds that aren't eating initially because it causes irritation at the site of injection.

There is also a chlortetracycline medicated feed (mash or pellet) available; however, it must be the only source of food and birds may not accept it very well.

Prevention and control

The following is an outline of prevention and control information provided in the Compendium of Measures to Control *Chlamydia psittaci* Infection Among Humans (Psittacosis) and Birds (Avian Chlamydiosis), 2017.[1]

- Protect persons at risk.
- Those people cleaning cages or handling infected birds, should wear protective clothing, gloves, a disposable surgical cap and a respirator.
- Maintain accurate records of all bird-related transactions to aid in identifying sources of infection and potentially exposed persons.
- Records should include date of purchase, species of bird, source of birds, leg band numbers, and describe any illness or deaths of birds.
- Avoid purchasing or selling birds that have signs of avian chlamydiosis.
- Isolate newly acquired birds.
  - New birds should be quarantined at least 30-45 days.
- Test birds before they are to be boarded or sold on consignment.
- Practice preventive husbandry.
- Do not stack cages. Keep cages and bowls clean and disinfect before adding new birds or using the cage for a different bird.
- Prevent the spread of infection.
  - Isolate birds and treat.
- Use disinfection measures.
- Chlamydial organisms can be killed with most commonly used disinfectants including a 1:100 dilution bleach solution for 10 minutes, 1% Lysol®, 70% isopropyl alcohol, or 1:1000 dilution of Roccal®. Gloves and respirators should be worn in an avian outbreak in any areas that have been exposed to the positive bird or its fecal matter (remember dried fecal matter is still contagious for several months). It is important to clean and remove all organic matter first and then disinfect.

Specific considerations

In most states avian chlamydiosis is reportable by veterinarians. In most states human chlamydiosis caused by organisms other than *C. trachomatis* is reportable by physicians (*C. trachomatis* is reportable by physicians in all states).

Salmonellosis

Taxonomy and nomenclature

The *Salmonella* genus comprises gram-negative rod-shaped bacteria. The serotypes that penetrate the mucosal barrier better are usually the pathogen, whereas the other serotypes are usually the secondary pathogens. Genus, species and variety are depicted in Table 1.

Table 1. *Salmonella* genus, species and variety.

Genus	Species	Variety
Salmonella	*typhimurium*	Many (e.g., var. Copenhagen)

Epidemiology

Salmonella has been reported in many avian families including Galliformes, Columbiformes, Psittaciformes, Passeriformes, Strigiformes, and Falconiformes. Generally, most birds with a cecum, such as Galliformes, commonly have *Salmonella* spp. in their gastrointestinal (GI) tract, but if it is isolated from any other tissue in these birds it is considered significant. Usually birds without cecae, such as Psittaciformes, do not have *Salmonella* sp. in their GI tract, and if cultured from the feces it is significant.

Pathogenesis

Transmission is usually through the fecal oral route, but may be from the aerosol route, after exposure to clinically or subclinically infected birds (i.e. asymptomatic carriers are common). Humans developing salmonellosis from birds usually have a history of being exposed to ducks (especially young children during Easter time), gulls or poultry. Infection is either acute, peracute, chronic or subclinical.

The incidence of salmonellosis in birds and people varies greatly with the host species and serotype. Factors that increase the likelihood of disease are exposure to high numbers of organisms, susceptibility of the host, and the serotype of the organism. Lories and African grey parrots seem to be highly susceptible to salmonellosis, even to non-invasive serotypes. *Salmonella* spp. can survive up to 28 months in avian feces.

Clinical signs

Clinical signs can include depression, lethargy, anorexia, weight loss, diarrhea (pasty vent), lameness, abscess formation and death. In people the clinical signs are primarily a hemorrhagic diarrhea and death. People under 5 years of age, the elderly, or those that are immunosuppressed are more likely to develop severe clinical signs and death.

Diagnosis

Salmonella is best grown on MacConkey's agar. *Salmonella* isolated from tissues other than the GI tract of any bird is significant, as well as if it is isolated from the feces or GI tract of birds without a cecum.

Treatment

Antibiotic sensitivity varies with the serotype isolated, therefore, use antibiotics known to include a gram-negative spectrum and change antibiotics accordingly when results of the sensitivity return. Birds whose normal fecal flora is known to have *Salmonella* sp. (those with cecae such as chickens) should not be treated because it is unnecessary, not possible to completely eliminate, and creates resistant organisms.

Prevention and control

Educate the public regarding the risk of salmonellosis from Easter ducks and chicks, waterfowl at parks that children feed and poultry. Free posters and downloadable handouts are available from the CDC website.

Specific considerations

In most states salmonellosis is reportable by veterinarians.

Mycobacteriosis

Taxonomy and nomenclature

Mycobacteriosis has been diagnosed in many species including Galliformes, Anseriformes, Gruiformes, Charadriiformes, Columbiformes, Falconiformes, Strigiformes, Ciconiiformes, Passeriformes and Psittaciformes. Domestic fowl are relatively resistant to infection and seem to be the primary host for many avian infections, whereas psittacine birds seem to be highly susceptible. Genus, species, and common hosts are depicted in Table 2.

Table 2. *Mycobacterium* genus, species and common hosts.

Genus	Species	Common host
Mycobacterium	*tuberculosis*	People
Mycobacterium	*avium*	Birds
Mycobacterium	*bovis*	Birds and people

Epidemiology and pathogenesis

Transmission is through inhalation or oral inoculation. Because of the long incubation period it is sometimes difficult to determine the source of people or bird exposure. There have been cases where *Mycobacterium avium* was isolated from both the people and a bird in a household, and it is unknown who had it first. The reported incidence of mycobacteriosis in wild populations varies between 4-40%.

Clinical signs

Avian mycobacteriosis generally causes chronic weight loss and unthriftiness. Radiographically or at necropsy granulomatous masses of the liver and intestines are most common. People generally develop tubercles in the lung. Gulls will present with mycobacteria related osteomyelitis in the limbs. Rarely, birds will present with a cutaneous form.

Diagnosis

Radiographs are suggestive. A significant monocytic leukocytosis is usually present. Demonstration of acid-fast organisms in feces is not definitive, since other non-pathogenic acid-fast organisms can be in the feces. Demonstration of acid-fast organisms in tissues is highly suggestive. A FITES acid-fast test is preferable to the standard acid-fast test since it is less destructive to the organism, especially for delicate mycobacteria species such as *M. genovense* or *M. kansasii*. Usually, *M. avian* is associated with large numbers of organisms in birds, whereas *M. tuberculosis* and *bovis* are not. The organism is difficult and dangerous to culture. The intradermal tuberculin test used in many mammals is not reliable in birds. There is a PCR test available.

Treatment

Treatment should be discouraged and euthanasia should be the first option discussed due to the difficulty of treating the *M. avium* in people. Veterinarians should probably have clients sign a statement outlining the human risks of treatment and to consult their physician. Various regimens have been described including the use of ethambutol which renders the cell wall permeable to other medications such as chlorambucil, azithromycin, enrofloxacin, among others.

Yersiniosis

Taxonomy and nomenclature:

Genus, species, common hosts, and geographic area for yersiniosis are depicted in Table 3.

Table 3. *Yersinia* genus, species, common hosts and geographic areas.

Genus	Species	Common host	Geographic area
Yersinia	*pseudotuberculosis*	People and animals	Northern Europe
Yersinia	*enterolitica*	People and animals	Worldwide

Epidemiology

Yersinia pseudotuberculosis has been reported in many species, whereas *Y. enterolitica* has been reported in geese, ducks, robins, canaries, pigeons and gulls. Some wild species are known to be refractory. The reservoir of *Yersinia* spp. is wild rodents and birds. In the author's experience toucans fed wild caught rodents at an aviary became infected and died of yersiniosis. Control rodents in an aviary to prevent spread of disease from rodents to birds.

Pathogenesis

Transmission is through the oral route from exposure to infected feces or urine.

Clinical signs

Clinical signs are usually of acute fatal sepsis, but clinical signs can include depression, lethargy, weight loss, diarrhea, respiratory distress and incoordination.

Diagnosis

Antemortem diagnosis is difficult. The organism is difficult to isolate in the laboratory.

Treatment

Treatment is difficult to start in time to be of benefit, but broad spectrum bacteriocidal antibiotics may be of benefit if started early in the course of the disease.

Campylobacteriosis

Taxonomy and nomenclature

Campylobater are gram-negative curved or spiral rods that cannot be distinguished by light microscopy from *Vibrio* spp. or *Helicobacter* spp. Genus, species and common hosts are depicted in Table 4.

Table 4. *Campylobacter* genus, species and common hosts.

Genus	Species	Common host
Campylobacter	*fetus*	Birds
Campylobacter	*intestinalis*	Birds and people
Campylobacter	*jejuni*	Birds and people

Epidemiology and pathogenesis

Campylobacter jejuni is the most common isolate from birds. The common bird host species are chickens, turkeys, pheasants, geese, pigeons, robins, finches, canaries, ducks, sparrows, starlings and blackbirds. Transmission is through the fecal-oral route and from contaminated or infected vectors. Eating contaminated poultry is a reported source of infection for humans.

Clinical signs

Clinical signs include lethargy, diarrhea and emaciation. Fledgling mortality can be high.

Diagnosis

Identification of curved rods on a fecal Gram's stain is suggestive. Culture or electron microscopy is definitive.

Treatment

Treatment for campylobacteriosis varies, but usually involves tetracyclines. Birds whose normal fecal flora is known to have *Campylobacter* sp. (those with cecae such as chickens) should not be treated as it is unnecessary and creates resistant organisms.

Influenza A

Overview and current update

There are two forms of avian influenza (AI): low pathogenic AI (LPAI) and high pathogenic AI (HPAI). Most AI encountered in the United States are LPAI, generally causing a mild respiratory disease.[18] There is a LPAI vaccine available, and is used in commercial poultry settings. Conversely, HPAI causes severe disease, is associated with a high mortality, and is a foreign animal disease in the United States with no available vaccine. The combined value of the U.S. production from broilers, eggs, turkeys, and the value of sales from chickens in 2020 was $35.5 billion. During the 2014/2015 HPAI outbreak, the U.S. Department of Agriculture (USDA) Animal and Plant Health Inspection Service (APHIS) spent $1 billion to eradicate HPAI and provide indemnity (payment) to those that lost birds and were National Poultry Improvement Plan (NPIP) certified. In that outbreak, about 49 million birds died or were depopulated and the United States lost 10% of its egg laying birds. HPAI is a threat to the U.S. economy, a threat to birds (backyard and commercial poultry, zoo birds, wild birds), and a possible threat to humans. Although a few cases have been shown to pass from birds to humans, and from birds to mammals (rats, mice, stone martens, civets, ferrets, pigs, cats, tigers, leopards, dogs, horses, coyotes, fox, raccoon, skunk, bear, dolphins, and more recently, harbor seals in the Netherlands, and harbor and gray seals in New England), the virus easily mutates, and could mutate into a strain that passes easily from person to person, creating a potential pandemic situation. The recent H5N1 clade 2.3.4.4b Eurasian strain that spread from Europe has now infected wild and domestic birds in North America, and South America, with devasting results. This strain is more pathogenic to wild birds (bald eagles, black vultures, waterfowl, pelicans, penguins, gulls) than past strains with high mortality events in many species and is trending towards becoming endemic in these new areas. The USDA is actively pursuing vaccine development and may change tactic in the near future to control HPAI from test and cull to vaccination.

Taxonomy and nomenclature

Influenza virus is in the Orthomyxoviridae family. There are three types of influenza, types A, B and C, but only type A is found in birds. There are two influenza A subtypes based on surface proteins, H for hemagglutinin, and N for neuraminidase. There are 15 known H subtypes and nine known N subtypes. All subtypes are found in birds, but only H1, H2, H3 and N1 and N2 are in humans. All avian virulent strains to date have been H5 or H7 subtype, but most H5 or H7 isolates have been of low virulence.

History

In 1918-1919 there was an influenza A pandemic in people called the Spanish flu that was found to be H1N1. This outbreak may have killed 20-50 million people worldwide, and 50% were young, healthy adults. Greater than 500,000 deaths were attributed to the flu in the United States.

In 1957-1958 the Asian flu, H2N2, caused approximately 70,000 human deaths in the United States. In 1968-1969, the Hong Kong flu, H3N2, caused 34,000 human deaths in the United States and still circulates today. In 1961, influenza A H5N1 was first isolated from terns in South Africa. In 1997, influenza A was shown to pass from birds to humans in an outbreak in Hong Kong where 18 people became sick and six died that had contact with infected birds. This outbreak was controlled by killing 1.5 million chickens.

Since December 2003, an Asian HPAI H5N1 virus has resulted in high mortality in poultry and wild birds in Asia, the Middle East, Europe, and Africa. In 2003, influenza A H5N1 was isolated from two human cases in Hong Kong, one of which died. Then the H5N1 subtype was confirmed in turkey poults in Cambodia, China

(Hong Kong had a single positive peregrine falcon), Indonesia, Japan, Laos, South Korea, Thailand and Viet Nam. Human deaths of influenza A H5N1 were reported in Thailand and Viet Nam after contact with infected birds or their excretions (nasal, saliva or feces). This H5N1 subtype isolated from humans has been genetically sequenced and all genes were of bird origin, but it was found to be resistant to two antiviral drugs, amantadine and rimantadine. It was still sensitive to oseltamivir and zanamivir. Also in 2003, influenza A H7N7 was isolated from poultry workers in the Netherlands. Over 80 cases were reported and one died, a veterinarian. People in the Netherlands outbreak exhibited eye infections with some respiratory infections. Another case occurred in 2003 involving one child in Hong Kong with influenza A H9N2 that recovered. In Canada in 2003, H7N3 was isolated.

In 2022, (January), HPAI (H5N1) was identified in the United States, initially in wild ducks in South Carolina and North Carolina. By March of 2022, HPAI had spread along the eastern seaboard in wild birds (AL, CT, DE, FL, GA, IA, IN, KS, KY, MA, MD, ME, MO, NE, NH, NJ, NY, NC, SC, TN, VA), and along the Atlantic and Mississippi migration flyways in commercial and backyard poultry (CT, DE, IA, IL, IN, KS, KY, MD, ME, MI, MO, NY, SD, VA, WI). Waterfowl act as a reservoir host and usually do not show any signs of infection, but the current outbreak is highly pathogenic to waterfowl and wild birds as well. The current outbreak is associated with migrating birds from Europe where the Eurasian strain of HPAI (H5N1) is essentially becoming endemic. Updated information can be found on the USDA-APHIS website regarding wild bird confirmations, and commercial and backyard flock confirmations, at the following websites, respectively:

- https://www.aphis.usda.gov/aphis/ourfocus/animalhealth/animal-disease-information/avian/avian-influenza/hpai-2022/2022-hpai-commercial-backyard-flocks

- https://www.aphis.usda.gov/aphis/ourfocus/animalhealth/animal-disease-information/avian/avian-influenza/hpai-2022/2022-hpai-wild-birds

Epidemiology and pathogenesis

Wild birds can be the natural hosts for influenza A, especially waterfowl, and with previous strains these bird species did not generally exhibit illness from infection, but with the currently circulating H5N1 wild birds including ducks, vultures, etc. are experiencing high mortality events. Domestic birds, particularly chickens are very susceptible to high mortality with influenza A. Animal to human transmission has occurred but is unusual at the current time, but a mutation in the virus could change this at any time. The chance for influenza A to develop a mutation that will allow it to be transmitted from human to human is very real and may happen in the near future. The virus is not efficient yet at human to human transmission.

The World Health Organization (WHO), as well as other organizations, are monitoring for human gene development. Currently, the goal is to stop the virus from spreading by culling any birds infected or exposed to pathogenic strains (H5 or H7). The OIE, an international organization based in France, and formed in 1928 by 28 countries, and now comprises 165 countries, is a clearinghouse for reported cases of animal disease from each of these countries. Their goal is to respond to multi-country outbreaks in a rapid manner.

Clinical signs

The current circulating HPAI (H5N1) is a foreign animal disease in the United States and is causing high mortality in both wild and domestic bird species. If clinical signs present prior to death, they are quite variable, so if you witness high avian mortality in any bird species, contact your state veterinarian immediately. AI is a highly contagious reportable foreign animal disease. You do not need to know or prove what is causing the high mortality event before you call them. The clinical signs of HPAI can be quite variable and may include: oculonasal discharge, infraorbital sinusitis, dyspnea, tachypnea, cyanosis, edema of the head, comb (swollen comb is common in chickens), and wattle, subcutaneous ecchymotic hemorrhages of the legs and feet, blood-tinged oral and nasal discharges, green diarrhea, torticollis, opisthotonos, and ataxia. Non-specific lesions are seen at gross necropsy, but a friable liver and/or splenic necrosis is commonly encountered.

Diagnosis

Many tests are available including the agar gel immunodiffusion (AGID) test, ELISA and RT-PCR, but these tests must be performed by a certified laboratory (usually state diagnostic laboratories or contact your state or USDA veterinarian). For PCR swabs it is recommended to obtain both a choanal and a cloacal sample (it can be on same swab), even in waterfowl. Not all birds develop demonstrable antibodies on the ELISA. Also available are hemagglutination and hemagglutination inhibition tests.

Treatment

Birds with H5 of H7 in the United States are required to be depopulated. The 2003 influenza A H5N1 isolated from people in Hong Kong, was genetically sequenced and found to be resistant to two antiviral drugs, amantadine and rimantadine. It was still sensitive to oseltamivir and zanamivir.

Prevention and control

A killed vaccine is available for the low pathogenic strains in the United States. There is a vaccine for the high pathogenic strains, but they are not currently used in the United States. Depopulation is the current recommended action taken by the USDA during an outbreak. The USDA is actively pursuing vaccine development and may change tactic in the near future to control HPAI from test and cull to vaccination.

Specific considerations

HPAI is a reportable foreign animal disease. Refer to the USDA website for current information on this rapidly changing outbreak. Indemnity from the federal government is available for birds that died after a diagnosis was made and if the flock is NPIP certified.

Newcastle Disease

Overview

Virulent Newcastle disease (VND) is a foreign animal disease and millions of dollars are spent to eradicate all exposed birds when there is an outbreak in the United States. VND affects many species of birds. The causative agent is a paramyxovirus. Technically it is considered a zoonotic disease since it can cause conjunctivitis in humans.

Taxonomy and nomenclature

Newcastle disease is caused by a paramyxovirus of which there are four pathotypes, lentogenic, mesogenic, neurotropic velogenic and viscerotropic velogenic (VVND). The latter, a foreign animal disease goes by many names and abbreviations including VVND, exotic Newcastle disease (END), and more recently as VND.

History

The non-exotic form of Newcastle disease was common before the use of vaccines for those forms. There is a vaccine available for exotic Newcastle disease, but its use is currently not allowed in the United States because the tactic to control the disease in this country is to test and cull any positive birds. In the early 1970s there was an outbreak of VND and millions of dollars were spent to destroy chickens in the outbreak to control the disease. In 2002, there was another outbreak of VND and again millions were spent to destroy infected and surrounding flocks to control the disease.

Epidemiology

The disease is spread by direct contact with viral particles from aerosolization of respiratory secretions or feces, or from fecal contaminated food, water or litter. Recovered birds are believed to shed the virus indefinitely.

Clinical signs

The clinical signs vary with the pathotype but include depression, diarrhea, anorexia, ruffled feathers, oculonasal discharge, conjunctivitis, dyspnea, ataxia, muscle tremors, paralysis and death. In people Newcastle disease causes a mild, acute granular conjunctivitis, general malaise and sinusitis that resolves within 7-20 days.

Diagnosis

Avian serum can be sent to a laboratory for PMV 1, 2 and 3 antibody testing. Other tests are available including the hemagglutination inhibition test, the ELISA, AGID, PCR, and virus isolation.

Treatment

For pet birds with mild lentogenic or mesogenic forms of Newcastle disease offer supportive care. Any bird, pet or otherwise that is positive for VND in the United States is to be reported to federal authorities and culled.

Prevention and control

Vaccines are available for control of lentogenic and mesogenic forms of Newcastle disease in the United States. VND is a foreign animal disease and millions are spent to eradicate all exposed birds when there is an outbreak in the United States (one was in California in 2002-2003; 3.5 million birds were culled). The latest outbreak in California started May, 2018 and was declared ended on June 1, 2020. At least 1.2 million birds were euthanized or depopulated. The cost is still undetermined, but over $100 million were spent to eradicate.

Specific considerations

Quarantine stations for birds entering the United States were developed to keep VND out, therefore any bird entering U.S. quarantine is tested serologically for VND (and also avian influenza). They are not tested for other diseases, not even chlamydiosis, although they are provided chlortetracycline impregnated food while in quarantine to prevent shedding of any possible chlamydial organisms to quarantine personnel while housed.

Viral Encephalitides

Taxonomy and nomenclature

Eastern and Western equine encephalomyelitis virus are known as EEE and WEE, respectively.

Epidemiology

Birds are the known reservoirs for these diseases. Many species of birds are involved. The virus is transmitted through a mosquito bite or in pheasants it has been shown to be transmitted through the broken skin created by pecking.

The mosquitos known to transmit EEE virus in the United States and Canada include *Culiseta melanura*, *Aedes* spp. and *Coqiullettidia* spp. Interestingly, *Culiseta melanura* feeds mainly on birds and rarely on people. The *A. sollicitans* and *A. vexans* mosquitos feed on both birds and people, but are less likely to be infected with the virus. Bird mites are known to be potential vectors. The mosquitos known to transmit WEE virus include *Culex tarsalis* in the western United States and Canada, and *Culiseta melanura* in the eastern United States.

People are an accidental host to the virus and if infected with EEE can have up to 80% mortality, and if infected with WEE can have up to 5-15% mortality. In Canada, a direct connection has been shown between the prevalence of the virus in wild birds and the number of human cases. During the 1962 outbreak in Saskatchewan, Canada the WEE virus was isolated from 22% of the wild bird population.

Clinical signs

Birds may show no clinical signs to ataxia, trembling, weakness, paralysis and death. People show varying degrees of neurologic signs and death.

Diagnosis

Antemortem diagnosis is difficult. Histopathologic lesions of lymphoplasmacytic encephalitis is suggestive.

Treatment

There is no specific treatment for viral encephalitis other than supportive care.

Rabies

All warm-blooded animals, including birds are susceptible to rabies virus. Avian infections are rare, but have been reported in chickens, waterfowl and raptors. A passive hemagglutination test in wild birds was seropositive in 23.1% of 65 predatory birds and 2.9% of 278 non-predatory birds. While no case has yet been documented of avian to human transmission of the rabies virus, preexposure rabies prophylaxis and preventing trauma while handling raptors is recommended.

West Nile Virus

Taxonomy and nomenclature

The causative organism is a flavivirus.

History

West Nile virus (WNV) is endemic in other countries, but in the late 1990's it was found within the eastern United States and has since spread across the country.

Epidemiology

Crows, jays and raptors and horses are susceptible species, whereas poultry are considered resistant. The WNV virus is spread by mosquitos. If people or dogs are affected they are usually older or immunosuppressed.

Clinical signs

Clinical signs range from none in resistant species such as poultry to neurologic signs (ataxia, circling, head tilt, and seizuring) and death in susceptible species. A CBC is usually normal or a lymphocytosis is present.

Diagnosis

There is a serum antibody test available.

Treatment

Treatment consists of supportive care. Recently in people the use of alpha interferon has seemed to result in better success.

Prevention and control

This disease has spread throughout the United States, so it is too late to prevent the disease in this country. There is a conditionally licensed vaccine available for use in horses that is currently being used intramuscularly in birds at the same or a reduced dose. Studies are underway evaluating the antibody response and protectiveness of this vaccine in some species of birds. Another vaccine specifically for use in birds is being developed.

Histoplasmosis

Taxonomy and nomenclature

Histoplasma capsulatum is an imperfect dimorphic fungus with a worldwide distribution.

Epidemiology and pathogenesis

Transmission is through the inhalation of highly infective spores. There is a wide host range including birds, mammals, and humans. The fungus proliferates particularly well in soil that is enriched with nitrogen from bird droppings or within accumulations of bird droppings (e.g., pigeon roosting sites are a common source of infection).

Clinical signs

In immunocompetent humans the infection is common, benign and self-limiting. In immunosuppressed humans, severe respiratory disease can result in chronic cavitary lesions or disseminated disease.

Diagnosis

Antemortem diagnosis is apparently difficult in humans since it mimics other less serious respiratory diseases, but thoracic radiographs and culture of sputum are recommended, as well as an antigen skin test to indicate sensitivity to the organism.

Treatment

Antifungals are used in the attempted treatment in people.

Cryptococcosis

Taxonomy and nomenclature

Cryptococcus neoformans is a saprophytic yeast-like fungus.

Epidemiology and pathogenesis

Cryptococcus infections have been reported in many avian species including swallows, pigeons, green winged macaw, canaries, thick-billed parrot, African grey parrot, triton cockatoo and Mollucan cockatoo. Cryptococcosis in people from exposure to infected birds is well documented and is particularly common after exposure to infected pigeon droppings which are high in nitrogen and encourage growth. The infectivity of *Cryptococcus* is thought to be lower in avians due to their higher core body temperature and the poor growth of this organism at higher temperatures such as 40 °C.

Clinical signs

Most birds show no clinical signs. Clinical signs in birds vary with the location of the growth of fungus but can include depression, weakness, anorexia, weight loss, acute diarrhea, incoordination, blindness, dyspnea, nasal discharge, oral masses and death. Generally, the respiratory lesions are the most common. Clinical signs in people usually include a meningitis and/or respiratory disease.

Diagnosis

Cryptococcal infections are associated with a gelatinous or granulomatous lesion usually in the air sacs. The organism has been isolated form intraocular sites, bone marrow, pectoral muscles, neck fascia, meninges, hepatic veins, air sacs, trachea, pararenal areas and lungs.

Treatment

Antifungal including amphotericin-B and flucytocine are recommended, and the success rate in people with this combination is 68%.

Summary

Several diseases can be transmitted from birds to people, but most are treatable if diagnosed early. This early diagnosis is enhanced by good communication between veterinarians and their clients and proper regulatory officials. Because avian veterinarians work in a field involving recently imported, recently domesticated or wild birds, they are exposed to disease entities that are not commonly encountered by the average veterinarian and should therefore be more vigilant in educating themselves and others to these diseases as well as keeping an open mind when observing clinical signs in their patients.

References

1. Balsamo G, Maxted AM, Midla JW, et.al. Compendium of measures to control *Chlamydia psittaci* infection among humans (psittacosis) and pet birds (avian chlamydiosis). *J Avian Med Surg.* 2017;31(3):262-282.

2. Chu J, Yarrarapu SNS, Vaqar S, et al. Psittacosis. In: StatPearls [Internet]. Treasure Island, FL: StatPearls Publishing; 2023.

3. Human psittacosis linked to a bird distributor in Mississippi, Massachusetts and Tennessee. *Morbidity and Mortality Weekly Report.* 1992, 41(42);794-797.

4. Chin J. *Control of Communicable Diseases Manual.* 17th Ed. Washington, DC: American Public Health Association; 2000.

5. Centers for Disease Control and Prevention. Epidemiologic Notes and Reports Psittacosis at a Turkey Processing Plant, North Carolina. 1989. Available at: http://www.cdc.gov/mmwr/preview/mmwrhtml/00001662.htm. Accessed June 1, 2023.

6. Centers for Disease Control and Prevention. Psittacosis in the United States. 1979. Available at: http://www.cdc.gov/mmwr/preview/mmwrhtml/00014688.htm. Accessed June 1, 2023.

7. Everett KD, Bush RM, Anderson AA. Emended description of the order Chlamydiales, proposal of Parachlamydiaceae fam. nov. and Simkaniaceae fam. nov., each containing one monotypic genus, revised taxonomy of the family Chlamydiaceae, including a new genus and five new species, and standards for the identification of organisms. *Int J Syst Bacteriol.* 1999;49(2):415-440.

8. Everett KD, Anderson AA. Identification of nine species of the Chlamydiaceae using PCR-RFLP. *Int J Syst Bacteriol.* 1999;49(2):803-813.

9. Centers for Disease Control and Prevention. The CDC's Division of Bacterial and Mycotic Diseases Disease Information Sheet on Psittacosis. Available at: http://www.cdc.gov/ncidod/dbmd/diseaseinfo/psittacosis_t.htm. Accessed June 1, 2023.

10. Centers for Disease Control and Prevention. Facts About Disease Case Definitions. Available at: https://www.cdc.gov/pneumonia/atypical/psittacosis/index.html. Accessed June 1, 2023.

11. Centers for Disease Control and Prevention. Morbidity and Mortality Weekly Report. Notice to Readers: Changes in National Notifiable Diseases List and Data Presentation. Available at: https://www.cdc.gov/mmwr/preview/mmwrhtml/mm5101a5.htm. Accessed June 1, 2023.

12. Federation of American Scientists. CDC Bioterrorism Agents. Available at: https://biosecurity.fas.org/resource/documents/CDC_Bioterrorism_Agents.pdf#:~:text=CDC%20Bioterrorism%20Agents%20Category%20A%20%C2%BE%20Anthrax%20%28Bacillus,Lassa%2C%20Machupo%5D%29%20Category%20B%20%C2%BE%20Brucellosis%20%28Brucella%20species%29. Accessed June 1, 2023.

13. Powers L, Flammer K, Papich M. Preliminary investigation of doxycycline plasma concentration in cockatiels (*Nymphicus hollandicus*) after administration by injection or in water or feed. *J Avian Med Surg.* 2000;4:23-30.

14. Flammer K, Whitt-Smith D, Papich M. Plasma concentrations of doxycycline in selected psittacine birds when administered in water for potential treatment of *Chlamydophila psittaci* infection. *J Avian Med Surg.* 2001;15:276-282.

15. Raso T, Junior AB, Pinto AA. Evidence of *Chlamydophila psittaci* infection in captive Amazon parrots in Brazil. *J Zoo Wildl Med.* 2002;33(2):118-121.

16. Ritchie BW, Dreesan DW. Avian zoonoses: proven and potential diseases. Part I. Bacteria and parasitic diseases. *Comp Cont Educ.* 1988;10(4):484-491.

17. Ritchie BW, Dreesan DW. Avian zoonoses: proven and potential diseases. Part II. Viral, fungal and miscellaneous diseases. *Comp Cont Educ.* 1988;10(6):688-695.

18. World Organisation for Animal Health. New OIE web portal on avian influenza. 2016. Available at: https://www.woah.org/en/new-oie-web-portal-on-avian-influenza/#:~:text=The%20new%20OIE%20portal%20on%20avian%20influenza%20brings,actions%20taken%20by%20the%20OIE%20and%20its%20partners. Accessed June 1, 2023.

19. Sachse K, Laroucau K. Two more species of *Chlamydia*-does it make a difference? *Pathog Dis.* 2015;73(1):1-3. doi.org/10.1093/femspd/ftu008.

20. Guo W, Li J, Kaltenboeck B, et al. *Chlamydia gallinacea,* not *C. psittaci*, is the endemic chlamydial species in chicken (*Gallus gallus*). *Sci Rep.* 2016;6:19638.

21. Centers for Disease Control and Prevention. CDC Multistate Psittacosis Outbreak among Poultry Plant Workers, 2018. Available at: https://www.cdc.gov/pneumonia/atypical/psittacosis/surveillance-reporting/outbreaks/2018-poultry-multistate-investigation.html. Accessed June 1, 2023.

Birds' Use of Technology for Enrichment and Socialization

Ilyena Hirskyj-Douglas, PhD,
Rebecca Kleinberger, PhD,
Jennifer Cunha, JD,
Susan Clubb, DVM

Session #3444

Affiliation: From University of Glasgow, Glasgow G12 8QQ, UK (Hirskyj-Douglas), Northeastern University, 360 Huntington Ave, Boston, MA 02115, USA (Kleinberger), Parrot Kindergarten, Inc., 601 Heritage Dr, Ste 424, Jupiter, FL 33458, USA (Cunha), Rainforest Clinic for Birds and Exotics, 3319 E Road, Loxahatchee, FL 33470, USA (Clubb).

Abstract: Technology has changed drastically in the past decade, creating new opportunities for enrichment and interactive devices for parrots and their caregivers. While recent advances have been made in parrot husbandry and care practices, the physical and mental health issues in captive parrots persist. While parrots in people's homes often do use technologies, such as tablets and smartphones, these technologies are designed by and for humans, who have vastly different physiology and cognitive processes than what parrots need. In this workshop, veterinarians and other experts in the parrot industry are invited to speculate on how technology can support parrots and caregivers to improve everyday lives and welfare, and increase sense of agency. In this workshop, the future of parrot technology is envisioned to critically explore and interrogate what emerging technologies can, and should, look like for birds. Using these visions, central themes of technology are distilled and important features of these systems are listed to create an expert-guided framework of key features that parrot-technology needs to address.

Introduction

Recent research has demonstrated that parrots (psittacine birds) are capable of high-level cognitive processing, including inference-by-exclusion,[1] counting,[2] music preference,[3] symbol use, and abstract concept processing.[4] Such cognitive stimulation further appears to have positive outcomes in terms of mental and physical health suggesting potential to use these tasks for enrichment. While research on non-human primates spans over a century, comparative research into avian cognition is much more recent and upcoming.

Comparing parrots to non-human primates (*Homo sapiens*), replications conducted on certain corvids (Corvidae) and parrots yield striking results. Güntürkün, Ströckens, Scarf and Colombo compared the performance of primates, corvids and pigeons (Columbidae) across a variety of tasks including short-term memory, object permanence, abstract numerical competence, and orthographic processing.[5] They discovered that pigeons and corvids demonstrated similarly high levels of performance when compared to non-human primates. Likewise, Goffin's cockatoos (*Cacatua goffiniana*), African grey parrots (*Psittacus erithacus*), and certain species of corvids have out-performed 7- to 10-year-old children and adult gorillas on some complex cognitive tests such as reasoning, memory, and mental manipulation tasks.[6-8]

Stereotypy and Enrichment

Perhaps due to their cognitive ability and unmet need for stimulation, captive birds often display stereotypic behaviors. Stereotypies are abnormal repetitive, unvarying, and apparently functionless behaviors that are often performed by captive and domesticated animals housed in barren environments.[9] They are believed to be in response to life in a captive environment. In birds, stereotypies include a variety of oral and locomotive stereotypies including feather-destruction , pacing, excessive vocalizations and hanging on cage bars.

Stereotypy rates in parrots are reported as high as 42% in cockatoos (*Cacatua galerita*) and 40% in African grey parrots[10] giving indication towards other parrot species. Some research has suggested a relationship between lack of enrichment and stereotypy.[9] For example, locomotive stereotypy (i.e., route-tracing and cage-hanging) may be due to lack of space and physical complexity and oral stereotypy may be related to lack of foraging enrichment. Orange-wing Amazon parrots (*Amazona amazonica*) raised in a barren environment had much higher rates of stereotypy than those raised with locomotion- and foraging-based enrichment environments.[11] Further, once the barren-environment raised parrots were relocated to enriched environments, their rates of stereotypy decreased.

Contemporary researchers have suggested a need for higher levels of cognitive enrichment for captive animals, beyond traditional methods of environmental stimulation.[11-13] Enrichment is traditionally composed of foraging opportunities, toys, variety in diet, destructible materials, and passive sensory stimulation (such as the radio or television). While these toys and devices may elicit natural behaviors, research suggests that these may not provide adequate mental stimulation with varied learning, motor, attentional and sensory components. This is attributed to the lack of cognitive challenges that increase in difficulty upon mastery of tasks, as may be demanded in wild settings.[12] These researchers argue that enhanced enrichment using non-traditional modalities including complex toys, puzzles, and even interactive technology including learning games or social interactions could additionally be utilized to meet the cognitive needs of captive animals.

Technology Enrichment

To meet the growing complex needs of animals, researchers have begun exploring how different technologies can support an animal's life and wellbeing. Nicky, et al. shows the growing relevance of interactive digital applications for captive primates and insists on the importance of giving control of the technologies to the animal.[14] Drawing from this, researchers have now deployed several technologies with animals in zoos, and various pets that aim to give animals agency by recognizing that they are sentient beings that benefit from controlling their environment.[15-17] Agency can be defined as the propensity of an animal to engage actively with the environment with the main purpose of gathering knowledge and enhancing its skills for future use.[18]

Likewise, offering parrots a similar kind of agency in shaping their environment, especially in turning systems on or off, may yield substantial benefits. Technological improvement in terms of sensing and recognition opens the door to various sorts of technologies for parrots.[16,19] Thus, while such systems should not be used in isolation to replace human care or co-specific interactions, there is potential for technologies to enrich, and connect birds to each other, their humans and extend their environment around them.

In terms of technological innovation, the field of animal-computer interactions has produced inspiring previous work in solo enrichment systems. Technology for animals is a growing billion-dollar industry.[17]

In the context of zoo-housed animals, computers and tablets have been used to provide primates with a variety of enriching applications, from face-matching games to "Tinder for orangutans" used in breeding research.[20]

In those examples, the main role of the human volunteers was to hold the devices. Keepers have expressed the need for enrichment systems that are hands-off since tablets often have to be held by caregivers, which is time-consuming and can cause safety issues. With the orangutans, frustration was observed because they were not able to hold the devices themselves.

Some of the most interesting works involving animals and digital technologies are the ones that acknowledge the key role played by humans in animal environments. Previous inspiring work from the human-computer interaction (HCI) community involving animals has acknowledged and curated the human's role and behavior to appeal to the specific animal or species, from a challenging play partner,[15] to an empathetic audience.[21] Such approaches have the potential to lead to higher levels of animal-human interaction, playfulness, and energy expenditure especially when the animal is provided with increased agency through the digitally mediated interspecies interaction.

Workshop Organization

The purpose of this workshop is to explore the design of technology that can improve the welfare of parrots. Although there have been efforts to develop technology for animals, most current technologies are not adapted to the unique needs of birds. Therefore, it is essential to develop technology based on parrot needs, with a good user interface and user experience for birds. The workshop is divided into three parts:

1. Overview of animal-computer interaction (ACI) and previous work on innovative experience design for parrots

   In this 30-minute overview, the field of ACI is introduced and inspiring previous work on technology for parrots is presented both by authors and other practitioners. This section proposes key considerations when working with animals and technology including agency, control, affordances, consent, and ethics.

2. Group design exercise

   In this 45-minute interactive section, participants gather in groups of four and are given a prompt to create an ideation around what they see the future of parrot-computer interfaces looking like. The method of design fiction is used to help move from current technology constraints to envision, explore and criticize future parrot-technologies. For this task, researchers support the groups and provide additional information on current technologies and take key notes on overarching themes that are present within the designs. The designs are focused on imaging technologies for parrot-to-parrot interaction, parrot-human interaction and parrots using technologies in general.

3. Interactive session with audience members

   After 45 minutes, all groups gather and present their thoughts on key features of their designs, any ethical concerns they have, any key opportunities that technology can take and any criticisms of future technologies. These thoughts are gathered and visualized in real-time to present key findings of future parrot-computer interfaces and how the future of technology can support parrots and things to be concerned about. This section is interactive with the audience to help identify the main contexts, needs, and senses of parrots where technology can be leveraged. Potential contexts are introduced in which technology can be used and audience feedback is sought through live polling that participants can access via their phones to help address the main needs and requirements that parrots have. The aim is to collect and reflect together quantitative and qualitative data based on participants observation and expertise and on the above exercise. The topics covered in this section include:

- Identifying and ranking main needs and requirements from parrots
- Identifying and ranking potential technologies that could be leveraged
- Identifying risks and ethical issues linked to using such technologies
- Identifying ways to alleviate those risks
- Reflecting on the role of human caretakers in deploying technologies for parrots
- Identifying rigorous ways to assess and evaluate the value/benefits of the systems on birds

Ethical Statement

All data collected is anonymized and following the conference copies of the overall findings are sent to participants desiring them. In encouraging the development of technologies for parrots, an emphasis is placed on not only ethical design features to create ease in interacting with the device but also in the deployment and training of animals to engage with the technologies in a force-free, uncoerced manner.

Outcomes

In this workshop, ACI is introduced and examples of currently available ACI technologies are provided. Participants are encouraged to develop innovative and practical technology solutions that benefit parrots and other animals.

This workshop gathers information from veterinarians and other avian experts regarding the ways technology might be developed to benefit parrots, as well as the unique physiology of parrots that need to be considered in technology. This is accomplished in several ways including via an online survey as well as with small group discussions.

With this information, data are presented to technology experts and developers to encourage the design and deployment of ethical technologies for animals. As a result of this information, the future will include improved welfare conditions, lower stereotypies, and higher rates of health for parrots in captivity.

Conclusions

The workshop looks at how technologies can support parrots, bringing together experts in parrot husbandry and veterinary science to explore the possibilities and limitations of future parrot-computer interfaces. Through group ideation and the design fiction method, participants are able to overcome current technological constraints and imagine potential technologies to support parrots and their caregivers. The workshop investigates key features of the designs, ethical concerns, opportunities for technology, and critiques of future technologies for parrots exploring how they can support the birds themselves, veterinarians and their human caregivers. The real-time visualization of the ideas provides a clear picture for attendees of the central themes that emerge from the workshop, and the expert-guided framework of key features for parrot technology serves as a valuable resource for researchers and practitioners in the field on the current state of the art technologies for animal welfare. The workshop emphasizes the importance of collaboration and interdisciplinary approaches to address the complex challenges in parrot care and welfare. The insights gained from this workshop will inform future research and development of technologies to support parrots and their caregivers, and ultimately contribute to improving the well-being of captive parrots.

References

1. O'Hara M, Auersperg AMI, Bugnyar T, Huber L. Inference by exclusion in Goffin cockatoos (*Cacatua goffini*). *PLoS ONE.* 2015;(10)8:e0134894. doi:10.1371/journal.pone.0134894.

2. Pepperberg IM, Carey S. Grey parrot number acquisition: the inference of cardinal value from ordinal position on the numeral list. *Cognition.* 2012;125:219-232.

3. Copping, J. The Business Insider. Parrots are Picky When it Comes to Music. Available at: https://www.businessinsider.com/parrots-have-specific-music-tastes-2012-12. Accessed June 1, 2023.

4. Pepperberg IM, Koepke A, Livingston P, et al. Reasoning by inference: further studies on exclusion in grey parrots (*Psittacus erithacus*). *J Comp Psychol.* 2013;127(3):272-281. doi.org/10.1037/a0031641.

5. Güntürkün O, Ströckens F, Scarf D, Colombo M. Apes, feathered apes, and pigeons: differences and similarities. *Curr Opin Behav Sci.* 2017;16:35-40.

6. Pepperberg IM, Gray SL, Lesser JS, Hartsfield LA. Piagetian liquid conservation in grey parrots (*Psittacus erithacus*). *J Comp Psychol.* 2017;131(4):370-383. doi.org/10.1037/com0000085.

7. Pepperberg IM, Gray SL, Mody S, et al. Logical reasoning by a Grey parrot? A case study of the disjunctive syllogism. *Behaviour.* 2019;156(5-8):409-445. doi.org/10.1163/1568539X-00003528.

8. Pepperberg IM, Pailin H. Evolution of mechanisms underlying visual working memory manipulation: when "bird-brain" is a compliment. *Proc Annu Conf Vision Sci Soc.* 2017.

9. Garner J, Meehan C, Famula T, Mench J. Genetic, environmental, and neighbor effects on the severity of stereotypies and feather picking in orange-winged Amazon parrots (*Amazona amazonica*): an epidemiological study. *Appl Anim Behav Sci.* 2006;96(1):153-168. doi:10.1016/j.applanim.2005.09.009.

10. Jayson S, Williams D, Wood J. Prevalence and risk factors of feather plucking in African grey parrots (*Psittacus erithacus erithacus* and *Psittacus erithacus timneh*) and cockatoo (*Cacatua* spp.) *J Exot Pet Med.* 2014;23:250-257. doi:10.1053/j.jepm.2014.06.012.

11. Meehan C, Garner J, Mench J. Environmental enrichment and development of cage stereotypy in orange-winged Amazon parrots (*Amazona amazonica*). *Dev Psychobiol.* 2004;44(4):209-218.

12. Bennett A, Bailoo J, Dutton M, et al. A model quantitative assessment tool for nonhuman primate environmental enrichment plans. *bioRxiv.* 2018;341206. doi.org/10.1101/341206.

13. Washburn DA. The four Cs of psychological wellbeing: lessons from three decades of computer-based environmental enrichment. *Anim Behav Cogn.* 2015;2(3):218-232. doi.org/10.12966/abc.08.02.2015.

14. Kim-McCormack NNE, Smith CL, Behie AM. Is interactive technology a relevant and effective enrichment for captive great apes? *Appl Anim Behav Sci.* 2016;185:1-8.

15. Hirskyj-Douglas I, Lucero A. On the internet, nobody knows you're a dog...unless you're another dog. *Proc Conf Human Factors Comput Syst.* 2019;1-12.

16. Cunha J, Rhoads C. Use of a tablet-based communication board and subsequent choice and behavioral correspondences in a Goffin's cockatoo (*Cacatua goffiana*). *Proc Conf Animal-Comput Interact.* 2020;1-9.

17. Kleinberger R, Cunha J, Vemuri M, Hirskyj-Douglas I. Birds of a feather video-flock together: agency-based parrot-to-parrot video-calling system for interspecies ethical enrichment. *Proc Conf Human Factors Comput Syst.* 2023. doi.org/10.1145/3544548.3581166.

18. Spinka M, Wemelsfelder F. Environmental challenge and animal agency. *Animal Welfare.* Wallingford, UK: CAB International; 2011:27-44.

19. Scheel B. Designing digital enrichment for orangutans. *Proc Conference Animal-Comput Interact.* 2018. doi.org/10.1145/3295598.3295603.

20. Westerlaken M, Gualeni S. Situated knowledges through game design: a transformative exercise with ants. *Proc Conf Philos Comput Games.* 2016.

21. Webber S, Marcus C, Sherwen S, et al. Kinecting with orangutans: zoo visitors' empathetic responses to animals' use of interactive technology. *Proc Conf Human Factors Comput Syst.* 2017:6075-6088. doi.org/10.1145/3025453.3025729.

Sedation in Birds

Christoph Mans, Dr med vet, Dipl ACZM, and
Grayson A. Doss, DVM, Dipl ACZM

Session #3486

Affiliation: From the School of Veterinary Medicine, University of Wisconsin-Madison, 2015 Linden Drive, Madison, WI 53706, USA.

Abstract: Sedation techniques provide a means of reducing physiologic stress in birds undergoing non-painful clinical procedures. In addition, sedation in birds provides easy restraint, increases the safety of many clinical procedures (e.g., blood collection, radiography, ultrasonography), and allows for a complete examination, which would otherwise only be achieved under general anesthesia. Using safe and effective sedative protocols in pet birds provides substantial benefits to the patients as well as the veterinarian and staff. Sedation should be considered for a variety of clinical procedures in avian patients.

Introduction

Sedation of dogs and cats in veterinary practice is a daily routine for a variety of procedures, such as radiographs and ultrasonography, or other non-painful but potentially stressful procedures. However, historically for avian patients, either manual restraint of conscious birds or general anesthesia is typically performed to complete most clinical procedures.[1] General anesthesia predisposes birds to cardiovascular and respiratory depression and may cause aspiration of gastric or crop contents and hypothermia. In contrast, manual restraint in conscious birds is simple to perform but can have negative consequences, including stress to the bird and/or handler, negative conditioning to the clinic environment (e.g., the person restraining or the towel used for restraint), hyperthermia, and the predisposition of trauma to the handler and/or bird. Several recent studies demonstrated that manual restraint of birds causes increased body temperature and respiratory rate.[1] In sick, old, or very stressed birds, acute collapse, and death secondary to manual restraint have been reported. Therefore, sedation techniques provide a useful alternative for reducing physiologic stress in birds undergoing non-painful clinical procedures. Furthermore, sedation in birds provides easier restraint, increases the safety of many clinical procedures (e.g., blood collection, radiography, ultrasonography), and allows for a complete examination, which would otherwise only be achieved under general anesthesia. Using safe and effective sedative protocols in pet birds provides substantial benefits to the patients as well as the veterinarian and staff and should be considered for a variety of clinical procedures.

Route of Drug Administration

Historically, sedative drugs have been most administered by intramuscular injection in birds. Recently, the intranasal administration of sedative drugs in birds has gained increased attention. Intranasal administration of midazolam in Hispaniolan Amazon parrots, ring-necked parakeets, budgerigars, and canary finches resulted in rapid onset of dose-dependent sedation, which is completely reversible with intranasal flumazenil.[1] Intranasal drug administration offers an alternative, non-invasive technique for drug administration in birds (Table 1). It is characterized by its ease of administration, high bioavailability, rapid onset of action, and reduced pain compared to intramuscular administration. Elevation of muscle enzymes in biochemistry panels secondary to intramuscular drug administration is avoided if intranasal administration is used instead. In addition, clients perceive the intranasal route as non-invasive, compared to intramuscular injection, which leads to better client compliance in cases in which sedation is recommended. The time of onset to sedation is rapid, typically within 3-5 minutes, which is comparable to intramuscular administration;[1-4] however, limitations of intranasal administration include incomplete drug delivery due to sneezing during administration, physiologically narrowed nostrils (e.g., cockatoos), or upper respiratory disease (e.g., blocked or stenotic nostrils). In some cases, in larger birds (e.g., macaws), the drug volume can also limit the effectiveness and produce excessive sneezing, therefore leading to incomplete drug delivery. Higher concentrated drugs (e.g., midazolam 50 mg/ml, Zoopharm, Windsor, CO) are available, but intramuscular administration might be more feasible in these cases.[1]

Table 1. Comparison of intramuscular and intranasal drug administration in pet birds.

	Intramuscular administration	Intranasal administration
Pro	Reliable drug delivery Faster administration	Non-invasive and not painful Higher client acceptance Reduced risk of needle stick injury No changes in the biochemical panel
Contra	Invasive Potentially painful Possible post-injection hemorrhage Changes in the biochemical panel	Risk of incomplete drug delivery Difficult to administer larger volumes May cause sneezing More difficult in birds with feathered nares

Drugs Used for Sedation

Midazolam

Midazolam is currently the most used drug for the sedation of pet birds and has a wide margin of safety. Midazolam has sedative, muscle-relaxing, anxiolytic, amnestic, and appetite-stimulating properties in birds.[1,5,6] Standard concentrated forms of injectable midazolam (midazolam hydrochloride (1 and 5 mg/ml); Hospira Inc, Lake Forest, IL) or a more concentrated form (50 mg/ml, Zoopharm, Windsor, CO) can be administered intranasally and intramuscular, without side effects.[1-3] Dosages commonly used in pet birds range from 0.5-3 mg/kg.[7] At the University of Wisconsin, we routinely use 2 mg/kg of midazolam in pet birds if administered intranasally and as the sole sedative agent. In smaller birds, such as finches or budgerigars, we routinely use 4-6 mg/kg of midazolam if administered alone (which is rarely indicated).

Diazepam

Diazepam is of similar efficacy as midazolam in birds following intranasal administration but has a slightly longer onset time and duration of action.[3,4] While less commonly used, diazepam represents a suitable alternative in cases in which midazolam might not be available. The intramuscular administration of diazepam should be avoided if possible due to the potential for delayed absorption and muscle irritation.[5] Dosages commonly used in pet birds range from 0.2-2 mg/kg if used as a sole sedative agent. Dosages as high as 10-15 mg/kg have been administered to finches and budgerigars without significant side effects, but this is not recommended in clinical settings, and instead, the benzodiazepines should be combined with butorphanol or dexmedetomidine.[3,4]

Butorphanol

Butorphanol is currently the most used opioid analgesic in birds. Besides its analgesic effects, butorphanol has sedative effects, which are potentiated by benzodiazepines (i.e., midazolam, diazepam). The combined administration of midazolam and butorphanol is recommended in birds for which midazolam alone provides only an insufficient level of sedation or where deeper sedation is needed for certain clinical procedures such as radiographic positioning. Butorphanol can be given intramuscularly as well as intranasally, drawn into a single syringe in combination with midazolam. No adverse effects of intranasal administration of butorphanol at a dose range of 1-3 mg/kg were seen in psittacine birds.[7] At the University of Wisconsin, we routinely use butorphanol at 2-3 mg/kg combined with 2-3 mg/kg of midazolam administered intranasally or by intramuscular injection in pet birds.

Dexmedetomidine

Historically, alpha$_2$-agonists have not been commonly used for sedation in birds; however, recently its safety and efficacy have been reported in budgerigars.[8] If combined with midazolam (3 mg/kg), dexmedetomidine (10 or 40 µg/kg IM) provided dose-dependent, moderate to deep sedation without obvious adverse effects. At the higher dexmedetomidine dose, positioning for radiographs was feasible. Reversal was quick following flumazenil and atipamezole administration.[8] The authors recommend 20 µg/kg IM as a starting dose in small psittacine and passerine birds instead of midazolam-butorphanol protocols, due to the superior sedation quality achieved. In birds with known or suspected cardiac disease, this medication should be avoided.

Reversal of Sedation

Reversal of sedation in pet birds will depend on the patient and the purpose of sedation. Sedation performed to facilitate physical examination, and diagnostic sample collection should always be reversed to have the patient return to normal behavior and food intake as soon as possible. It is important not to discharge sedated patients, as owners are often concerned about caring for a partially sedated bird, which might be ataxic, sleepy or refuse to eat. Birds that were sedated for e-collar or bandage placement or control of seizures should not be reversed. In these cases, birds should be carefully monitored, and reversal should be considered if the level of sedation is perceived as too deep or the duration of sedation is prolonged and might interfere with physiological behavior, particularly food intake.

Flumazenil

Flumazenil is a benzodiazepine antagonist and is used to reverse the sedative effects of midazolam and diazepam in birds.[1,2,7] The injectable form of flumazenil (flumazenil hydrochloride (0.1 mg/ml), Abaxis Pharmaceutical Products, Schaumburg, IL) can be administered intranasally, intramuscularly, or intravenously, without side effects. The recommended dosages range from 0.01-0.1 mg/kg. Alternatively, a flumazenil to midazolam ratio of 13:1 has been recommended, but it has been shown that lower doses of flumazenil achieve complete recovery from midazolam-induced sedation in birds.[1] The author prefers to administer 0.05 mg/kg initially. If the reversal is deemed unsatisfactory 10 minutes after drug administration, then the same amount of flumazenil can be administered repeatedly. Recovery from sedation is usually complete within 10-15 minutes. Reversal with flumazenil is recommended in all outpatient cases, to avoid injuries from falling off perches and negative client perceptions of interacting with their abnormally behaving bird. Therefore, it is important to ensure that complete reversal has occurred prior to discharge from the hospital by assessing complete recovery of the sensory and motor function (e.g., approaching the bird, rotating finger/perch to evaluate balance).

Atipamezole

Sedation protocols using dexmedetomidine or medetomidine should always be reversed using atipamezole. Atipamezole should be dosed at 10× the dose of dexmedetomidine in milligrams or 5× the dose of medetomidine in milligrams. Redose together with flumazenil if recovery is considered insufficient.

Step-by-Step Procedure

1. Educate the bird owners, if present during the sedation event, about what to expect regarding the sedation of their bird. Sedation using published protocols is safe in healthy pet birds, but birds will appear sleepy, which owners are not used to. Inform them that in some cases, regurgitation might occur, particularly in macaws.

2. Manually restrain the bird using a towel and immobilize the head.

3. Administer the sedative drugs either by intramuscular injection or intranasally. For intranasal administration, administer into the nostrils over 5-10 seconds and split the total dose between both nostrils.

4. Release the bird from manual restraint and allow 7-10 minutes to reach the maximum sedative effect. Do not allow the bird to remain high off the ground; instead, place it on the floor or in its carrier.

5. Manually restrain the sedated bird. Monitor respiration throughout the restraint period, as well as heart rate. Perform necessary clinical procedures (e.g., physical examination, blood collection, radiographs).

6. If reversing sedation, administer flumazenil IM or intranasally (split total volume between nostrils).

7. Place the bird on the floor or in its carrier and monitor recovery. Do not allow the bird to remain high off the ground until it has completely recovered to avoid trauma from falling. Most birds will recover within 10-15 minutes. Assess recovery before discharging the bird. Administer a second dose of reversal flumazenil if recovery is incomplete. Cockatoos often recover slowly and, in most cases, need a second dose. Do not discharge an incompletely recovered bird.

Useful Tips for Sedation of Birds

Below are tips for use during sedation of birds.

- Each bird requires an individual assessment prior to choosing a suitable sedative drug protocol. Birds which have no previous experience with manual restraint may require less or no sedation compared to birds that had previous negative experiences and become readily stressed or fearful in a veterinary clinic environment.

- Macaws and grey parrots usually require a combination of midazolam and butorphanol to achieve a sufficiently deep level of sedation suitable to perform a variety of clinical procedures.

- Smaller psittacines, passerines and other non-psittacines often need higher doses of midazolam-butorphanol and often the induced sedation is not satisfactory. Therefore, consider using dexmedetomidine-midazolam protocols instead.

- Even though a bird might not appear sufficiently deeply sedated while under manual restraint for a clinical procedure, it might become more sedated once stimulation is discontinued, making it unsuitable to be discharged from the hospital. Therefore, every bird, which is intended to be discharged shortly after completion of the procedures performed under sedation, should receive flumazenil reversal.

- Cockatoos often do not recover completely after the initial dose of flumazenil. A second dose of flumazenil is typically necessary.

- Record the sedation protocol used and the achieved level of sedation, indication for sedation and adverse effects in the patient's medical record, to better guide sedation protocol selection for future visits.

References

1. Doss GA, Mans C. Avian sedation. *J Avian Med Surg.* 2021;35:253-268.

2. Mans C, Guzman DSM, Lahner LL, et al. Sedation and physiologic response to manual restraint after intranasal administration of midazolam in Hispaniolan Amazon parrots (*Amazona ventralis*). *J Avian Med Surg*. 2012;26:130-139.

3. Sadegh AB. Comparison of intranasal administration of xylazine, diazepam, and midazolam in budgerigars (*Melopsittacus undulatus*): clinical evaluation. *J Zoo Wildl Med.* 2013;44:241-244.

4. Vesal N, Zare P. Clinical evaluation of intranasal benzodiazepines, alpha-agonists and their antagonists in canaries. *Vet Anaesth Analg.* 2006;33:143-148.

5. Scagnelli A, Titel C, Doss G, Mans C. Effects of midazolam and lorazepam on food intake in budgerigars. *J Exot Pet Med.* 2022;41:42-45.

6. Martel A, Doss G, Berg C, Mans C. Effects of midazolam on food intake in budgerigars. *J Avian Med Surg.* 2022;36:53-57.

7. Mans C, Braun J, Feeny M. Intranasal sediation with midazolam or midazolam-butorphanol in psittacine birds: a retrospective evaluation of 114 cases (2010-2012). *Proc ICARE.* 2013.

8. Mumm L, Mans C. Evaluation of dexmedetomidine-midazolam sedation in budgerigars. *J Am Vet Med Assoc.* 2022;260:1194-1199.

Section 6

AAV Masterclasses Track 2

Michael Huffaker, DVM;
Seth Oster, DVM
Session Chairs

Avian Wound Management

Heather Wilson Barron, DVM, DABVP (Avian Practice), and
Rebecca Duerr, DVM, MPVM, PhD

Session #3220

Affiliation: From Loggerhead Marinelife Center, 14200 US HWY 1, Juno Beach, FL 33478, USA (Barron), International Bird Rescue, 4369 Cordelia Rd, Fairfield, CA 94534, USA (Duerr).

Abstract: The purpose of this masterclass is to provide in-depth understanding of wound management of birds. Techniques are applicable to all bird species and case examples are given. The course includes details on practical applications of things like bandaging, with clear guidelines on how and when to use bioactive, hydrocolloid, hydrogel, and other dressings. Further discussion involves topical medications, antibiotic impregnated beads, regional limb perfusion, bioactive glass, and surgical management options like autografting, advancement flaps, and closure of severely contaminated wounds when appropriate due to species concerns. In addition, vacuum-assisted closure is discussed. Videos are used to ensure understanding of more complex procedures. Poll questions are used to ensure audience engagement and understanding.

Introduction

By definition, a wound disrupts the normal skin barrier, which is the body's first line of defense against invading microorganisms. Wounds are a common and debilitating problem in domestic and companion birds, as well as being a common reason that wild birds are presented for care. Understanding and delivering modern, appropriate wound care for birds can present a unique challenge for the avian practitioner. Some of the special anatomic features, physiology, or behavior of the myriad of bird species seen can engender a need for creative thinking. When wounds do not heal in a timely manner, there is an increase in discomfort and disease that interferes with health and quality of life, and in wild species that present in an extremely debilitated state, or with unfamiliar behavioral or physiologic needs, delays in healing can lead to the development of new problems and eventual euthanasia. There is a vast array of options for wound treatment and it is important that the avian practitioner remains current with this knowledge to offer the best patient care. Wound care in companion birds is well-covered in other sources, hence the options presented here largely focus on techniques used by the authors in their roles as wild bird veterinarians.

Avian Integument

One thing that clinicians may appreciate when attempting to suture bird skin is that it is less elastic than mammalian skin, drier, and much thinner (only ~4-10 cells thick), although grossly there is much variation due to species. For example, duck skin seems thick, fatty, and friable while songbird skin may resemble tough wet tissue paper. Other variations include air cells as in pelicans, dermal ornamentations such as wattles, combs, and beards, species-dependent anatomic locations of subcutaneous fat deposits, and changes due to seasonal transitions or loss of body condition.

The other thing that is obviously unique to birds is the presence of feathers. Be very conservative when removing feathers from wild birds, especially flight or tail feathers, lest you inadvertently prolong the bird's captivity.

Some species may not be able to stay in captivity long enough to regrow feathers before developing sometimes-lethal, captivity-related problems. Larger species may take three or more years to molt their entire set of flight feathers, for example.

The dermis is the site of feather development and is also quite thin. Therefore, be cautious in the application of topical medications as they may be more easily absorbed transdermally. Further, caution should always be used in the application of any ointments or creams in birds topically (unless well-bandaged) because these products may damage the plumage and render it ineffective in keeping the bird warm and waterproof. This holds special importance during wound care in wild birds because some species require rapid transition into water-based housing, and if dirtied by topical products, birds may require washing. Birds have thin skin and limited soft tissue covering the distal extremities and thus a reduced vascular supply to these areas, which means wounds in this area may be slow to heal and simple closure of skin defects may be difficult. Birds also lack myeloperoxidase and tend to have very caseous pus and thus drains are typically not used, although because birds are so diverse, there are exceptions to every rule. One author (RD) has successfully used drains on several species with persistent liquid pus production from wounds, particularly those involving aquatic birds with severe fishing hook or line injuries (including pelicans, shorebirds, gulls, geese, others). Finally, birds lack biliverdin reductase and hence bruising in avian skin and muscle may appear green.

Tissues deep to the skin vary by anatomic location and may be merely exposed and subject to desiccation or infection, or the deeper tissues may be involved in the injury itself. Muscle, tendons, ligaments, bones, nerves, and other tissues may be directly involved, and the fate of these tissues may drive the outcome of the case. With wild birds that must have appropriate function and athleticism for their wild lifestyle after recovery, permanent disability may become an indication for euthanasia. Once in a good nutritional plane, most wild birds have an excellent ability to heal even severe wounds with appropriate care, as long as the injury has not ruined a non-expendable structure such as a joint vital for flight, or a toe needed for killing prey.

Anesthesia and Analgesia

Anesthesia or sedation may be necessary for adequate wound care. Analgesics are an important part of treatment and general anesthesia may be indicated. One author (HB) generally prefers buprenorphine or hydromorphone for analgesia in most species of wild birds. Avian species responses to these drugs are highly variable, even with closely related species The other author (RD) commonly uses butorphanol, meloxicam, and gabapentin, again with awareness of species differences and an eye out for beneficial and adverse responses to therapy. There are concerns regarding the use of NSAIDS in poorly studied taxonomic groups of birds, hence staying abreast of recent developments in avian pharmacology is prudent and more work in this area is needed. Clinical presentation including species considerations may help dictate anesthetic protocols. The patient should be ideally warm and stable prior to anesthesia or aggressive wound management. Until then, simple bandaging may provide temporary fracture stabilization and wound coverage.

Wound Care

Basic care for most wounds necessitates fully identifying the nature of the injury. Begin with gross removal of debris from the wound, and carefully pluck a narrow margin of healthy skin around the edge. Pluck towards the wound whenever possible to avoid creating a larger defect by tearing the skin or hold flat-tipped forceps on the skin at the base of each feather to reduce tension. Follow the wound margin to fully identify the extent of the lesion. This can be quite time consuming in aquatic birds with dense plumage, as necrotic skin may be covered with relatively normal-looking feathers. Do not remove flight (remiges) or tail (rectrices) feathers if at all possible, as it is preferred to allow soft tissues to heal with a feather in the follicle.

Once gross debris has been manually removed, follow with lavage of large amounts of warm sterile physiologic fluid with or without dilute chlorhexidine, betadine, colloidal silver solutions, or Tris EDTA as appropriate. Irrigation pressure is ideally around 8-10 PSI to adequately clean and debride the wound. Lavage is contraindicated in wounds that include an opening into the bird's air sac system. If uncertain whether flushing a wound will risk drowning the bird, it is helpful to have the bird anesthetized and intubated, and have the anesthetist give the bird a nice big breath to check for leakage of air out the wound. If the wound does communicate with the respiratory system, cleansing the area repeatedly with sterile gauze moistened with sterile flushing fluids may suffice.

Debridement may be done as needed surgically or with wet to dry bandages over the first few days. Lavage may be done once daily and then gradually less often. Wet-to-dry bandages can be very helpful in removal of debris and in moistening dried out jerky-type tissue that needs removal. Coagulation necrosis is common in wounded avian tissue, and these necrotic tissues must be removed for full healing to occur. Sterile hydrogel-impregnated gauze is also useful for softening necrotic tissues prior to debridement. During debridement it may become apparent that an important structure is involved in the wound. Prognosis and outcomes are highly dependent on what structures are involved. There are some tendons that can be lost without a perceptible effect on function but losing a large bundle of tendons that completely control, for example, flexion or extension of the hock joint, is not compatible with being a releasable (or even comfortable in captivity) wild bird. Quality of life must be assessed for captive birds with similar loss of function. The outcome of these cases is usually dependent on what function has been lost, but it is often worth attempting treatment to see what functionality is retained.

Muscle tissue in birds has a tremendous capacity for repair. Large wounds in muscle, even when a large portion of the muscle is essentially destroyed, may heal well and result in a full return to function. Tendons are, however, more delicate, and tend to desiccate, die, and rupture if allowed to remain exposed for any length of time. Freshly exposed tendons that are shiny and white should be prioritized for treatment, as they are still healthy and need help to stay that way. As might be expected from a slippery structure designed to move freely and not adhere to surrounding tissue, tendon can be difficult to heal over with second intention healing and requires primary closure for best healing. See Duerr[1] for information on the use of autografts in providing coverage when locally available skin is not adequate. Tendon that has been exposed for long periods that looks yellowish or tan or has fibrinous debris covering it has a poor prognosis. Even if the unhealthy tendon appears intact and continues to function for days to weeks, it is not unexpected that it may rupture after a few weeks in care. Dead tendons can continue to perform their mechanical functions until rupture occurs. These necrotic tendons typically feel like firm tubular structures on palpation and appear denser than surrounding soft tissue on radiography. Careful regular palpation of tissues surrounding wounds can identify these problems prior to occurrence. In the author's (RD) experience, if the tendon feels thickened but springy in texture, it may merely be inflamed and may recover; monitor it regularly to assess whether the inflamed zone is increasing or decreasing in size. But if it is firm with any sort of lumpy character, it may be dead or dying and may require removal or may suddenly extrude itself from a nearby wound being treated when the area is manipulated.

One advantage that birds have over mammals is that bone exposure in wounds is usually much less concerning, and with appropriate care these areas heal very well. See Duerr[1] for repair of degloving injuries to the neck and scalp, and Duerr[2] for information regarding repair of keratin damage exposing pelican mandible, which is also applicable to bone exposure in any species near the feathered skin-bill keratin border. Open joints are always of serious concern and healing that results in a fully functional joint is uncommon.

Skull, synsacrum, or other bone may be laid bare by skin damage. Small areas no larger than the skin is thick do not usually require treatment, but larger exposures usually do. In wild birds these injuries may not present until chronic. Left to its own devices, the skin at the edges of the defect adheres to the bone and begins to granulate in place, while cells are unable to cross the defect because it cannot heal from below. A raised edge of granulating skin surrounding a flat plain of eroding bone is often the result. In a radiograph that catches a cross section of these areas on long bones, it may be apparent that the exposed area is eroding while the inside of the medullary cavity is becoming reinforced. This at least means that the bone is rarely penetrated by the erosion, but the site is a perpetual source of discomfort from the tissue layers being stuck together (which may impede movement such as flight) and the open wound remains a source of potential infection. Exposed skull from degloving injuries in ducks and geese tends to develop a leathery scar on the bone and wound contracture may cause the palpebral fissure to become displaced from the globe, leading to visual impairment. One author's (RD) preferred resolution for these injuries is to meticulously debride the surface of the bone and associated crevices by scraping with a scalpel blade, then removing the granulated edge to create a fresh skin margin, undermining the adjacent tissue and then closing the gap across the bone with as little tension as possible. If this is occurring on an exposed skull, this closure becomes an advancement graft from the cranial neck skin, which is placed onto the crown like a hood. Excess tension or inadequate debris removal may result in graft necrosis or dehiscence. Ducks and geese that have been hit by cars may present with macerated skin with areas of exposure of the synsacrum. These areas often granulate well with minimal treatment after simple debris removal. Periodic debridement may be useful; autografts are usually not needed.

Topical medications

The use of topical antibiotics is indicated when treating infected wounds. Ointments may initially help to keep tissues moist but can delay wound healing by impeding the migration of fibroblasts into the area. These agents most often are indicated when applied to small, recent wounds for short periods of time. Silver sulfadiazine cream is effective against most gram-positive and gram-negative bacteria as well as most fungi. It is often used for thermal burns, as in electrocution cases, because of its ability to penetrate eschar and necrotic tissue.[3,4] But it may have a negative effect on fibroblasts and thus may impede wound contraction and so is used infrequently by an author (HB).[5,6] Medical grade honey and silver impregnated hydrogels have been documented to be effective in a variety of avian wounds and are generally both authors' empiric preferences.[5] Honey and granular sugar also have uses in reducing the gross contamination of wounds that are producing a lot of discharge and debris; however, be aware that hyperosmotic dressings may cause tissue such as exposed tendons or bare bone to desiccate. Whenever possible, bone and tendon should be covered with vital tissue, even if it means only partially closing a wound.

Some species of birds tend to quickly close skin wounds over pockets of contaminated debris; hence keeping an abscessed area open for dressings and debridement may be best while a wound heals from below. Routine palpation of deeper tissues as a wound heals allows monitoring for abscess or granuloma formation. In birds, the skin should remain largely loose in relation to deeper tissues—once the skin has healed, an area of adhesion may indicate a problem such as a pocket of debris.

Petrolatum impregnated dressing works well to keep the tissues moist as they heal, although oily dressings of any kind are not suitable for use in species of birds that will require waterproof plumage in the very near future (loons, grebes, diving ducks, and seabirds to name a few). All wild birds must be waterproof prior to release lest they become soaked and chilled the first time it rains. Washing a wild bird simply for ointment removal adds stress to an already-stressed patient.

Vacuum-Assisted Closure

Once hemorrhage is controlled and a clean wound bed established, vacuum assisted closure may reduce wound healing time. In one study in humans, vacuum-assisted closure (VAC)-treated wounds decreased in size by 78% compared to a 30% size reduction in wounds treated with wet-to-dry bandages over the same time frame. There are few publications describing the adaptations for this technique in birds.[6] The VAC is usually left in place until a healthy bed of granulation tissue is noted. Bandage changes generally start out being done more frequently and then may stretch out to once every few days once the wound is progressing well. Aquatic species may need to be alternated in water and dry docking with the VAC if waterproofing isn't possible. Antimicrobials commonly used by the author (HB) with the VAC include water-soluble products, such as honey or colloidal silver hydrogels. Do not use topical antimicrobials that will impair the VAC, such as creams or ointments.

External Coaptation and Physical Therapy

Keep a bandage box with appropriately sized bandage materials for small patients available to help facilitate rapid application in emergent situations. A figure-of-8 bandage may be indicated in birds for temporary stabilization of fractures of the radius, ulna or both; fractures or luxations of the elbow or carpus; or soft tissue injuries to the brachium or antebrachium.[7] Generally, cast padding and an outer layer of vet wrap will suffice for most wild birds. Usually no more than 2-3 circuits around are needed; otherwise the bandage is needlessly bulky. A body wrap is indicated for shoulder or humeral injuries. A correctly placed bandage will have the primary feathers sitting **under** the secondary feathers, as seen in the normal perching pose. Be sure to downsize bandaging materials appropriately for small patients.

Generally physical therapy is indicated in wing wraps to prevent patagial membrane contracture and keep a normal range of motion (ROM) in the joints. Physical therapy (PT) may begin as soon as 3 days post-admission or fixation of the limb. Initially, PT should be done daily or every other day. If ROM is good, this may be extended to every third day. Once contracture has occurred, it is more difficult to reverse it. So, it is better to do PT more often. Initially, it may be necessary to do PT under general multi-modal anesthesia with injectable pain medications plus inhalant anesthesia. After some initial healing, it may be possible to do PT awake in sedated animals given injectable analgesic drugs. Eventually, it may be possible to do PT using oral pain medication with or without sedation. The author (HB) tries to minimize the use of intramuscular (IM) injectable medications where possible and puts medications into food items to avoid the pain and trauma that repeated IM injections may cause. Generally, external coaptation devices are removed between 10-21 days if they are the only method of fixation. Joints should never be bandaged for longer than necessary and timely PT is always required for a good outcome. In species with calmer demeanors, the author (HB) may keep full coaptation on for shorter time periods, gradually reduce, and try strict cage rest with anxiolytics and analgesics to cut down coaptation time.

Functions of tendons and joints are assessed by manipulation of the reciprocal apparatus of the wings and legs. For example, in the wings, the carpus is depressed caudally while the elbow is depressed cranially. These combined forces should cause the elbow and carpus to pivot open and spread the wing utilizing the tendons primarily. See Freshman and Duerr[8] for more information on developing physical therapy plans for wild birds.

Goniometer measurements should be used to assess ROM. Heat packing prior to PT exercises may facilitate better ROM. While there is some anecdotal information that low level laser therapy may also be effective, there are very few published studies demonstrating clear efficacy in birds.[9] In one study in green iguanas, laser therapy did not result in substantial improvement in histologic evidence of wound healing, compared with results for traditional wound management.[10]

References

1. Duerr RS. Surgical repair of keel lesions and lacerations in aquatic birds. *Proc ExoticsCon.* 2016;83-88.

2. Duerr RS. Successful surgical management of pouch and bill injuries in pelicans. *Proc ExoticsCon.* 2018;363-367.

3. Grunkemeyer V. Management of avian skin wounds. *Proc Annu Conf Assoc Avian Vet.* 2009;57-62.

4. Burke HF, Swaim SF, Amalsadvala T. Review of wound management in raptors. *J Avian Med Surg.* 2002;16:180-191.

5. Mans C, Sunohara-Neilson J, Higginson G, et al. Wound management in a trumpeter swan using honey and a sustained release ionic silver hydrogel. *Exot DVM.* 2006;8(5):21-23.

6. Knapp-Hoch H, de Matos R. Clinical technique: negative pressure wound therapy-general principles and use in avian species. *J Exotic Pet Med.* 2014;23(1):56-66.

7. Duerr RS, Purdin GJ, eds. *Topics in Wildlife Medicine: Orthopedics.* Vol. 4. St. Paul, MN: National Wildlife Rehabilitator's Association; 2017.

8. Freshman J, Duerr R. Physical therapy. In: *Topics in Wildlife Medicine: Orthopedics*. Vol. 4. St. Paul, MN: National Wildlife Rehabilitator's Association; 2017:6-14.

9. Hippólito AG, Stanski G, Fernandes LL, et al. Evaluation of topical medications, lavage solutions, and laser therapy for treating open wounds induced in blue-fronted Amazon parrots, *Amazona aestiva*. *J Exot Pet Med.* 2019;29(0):1-9.

10. Cusack LM, Mayer J, Cutler DC, et al. Gross and histologic evaluation of effects of photobiomodulation, silver sulfadiazine, and a topical antimicrobial product on experimentally induced full-thickness skin wounds in green iguanas (*Iguana iguana*). *Am J Vet Res.* 2018;79(4):465-473.

Vulture Medicine and Conservation

John Chitty, BVetMed, CertZooMed, MRCVS,
Michael Jones, DVM, DABVP (Avian Practice),
Julia Ponder, DVM, MPH

Session #3212

Affiliation: From Wombourne, Allington, Salisbury, SP4 0DD, UK (Chitty), The American Eagle Foundation, 2957 Veterans Blvd.,Pigeon Forge, TN 37863, USA (Jones), College of Veterinary Medicine, University of Minnesota, 1365 Gortner Avenue, St. Paul, MN 55108, USA (Ponder).

Abstract: Old World and New World vulture species face threats worldwide and are of major conservation significance. Many of these species have previously been extremely common but have suffered major population crashes. Veterinarians working with these species—whether in zoos, rehabilitation, or as part of conservation projects—need an understanding of the unique biology of these birds to provide successful treatment and outcomes. Their intelligence and complex social structures require an understanding of their behavior and behavioral diseases they may experience. Differences are highlighted by the sensitivity of some of these species to commonly used drugs, especially non-steroidal anti-inflammatory drugs (NSAIDs). This masterclass provides information about the natural history as well as rearing and husbandry of Old and New World vultures encountered in managed care, handling and examination, behavioral conditions (i.e., self-mutilation syndromes), common conditions, diagnostic techniques and clinical pathology, therapeutics, analgesia, anesthesia and surgical techniques, all while highlighting similarities and differences between vultures and other raptors.

Vulture Taxonomy

Taxonomy and classification of raptors has gone through several significant changes in the last decade. Currently, four avian lists are available; however, they may not completely agree, which some would argue is detrimental to avian conservation.[1] In particular, species level taxonomy is often the basis for decisions for conservation prioritization.[1] The four lists available include the International Ornithological Community (IOC) World Birds List, the Howard and Moore Checklist of the Birds of the World, the eBird/Clements Checklist of Birds of the World, and the Handbook of the Birds of the World and Birdlife International Digital Checklist of the Birds of the World. These lists differ not only in the scientific names they use but also in the groups of individual organisms to which the scientific name refers.[1] Previously, diurnal birds of prey were included in the Order Falconiformes under which five families were noted to include Sagittariidae (Secretary birds), Pandionidae (ospreys), Accipitridae (Old World Vultures (OWVs), harriers, hawks, eagles, and buzzards), Falconidae (falcons and caracaras) and the family Cathartidae (New World vultures). Nocturnal birds of prey were listed under the Order Strigiformes with two families, the Strigidae (most owls) and Tytonidae (barn and bay owls). Subsequently, diurnal birds of prey were divided into three different Orders including Falconiformes, Accipitriformes, and Cariamiformes to include the five families listed previously but with the addition of the family Cariamidae (Seriemas). Strigiformes remained the same. At one point, Cathartidae were considered part of the Order Ciconiformes along with storks. Currently, while raptor classifications are still in a flux, there appears to be some agreement that there are five taxonomic Orders of birds of prey.[1] These include the Orders Acciptirformes, Falconiformes, Cariamiformes and Cathartiformes for diurnal birds of prey, and Strigiformes for nocturnal birds of prey. Interestingly, OWVs are classified with Accipitriformes despite their natural history as obligate scavengers compared to all other member of the Order.

Vulture Conservation

Worldwide, vulture populations are under pressure with 61% of vulture species threatened with extinction. Of the 16 African-Eurasian species of OWVs, eight are critically endangered (four in Africa and four in Asia), two are endangered, one is vulnerable and three are near-threatened (IUCN Redlist). Most of the seven New World vulture species are categorized as least concern although the California condor (*Gymnogyps californianus*) is critically endangered and the Andean condor is in rapid decline. While turkey vulture and black vulture populations are on the rise, most of the rest are in decline.[2]

For some Old OWV species, the current population problems can be traced to the 1990's, when three species of *Gyps* vultures experienced 95% population reductions over the course of one decade through exposure to diclofenac, a non-steroidal anti-inflammatory drug used by veterinarians in treating cattle in Southeast Asia. The exquisite sensitivity of vultures to this drug as well as the communal feeding strategies led to massive die-offs throughout the region.

There are many reasons for the on-going vulture population declines, with poisoning and persecution at the top of most (if not, all) lists. As the only known obligate scavengers outside of the ocean, vultures have a high-risk feeding strategy that increases the likelihood of exposure to contaminants and toxicants. They often feed communally, increasing the risk that significant numbers of individuals will be impacted by a single carcass. In addition, they are long-lived, high trophic level feeders, which increases the risk for bioaccumulation of toxicants. These shared characteristics of vulture populations put them at high risk for poisoning, whether the exposure is intentional or unintentional.[3] Banning of the use of diclofenac for cattle in several counties has led to some populations stabilizing and recovering. In Pakistan, a population of colony of long-billed vultures was shown to have a significant decline in reproductive parameters in the three years before the 2006 ban of diclofenac and an increase in the same parameters for the two years after the ban.[4]

For OWVs, the number one threat to vultures is probably unintentional poisoning secondary to deliberate poisonings of carnivores to protect game and livestock. These poisonings may be efforts to reduce predators of game animals or in retaliation for carnivore predation on livestock.[5] Additionally, vultures have been intentionally poisoned to prevent their signalling poachers of the presence of an elephant carcass, as well as for collection of body parts for belief-based use and trade. Between September 2019 and March 2020 across Guinea-Bissau, West Africa, greater than 2000 hooded vultures were killed by poisoning with methiocarb, a carbamate pesticide, and their heads were collected for cultural use.[6]

New World vulture populations also face many problems. The main threats for California Condors, a highly endangered population with approximately 300 free-flying individuals, are lead poisoning from spent hunting ammunition; utility line-related mortality (largely mitigated) and vulnerability to catastrophic losses secondary to forest fires and changing climate. One study demonstrated the risk of exposure to second-generation anticoagulant rodenticides (SGAR) with 42% of condors tested being positive for residues and evidence of prolonged clotting time in 16% of free-flying condors.[7] For Andean condors, persecution is the greatest population threat with poisoning (sodium monofluoroacetate and carbofuran) on top of the list with shooting second.[8] Other threats faced by vulture populations solidly reflect the intersections with human activity and include collision and electrocution at energy infrastructures; reduced food supplies; habitat destruction and degradation; and human disturbances.

While many of the threats are based in beliefs that vultures cause problems for humans, they need to be balanced by the ecologic, economic and social case for conservation. Vultures provide many ecosystem services. They clean up carcasses through scavenging, promoting the flow of energy through the food web and preventing disease outbreaks. They act as sentinels for other predators in locating food sources. And their presence can contribute to ecotourism.[9]

Literature Review of Infectious Diseases in Vulture Species

Given their natural history as obligate scavengers that consume carcasses in various stages of decomposition, vultures are exposed to many infectious disease-causing organisms. In a systemic global review of available scientific literature, infectious diseases account for 16% of causes of morbidity and mortality in free-living vulture species worldwide.[10] Other reports of opportunistic infectious diseases occurring in vultures in managed care are available but are also limited in number, suggesting a gap in our understanding of infectious disease in vulture species world-wide. In many instances, susceptibility to infectious diseases may be a consequence of co-morbidities (i.e., toxin exposure, malnutrition) that inhibit an appropriate immune response in affected vultures.

Fungal diseases

Ives et al (2022)[10] reported fungal diseases, as the most commonly reported infectious disease in free-living Old World and New World vultures. *Aspergillus* spp. was the most commonly reported in the literature; however, *Candida* spp. affected more individual free-living vultures and was also reported more in nestling and fledglings.[10] Chege et al. (2013)[11] reported on opportunistic *Aspergillus* spp. infection and mixed bacterial infection in Cape vultures (*Gyps coprotheres*) in managed care.

Bacterial diseases

Bacteria reported to cause disease in vulture species include *Mycoplasma* spp., *Mycobacterium* spp., *Escherichia coli*, *Salmonella* spp., and *Staphylococcus* spp.

Several reports describe disease caused by *Mycobacterium* spp. For example, granulomatous disease associated with *Mycobacteria* spp. has been described in an adult male white-rumped vultures (*Gyps bengalensis*) in Pakistan.[12] In addition, upper respiratory tract disease associated with *Mycobacteria* spp. has also been described in other vultures.

It is difficult to determine the significance of *Mycoplasma* spp. infections in raptors as the presence of the organism does not always correlate with a disease process; however, Panagala (1993)[13] described a new species, *Mycoplasma corogypsi* sp. associated with footpad abscesses in a black vulture (*Coragyps atratus*). Similarly, Ruder et al (2009)[14] were the first to report polyarthritis diagnosed in association with *Mycoplasma corogypsi* in a black vulture and documented an acute fibrino-heterophilic polyarthritis. The vulture also suffered from lead toxicosis. Oaks et al (2004)[15] described mild heterophilic inflammation in the trachea and bronchi associated with a novel *Mycoplasma* spp. in an Oriental white-backed vulture. This vulture was found dead and demonstrated renal failure and visceral gout related to diclofenac intoxication.[15]

Other bacterial species such as *Salmonella* spp. and *E. coli*, while reported as occurring in vulture species, appear to be associated more with intense agricultural practices, antimicrobial resistance and concerns of carrier or reservoir status in vultures and may not represent actual diseases in these species. Interestingly, it is suspected that the microbiome of vultures may make them more resistant to enteric diseases.

Viral diseases

Viral diseases reported in vulture species world-wide include West Nile virus (WNV), Influenza A, and poxvirus. The Centers for Disease Control and Prevention reports detection of WNV in all bird species from 1999 to 2016. In this report, WNV was detected in a free-living black vulture, a turkey vulture and a cinereus vulture in managed care.[16] Plaza et al (2020)[17] reviewed recent literature and reported the presence of WNV in California condors, bearded vultures (*Gypaetus barbatus*), and a cinereous vulture (*Aegypius monachus*). In this report, the presence of inflammation and hemorrhage produce by WNV was only reported in a California condor.[17] The other species in which WNV was detected either had no disease condition reported or had concomitant conditions such as oral lesions associated with *Candida* spp. in a cinereous vulture and footpad abscesses in a black vulture.

Highly pathogenic avian influenza A (HPAI), specifically the Eurasian H5N1 lineage, has caused severe disease and large-scale die-offs in vultures in the United States, especially in Georgia, Kentucky, New Jersey, and Tennessee over the past 1 to 2 years, The largest die-off of 700 black vultures occurred at a Sanctuary in Georgia. As of April 13, 2023, the United States Fish & Wildlife Service reported the deaths of 18 California condors in the Arizona/Utah flock; HPAI has been confirmed for six of those condors. Certainly, this is a major concern for the population of condors should the disease spread to other regions, especially to California. To protect the species, biologist are considering capturing wild condors to protect them from infection until a vaccine can be developed.[18] Interested parties can follow detection of HPAI in the United States by reviewing the U.S. Department of Agriculture Animal and Plant Health Inspection Service site.[19] Certainly, HPAI infections in vultures have been described in other countries many years prior to 2022. Vultures are most likely to become infected by consuming infected migratory bird species (waterfowl and shorebird), and cannibalism of dead infected vultures may also contribute to the spread of this disease. Characteristic neurologic clinical signs associated with avian influenza are similar to those reported in other raptor species along with pathologic lesions demonstrating multiorgan necrosis.

Poxvirus that affects raptors are distributed worldwide, and infections are likely underreported in vulture species. This author (MPJ) and others in the United States previously noted what appeared to be poxvirus infections in turkey vultures (*Cathartes aura*). Poxvirus also has been detected in a turkey vulture in Chile by Gomez et al (2014)[20] and by Mora-Carreño et al (2019)[21] in Venezuela. The raised and sometimes ulcerated appearance of poxvirus lesions noted on unfeathered areas of the feet and the head of vultures are similar to lesions noted in other raptor and avian species.

Parasitic and protozoal diseases

Similar to other raptors, parasitic and protozoal diseases are reported in vulture species. Martinez-Herrero, et al (2014 and 2019)[22,23] documented oropharyngeal trichomonosis with a novel Trichomonas spp., *Trichomonas gypaetinii*, in Griffon (*Gyps fulvus*), Egyptian (*Neophron percnopterus*), and Cinereous vultures. The reported lesions were similar to those reported in other raptor species and trichomonad trophozoites were detected from wet mount preparations of the lesions.

In general, hemoparasites are considered non-lethal at lower levels of infection in most avian species. Webb SL et al (2005)[24] performed a survey of blood parasites in black vultures and turkey vultures from South Carolina, USA and noted that while blood parasites have been reported in black vultures, no blood protozoans were observed on blood smears of black vultures with the exception of a microfilaria found in one black vulture. In another study, Yabsley et al. (2018)[25] opportunistically obtained blood and/or tissue samples from 162 turkey vultures and 95 black vultures to identify haemosporidian parasites in these species. Similar to the previously noted study, no parasites were detected in the black vultures by blood smears or PCR testing of their samples.

Yabsley et al. (2018)[25] also provide a brief review of hemoparasites documented in New World and OWV species. *Hemoproteus elani*, *Haemoproteus janovyi*, *Leucocytozoon toddi* and *Plasmodium fallax* have been documented in OWVs, and *Haemoproteus catharti*, *Leucocytozoon toddi* and *Plasmodium elongatum* have been reported in New World vultures. Yabsley et al. (2018)[25] further asserts that there are numerous records of *Haemoproteus* spp., *Plasmodium* spp., and *Leucocytozoon* spp. that have yet to be morphologically or genetically characterized in New World vultures.

New World Vulture Rehabilitation and Ethics

Vultures are frequently seen in wildlife rehabilitation facilities. As with any injured or ill animal admitted for treatment, they present with a wide variety of issues, with miscellaneous trauma and toxicities being two of the more common presenting problems. While the goal of wildlife rehabilitation is always the treatment and release back to the wild of the patient, there are a series of decisions from triage through final outcome that determine the bird's future and require thoughtful and ethical considerations that prioritizes the bird's welfare. There are three possible outcomes for a vulture admitted to a wildlife rehabilitation facility: successful treatment and release, euthanasia and placement in permanent captivity. While the veterinarian's role in clinical treatment and release is the most obvious, the DVM also plays a key role in each of the other two outcomes. With the high bar of a full return to normal being required for release, the veterinarian's experience is helpful in determining a prognosis. When that bar cannot reasonably be expected to be reached, the veterinarian's access to humane and acceptable methods of euthanasia is a critical support for wildlife rehabilitators.

Vultures present some unique concerns when considering the option of placement in permanent captivity. As with all animals, it is important to consider the physical, affective and environmental states for the bird when determining if it is a candidate for placement. Physically, the bird's injuries must be completely healed, the illness/injury should be expected not to progress (no future complications), the bird should have no imbalance issues (such as those associated with amputations, pelvic/leg issues or nerve damage), and the bird's permanent injury must be compatible with its captive purpose. In the United States, a wild native bird cannot be kept in permanent captivity unless it is deemed non-releasable by a veterinarian, who must also attest to the fact that the bird meets the other conditions listed above.

Links to U.S. Fish and Wildlife Service legal requirements and documents

- Guidance for veterinary certification of condition that renders migratory bird/eagle as non-releasable https://www.nwrawildlife.org/resource/resmgr/symp_2020/handouts/Handout_Template_Vet_Letter.pdf
- Euthanasia requirements (50 CFR 21.76(e)(4)(iii)) https://www.ecfr.gov/current/title-50/chapter-I/subchapter-B/part-21/subpart-C/section-21.76#p-21.76(e)(4)(iii)
- Requirement of commitment of veterinarian to provide medical care of a placed bird (50 CFR 21.76(e)(4)(iii)(A)) https://www.ecfr.gov/current/title-50/chapter-I/subchapter-B/part-21/subpart-C/section-21.76#p-21.76(e)(4)(iii)(A)

Beyond these physical requirements, the consideration of the affective or mental state of a vulture is a key consideration for placement. Vultures are highly social animals, extremely intelligent and very long-lived. While young vultures are the most likely to adapt well to captivity, they also require considerable social learning from conspecifics. Single vultures housed alone present a welfare consideration as it is likely detrimental to their mental state. These social animals tend toward neophobia when imprinted or taken out of their social context, which impacts the ability to adapt. It can require extensive knowledge, time, resources and skills to integrate them into an established community. Confidence-building is key to avoiding behavioral issues, which are often seen as aggression, which covers for neophobia. Conspecifics are critical for this confidence-building. With enough space, non-imprinted adults may integrate well into a display situation as long as there is no need for constant management or medical treatment.

Finally, when considering placement, the longevity of vultures must be taken into consideration. It is extraordinarily rare that a captive vulture will have the same curator/handler or housing situation for its whole captive life and there is a significant probability that there will be need for extended management of chronic disease as the bird ages. All of these issues should be taken into consideration by the veterinarian when determining whether to sign the official statement saying the bird is appropriate for captive placement. The question is not "can the animal be placed," but "should the animal be placed?"

Husbandry and Medicine of Old World Vultures

Enclosure design and construction

The size and weight of OWVs dictates several recommended design parameters. In addition, they are generally social birds so aviaries should be large enough to hold multiple birds (they are also suitable for mixed vulture species exhibits) with sufficient space and nesting areas to allow pairs to break away from the main group in breeding season.

Height is important and should be a minimum of 3 meters (10 feet). Overall dimensions for a pair of birds can be in the range of 8.5-10 meters (25-33 feet) long and 6-8 meters (20-24 feet) wide, although larger enclosures are possible. A mixture of covered and uncovered areas is recommended, as birds will often deliberately sit unprotected in the rain and will actively sunbathe with wings outstretched but need the option of shelter from rain and summer sun.

A suitable enclosure will ideally be positioned to allow at least 2-3 hours of winter sun daily, with protection from prevailing winds in the form of a solid barrier wall. Heating is not generally necessary provided adequate shelter is available. The provision of mild heat to nesting areas is a possibility, particularly in Europe where some species will breed during the winter months (November to March).

In winter months, birds should be observed regularly for wing drooping and lameness or discoloration of the feet-peripheral vascular changes are common (especially in older birds) and can result in loss of digits or wing-tip edema syndrome. It is likely these may be linked to underlying cardiovascular or atherosclerotic conditions but may be triggered in cold weather.

Enclosure materials should be robust with consideration given to potential for injury. Chicken wire is totally unacceptable. Welded mesh is a superior enclosure medium and should be of sufficient gauge (40-50 millimeters) to avoid abrasion.

Bathing, preening and sunbathing constitute a significant proportion of the daily activity of OWVs. Fresh water is essential, and should be available in a bath at least 1 meter in diameter and 10-15 centimeters deep. Vultures will often only bathe in or drink water that is changed virtually every day.

The type of enclosure substrate used is variable and can consist of sand, pea gravel, grass or a combination of these. Woodchips should be avoided, particularly in enclosed or confined areas due to risk of aspergillosis although, in general, OWVs are fairly resistant to this infection. They do not appear overly prone to geosedimentation problems when housed on sand and rarely deliberately eat stones.

Perch sizes should be varied. Although OWVs can land and sit on surprisingly narrow perches, an enclosure should contain a number of perches 10-15 centimeters in diameter sufficient for the number of birds enclosed. Care is required to ensure perch positions do not allow feces to reach water baths or feeding areas, and perches should be spaced throughout the enclosure with sufficient in sheltered areas that subordinate individuals are not pushed out of the shelter. Large vultures kept on small, hard perches are more prone to bumblefoot as are those where a significant part of the flooring is concrete. Astroturf™ or similar artificial materials can aid weight dispersal on perches, although some vultures will destroy this material rapidly.

Old World Vultures should not be tethered, even when being used regularly in flying demonstrations. They do not tolerate well the actual tethering, or the equipment used, and will often remove items attached to their legs. Even where equipment is tolerated, tethered birds often become depressed and fail to thrive. Where part of an educational programme, demonstration birds should therefore be flown directly from an enclosure, without equipment.

Feeding

The gregarious feeding behaviour of many OWVs needs to be taken into account for successful captive management. Interactions between birds during feeding is important for general social interaction, the establishment of a colony pecking order and for pair formation.

Food type and presentation is important, and ideally consists of large animal parts (e.g., jointed beef) or small to medium sized mammals (e.g., rabbit). It is important that bone is given and that large bones are smashed to enable swallowing. Nutritional osteodystrophy has been recorded in these birds when fed meat alone or when given bone in a form they cannot ingest. Calcium supplementation also is recommended for breeding and growing birds in the form of a calcium and vitamin D_3 powder (e.g., Nutrobal: Vetark, Winchester, UK).

Feeding regimes should attempt to mimic wild conditions. At least one fasting day per week is recommended and should follow a sizeable feed the previous day. It is not unusual for vultures to still exhibit a protruding crop more than 24 hours after a large feed. Within reason, carcass remains should be left in aviaries for the day. Fossicking of carcass remains and the consumption of small food pieces around a feeding area is an important behavioural characteristic and is an important activity for young or subordinate birds held within a colony.

Care should be taken that birds do not become obese. If food is regularly found uneaten after more than a day, then overfeeding should be suspected and body condition monitored.

Behavioral characteristics and breeding

Females generally take main responsibility for nest building, although males will also bring material and add to the nest. Nest areas should have a moderate depth (5-10 centimeters or 2-4 inches) of substrate, such as a mixture of sand and wood shavings upon which the birds can build a nest. Care should be taken that nesting material is not rotten as aspergillosis may result from inappropriate material.

The nest, varying between 50 centimeters and 180 centimeters in diameter, is made of sticks and twigs. In the wild, some OWVs are tree-nesting species (e.g., *Gyps africanus*, *G. bengalensis*, *Aegypius*); however, the option for providing large trees for captive birds is often limited. Captive birds are traditionally offered two main nesting areas: platforms and caves. The use of caves by tree-nesting OWVs in captivity is worthy of note, given that they tend to avoid cliffs and/or cliff cavities for nesting in the wild.

In captivity, courtship displays vary from phlegmatic or non-existent to preening and head-turning (especially *Aegypius* and *Torgos*). Mating takes place either in the nest or on a preferred perch nearby. Copulation can be prolonged (compared with other avian species) and lasts between 10 and 30 seconds. The male will often bite the nape of the female's neck during copulation.

Among the species, OWVs generally lay one egg (rarely two). The incubation period ranges between 42 (*Neophron*) and up to 58 days (larger *Gyps*). Both the male and female will perform incubation duties. While OWVs will breed successfully as a single pair kept in an enclosure, the colonial nesting nature of some species in the wild makes it possible for a pair of birds to breed in an enclosure with many specimens. Interference from other birds at the nest and during incubation is limited, as breeding birds are defensive of the small area surrounding the nest.

Restraint

Compared with other raptors the talons of OWVs are relatively weak for their size; however, all large OWVs have a powerful grip, and although significant injury to a keeper is rare, due caution must be exercised. In any event, the handler should primarily restrain the head due to the strength of the bite and sharpness of the bill.

In general, two people are required to capture and restrain a large vulture. The first person should secure the head and neck immediately prior to the second person securing the wings, body and legs. A vulture that is aware of the fact it is to be captured will usually vomit, and this reaction is likely to continue during restraint. Care must be taken that the bird does not aspirate regurgitated food. Where handling is planned, the bird should be starved for 24 hours before the event to ensure an empty crop. Similarly positive reinforcement training of young birds to accept handling will make the whole less stressful for all concerned, especially as repeated handlings over many days (e.g., for long term veterinary treatment) can result in loss of body condition due to repeated regurgitation.

Blood sampling

Ulnar or medial metatarsal veins are suitable. The right jugular vein may also be used but generally is more difficult due to the strength of these birds and difficulties in head and neck restraint.

If sampling zoo birds, it is generally advised to fast birds for 24 hours pre-sampling to reduce post-prandial effects, especially when looking at renal and lipid parameters. Otherwise, blood values appear similar to other raptor species.

Imaging

Imaging is frequently required as part of a clinical investigation. Radiography is the mainstay of this but computed tomography should be considered especially when looking at suspected osteoarthritis of coxo-femoral joints as these are quite difficult to adequately evaluate using conventional radiography. New World Vulture species appear prone to laying down considerable amounts of new bone around osteoarthritic joints.

Anesthesia

OWVs should be anesthetised using isoflurane with induction by restraint and open mask. Breath holding is common making stabilization difficult on occasion. They should therefore be intubated and placed on mechanical ventilation as soon as possible after induction.

Pain management

Old World Vultures are sensitive to the renal side-effects of NSAIDs and this author (JRC) will avoid their use until the post-anaesthetic period when blood pressure is likely to have been restored. Fluid therapy should be given during all procedures in OWVs to assist with renal perfusion.

Analgesia in the pre- and peri-operative period should be based on opiate use, typically buprenorphine at 0.5 mg/kg IM q12h. For longer term management (e.g., for osteoarthritis) meloxicam may be used but this author (JRC) will generally use this at lower dosage and frequencies compared to other avian species. A dose of 0.2 mg/kg PO q24h generally appears effective and well-tolerated. Use of management techniques to assist movement, and complementary agents (e.g., glucosamine or poly-sulphated glycosaminoglycans) may assist in reducing NSAID use.

Fenbendazole

Old World Vultures have been shown to be sensitive to the hepatotoxic and radiomimetic side-effects of fenbendazole. The drug may be used for deworming OWV species, but lower doses should be used (generally 20 mg/kg PO q24h for 3-5 days) and the suspension should always be well-shaken both before dispensing from a larger bottle, and before dosing in order to reduce the chances of accidental overdosage due to settling out of the suspension.

Therapeutic dosing

Due to the risks of regurgitation, oral drugs should generally be given hidden in part of the daily ration (e.g., in body cavity of day-old cockerel or rat) and not by direct administration into the esophagus or crop. Handling interventions for birds on medication should be performed before feeding/ dosing rather than after. For debilitated birds, placement of a proventricular feeding tube should be considered where repeated oral dosing and/or supportive feeding is needed.

References

1. McClure CJ, Lepage D, Dunn L, et al. Towards reconciliation of the four world bird lists: hotspots of disagreement in taxonomy of raptors. *Proc Royal Soc. B.* 2020;287(1929):20200683.

2. Safford R, Andevski J, Botha A, et al. Vulture conservation: the case for urgent action. *Bird Conserv Int.* 2019;29(1):1-9.

3. Ogada DL, Keesing F, Virani MZ. Dropping dead: causes and consequences of vulture population declines worldwide. *Ann NY Acad Sci.* 2012;1249(1):57-71.

4. Chaudhry MJI, Ogada DL, Malik RN, et al. First evidence that populations of the critically endangered long-billed vulture *Gyps indicus* in Pakistan have increased following the ban of the toxic veterinary drug diclofenac in south Asia. *Bird Conserv Int.* 2012;22(4):389-397.

5. Ogada D, Shaw P, Beyers RL, et al. Another continental vulture crisis: Africa's vultures collapsing toward extinction. *Conserv Lett.* 2016;9(2):89-97.

6. Henriques M, Buij R, Monteiro H, et al. Deliberate poisoning of Africa's vultures. *Science.* 2020;370(6514): 304-304.

7. Herring G, Eagles-Smith CA, Wolstenholme R, et al. Collateral damage: anticoagulant rodenticides pose threats to California condors. *Environ Pollut.* 2022;311:119925.

8. Restrepo-Cardona JS, Parrado MA, Vargas FH, et al. Anthropogenic threats to the vulnerable Andean condor in northern South America. *PLoS One.* 2022;17(12):e0278331.

9. Carucci T, Whitehouse-Tedd K, Yarnell RW, et al. Ecosystem services and disservices associated with vultures: a systematic review and evidence assessment. *Ecosyst Serv.* 2022;56:101447.

10. Ives AM, Brenn-White M, Buckley JY, et al. A global review of causes of morbidity and mortality in free-living vultures. *EcoHealth.* 2022;19:40-54.

11. Chege S, Howlett J, Al Qassimi M, et al. Opportunistic infection of *Aspergillosis* and bacteria in captive Cape vultures (*Gyps copreotheres*). *Asian Pac J Trop Biomed.* 2013;3(5):401-406.

12. Gilbert M, Virani MZ, Watson RT, et al. Breeding and mortality of Oriental white-backed vulture *Gyps bengalensis* in Punjab Province, Pakistan. *Bird Conserv Int.* 2002;12:311-326.

13. Panagala VS, Stingfellow JS, Dybvig K, et al. *Mycoplasma corogypsi* sp. nov., a new species from the footpad abscess of a black vulture, *Coragyps atratus*. *Int J Syst Bacteriol.* 1993;43:585-590.

14. Ruder MG, Feldman SH, Wünschmann A, et al. Association of *Mycoplasma corogypsi* and polyarthritis in a black vulture (*Coragyps atratus*) in Virginia. *J Wildl Dis.* 2009;45(3):808-816.

15. Oaks JL, Donahoe SL, Rurangirwa FR, et al. Identification of a novel *Mycoplasma* species from an Oriental White-backed vulture (*Gyps bengalensis*). *J Clin Microbiol.* 2004;42(12):5909-5912.

16. Centers for Disease Control and Prevention. West Nile Virus and Dead Birds. Available at: https://www.cdc.gov/westnile/dead-birds/index.html. Accessed May 8, 2023.

17. Plaza PI, Blanco G, Lambertucci SA. Implications of bacterial, viral, and mycotic microorganisms in vultures for wildlife conservation, ecosystem services and public health. *Ibis.* 2020;162:1109-1124.

18. U.S. Fish and Wildlife Service. California Condor Recovery Program. Available at: https://www.fws.gov/program/california-condor-recovery/southwest-california-condor-flock-hpai-information-updates-2023. Accessed April 17, 2023.

19. U.S. Department of Agriculture Animal and Plant Health Inspection Service. 2022-2023 Detections of Highly Pathogenic Avian Influenza in Wild Birds. Available at: https://www.aphis.usda.gov/aphis/ourfocus/animalhealth/animal-disease-information/avian/avian-influenza/hpai-2022/2022-hpai-wild-birds. Accessed May 8, 2023.

20. Gomez O, Lopez R, Naveda-Rodriguez A. Avian pox in turkey vulture (*Cathartes aura*). *Vulture News.* 2014;66:42-48.

21. Mora-Carreño M, Guerra-Correa C, Moroni M, Paredes E. Avian pox in a turkey vulture (*Cathartes aura*) from northern Chile. *Austral J Vet Sci.* 2019;51:41-43.

22. Martinez-Herrero MC, Sansano-Maestre J, López Márquez I, et al. Genetic characterization of oropharyngeal trichomonad isolates from wild birds indicates that genotype is associated with host species, diet and presence of pathognomonic lesions. *Avian Pathol.* 2014;43(6):535-546.

23. Martinez-Herrero MC, González-González F, López-Márquez I, et al. Oropharyngeal trichomonosis due to *Trichomonas gypaetinii* in a cinereous vulture (*Aegypius monachus*) fledgling in Spain. *J Wildl Dis.* 2019;55(1):153-157.

24. Webb SL, Fedynich AM, Yeltatzie SK, et al. Survey of blood parasites in black vultures and turkey vultures from South Carolina. *Southeast Nat.* 2005;4(2):355-360.

25. Yabsley MJ, Vanstreels RET, Martinsen ES, et al. Parasitaemia data and molecular characterization of *Haemoproteus catharti* from New World vultures (Cathartidae) reveals a novel clade of *Haemosporida. Malar J.* 2018;17:12.

Avian Orthopedic Surgery: From Theory to Practice

Shangzhe Xie, BSc/BVM, MVS, PhD, DABVP (Avian Practice), DACCM

Session #3104

Affiliation: From Mandai Wildlife Group, 80 Mandai Lake Road Singapore 729826.

Abstract: Basic principles of avian orthopedics are well-established and well-described in several review papers and textbook chapters. The first part of this masterclass covers thebasics of avian orthopedic surgery including the suitability of each fracture fixation method (e.g., intramedullary pin, external skeletal fixator, tie-in fixator, coaptation) for different fracture locations and types. Given the wide range of wing and limb sizes, shapes and conformations, some principles cannot simply be extrapolated across all avian species. The second part of the masterclass utilizes real case studies to discuss the need for adjustment of these techniques and methods for patients of various avian species, including Malagasy turtle dove, honey buzzard, olive-backed sunbird, sun conure, rainbow lorikeet and Vietnamese pheasant. In particular, the honey buzzard case includes a discussion on the level of physical therapy and rehabilitation required for cases to regain full flight and ability to thrive back in the wild, as backed by post-release tracking data.

Introduction

Avian bones are the epitome of oxymorons: they are lightweight, yet strong; they are easy to break, yet quick to heal; they are simple in structure, with components similar to mammalian bones, yet complex in function, with physiologic integration into other organ systems such as the respiratory and reproductive systems; the cortices are thin, yet can withstand the forces associated with flight. The powerful flight muscles that provide some protection to the bone can also cause rotational deformity of fragments of long bones after they are fractured. These unique features of avian bone result in fracture presentations that may differ from mammalian fractures. For example, compared to mammalian fractures, avian fractures are more frequently open and comminuted due to the lack of soft tissue protection.

Clinically Relevant Avian Musculoskeletal Anatomy and Physiology

Very few fractures are orthopedic emergencies. Taking time to stabilize the patient from life-threatening injuries resulting from the same traumatic event that caused the fracture will improve overall survival and outcome of fracture repair attempts.

It is important to understand the clinically relevant anatomic and physiologic details of each patient's fracture, which will differ according to patient species, size, fracture location and type etc. The periosteal, medullary, metaphyseal and epiphyseal blood vessels supply blood to avian bone. The periosteal blood supply is often the most important component in callus formation, especially in pneumatic bones such as the humerus and femur.[1] Therefore, preserving as much of the soft tissue around the bone as possible during fracture repair will ensure that bone healing occurs at the optimal rate. Bone fragments in a comminuted fracture that have been displaced far from soft tissue and therefore lost vascular supply are potentially likely to become sequestra.[2] Management of sequestra depends on their size and location. A large sequestrum that may be important in contributing to the overall stability of the repaired bone could be better left in place until a stable callus has formed,[1] unless it starts acting as a nidus for infection. Smaller sequestra that serve no function in the bone healing should be removed.[2]

Avian bones can heal in the combination of a few ways:

1. Primary healing through the Haversian system when there is good fixation and bone apposition. There is usually minimal callus formation if this is the main way of healing, however fracture healing time will be greater than with secondary healing.

2. Secondary healing, of which there are two types:

    a. Endosteal callus formation between well-aligned bone fragments, which is often the most realistic objective of fracture repair.

    b. Periosteal callus formation when fractures are not aligned, with movement at the fracture site.[1]

Most fracture healing is a combination of these two.

The type of healing affects the rate of healing; however, the general rate of healing is faster in a bird than in a mammal for a similar fracture type. Depending on factors such as age, size, weight, activity level, fracture location, and specific bone (i.e., humerus, ulna, metatarsus etc.), fracture healing may be as short as 2 weeks and as long as 9-18 weeks. In general, more distal bones heal more slowly than those close to the body. Clinical stability of a fracture (i.e. a firm fibrous callus) at 2-3 weeks often precedes radiographic evidence of bone healing, which often takes 4-8 weeks.[1] Therefore, although radiographs taken at 10-14 days post-op are important for evaluation of adequacy and integrity of fixation, evidence of osteomyelitis, sequestrum formation, and other issues, the need for such radiographs should be assessed. For example, if the callus is palpable and solid and return to function of the patient is progressing as expected, then there is less need for stressful restraint and anesthesia for early radiographic reassessment.

Ligaments surrounding avian joints and tendons that are immobilized for long periods of time can become contracted, resulting in a permanently reduced range of motion.[1,2] Similarly, fracture callus formation and ligament/tendon adhesions can also affect the joint range of motion. As formation of hematomas and cartilaginous bridges delays fracture healing, there should be balance between excessive movement at the fracture site versus maintenance of the range of motion through early and regular physiotherapy. Without considering and addressing these factors, complete healing of the bone may not correlate with complete return to flight or weight-bearing capabilities of the patient.

Principles of Fracture Management

Many good resources exist that describe different types of surgical techniques used to repair avian fractures, including the most appropriate fixation type for each anatomic location.[1,2] In addition, considerable assessment and judgement must be made by the clinician in choosing a management option. The objectives of fracture management are to achieve alignment of the bone fragments with minimal loss of bone length and alignment, while minimizing soft tissue damage. Depending on the type of patient, full return of function may be important, especially for patients that are intended for release back into the wild, or into a large flight aviary. The skill and experience of the clinician with different techniques may also be an important factor, as surgical fixation requires skills that are developed through training and experience. Lastly, the financial cost of repair should also be considered, as an appropriately charged surgical fracture repair will cost a lot more than external coaptation alone. However, the number of follow-up revisits required for maintenance of a splint and the higher likelihood of failure may also contribute significantly to the overall cost.

External coaptation

After assessment of a fracture, including well-positioned radiographs, external coaptation is usually the first form of temporary stabilization of the fracture. This is important to prevent further damage to the soft tissues from excessive movement of mobile bone fragments. Good bone alignment can sometimes be achieved with this alone, and it can be an appropriate sole treatment for some fractures and patients. It is also the only option for patients that are too small for surgery. Bone alignment is best assessed on radiographs, so radiographs after external coaptation should be taken wherever possible. Examples of external coaptation techniques include figure of eight bandages for antebrachial and metacarpal fractures; wing to body wraps for humeral fractures; Robert Jones bandages, Altman splints, Schroeder-Thomas splints, or stirrup splints for tibiotarsal and tarsometatarsal fractures.[1] Pressure necrosis is a likelihood from bandages that are applied too tightly, so frequent monitoring is required when coaptation is selected. Some patients also tolerate bandages poorly, and the frequent movement and self-traumatization associated with them may also diminish the likelihood of a favorable outcome.

Surgical Repair Techniques

Intramedullary pins

Intramedullary (IM) pins can be inserted normograde or retrograde, depending on the nature of the fracture. Retrograde insertion is generally more straightforward as the fracture fragments can be visualized directly and manipulated into position; however, unless the fracture was already open, the surgical approach should be well-planned to ensure minimal soft tissue disruption, including avoiding joint penetration wherever possible. Anatomic guides such as *Avian Surgical Anatomy: Thoracic and Pelvic Limbs*[3] are useful in planning the most direct approach based on the fracture location. Normograde pin placement requires practice and generally works best for fractures that are not excessively displaced. IM pins are very good for maintaining alignment and length of the bone, but lack rotational stability.[1,2] Stack pinning using multiple smaller pins of the humerus and femur can provide further rotational stability, but in general, combining an IM pin with an external skeletal fixator (i.e., tie-in fixation, or with external coaptation is utilized for this purpose).[1]

External skeletal fixator

External skeletal fixators (ESF) are light, strong, inexpensive and adaptable to many fracture types ESFs also allow easier access to the soft tissue wounds in open fractures, if those need to be managed by secondary intention healing. ESFs may be the preferred fixator type if the anatomy precludes placement of an IM pin (e.g., if the pin has to go through joints, potentially affecting long-term function). Type 1 ESF, with pins connected on one side of the limb only may be the only type possible in some anatomic locations due to the lack of space, compared to Type II ESFs, with pins through the bone and connected on both sides of the limb.[1] Maintaining alignment of the bones until the ESF is completely placed and stabilized can be challenging because the bone fragments are not held in place unlike in IM pinning, so a surgical assistant or the use of temporary jigs is essential.

IM pin-ESF tie-in fixator

Combining an IM pin and ESF overcomes the disadvantages of each. It may not be suitable for the tarsometatarsus as there is no marrow cavity into which to insert an IM pin, but is otherwise a very versatile technique. Technically, this is just a combination of each of the two techniques with some minor adjustments (e.g., a smaller IM pin, 50-65% of the medullary cavity) to allow the cross pins for the ESF to be placed within the same cavity.[1] A connecting bar can be either a separate pin, or the protruding end of the IM pin can be bent back 180° to act as the connecting bar.

Bone plates

Bone plating is not as well-described as the above techniques in literature, and is not as well-practiced as IM pinning. The cortices of avian bones were traditionally thought to be too thin to engage the thread of bone screws. Also, given how well IM pins and tie-in fixators work for the majority of avian fractures, the theoretical advantages (e.g., quicker return to function, superior rotational stability) of bone plates may not result in a better outcome for avian fractures.

Case Studies

Although the methods discussed above form a sound basis for managing many avian fractures, these methods have historically been applied to a limited range of avian species (e.g., raptor and psittacine species). Some of these methods may not be readily extrapolated to some avian species and factors such as life history and patient size, among others, need to be considered for each individual patient, especially if the alternative to an imperfect fracture repair is euthanasia. A few case studies are presented here, demonstrating a combination of uncommon fractures in common species, common fractures in uncommon species, typical fracture management plan with setbacks etc. The common theme with these cases is that assessment by the clinician of the overall patient circumstances resulted in modification of standard methodology to create a good outcome for the patient.

Malagasy turtle dove (*Nesoenas picturatus*)

A Malagasy turtle dove suffered a right comminuted humeral fracture, which was repaired with IM pinning and figure of 8 bandage. As the fracture was comminuted and the patient was small, surgical repair was not straightforward and bone length was not fully restored; however, with staged physiotherapy, starting with passive range of motion exercises and progressing to increasingly large cages for flight exercises, the patient regained sufficient flight capabilities for release into a large flight aviary after approximately 8 weeks.

Oriental honey buzzard (*Pernis ptilorhynchus*)

A wild oriental honey buzzard presented with an oblique fracture of the left humerus that was repaired with IM pinning and figure of 8 bandage. Aesthetic complications precluded the additional surgical time needed to add an ESF tie-in. Bone alignment was good, but not perfect, and healing was slightly delayed, likely due to excessive mobility due to the temperament of the patient and the need to restrain it regularly for supportive feeding. The patient was finally released 16 weeks post-surgery, with 10 weeks of active rehabilitation, including passive range of motion exercises, flight exercises in large flight cages and creance flying.

Olive-backed sunbird (*Cinnyris jugularis*)

An olive-backed sunbird presented with a left radius/ulna fracture. Due to the small size of the patient, surgical fixation was not possible, and a tape splint was applied. The splint was removed 2 weeks later, by which time a bony callus could be palpated. However, the alignment of both bones was not perfect. The patient was placed in a small cage while waiting to see if it could be placed permanently in captivity, but it seemed to start flying well in this space during this time. A decision was made to move the patient to a larger space to further assess its flight capabilities and it continued to demonstrate good flight. Eventually, 6 weeks post presentation, it was released into the wild in a location with abundant food resources.

Sun conure (*Aratinga solstitialis*)

A sun conure suffered a traumatic injury resulting in concurrent closed, mid-shaft fractures of the right femur and right tibiotarsus. IM pins were placed in both the right femur and right tibiotarsus. Both pins exited at the stifle joint and were connected just in front of the stifle joint externally using acrylic. Both pins were removed 2 weeks post-surgery. The right leg was fully functional 3 weeks post-surgery (and no apparent morbidity to the stifle joint).

Rainbow lorikeet (*Trichoglossus moluccanus*)

A rainbow lorikeet was presented acutely with left leg lameness. Radiographic assessment revealed a midshaft, mildly displaced femoral fracture. The patient developed respiratory problems during this time that made it an unsuitable anaesthetic candidate. As the focus shifted to management of its respiratory problems, the femoral fracture healed during this 2-week period, with the left leg returning to full function without any form of fixation, even though the bone alignment was not perfect.

Vietnamese pheasant (*Lophura edwardsi*)

A juvenile Vietnamese pheasant presented acutely with left leg lameness. Radiographic assessment revealed a midshaft, mildly displaced left femoral fracture. As time was taken to plan for potential surgical fixation while avoiding damage to the growth plates, the patient's lameness improved significantly over the next 3 days. Due to this rapid improvement, a decision was made to forego surgery altogether. This patient is still being monitored at the time of writing but is expected to return to full function.

References

1. Doneley B. *Avian Medicine and Surgery in Practice: Companion and Aviary Birds*. 2nd Ed. Boca Raton, FL; CRC Press; 2016.

2. Redig PT, Ponder J. Orthopedic surgery. In: Samour J, ed. *Avian Medicine.* 3rd Ed. St. Louis, MO: Elsevier; 2016:312-358.

3. Orosz SE, Ensley PK, Haynes CJ. *Avian Surgical Anatomy: Thoracic and Pelvic Limbs*. Philadelphia, PA: WB Saunders; 1992.

Section 7

AAV Masterclasses Track 3

Shangzhe Xie, BSc/BVMS, MVS, PhD, DABVP (Avian Practice), DACCM;
Hannah Attarian, DVM
Session Chairs

Management of Reproductive and Behavioral Diseases of Captive Psittacine Birds

Amber Lee, BVSc, DABVP (Avian Practice)

Session #3186

Affiliation: From The Unusual Pet Vets, Peninsula, 210 Karingal Drive, Frankston, VIC, 3199, Australia.

Abstract: Captive psittacine birds commonly present for reproductive problems; often including a spectrum of medical, surgical and behavioral disorders. This stems from a birds' strong innate drive to reproduce, and oftentimes the conditions in captivity are perfect. There are often links between behavior changes and the onset of reproductive activity, of which many owners misunderstand or are unaware of their significance. Similarly, management of many of these conditions requires both medical therapy and behavioral modification to ensure that the bird has no further problems in the future. This masterclass covers the importance of the avian behavioral exam, how to recognize and diagnosis reproductive disease, and case examples for management and treatment options. The medical and surgical management of egg binding, dystocia and cloacal prolapse are highlighted.

Introduction

Reproductive disorders are a common occurrence in avian practice. Birds frequently present for evaluation of chronic egg laying, egg binding or dystocia, egg yolk coelomitis, gonadal neoplasia and cloacal prolapse. Avian veterinarians must have a solid understanding of reproductive anatomy, physiology and normal breeding behaviors to be able to diagnose and treat reproductive disease appropriately. Similarly, it is important for veterinarians to be able to recognise the role that environmental, nutritional and behavioral modification play in the management of these conditions.

There can often be an association between the development of these reproductive diseases and behavioral disorders for many avian patients. These can include territorial aggression, inappropriate bonding, redirected aggression, attention seeking behavior, excessive screaming, masturbation and feather destructive behaviors. Understanding these problem behaviors requires detailed history acquisition to ascertain the bird's hand-rearing status or degree of early learning and development, its current household relationships as well as its health, diet, husbandry and environmental cues.[1]

Behavior as Part of a History and Why It is Important

Pet birds lack domesticity, and as a result require the need for daily maintenance behaviors of their wild parents or grandparents. Often, the husbandry provided and the human relationships that result are not appropriate or sufficient for the engagement of these normal behaviors.[2] When owners fail to understand normal parrot behavior, they anthropomorphise and expect the bird to fit in with their human expectations. This then leads to the lack of appropriate learning opportunities for the bird to engage in normal flock interactions such as conflict resolution, self-maintenance and other social behaviors. This can lead to the development of many behavioral problems including the inappropriate pair bonding that occurs with many birds and their owners. Most bird owners are not cognizant of the fact that physical touching, cuddling or sharing food with their bird is perceived by their bird as mating behaviors.

Captive psittacine birds, particularly those that are hand-reared often do not have the opportunities to learn normal social behaviors. If a bird is not appropriately socialized, attention-seeking behaviors such as begging and screaming or displacement behaviors such as biting, phobias and feather destructive behavior can result. As these antisocial behaviors develop, the bird often becomes more isolated, and therefore the behaviors can worsen as the bird is positively or negatively reinforced.

Changes in behavior can indicate underlying reproductive disease. For example, broody behavior without oviposition may indicate cystic ovarian changes.[3] Veterinarians can work with owners to modify any unwanted or problem behaviors that could lead to conditions such as chronic egg laying, dystocia, coelomitis and feather destructive behavior.

One of the most common causes of relinquishment of birds, especially for larger species like macaws and cockatoos, is for behavioral reasons.[4] When a bird is relinquished to a rescue the effects are threefold. First, there is disruption for the bird. Second, there is disappointment for the owner. Third, there is a loss of clientele and income for the veterinarian. Finally, it is important to consider difficulties handling many birds in the consultation room, the use of Fear Free techniques for the welfare of the bird, and the safety of the staff.

Normal Sexual Behaviors of Wild Psittacine Species and How They Become Problems in Captivity

It is imperative to understand normal sexual behaviors of wild psittacine birds if veterinarians are going to be able to prevent the development of problem sexual behaviors in captive counterparts. It is important to be able to recognize normal sexual behaviors, environmental triggers of wild parrots, and how they occur or are expressed in the home setting.[5]

Reproductive behaviors of wild psittacine birds include pair bonding, courtship regurgitation, nest seeking, nest building, territorial aggression and copulation. The display of these behaviors is regulated by environment (e.g., photoperiod, rainfall, food supply, presence of a nest, presence of a mate), hormonal and social cues (e.g., pair bond formation and social hierarchy of the flock).[6] The following table illustrates the above reproductive stimuli and addresses management of these factors from the perspective of pet bird owners (Table 1). Prevention, rather than intervention, is key as is understanding the unique natural history and species-specific differences in breeding strategies exhibited by captive parrots.

Table 1. Therapeutic considerations for treatment of reproductive behaviors.[1,2,5,26-28]

Stimulus	Explanation	Management
Environmental		
Presence of a nest site Presence of nest material Maintenance of a territory	Presence and maintenance of a nest site in the wild helps stimulate reproductive drive.	Do not provide a nest or happy hut. Deny access to drawers, underneath furniture, other dark cavities. Deny access to the newspaper at the bottom of the cage, move the paper to underneath the grate. Frequently rearrange perches and cage furniture to create some (gentle) environmental instability. Provide opportunities for exercise, especially flight.
Long day length	Increasing day length can signal the start of breeding season for many species.	Ensure birds gets at least 12 hours of uninterrupted sleep. Ideally, sleeps in a room that no one uses after the bird goes to bed.
Rainfall	Rain can indicate that the availability of fresh food is about to increase, which is ideal for reproduction.	Rainfall cannot be controlled but owners can limit the amount of shower time or misting a bird receives.
Loss of eggs	Females may try to replace eggs lost from clutch.	Once a hen has started laying a clutch, allow it to brood the clutch for at least 3 weeks (instead immediately removing eggs).
Nutritional		
High energy diet	A high-energy diet facilitates egg production and chick rearing.	Birds fed a seed-only diet or predominantly seed diet should be converted to a pellet-based diet with low energy vegetables and only a small amount of fruit, seed and nuts.
Easily accessible food	Accessible food makes feeding a mate or chicks easier. Wild birds spend 60-80% of time foraging, but most captive birds meet nutritional requirements in less than 20 minutes over the course of the day.	Birds should be required to work and expend energy (forage) for their food rather than receive it freely. Dry food can be mixed with recycled cat litter and fruit and vegetables can be wrapped in paper and cardboard to increase difficulty of procurement. Various foraging toys are available for purchase or can be made by hand by owners.
Social		
Interactions with a real or perceived mate	In the wild, courtship behavior between a bonded pair is a strong driver for nest building and reproduction. In captivity, a bird may pair-bond with another bird, a human or an inanimate object.	If the bird is bonded to another bird, denying visual and auditory contact with that bird may reduce reproductive drive in the long term, but is likely to cause significant distress and therefore is generally not recommended. If the bird is bonded to a person, reduce the reproductive behaviors performed by that person. Keep all physical contact to the head, avoid touching the back or under the wings. Do not share food with the bird or feed the bird directly from your mouth. Do not allow the bird to masturbate on you. Foster platonic interactions, such as clicker training, as an alternative to sexual interactions. Recommendations for different species vary but keep birds of the same sex of the same or similar species together. Allow other members of the household the opportunity to interact positively with the bird. If the bird is bonded to an object (e.g., food bowl, mirror, bell), consider removing that object. Do not purchase unweaned baby birds! These birds are more likely to be poorly socialized and develop reproductive problems.
Innate		
Genetic	Some species, or individuals within a species are more prone to excessive reproductive drive.	Small parrots including cockatiels, budgerigar and lovebirds. Seasonal aggression of Amazon parrots. Peach-faced lovebirds tuck long pieces of nest material into the feathers of their back for transportation.

Formation of pair bonds

The formation of pair bonds is not well understood but appears to be the result of increased associative interactions, including spending more time together, allopreening, and reduced aggression towards pair mate than to non-mates.[7] Many pair mates remain together almost constantly, performing many daily activities together, with some species exhibiting social synchrony (mimicking each other's behaviors). The main advantage of forming a pair bond for psittacine birds is in the ability to undertake the huge task of hatching and rearing altricial offspring, that would not be possible for an individual bird. Often these pair bonds exist year-round due to the advantages that pair bonds provide including increasing social standing of the pair within the flock.

Vocalization

Unique vocalizing and mating calls of various species signal their sexual readiness. Male cockatiels can easily be led to start singing their mating call with encouragement from owner vocalizations (author's personal experience). Female cockatiels will cluck softly or chatter when ready to mate. Pair bonded captive psittacine birds will often vocalise when their caregiver leaves the room or house. Their wild counterparts will contact call to keep in communication with a mate, so excessive screaming may be escalation of contact call if the bird is distressed by their mate's absence.

Courting behavior

Psittacine birds exhibit several different moves in performing courtship behavior including hopping, bowing, strutting or tail wagging. Many species show pinning of the pupils (rapid dilation and constriction) or blushing of facial skin areas (macaws and Palm cockatoos). Cockatoos will display movements of the crest feathers on the head. Pair bonded birds will bow their head to solicit preening from a mate. Some species usually remain physically distant and only come together for breeding. Thus, attention paid by owners and excessive physical contact in the form of petting (preening) can encourage chronic reproductive behavior.

Nest seeking

Nest cavities are often a limiting factor for breeding for many Neotropic species, and in captivity the provision of a cage (nest), nesting material (newspaper), or free ranging through the house allows cavity-seeking behavior (e.g., under furniture, dark spots or corners). Fleece huts are a particular problem as many owners think that their birds require a nest to sleep in at night and inadvertently provide, a dark cozy place for egg laying or nesting behaviors.

Copulation

The drive to mate is strong for pair bonded birds. In captive birds, this desire may result in masturbation on the owner, or on perches or toys within the cage.

Laying eggs

Generally, larger species lay smaller clutches compared to smaller species. For cockatiels, budgerigars and lovebirds, the development of chronic egg laying in captivity is very common. Some species share incubation duties (e.g., cockatoos), while female macaws incubate and it is the male's job to ensure she is fed through regurgitation. Thus, excessive regurgitation may present a problem in captivity as the bird tries to feed their owner as their perceived mate.

Aggression

Territorial aggression towards non-mates for protection of the nest or guarding of a mate is common in wild psittacine birds. In captivity this may lead to cage aggression, mate-related aggression towards competing human family members, or even redirected aggression towards a favored human if that person is approached by another person. This is commonly seen in Amazon parrots, cockatoos, (miniature) macaws, grey parrots, conures, quaker parrots and lovebirds.[8]

Feather destructive behavior

Chronic hormone stimulation and inappropriate bonding can be an underlying cause for feather destructive behavior (FDB) in cockatoos, macaws, conures, and grey-cheeked parakeets.[3] Sexual frustration can result from elongated photoperiod, high-fat diet, loss of a mate or inappropriate mate selection, or the presence of mating pairs in the environment. Some reproductive disorders can lead to pain or neuropathy which may result in FDB.

Female Reproductive Diseases

Disorders of the reproductive tract in captive psittacine birds involve a complete spectrum from mild and self-limiting to severe and life-threatening. Disease can involve any part of the reproductive tract, and many diseases can present with non-specific signs so it is important to determine whether the reproductive tract is involved from history, clinical signs and diagnostics. The following is a review of the general approach to diagnosing reproductive disease and general treatment, with a specific focus on egg binding, dystocia and cloaca prolapse.

History

Confirmation of the sex of the bird is important. Changes in behavior can come with puberty (6 months to 4 years depending on species) with changes including screaming, irritability, preference of a caregiver, territorial aggression, and overt sexual behaviors (regurgitation, masturbation). Nest seeking or building, along with recent oviposition are indicators of reproductive activity. Changes in reproductive cycle with oviposition and then cessation of lay may suggest egg binding or dystocia along with the laying of abnormal or soft-shelled eggs.

Clinical signs

Often signs are non-specific including lethargy, dyspnoea, coelomic distension, lameness, or feather destructive behavior. Specifically, there may be a palpable egg or mass in coelom, straining, tenesmus, tail wagging, wide-based stance or drooped wings. Depending on the chronicity of disease, the bird may present severely emaciated and weak. With hypocalcemia there maybe pathologic fractures or tremors. Cloacal prolapse can occur secondary to straining from dystocia, space-occupying masses or behaviorally in male cockatoos.

Diagnosis

A diagnosis can be difficult; therefore, it is important to obtain clues from history, physical examination and diagnostics. Blood work can be completely normal. A complete blood cell count may show anemia (sometimes non-regenerative), or evidence of infection or inflammation. Biochemical changes include elevation of alkaline phosphatase, elevated creatine phosphokinase, hyperglobulinemia, hypercalcemia or hypocalcemia, hypercholesterolemia with lipemic serum.[9,10] Radiography often will help to detect an intact egg or shell fragments, loss of serosal detail (e.g., free fluid), mass effect in the region of the reproductive tract, and medullary hyperostosis.

Ultrasonography provides a more sensitive assessment of the reproductive tract and can help identify neoplasms, unmineralized eggs, ectopic ova, free fluid, and cystic tissue associated with the ovaries or oviduct. Ultrasound-guided coelomocentesis can be both therapeutic to help relieve dyspnoea and diagnostic as aspirated fluid can be submitted for cytology and culture and sensitivity. The use of endoscopy can further help elucidate the type of reproductive disease present with direct visualization of reproductive organs and biopsy of lesion at the same time. For definitive diagnosis of some conditions, exploratory laparotomy may be the only way to confirm and/or treat the disease.

Treatment

Generally, treatment is dependent upon the diagnosis (or suspected diagnosis) and the condition of the patient. While birds may present acutely, often disease may actually be chronic in nature, and the patient must be stabilized and any secondary complications treated. Once the reproductive disease has been identified and treatment initiated, the caregivers of the bird must be counselled on environmental, behavioral and dietary modifications to eliminate the risk of future egg laying or reproductive stimulation. This includes altering the relationship to include less intimacy; less petting, sitting on the shoulder, physical touch or sharing of food. Positive reinforcement training is recommended to get the bird to step-up readily, and only use the hand to transfer to and from the cage with the use of stations and play stands in the environment. Any high fat or high sugar foods should be eliminated from the diet, as well as addressing photoperiod (12-14 hours of darkness) and removal of any potential nesting sites or nesting material.

The pharmacologic use of hormonal therapy for control of chronic hormone stimulation and reproductive disease has been largely through the use of gonadotropin releasing hormone (GnRH) agonists leuprolide acetate (e.g., Lucrin; AbbVie Pty Ltd, Mascot NSW 2020) or deslorelin acetate (e.g., Suprelorin; Virbac (Australia) Pty Limited, Milperra NSW 2214). Previously used agents included medroxyprogesterone acetate, human chorionic gonadotropin and testosterone have since fallen out of use due to severe and potentially fatal side effects.[3] GnRH agonists initially stimulate the secretion of luteinizing hormone (LH) and follicle-stimulating hormone (FSH) from the pituitary gland, which results in oestrus and ovulation as well as the production of GnRH receptors.[11] When administered long term or as sustained release they lead to the downregulation of GnRH receptors via negative feedback.[12]

Leuprolide acetate is a short-acting depot formulation that has been studied in several species.[13-17] Length of efficacy varies between species with the average length of cessation of egg laying in cockatiels being 12-19 days[13] and the effect shorter in budgerigars at the same dose, only lasting for 12-14 days.[16] There is a wide range of doses being recommended from 400-1,000 µg/kg IM every 2-3 weeks.[12] There is also anecdotal evidence that the efficacy of Lupron reduces with long-term usage.

Deslorelin acetate is a longer-acting, subcutaneous, controlled release implant that is available in a 4.7-mg or 9.4-mg dose. Deslorelin acetate, like leuprolide acetate has been used successfully to reduce reproductive behaviors and egg laying in birds, but additional uses have been identified in the management of ovarian neoplasia in cockatiels,[18,19] and budgerigars with Sertoli cell tumors.[16] There is also evidence in chickens that deslorelin acetate has chemoprotective properties and prevents against the development of ovarian neoplasia.[20] The only prospective studies of deslorelin acetate have been done in chickens,[21] quail[22-24] and pigeons.[25] The length of treatment varies with species and the type of disease being treated. Recommended dosing intervals are difficult to ascertain, and indicators such as return of reproductive behaviors, egg laying or both, have been used as indicators for reimplantation.[12] In psittacine birds, the most common dosing frequency is every 3 months, with a range of 2-5 months in addition to behavioral and environmental modifications.[12]

Egg binding and dystocia

Birds presenting with egg binding or dystocia can present in various stages of shock. Patient stability will determine when further imaging (e.g., radiographs or ultrasound) is performed to locate the egg and determine its position. If the bird is obstructed (i.e., dystocia) and not able to pass feces, then immediate stabilization and removal of the egg must occur.

In general, provide supportive care with supplemental heat 84-90 °F (29-32 °C), fluid therapy, analgesics, parenteral nutrition and parenteral calcium. Antibiotics can be provided if there is suspicion of ruptured oviduct (i.e., egg no longer within oviduct, secondary coelomitis). Some eggs can pass with supportive care alone, however if the patient declines in condition, then further intervention is warranted. The use of prostaglandin, oxytocin and hormone therapy can be considered if there is no suspicion of oviduct obstruction, adhesion or perforation and the uterovaginal sphincter is patent.[3] Parenteral calcium must be administered prior as these treatments potentiate muscle contraction.

If these treatments are ineffective then gentle massage and manipulation of the lower coelom under anesthesia to facilitate vaginal opening can be attempted. Once the uterovaginal sphincter is dilated, then direct pressure is applied above and partially behind the egg to move it caudally into oviposition. Ovocentesis is then performed if the above treatments have failed. Transcloacal ovocentesis is preferred over transcoelomic due to the risk of trauma. Ideally, crown the egg to visualise through the sphincter and then use a needle and syringe to collapse and drain the contents. A vaginal speculum or lone star retractor can be used to help open the lips of the vent and identify the uterovaginal sphincter. If the egg is not visualized, then it is likely not in the cloaca and the procedure should be converted to a coeliotomy. Prepare owners that many hens may lay another egg within the 72 hours following the passage of an egg. In our practice, we usually administer leuprolide acetate at the time of egg removal, and then wait a few days before administering long-acting deslorelin acetate. The owner is then counselled on environmental, behavior and nutritional factors to help prevent further egg laying.

Cloacal prolapse

Prolapse of cloacal tissues, vagina, uterus or oviduct are a common emergency presentation and can result from dystocia, normal oviposition, manual delivery of an egg, malnutrition, obesity and general debilitation from chronic egg laying.[26,27] General prolapse treatment principles follow.

- Prolapsed tissue must be protected from further damage and desiccation.
- Supportive care is provided to the patient, which may include shock therapy, fluid resuscitation, antibiotics, analgesia and nutritional support.
- Resolution of the prolapse is based upon identification and treatment of the underlying cause.

At presentation, the stability of the patient must be carefully evaluated as well as history acquisition to suggest the potential for underlying reproductive disease as the cause for cloacal prolapse. If the patient is bright, alert and responsive (with a smaller prolapse), initial interventions may include warmth, fluids and gentle cleansing and lubrication of prolapsed tissues. For those patients presenting in poor condition due to septicemia, blood loss or dystocia, more aggressive emergency and shock therapy measures are required.

If the prolapse is small and no egg or mass is palpated, then it can be replaced into the cloaca with stay sutures, while further diagnostics and definitive therapy is planned. Two lateral horizontal mattress sutures are placed individually either side of the vent, while a cotton-tipped applicator is used to hold the replaced tissue within the cloaca. The resulting opening must be small enough that the tissue does not re-prolapse but large enough that urofeces can still pass normally.

For those prolapses where a large amount of tissue is present, or there are signs of devitalization, replacement of necrotic tissue is not recommended due to risk of coelomitis and sepsis which can result in death. These cases require further advanced coelomic surgical resolution (i.e., salpingohysterectomy) and necessitate referral to an avian specialist. If an egg is present within the prolapsed tissue, it may be gently manipulated out of oviductal tissue after flushing and lubrication. The egg should crown through the opening of the uterovaginal sphincter. If the egg is unintentionally passed through the wall of the oviduct or cloaca, then the defect must be repaired.

Cloacal prolapses can occur in cases with chronic hormone stimulation and/or masturbation as a result of chronic straining. It is important that all other underlying causes of prolapse are ruled out in these cases. In chronic cases, the cloacal wall and sphincter may become flaccid and this loss of tone leads to soiling of skin/feathers with malodorous urofeces and repeated prolapse. These cases require reconstructive surgery with cloacopexy or ventplasty as well medical and behavioral therapy. Diagnosis of underlying reproductive disease in cloacal prolapse cases is imperative as in many cases if the bird resumes oviposition, then the prolapse is likely to recur. Initial diagnostic work-up can include complete blood cell count, biochemistry panel and radiographs in two views. The use of ultrasound may be superior to radiographs in visualizing eggs or egg material (non-mineralized) within the oviduct or coelom (egg yolk coelomitis) as well as aide in the diagnosis of salpingitis or metritis and differentiate oviductal masses from caudal coelomic masses. The examination of the cloaca itself can be helpful. This can be achieved with manual eversion with a lubricated cotton-tipped applicator in larger birds. Cloacoscopy is ideal to evaluate for underlying masses, and to collect mucosal biopsies.

While it is important to protect and replace the prolapsed tissues, it is imperative in the management of prolapses that the underlying disease process is identified and addressed. Then all attempts to decrease hormone stimulation and discourage reproductive behavior must also occur to prevent recurrence. These include the aforementioned husbandry, behavioral,[28] environmental and hormonal therapies.

Male Reproductive Diseases

Diseases of the male reproductive tract include infectious or neoplastic etiologies. Orchitis is inflammation associated with the testicular tissue and can be due to local disease, extension from the cloaca or hematogenous spread of an infection. Clinical signs can be vague with decreased fertility, lethargy, decreased appetite or pain.

The most common neoplasms of the male gonads include Sertoli cell tumors, seminomas, Leydig (interstitial) cell tumors, adenocarcinoma, leiomyosarcoma, and lymphosarcoma. These can present with the above signs, but due to space-occupying effects also present with lameness, paresis, or paralysis of the hindlimbs and coelomic distension.

Diagnosis is as above for with female reproductive diseases, with ultrasound or computed tomography scan likely providing greater detail and more accurate images over convention radiography. Coelioscopy can provide visualization of the testes and the ability to biopsy affected tissues.

Treatment is aimed at the underlying disease with antibiotics based on culture and sensitivity for orchitis. For testicular neoplasms the use of GnRH agonists may aide in reduction of size, along with regular coelomocentesis for fluid accumulation, and general supportive care. Removal of neoplasms is ideal, but difficult in many small avian patients and there is limited information regarding chemotherapeutic options.[3]

References

1. Doneley B. Reproduction. In: Chitty J, Monks D, eds. *BSAVA Manual of Avian Practice: A Foundation Manual.* Gloucester, UK: British Small Animal Veterinary Association; 2018:54-67.

2. Stelow E. Avian behavior consultation for exotic pet practitioners. *Vet Clin North Am Exot Anim Pract.* 2021;24(1):103-117. doi:10.1016/j.cvex.2020.09.006.

3. Scagnelli AM, Tully TN. Reproductive disorders in parrots. *Vet Clin North Am Exot Anim Pract.* 2017;20(2):485-507. doi:10.1016/j.cvex.2016.11.012.

4. Clubb S. Parrot relinquishment in the US: why are birds losing their homes? *Proc Annu Conf Assoc Avian Vet.* 2011:279-284.

5. Van Sant F. Problem sexual behaviors of companion parrots. In: Luescher AU, ed. *Manual of Parrot Behavior.* Ames, IA: Blackwell Publishing; 2006:233-245.

6. Seibert LM, Sung W. Psittacines. In: Tynes VV, ed. *Behavior of Exotic Pets.* Hoboken, NJ: Wiley-Blackwell; 2010:1-11.

7. Spoon TR. Parrot reproductive behavior, or who associates, who mates, and who cares? In: Luescher AU, ed. *Manual of Parrot Behavior.* Ames, IA: Blackwell Publishing; 2006:67-82.

8. Welle KR, Luescherm Andrew U. Aggressive behavior in pet birds. In: Luescher AU, ed. *Manual of Parrot Behavior.* Ames, IA: Blackwell Publishing; 2006:211-217.

9. Bowles HL. Reproductive diseases of pet bird species. *Vet Clin North Am Exot Anim Pract.* 2002;5(3):489-506. doi:10.1016/S1094-9194(02)00008-7.

10. Joyner KL. Theriogenology. In: Ritchie BW, Harrison GJ, Harrison LR, eds. *Avian Medicine: Principles and Application.* Lake Worth, FL: Wingers Publishing; 1994:748-775.

11. Johnson JG. Therapeutic review: deslorelin acetate subcutaneous implant. *J Exot Pet Med.* 2013;22(1):82-84. doi:10.1053/j.jepm.2012.12.013.

12. Petriz OA. Advancements in methods for decreasing reproductive success. In: Speer BL, ed. *Current Therapy in Avian Medicine.* St. Louis, MO: Elsevier; 2016:446-454.

13. Millam J, Finney H. Leuprolide acetate can reversibly prevent egg laying in cockatiels. *Zoo Biol.* 1994;13:149-155.

14. Millam JR. Reproductive management of captive parrots. *Vet Clin North Am Exot Anim Pract.* 1999:93-110.

15. de Wit M, Westerhof I, Pefold L. Effect of leuprolide acetate on avian reproduction. *Proc Annu Conf Assoc Avian Vet.* 2004:73-74.

16. Straub J, Zenker I. First experience in hormonal treatment of Sertoli cell tumors in budgerigars (*M. undulatus*) with absorbable extended-release GnRH chips (Suprelorin). *Proc ICARE.* 2013:299-301.

17. Klaphake E, Fecteau K, DeWit M, et al. Effects of leuprolide acetate on selected blood and fecal sex hormones in Hispaniolan Amazon parrots (*Amazona ventralis*). *J Avian Med Surg.* 2016;23(4):253-262.

18. Keller KA, Beaufrère H, Brandão J, et al. Long-term management of ovarian neoplasia in two cockatiels (*Nymphicus hollandicus*). *J Avian Med Surg.* 2013;27(1):44-52. doi:10.1647/2012-006.

19. Nemetz L. Deslorelin acetate long-term suppression of ovarian carcinoma in a cockatiel (*Nymphicus hollandicus*). *Proc Annu Conf Assoc Avian Vet.* 2012:37-42.

20. de Matos R. Investigation of the chemopreventative effects of deslorelin in domestic chickens with a high prevalence of ovarian cancer. *Proc ICARE.* 2013:90.

21. Noonan B, Johnson P, de Matos R: Evaluation of egg-laying suppression effects of the GnRH agonist deslorelin in domestic chickens. *Proc Annu Conf Assoc Avian Vet.* 2012:321.

22. Petritz OA, Sanchez-Migallon Guzman D, Paul-Murphy J, et al. Evaluation of the efficacy and safety of single administration of 4.7-mg deslorelin acetate implants on egg production and plasma sex hormones in Japanese quail (*Coturnix coturnix japonica*). *Am J Vet Res.* 2013;74:316-323.

23. Petritz O, Guzman DS, Hawkins M et al: Evaluation of deslorelin acetate implant dosage on egg production and plasma progesterone in Japanese quail (*Coturnix coturnix japonica*). *Proc Annu Conf Assoc Avian Vet.* 2013:17.

24. Schmidt F, Legler M, Einspanier A et al: Influence of the GnRH slow release agonist deslorelin on the gonadal activity of Japanese quail (*Coturnix coturnix japonica*). *Proc ICARE.* 2013:501-502.

25. Cowan ML, Martin GB, Monks DJ, et al. Inhibition of the reproductive system by deslorelin in male and female pigeons (*Columba livia*). *J Avian Med Surg.* 2014;28:102-108.

26. Lennox A, Lee A. Cloacal, uterine and rectal prolapse. In: Chitty J, Monks D, eds. *BSAVA Manual of Avian Practice: A Foundation Manual.* Gloucester, UK: British Small Animal Veterinary Association; 2018: 358-365.

27. Curd S. Egg retention. In: Chitty J, Monks D, eds. *BSAVA Manual of Avian Practice A Foundation Manual.* Gloucester, UK: British Small Animal Veterinary Association; 2018:350-3537.

28. van Zeeland YRR, Friedman SG, Bergman L. Behavior. In: Speer BL, ed. *Current Therapy in Avian Medicine and Surgery.* St. Louis, MO: Elsevier; 2016:177-251.

Veterinary Care of Breeding Collections of Birds of Prey and Psittacine Species: Tips and Tricks to Help Your Client

Michael Lierz, DZooMed, DipECZM (Wildlife Population Health), DipECPVS

Session #3108

Affiliation: From Clinic for Birds, Reptiles, Amphibians and Fish, Faculty of Veterinary Medicine, Justus Liebig University Giessen, Frankfurter Str. 114, D-35392 Giessen.

Abstract: Captive breeding has contributed to the successful restoration of many species of birds of prey and psittacine birds. Avicultural breeding techniques include double clutching, direct fostering, cross fostering, hatch and switch, hacking, imprinting and assisted reproduction techniques such as semen collection and artificial insemination; however reproductive failure is a common occurrence. Failures are mainly related to management problems, including hygiene measures, food quality issues, breeding collection structure, or to individual health issues of breeding birds. These issues may result in non-egg laying females, low quality eggs or infertile eggs due to male infertility. Therefore, veterinary care of breeding collections is important. This masterclass provides an overview of veterinary involvement in birds of prey and psittacine bird breeding projects. Insights into the most common management problems are provided, basic aviculture techniques are explained and state-of-the-art as well as cutting-edge assisted reproduction techniques in avian reproduction are described. Information from this veterinary field allows the veterinarian to expand expertise and service to clients. The latest developments in prophylactic health measures, preparation of birds and cages, semen collection, evaluation and cryopreservation techniques, artificial insemination and ex-ovo fertilization are covered.

Manuscript previously published in: Lierz M. Veterinary care of breeding flocks of birds of prey and psittacine species: tips and tricks to help your client. Proc ExoticsCon. 2022;57-62.

Introduction

Many raptor and psittacine species are endangered or threatened with extinction and species conservation programs exist. In addition, commercial breeding is common. Reproduction attempts in birds of prey or psittacine birds in captivity regularly fail or the reproduction rate is low. The reason for this is multifaceted. Management-related problems are the most common reason, followed by diseases and infertility. To solve problems in a breeding collection, a detailed investigation is necessary. The veterinarian plays the most important role in this process as s/he has the capability to perform all necessary detailed investigations. A veterinarian in this role needs to understand how to manage a breeding collection and which aspects are of special importance. This masterclass provides a basic introduction. For more detail the reader is directed to further literature.[1-16]

Basics of Bird of Prey Reproduction

Birds of prey are monogamous and reach their sexual maturity usually between 2-6 years. Pairs are very territorial and the bond to their nesting site is strong. The male is around one third smaller than the female and provides the food for the brooding female. Males brood to a much lesser extend compared to the female. Birds of prey have usually a sibling sexual imprint starting around day 8 after hatch. Purely human hand-reared single birds of prey are unable to reproduce naturally, but can be stimulated by man and artificially inseminated. This is usually used for to produce species hybrids. Birds of prey hand-reared in groups may be used for natural breeding and sexual stimulation by humans, but parent-reared birds are best for natural reproduction.

Birds of prey usually lay two to four eggs in one clutch. By removing the eggs before the clutch is completed, the female may be stimulated to lay more eggs (egg pulling). If the eggs are removed within 2-3 weeks after clutching, a second clutch will be laid.

Breeding birds should be kept in flying cages with a nesting site. In cases where assisted reproduction techniques are used, imprinted semen donors can be kept by falconry methods.

Basics of Psittacine Bird Reproduction

Compared to birds of prey, psittacine birds have multiple reproductive strategies, from strongly monogamic, to group nesting and polygamic. Imprinting birds for reproduction purposes is not used in this avian group as imprinting causes usually behavioral problems preventing natural breeding or even nesting. Also artificial insemination is not routinely used as semen collection is much more difficult in this group compared to birds of prey. Breeders usually use double clutching to increase the amount of offspring. Psittacine reproduction usually relies on natural pair breeding.

Assisted Reproduction Techniques

In falconry outside Germany, hybrid falcons are commonly used. Those hybrids are mixed breeds of two different falcon species, which under natural circumstances, do not reproduce together. Therefore, artificial insemination techniques are used. Breeding birds are imprinted to humans and when they reach sexual maturity the breeder can stimulate the bird. Semen is collected from the males by massage methods or by voluntary semen donation. This semen is transferred into the cloaca or the oviduct of a female immediately prior to egg laying of the first egg and shortly after each subsequent egg laid.[12,16]

In psittacine species, artificial insemination techniques have only recently been introduced, as semen collection is more difficult compared to birds of prey. These techniques are usually used in species conservation programs, but commercial breeders begun to include them in the management. More and more parrot owners ask for species hybrids to produce uncommon colors, but there is really no argument to support this change.

In several psittacine or bird of prey breeding facilities, eggs are removed and placed in an incubator. Natural incubation in captivity can be risky, especially with inexperienced females as they might leave the nest or damage the eggs. Artificial incubation of eggs is not easy and each egg needs to be treated individually. Apart from the temperature (37.2-37.4 °C), the humidity is an important key factor. Each egg should lose between 14-16% weight during the complete incubation period. Therefore, the egg needs to be weighed and humidity adjusted to reach those loss rates. If alteration is insufficient, the eggshell needs to be manipulated to increase or decrease the water loss through the shell. In addition, vitality of the embryo should be monitored, either by candling or by a device that is able to monitor the heart rate (Buddy ® Digital Egg monitor).[11]

Chicks break the eggshell from inside (i.e., pipping) at the end of the species-specific incubation period. The total incubation period is highly variable between species. After pipping it still takes 48-50 hours until hatch is completed and during this time no manipulations should be done. Chicks are then fed at the earliest 6 hours after hatch and usually are placed back to the parents between day 6-10 of life (birds of prey), or purely hand reared, which must be done in groups in psittacine birds to reduce risk of behavioral problems later in life.

Veterinary Considerations of Bird of Prey and Psittacine Breeding

The process of breeding birds includes many risk factors that need to be considered. Pairs are regularly in large cages with no direct control of the individual. Especially during the reproduction period, disturbances usually interfere with breeding activity. In birds of prey, imprinted single breeding birds are much better to control from the health point of view; however, repeated inseminations do carry an increased risk of diseases of the reproductive organs. Finally, reproductive behavior and egg laying is an increased burden to the birds. In Europe, psittacine and raptorial birds are usually kept in outside aviaries where those birds have contact with free- ranging populations of various birds. Also, partners are usually placed together without an intensive health check. Breeders regularly place eggs from other breeders in their incubator or place chicks of different sources together, even to the same pair. All this significantly increases the risk of infectious disease.

Problems in Reproduction

Infertile clutches

The most common problem in bird of prey and parrot breeding facilities is infertile eggs. Pair behavior is not always ideal and despite the owner not seeing a problem because birds demonstrate display behaviour, copulation does not occur successfully. Psittacine birds are highly social birds with a complex partner selection system. Forced pairing might regularly lead to insufficient partner acceptance. In birds of prey, the male, as the smaller partner, must approach the larger female and must fly onto her back. Females might interpret this as aggression and hurt the male. Compared to nature, males cannot escape aggressive behavior because they are in an aviary. Therefore males must approach the female carefully, offer her food and stimulate her for copulation. Even after an attack from the female, males need to be brave enough to try copulations repeatedly. Many males fail this, offer food, demonstrate reproductive behavior, but do not try to copulate again after one poor experience. Therefore, the male represents the key factor in natural reproducing pairs. As food-offer behavior from the male to the female is important, males must be supported by the breeder. During reproduction period, the pair should be fed several times during the day with high-quality food.

In psittacine birds as well as birds of prey, infertility of the male is a common reason for poor breeding performance. Infections (e.g., *Salmonella* sp., *Chlamydia* sp., *Mycoplasma* sp.) involving the reproductive tract, environmental toxic agents (e.g., wood protecting paint, heavy metals, pesticides, mycotoxins), medications (e.g., coccidiostats given in animal feed and consumed by raptors) or vitamin deficiencies (e.g., vitamin E, A) are possible causes. In cases where infertility of a male is suspected, a detailed spermatologic investigation (e.g., motility, pH, live/death ratio, spermatozoa morphology) should be performed. The semen of falcons contains so-called round bodies (RB), which are a normal structure, but not found in all avian species. In falcons the number of RB can be high despite that their function is presently unknown.[5] Reference values are available for the different falcon species,[4] but not for all birds of prey species and a very rare in psittacine birds.[2] If reference values are not available, the comparison of ejaculate parameters from the investigated male with proven breeding males is useful. To date, spermatologic investigations can be supported by computer assisted semen analysis (CASA) which provides additional very useful information.[7] Functional spermatologic tests, such as the perivitelline membrane penetration test,[9] are very helpful in investigating infertile males that demonstrate motile semen. Electron microscopic investigations of individual spermatozoa[15] assist in the determination of the reason for infertility and might be used to complete the evaluation of the semen.

In addition, a detailed endoscopic investigation of the reproductive tract is recommended. It should include a testicle biopsy in case the spermatologic values or the endoscopic appearance of the testicle are altered.

Finally, infertile clutches occur after artificial insemination with contaminated semen. This is especially true when urate or fecal contamination decreases spermatozoa vitality significantly.

Reduced egg quality

Alterations of the eggshell occur if the female faces stressful events. These stresses might come from outside or from disturbed pair behavior. Thin eggshells might also be caused by calcium deficiencies. Causes include low calcium content or imbalanced calcium:phosphorus ratios in the food, vitamin-D deficiency and intestinal or kidney disorders. Diseases of the female reproductive tract, such as oophoritis, salpingitis or cloacitis need to be considered.

Reduced hatching rate

A reduced hatching rate is caused by incubation management failure. Incorrect temperatures (calibrate the thermometer!) and not controlling the weight loss of each egg are causes. In addition, poor incubation hygiene and malpositioning of embryos in the eggs also reduce hatching rate.

An egg's content is a closed environment. Once the egg is laid, nothing can be added or removed. Therefore, feeding management of the parents, especially the female, is of major importance for a satisfactory hatching rate. The egg follicle starts growing approximately 6 weeks prior to ovulation and the egg yolk contains most of the nutritional ingredients for the growing embryo. High quality, vitamin-enriched (E, B_2, D_3) food must be supplied months before the first reproductive activity is visible. This is in many cases not considered by the breeder.

Vertically transmitted pathogens need to be considered. Infected females often are clinically healthy, but carriers of pathogens and their vertically infected embryos may die. The same occurs with environmental toxic agents (e.g., pesticides, mycotoxins), which may not affect the female clinically, but are transmitted via the egg yolk and affect the embryo severely.

The investigation of a decreased hatching rate involves a detailed assessment of the breeding and feeding management and a postmortem examination of the egg, including evaluation of the time of death of the embryo. Common reasons for early embryonic death are inadequate egg transport, egg storage prior to incubation, severely inappropriate incubation temperature, microbiologic egg contamination or environmental toxic agents. Late embryonic death usually occurs after malnutrition of the parents, vertically transmitted pathogens, failure in incubation management such as deficient egg turning, slightly wrong incubation temperature, insufficient or increased egg weight loss.

The incubator needs to be tested and calibrated so all electronics and mechanisms work correctly. Swabs for microbiologic investigations should be taken to assess incubation hygiene. Investigation of the egg and embryos that failed to hatch should involve a microbiologic examination of the egg yolk. The egg yolk membrane and the trachea of the embryo should be taken to investigate for mycoplasma, the eggshell for detection of *Salmonella*, liver and kidney of the embryo should be stored for virologic and molecular biologic examination and the complete embryo should then be placed in formalin and sent for histopathologic examination.

Veterinary Management of Breeding Birds

Breeding birds demand special attention, not only because of their high value, but also due to their increased health risk (see above). On a regular basis (e.g., every 3-6 months) a fecal sample should be taken and examined for the presence parasites and *Salmonella* spp. At least once per year the birds should be evaluated in detail. This involves a swab for detection of *Chlamydia* sp. and a stress test[3] to assess the function of the respiratory tract. In cases where the stress test demonstrates alterations, an endoscopic examination of the celomic cavity of the bird should be performed. A blood sample should be taken to investigate the antibody level against paramyxovirus-1 and avian influenza virus H5 and H7, blood chemistry to investigate liver and kidney function, hematology to detect chronic diseases and determination of the vitamin and trace element level, especially if a nutritional imbalance is suspected. In psittacine birds, samples to check for circo-, polyoma-, borna-, herpes-, and adenovirus should be taken, including antibody screening, especially for antibodies directed against avian bornavirus. Treatments (e.g., against parasites) or prophylactic measures such as vaccinations (e.g., in birds of prey: paramyxovirus-1 [PMV-1], avian influenza [AIV, H5/H7] and West Nile virus) can be done at that time. In birds of prey, it has been demonstrated in vaccination and challenge studies, that vaccination against H5 using inactivated vaccine protects the birds completely from clinical disease[10] and partial protection against West Nile virus can be achieved.

Conclusions

Breeding collections of birds of prey and psittacine birds in captivity are complex systems. The basics of management and behavior of the birds are not fully understood by many owners. Reproductive success rates of 40-50% (hatched chicks to laid eggs) are considered successful; however, this is well below natural reproduction rate (> 80%), not to mention pairs that do not lay at all. The role of a veterinarian in a breeding facility is critical. Increasing success rate is usually possible. The veterinarian plays a key role in increasing the success of such facilities and the health of the breeding birds.

References

1. Bailey T, Lierz M. Veterinary aspects of bird of prey reproduction. *Vet Clin North Am Exot Anim Pract.* 2017;20(2):455-483.

2. Bublat AD, Fischer S, Bruslund H, et al. Seasonal and genera-specific variations in semen availability and semen characteristics in large parrots. *Theriogenology.* 2017;5(91):82-89.

3. Chitty J, Lierz M, eds. *BSAVA Manual of Raptors, Pigeons and Passerine Birds.* Gloucester, UK: British Small Animal Veterinary Association; 2008.

4. Fischer D, Garcia De La Fuente J, Wehrend A, Lierz M. Semen quality and semen characteristics of large falcons with special emphasis on fertility rate after artificial insemination. *Reprod Domest Anim.* 2011;46:14.

5. Fischer D, Garcia De La Fuente J, Wehrend A, Lierz M. Celibate birds needs assistance-spermatology and artificial insemination in falcons. *Proc 4th Int Giessen Grad Centre Life Sci.* 2011;127.

6. Fischer D, Neumann D, Purchase C, et al. The use of semen evaluation and assisted reproduction in Spix´s macaws in terms of species conservation. *Zoo Biol.* 2014;33(3):234-244.

7. Fischer D, Neumann D, Wehrend A, Lierz M. Comparison of conventional and computer-assisted semen analysis in cockatiels (*Nymphicus hollandicus*) and evaluation of different insemination dosages for artificial insemination. *Theriogenology.* 2014;82(4):613-620.

8. Fischer D, Angenvoort J, Ziegler U, et al. DNA vaccines encoding the envelop protein of West Nile virus lineages 1 or 2 administered intramascularly, via electroporation and with recombinant virus protein induce partial protection in large falcons (*Falco* spp.). *Vet Res.* 2015;46:87.

9. Krohn J, Meinecke-Tillmann S, Schneider H, et al. Technical note: investigations on the accomplishment of the avian perivitelline membrane-penetration-assay. *Reprod Domest Anim.* 2012;33.

10. Lierz M, Hafez HM, Klopfleisch R, et al. Protection and virus shedding of falcons vaccinated against highly pathogenic avian influenza A virus (H5N1). *Emerg Infect Dis.* 2007;13(11):1667-1674.

11. Lierz M, Gooss O, Hafez HM. Non-invasive heart rate measurement using a digital egg monitor in chicken and turkey embryos. *J Avian Med Surg.* 2006;20(3):141-146.

12. Lierz M. Raptors: reproductive disease, incubation and artificial insemination. In: Chitty J, Lierz M, eds. *BSAVA Manual of Raptors, Pigeons and Passerine Birds.* Gloucester, UK: British Small Animal Veterinary Association; 2008:235-249.

13. Lierz M, Reinschmidt M, Müller H, et al. A novel method for semen collection and artificial insemination in large parrots (Psittaciformes). *Sci Rep.* 2013;3:2066.

14. Lierz M. Reproduction: advancements in methods for improving reproductive success. In: Speer B, ed. *Current Therapy in Avian Medicine and Surgery.* St. Louis, MO: Elsevier; 2016:433-445.

15. Schneider H, Fischer D, Bergmann M, et al. Investigations on crypreservation of psittacine semen. *Proc ICARE.* 2013;386-387.

16. Temple SA. Artificial insemination with imprinted birds of prey. *Nature.* 1972;237:287-288.

Cloacal Prolapses in Cockatoos

Brynn McCleery, DVM, DABVP (Avian Practice)

Session #3136

Affiliation: From Red Bank Veterinary Hospital, 197 Hance Avenue, Tinton Falls, NJ 07722, USA.

Abstract: Cloacal prolapse is the most common reported disease condition of the cloaca in companion parrots with cockatoo species over-represented. Cloacal anatomy, causes of prolapse, signalment, clinical signs, and diagnostic techniques utilized in cases of cloacal prolapse are covered in this masterclass. The focus of the masterclass is on medical and surgical management options for cases of chronic, recurrent cloacal prolapses.

Introduction

Chronic cloacal prolapse of psittacine birds is a common and frustrating disease process for both owners and veterinarians. Cockatoos are overrepresented amongst birds suffering from cloacal prolapse, and prolapsing in this species may be due to a hypersexual or idiopathic syndrome, making treatment more challenging and recurrence more likely. Chronic cloacal prolapses in cockatoos almost always must be treated with a combination of medical, behavioral, and surgical therapies.

Cloacal Anatomy

The cloaca serves as a common termination point for the gastrointestinal, urinary, and reproductive tracts in birds.[1-4] It consists of three compartments (in oral to aboral positioning): the coprodeum, urodeum, and proctodeum.[1-4] The most proximal compartment, the coprodeum, is the largest of the three and connects with the rectum.[1,3,4] In psittacine birds, the rectum attaches to the left side of the coprodeum at an approximately 80-90 degree angle.[3] It is separated from the urodeum via the sphincter-like coprourodeal fold, which is everted from the vent during defecation to prevent fecal contamination of the other cloacal compartments.[1-4] The urodeum is the middle compartment and the smallest of the three.[1,3] It contains the termination of the ureters in the dorsolateral walls.[1-4] The urodeum also connects to the reproductive system of males and females.[1,3,4] In female birds, the oviduct has a rosette-like opening into the urodeum on the left ventrolateral wall, while in males the ductus deferens exits into the ventrolateral wall bilaterally on symmetric, raised papillae.[1,2] The dorsal semicircular uroproctodeal fold separates the urodeum from the most distal compartment, the proctodeum. In juvenile birds, the dorsal wall of the proctodeum contains the opening of the Bursa of Fabricius, an important side of B-lymphocyte differentiation.[1,3,4] This opening regresses into an ostium after the involution of the cloacal bursa as birds mature.[1,3] In some avian species, such as ratites and waterfowl, a copulatory structure is also present within the proctodeum.[2] Vasa parrots (*Coracopsis vasa*) can form an eversion of the cloacal for copulation.

The muscular folds between the chambers of the cloaca create both a physical and functional separation and facilitate many of the important functions of the cloaca. In addition to defecation, as mentioned above, the folds assist with urination, retroperistalsis of urine into the colon for water and solute resorption, copulation, oviposition, and exposure of the Bursa of Fabricius to environmental antigens.[3,4]

The vent is composed of dorsal and ventral lips with right and left commissures and serves as the outer opening of the cloaca, analogous to the anus of mammals.[3-5] Opening and closing of the vent is controlled by the sphincter, transverse, and levator muscles of the cloaca.[4]

Vascular supply to the cloaca is provided by the pudendal artery and vein. The pudendal nerve provides innervation, along with the connexus caudalis and caudal coxal nerves. This pudendal nerve travels parallel to the ureter to the dorsolateral cloacal wall where a plexus is formed.[1,3]

Cloacal Prolapse

Cloacal prolapse is a common disorder reported in psittacine birds and can occur from primary disease within the cloaca or secondary to diseases involving structures that terminate in or near the cloaca.[5-7] Cloacal prolapses may occur in conjunction with prolapses of the oviduct or rectum.[3]

Causes

Cloacal prolapse is ultimately caused by excessive straining and/or loss of normal vent sphincter tone. Confirmed or theorized causes of cloacal prolapse in psittacine birds include: [3,5-9]

Primary: infectious or inflammatory cloacitis (bacterial including mycobacteriosis, fungal [pathogenic yeasts, especially *Candida albicans*[3]], viral [Herpesvirus], parasitic [*Sarcocystis* spp., *Cryptosporidium* avian genotype V[19]]), trauma, neoplasia (adenocarcinoma, leiomyoma, lymphosarcoma, squamous cell carcinoma, fibrosarcoma), hematoma, stricture (especially after treatment for papillomatous masses), adenomatous polyps, viral induced or nonviral papillomatous masses[18]

Secondary: chronic behavioral maladaptation, inappropriate bird-human bond, dystocia, other reproductive diseases (salpingitis, metritis, cystic ovaries), gastrointestinal disease (megacloaca, infectious enteritis, heavy metal gastroenteritis, giardiasis, helminths), cloacolithiasis, ureterolithiasis, neurogenic or neuromuscular deficits resulting in reduced sphincter tone or cloacal atony (from chronic vent trauma or spinal cord disease), coelomic masses/space occupying lesions (organomegaly, tumors, granulomas), ascites

Cloacal prolapse from excessive straining associated with secondary causes is considered more common than primary cloacal disease.[5]

In a study of > 500 cloacal biopsies from avian patients, the three most common diagnoses included: cloacitis (40%), cloacal papillomas (20%), and neoplasia (14%).[5] Herpesvirus is one of the more commonly reported causes of cloacitis in new world parrots, especially green-winged macaws.[3,5] Bacterial cloacitis has been reported due to *Pasteurella* spp., *Salmonella* spp., *E. coli*, and *Staphylococcus* spp.[3] Cloacal prolapses suspected to be from reproductive, behavioral, or idiopathic causes may be less likely to be biopsied compared to other cloacal diseases, and the results of the above retrospective are unlikely to be reflective of the most common causes of cloacal prolapse.

Reproductive tract disease is the most common cause of prolapsed tissue in female birds and can include infection or inflammation (salpingitis, metritis), neoplasia, and recent oviposition. Nutritional deficiencies, obesity, and debilitation from chronic egg laying are risk factors associated with prolapse in these birds.[5] In male birds, masturbation and other sexual behaviors are a common cause of cloacal prolapsing.[10]

Syndromes of cloacal prolapse in cockatoos

Coprodeal prolapse syndrome has been reported cockatoos (and a few other species). This is sometimes considered a single syndrome, while other authors have suggested two separate syndomes.[3,6,11,12]

1. *Secondary to excess hormone stimulation and hypersexual behavior in imprinted parrots (Hypersexual):* This syndrome is reported as more common in cockatoos and affects both males and females. Chronic hormone stimulation leads to anatomic changes to the cloaca and vent.

2. *Progressive idiopathic coprodeal prolapse and straining (Idiopathic):* This syndrome generally starts around sexual maturity (5-8 years in most cockatoos), but the birds do not display normal sexual behaviors—which is considered the distinguishing feature from the other syndrome. This syndrome is most reported in cockatoos (including umbrella and Moluccan cockatoos) but has also been reported in African grey parrots and blue and gold macaws. Males are more commonly affected. Speculated risk factors include living in a single parrot household, being raised without conspecifics, holding feces for prolonged periods, inappropriate bond to one human, and a prolonged weaning period.

Prevalence and signalment

In a retrospective study of companion parrots, prevalence of cloacal disease overall was 3.8%. Prolapse (cloacal, oviductal, and rectal) was diagnosed most often (65.1%), followed by masses and cloacoliths. Cloacitis was identified in approximately one-third of cases (34.9%) and was considered a secondary disease condition in each instance.[4]

Although cloacal prolapse has been reported in a multitude of psittacine bird species, cockatoos are overrepresented.[5] Of the 61.5% of parrots with cloacal prolapse in the study mentioned above, 60.7% were cockatoos, despite cockatoos making up only 10% of the parrot population evaluated. All of the cockatoos that had prolapses were umbrella cockatoos (*Cacatua alba*). Of the cockatoos included in the study, 94.4% were presented for prolapse (cloacal, oviductal, rectal). Sex distribution of the cockatoos with prolapses was 67.9% female, 25% male, and 7.1% of unknown sex. Almost half of the birds in the study (46.4%) with cloacal disease had a history of inappropriate sexual behaviors (excluding egg laying); 76.9% were cockatoos and 84.6% were females. This contradicts previous reports of male cockatoos being overrepresented for cloacal prolapse. The authors theorized that females may have more commonly prolapsed secondary to hypersexual behavior and increased hormones while males had prolapses more commonly caused by idiopathic prolapse syndrome. Sexual behavior was commonly noted in cockatoos and females; however, male cockatoos and a single African grey parrot made up the group of birds for which no underlying cause of prolapse was identified.[4]

Cockatoos were also one of the most common bird groups for which cloacal biopsies were submitted to a pathology lab, along with Amazons and macaws. Cloacitis was commonly diagnosed on histopathologic diagnosis of the biopsy samples from cockatoos. In that study, prolapse was rare with only 10/712 (1.4%) samples reported to be from prolapsed tissue; however, underreporting may have affected those numbers due to the retrospective nature of the study.[13]

Clinical Signs

Clinical signs associated with cloacal prolapse can vary immensely. Birds may show no other apparent signs of disease aside from the prolapse or may present in critical condition.[5] Severity of symptoms depends on the length of time the prolapse has been present, the amount of tissue prolapsed, and the health of the prolapsed tissue.

Symptoms reported by the owners or seen at presentation can include a permanent or intermittent prolapse of variable size, diarrhea, reduced fecal production, tenesmus, dyschezia, haematochezia, malodorous droppings, flatulence, alterations in respiratory rate and/or effort, self-mutilation of the vent, a soiled pericloacal area/vent gleet, infertility, change in perching posture, and pelvic limb paresis or paralysis.[4,5,8] The owners may also report recent oviposition.

Sequelae of Prolapse

The prolapsed tissue is at risk of trauma and subsequent hemorrhage and/or inflammation, infection (cloacitis), desiccation, ischemia, and necrosis.[6,14] Cloaca prolapses can lead to impaired excretion of urinary and gastrointestinal waste products, ascending urogenital infections, cloacitis, enteritis, impaired water resorption, dehydration, egg binding, dystocia, oviductal impaction, gastrointestinal intussusception, and cloacal rent formation.[6,9] With chronic prolapses especially, there may be loss of cloacal wall and vent sphincter muscle tone.[9]

Vegetative endocarditis due to *Staphylococcus aureus* was also reported in an umbrella cockatoo secondary to suspected translocation from the owner's skin during digital reduction of a chronic cloacal prolapse over the course of 1 year.[15]

Diagnosis

A full diagnostic work-up is recommended in cases of cloacal prolapse to try and identify the underlying cause and to guide therapy. Even if chronic hormone stimulation and masturbation is suspected, other causes of prolapse should be ruled out. Unfortunately, in many cases, a cause may not be identified.[7,9]

Physical exam findings may help elucidate a cause of the prolapse, especially if a coelomic mass is palpated or effusion is suspected. During physical examination, the prolapse should be evaluated and cloacal tissue may need to be differentiated from prolapsed rectum or oviduct. Prolapsed cloacal tissue should be carefully evaluated for any areas of discoloration, necrosis, ulceration, or other trauma. The prolapse itself, as well as the feathers around the vent, may be heavily soiled with feces, urates, and blood and should be gently cleaned with sterile saline for evaluation.

Physical examination should include cloacal evaluation. On awake exam, the vent and a small portion of the proctodeum can be evaluated by gentle eversion using a cotton tipped applicator. Further cloacal evaluation can be performed via indirect visualization with an otoscope and sterile cone; however, this technique is still limited and can cause trauma without sedation or anesthesia.[3,5] The preferred method of cloacal evaluation is through cloacoscopy.[5] Saline-infused cloacoscopy is preferred due to improved visualization and maneuverability.[3,5,6,16] This procedure can be performed with a 2.7 or 1.9 mm rigid telescope with a diagnostic sheath for sterile saline insufflation.[3,5,16] Cloacal lesions should be biopsied for histopathology and/or culture. Care must be taken to only sample superficial tissue (avoid full thickness biopsies) and to avoid pertinent anatomic structures.[3]

A minimum database of a complete blood cell count and biochemistry panel is recommended for patients presenting with a cloacal prolapse.[5,6] Blood work will help evaluate for anemia, systemic infection, renal disease, liver disease, and changes associated with ovulation and egg laying.[5] Protein electrophoresis may also be indicated. Fecal evaluation should be performed including fecal cytology (evaluated especially for sporulated rods), fecal parasite evaluation (direct, floatation), and possibly fecal culture.[6] Malodorous fecal production with gas may be associated with *Clostridium* spp. or *E. coli* cloacitis.[3-5]

Orthogonal radiographs are recommended to evaluate for reproductive changes in females (e.g., eggs, polyostotic hyperostosis or physiologic osteomyelosclerosis), organomegaly, and coelomic masses. Contrast gastrointestinal radiographs or a cloacal contrast study may help delineate diseases of the GI tract causing secondary prolapse or cloacal luminal masses.[5] Alternatively, fluoroscopy can be utilized to assess GI motility and evaluate for luminal abnormalities/masses. Advanced imaging (such as abdominal ultrasound, CT, or MRI) may also be indicated. Ultrasound and CT are superior for the detection of unmineralized eggs or egg products, changes to the oviduct (salpingitis, neoplasia), and differentiation of coelomic masses compared to radiographs.

Additional testing may be beneficial in some cases including specific pathogen testing (e.g., avian bornavirus, avian antiganglioside antibody, herpesvirus) and coelioscopy. Biopsy for cytology or histopathology is required for definitive diagnosis of cloacal or coelomic masses and may be performed with ultrasound guidance, endoscopically, or surgically.

Treatment

If left untreated, cloacal prolapse can potentially be fatal.[6] Initial treatment should be centered around supportive care and may include providing appropriate warmth, fluid resuscitation, and cleaning of the prolapsed tissue with subsequent lubrication to present further desiccation or damage.[5]

Although most chronic cloacal prolapses in cockatoos are mild (and often intermittent), some patients may present with severe prolapses and be in shock secondary to blood loss/hypovolemia, underlying disease, compression of coelomic nerves or vasculature (secondary to a space occupying lesion, mass, or egg), and/or septicemia. These patients should receive emergency management including shock therapy.[5]

Appropriate therapeutic intervention will depend on the underlying cause of the prolapse. Although many theorized or reported causes of cloacal prolapse were listed above for psittacine birds, many of these causes have not been identified in cockatoos or are considered extremely rare. For this reason, specific management of papillomatosis, cloacolithiasis, and ureterolithiasis are not covered in these proceedings. Diagnosis and treatment of cloacitis should be guided by biopsy and culture results. Similarly, if neoplasia is diagnosed within the cloaca or coelom, treatment will depend on diagnostic results (e.g., advanced imaging, cytology, histopathology) and may include complete surgical debridement vs. debulking and adjunct therapies (e.g., chemotherapy, radiotherapy).

Summary of treatment goals

- *Protect the tissue:* Prolapsed tissue must be protected against trauma and desiccation. The tissue should be gently cleaned with sterile saline and lubricated. Hypertonic saline or dextrose can be applied if the tissue appears edematous.

- *Patient stabilization:* Supportive care including warmth, fluid resuscitation, analgesia (topical anesthetics, systemic opioids, nonsteroidal anti-inflammatory medication), antibiotics if indicated for primary or secondary infection, and nutritional support should be provided prior to diagnostics or any surgical therapies.

- *Reduction:* Prolapse reduction can be performed digitally with a gloved hand, utilizing cotton-tipped applicators as needed. Tissue must be replaced in normal anatomic positioning. Transverse interrupted or horizontal mattress vent sutures may be needed to maintain position of the tissue during diagnostic work-up and definitive therapy.
- *Resolution:* This will be based on identification and treatment of the cause of prolapse. It is important to inform owners that a simple reduction with transverse sutures is unlikely to solve the problem in most cases.

Prolapse related to egg laying

In cases of confirmed egg binding or dystocia, the egg should be removed manually or through transcloacal ovocentesis under heavy sedation or general anesthesia. If the egg is located proximally, cannot be safely removed from prolapsed tissue, or the tissue is necrotic, surgery is indicated and may include a salpingohysterectomy and/or partial cloacal resection. Replacement of necrotic tissue puts the bird at risk of disease and death secondary to coelomitis and septicemia.[5] Procedures for ovocentesis and salpingohysterectomy have described in detail elsewhere.[17]

Hormone therapy can be utilized in birds with a suspected or confirmed reproductive cause of cloacal prolapse (e.g., recent egg binding or dystocia, history of oviposition prior to prolapse). Long-acting gonadotropin releasing hormone analogues (e.g., leuprolide acetate, deslorelin acetate) are used commonly.[12] Although these medications initially stimulate release of leutinizing hormone (LH) and follicle stimulating hormone (FSH) from the anterior pituitary gland, downregulation of GnRH receptors in cells producing these hormones follows through a negative feedback mechanism on both the hypothalamus and pituitary gland.[5] Chronic use of GnRH agonist medication may cause the medications to be less effective over time, and an increased dose or dosing frequency may be required.[12]

Surgical Intervention

Many cockatoos will present with chronic, intermittent, or persistent prolapses. These cases typically require surgical intervention. Each of these surgical procedures only addresses the clinical signs associated with prolapsing. The underlying cause must still be identified and addressed or recurrence is likely. These procedures can often be performed concurrently to address chronic changes to the anatomy (e.g., atony of the cloaca, atony of the vent).

Transverse sutures

After prolapse reduction, transverse simple interrupted or horizontal mattress sutures can be used to reduce the vent opening to prevent prolapse recurrence while treating the underlying cause and to allow time for reduction of tissue inflammation. The vent opening must be reduced to a size small enough to prevent prolapse recurrence but large enough to allow normal passage of feces.[3,5] It is rare for simple reduction as a sole therapeutic modality to provide permanent resolution of cloacal prolapses, and owners should be properly informed. In chronic cases, the cloacal wall and vent sphincter may be damaged, leading to fecal incontinence, secondary infections, and repeated prolapsing without more aggressive surgical intervention (i.e., reconstructive surgeries).

Purse strings in the vent are not recommended due to possible damage to the vent sphincter or its innervation and impedance of dorsal and ventral movements of the vent during defecation, predisposing to obstipation.[3,9] Placement and periodic replacement of stainless steel transverse sutures has also been reported.[3,6]

Cloacotomy

Direct incision into the cloaca through a coelomic approach can be performed for removal or biopsy of masses or abnormal tissue.[5] This is performed through a ventral midline approach to the caudal coelom.[17] Avoidance of important anatomic structures, including the ureters and reproductive tract openings, is essential.[5]

Cloacopexy

This procedure creates a permanent adhesion between the cloaca and body wall to help prevent further prolapse. Several approaches have been reported in psittacine birds. Intracoelomic approaches involve performing a coeliotomy via a vertical or horizontal skin incision made over the cloaca and reflection or removal of the fat pad over the ventral cloaca.[17] A cotton-tipped applicator placed in the cloaca prior to the below procedures can help confirm placement of the cloaca within the coelom.

- *Percutaneous:* Multiple (two to three) sutures are placed blindly through the skin and cloacal wall into a cotton-tipped applicator passed retrograde into the cloaca from the vent. The cotton-tipped applicator is removed from the vent and the cotton is teased off the sutures, which are then tightened. This procedure is more likely to cause inadvertent perforation of other structures (ureter, rectum, duodenum, pancreas), cause GI or ureteral entrapment, or cause functional colonic obstruction (if sutures placed extremely cranial) than an open procedure and is not recommended by the author.[6,9]

- *Circumcostal:* Sutures are placed around the most distal rib (at the junction of the sternal and vertebral portions) and passed full thickness through the craniolateral cloacal wall. This procedure is performed bilaterally and can be combined with a ventral body wall cloacopexy during closure.[6,9,17]

- *Sternal:* Anchoring to the pubic bones and sternum has also been reported. For sternal cloacopexy, two sutures are passed through the cartilaginous border of the sternum, then full thickness through the cloaca.[17]

- *Incisional:* The serosal surface can be incised on the coprodeum (exposing the subserosa) and sutured to an incision in the abdominal wall, similar to a gastropexy procedure in canine patients.[6,17] To perform this procedure, a 2- to 5-mm incision is made in the serosa of the coprodeum longitudinally, approximately 5 to 10 millimeters from the midline. An opposite incision is created in the paramedial body wall at a position that results in slight inversion of the vent when the cloaca is held up to it. Sutures are preplaced (three to four) then tied. During closure, a ventral midline cloacopexy can also be performed.[17]

- *Ventral midline*: This can be performed on its own or with one of the above cloacopexy procedures. To encourage adhesion formation, the ventral surface of the cloaca should be scarified or incised. Full thickness bites through the cloaca are taken during closure.

Possible and reported complications of cloacopexy include adhesion formation, colonic entrapment (reported in two cockatoos),[9] infection, intraoperative trauma to the cloaca/oviduct GI tract, failure or tearing of the cloacopexy site, and obstructions.[5,9,17]

Traditional ventplasty

Ventplasty reduces the vent size permanently in an effort to prevent cloacal prolapse.[10,17] This procedure should be reserved for chronic cloacal prolapses where cloacal muscles and supporting structures are permanently stretched and nonfunctional, leading to vent atony.[6,10] The amount of tissue resected depends on the extent of the vent dilation. One section can be removed in cases of mild to moderate dilation, while two sections are typically removed in cases of severely dilated vents using a traditional approach.[10,17] The goal is to return to vent to a more normal diameter by reducing the opening by approximately 25-30%.[6,10] Triangular incisions are made in both approaches with the base of the triangle along the leading edge of the vent with the point away from the vent.[6,17] Excision of the epidermis and dermis is performed in each approach with effort made to spare the sphincter and transverse cloacal muscle, if visible. The dermis can typically be bluntly dissected from the underlying muscular and submucosal layers. After resection of the triangular piece of tissue, gentle manual apposition of the new epidermal edges will form the vent diameter. If insufficient, additional epidermal and dermal tissue can be resected as needed until the desired vent diameter is achieved. During closure, a simple interrupted pattern is used to appose the submucosa, dermis, and epidermis in separate layers using absorbable sutures.[6,17]

- *Singular incision:* A single incision over the cranioventral aspect of the vent can be performed in cases of mild to moderate dilation.[17]

- *Bilateral approach:* Bilateral triangular wedge resections on the left and right lateral aspects of the vent can be performed in more severe cases.

It is important to note that ventplasty alone is likely to fail if the underlying cause of straining is not identified and addressed (as with cloacopexy alone).[10] Furthermore, prolapse of tissue through a more narrowed opening puts the tissue at increased risk of vascular compromise, inflammation, and necrosis. In females, egg laying must be controlled through surgery (salpingohysterectomy), behavioral modification, or hormone therapy, otherwise dystocia or rupture of the ventplasty sutures may occur.[10]

Asymmetric ventplasty[6]

During this procedure, the mucocutaneous junction of the vent is superficially incised in a C-shape on the right side with the incision encompassing approximately 50% of the vent circumference. A width of 1 to 2 millimeters of mucosa is removed with tenotomy scissors, and the mucosa is subsequently apposed with horizontal mattress sutures (5-0 PDS) placed on the vent sphincter in order to maximize contact between the dorsal and ventral vent lips. Skin edges are oversewn with simple interrupted sutures (6-0 PDS) to prevent inward curling. Finally, an additional tension-relieving mattress suture (4-0 PDS) is placed cranial to the cloacal sphincter as the luminal edge of the incision and removed 2 weeks later. The tension-relieving suture helps protect the area most susceptible to stretching leading to dehiscence.

This procedure reduces incision tension secondary to midline straining that occurs post-operatively. The asymmetric positioning creates a physical barrier to coprodeal tissue during straining but still allows defecation in parrots due to left-sided positioning of the rectal opening into the coprodeum. As with traditional ventplasty, this procedure is used in chronic cases associated with chronic reproductive conditions or idiopathic prolapsing leading to cloacal atony.

Reported acute complications include dehiscence, transitory straining, transitory mild fecal staining, and high velocity and lateralized fecal output. Possible long-term complications include dystocia, cloacolith formation or cloacitis secondary to fecal/urate stagnation.

Colopexy[7]

Colopexy was reported as a treatment for chronic recurrent colocloacal prolapse in an 8-year-old male sulphur-crested cockatoo (*Cacatua galerita*). The prolapses often occurred in association with hypersexual behaviors including masturbation. A ventral U-shaped skin incision was made, and the distal colon was sutured to the left abdominal wall using the same technique reported in dogs and cats. The seromuscular layer of the colon was incised longitudinally along the antimesenteric side. A corresponding incision was made in a craniocaudal direction in the opposing left transverse abdominal muscle. The dorsal and ventral edges of the seromuscular incision were sutured to the corresponding edges of the abdominal wall incision with simple continuous sutures with the colon retracted cranially under slight tension. Although technically more challenging than cloacopexy due to the relatively thin colon wall and risk of perforation, this procedure provides cranially directed tension on the cloaca with the intestines fixed in their normal anatomic position with the pexy site on the left body side, thereby reducing the risk of intestinal entrapment.

Possible complications of colopexy include infection/coelomitis perforation or suture penetration of the lumen, fistula formation, and stricture.

Environmental Changes

Environmental factors need to be addressed to help prevent recurrence of chronic cloacal prolapse in cockatoos, as the vast majority of patients are exhibiting excessive reproductive behaviors (e.g., hypersexual, chronic egg laying) or have idiopathic prolapse with environmental/behavioral risk factors. Information regarding the diet, housing, reproductive status, exposure to other birds, acquisition, and behavior at home should be obtained from the owner.[5]

Environmental changes and enrichment should include dietary improvement (e.g., conversion to a formulated diet, limiting high-calorie foods, switching to meal feeding if obese), increasing exercise, and encouraging foraging.

If a reproductive cause is identified or confirmed, special attention must be placed on at-home reduction of hormonal stimulation and discouragement of reproductive behavior. This can include reduced photoperiod, removal of nesting material, removal or preventing access to actual or perceived nesting sites, separation from other birds in the home, removal of toys stimulating masturbation or other reproductive behaviors, and dietary changes to reduce high fat, warm, or mushy foods (mimics courtship feeding).[5,12] The owner-bird bond should be evaluated for behavior contributing to reproductive activity/hormonal stimulation, such as petting below the head/neck (mimics mating), kissing (mimics courtship feeding), and cuddling (mimics nesting).

Applied behavioral analysis should be utilized to identify and correct undesirable behaviors that may be contributing to prolapsing, or the prolapsing itself if intermittent and the patient appears to have some control.

Behavioral Pharmacology

Behavioral pharmacology may be beneficial in some patients. Clomipramine hydrochloride has been anecdotally reported to contract the vent orifice and help prevent cloacal prolapse; however, these claims have not been validated in scientific studies.[20] Clomipramine is a tricyclic antidepressant and serotonin-norepinephrine reuptake inhibitor (SNRI) that has been used for the treatment of obsessive-compulsive disorders in humans.[21] It was evaluated in a double-blind, placebo-controlled trial in cockatoos and was showed to reduce feather picking.[22] Pharmacokinetic studies in avian species are lacking, and dosing is based on extrapolation or progressive titration from low doses. Side effects have been reported and are typically mild in birds including drowsiness, regurgitation, transient ataxia, and suspected hallucinations (from paradoxical anxiety). More severe extrapyramidal side effects were reported in a blue and gold macaw treated concurrently with haloperidol.[21] The benefit of clomipramine or other behavioral modification medications in cockatoos with chronic cloacal prolapse is unknown.

Prognosis

Management of chronic cloacal prolapse in cockatoos requires a dedicated owner. Full resolution can be difficult to achieve and often requires a combination of medical, surgical, environmental, and behavioral interventions. Even with surgical management, recurrence is possible, and repeated surgical procedures may be necessary.

References

1. Graham JE, Tell LA, Lamm MG, Lowenstine LJ. Megacloaca in a Moluccan cockatoo (*Cacatua moluccensis*). *J Avian Med Surg.* 2004;18;41-49.

2. Sanmeier P. Anatomy and physiology. In: Chitty J, Monks D, eds. *BSAVA Manual of Avian Practice: A Foundation Manual.* Gloucester, UK: British Small Animal Veterinary Association; 2018;14-34.

3. Taylor WM. Clinical significance of the avian cloaca: interrelationships with the kidneys and the hindgut. In: Speer BL, ed. *Current Therapy in Avian Medicine and Surgery.* St Louis, MO: Elsevier; 2016:329-344.

4. Gill KS, Helmer PJ. Clocal diseases in companion parrots: a retrospective study of 43 cases (2012-2018). *J Avian Med Surg.* 2020;34(4):364-370.

5. Lennox A, Lee A. Cloacal, uterine, and rectal prolapse. In: Chitty J, Monks D, eds. *BSAVA Manual of Avian Practice: A Foundation Manual.* Gloucester, UK: British Small Animal Veterinary Association. 2018; 358-365.

6. Zaheer O, Beaufrere H, Jajou S, Gardhouse S. Asymmetrical cloacoplasty for the treatment of chronic cloacal prolapse in Psittaciformes: a case series. *J Avian Med Surg.* 2020;34(2):172-180.

7. van Zeeland YRA, Schoemaker NJ, van Sluijs FK. Incisional colopexy for treatment of chronic, recurrent colocloacal prolapse in a sulphur-credted cockatoo (*Cacatua galerita*). *Vet Surg.* 2014;43:882-887.

8. Detweiler DA, Carpenter JW, Kraft SL et al. Radiographic diagnosis: avian cloacal adenocarcinoma. *Vet Radiol Ultrasound.* 2000;41(6):539-541.

9. Radlinsky MG, Carpenter JW, Mison MB, Armbrust LJ. Colonic entrapment after cloacopexy in two psittacine birds. *J Avian Med Surg.* 2004;18(3):175-182.

10. Echols MS. Basic surgery. In: Chitty J, Monks D, eds. *BSAVA Manual of Avian Practice: A Foundation Manual.* Gloucester, UK: British Small Animal Veterinary Association. 2018;242-268.

11. Beaufrere H, Taylor WM. Case series: long-term management in five umbrella cockatoos with idiopathic cloacal prolapse syndrome. *Proc Annu Conf Eur Assoc Avian Vet.* 2009:163-169.

12. van Zeeland YRA, Friedman SG, Bergman L. Behavior. In: Speer BL, ed. *Current Therapy in Avian Medicine and Surgery.* St Louis, MO: Elsevier; 2016:177-251.

13. Lee A, Lennox A, Reavill D. Diseases of the cloaca: a review of 712 cloacal biopsies. *Proc Annu Conf Assoc Avian Vet.* 2014:51-60.

14. Dutton TAG, Forbes NA, Carrasco DC. Cloacal prolapse in raptors: review of 16 cases. *J Avian Med Surg.* 2016;30(2):133-140.

15. Steinagel A, Quesenberry K, Donovan T. Vegetative endocarditis due to *Staphylococcus aureus* in an umbrella cockatoo (*Cacatua alba*). *J Avian Med Surg.* 2019;33(4):419-426.

16. Taylor M. Examining the avian cloaca using saline infusion cloacoscopy. *Exot DVM.* 2001;3(3):77-79.

17. Rubin JA, Runge JJ, Mison M, et al. Surgery. In: Speer BL, ed. *Current Therapy in Avian Medicine and Surgery.* St Louis, MO: Elsevier; 2016:631-667.

18. Gartrell BD, Morgan KJ, Hoe L, et al. Cloacal papillomatosis in the absence herpesvirus and papillomavirus in a sulphur-crested cockatoo (*Cacatua galerita*). *N Z Vet J.* 2009;57(4):241-243.

19. Curtiss JB, Leone AM, Wellehan JFX, et al. Renal and cloacal cryptosporidiosis (*Cryptosporidium* avian genotype V) in a Major Mitchell's cockatoo (*Lophochroa leadbeteri*). *J Zoo Wildl Med.* 2016;46(4): 934-935.

20. Bowles H, Lichtenberger M, Lennox A. Emergency and critical care of pet birds. *Vet Clin North Am Exot Anim Pract.* 2007;10:345-394.

21. Starkey SR, Morrisey JK, Hickman HD, et al. Extrapyramidal side effects in a blue and gold macaw (*Ara ararauna*) treated with haloperidol and clomipramine. *J Avian Med Surg.* 2008;22(3):234-239.

22. Seibert LM, Crowell-Davis SL, Wilson GH, Ritchie BW. Placebo-controlled clomipramine trial for the treatment of feather picking disorder in cockatoos. *J Am Anim Hosp Assoc.* 2004;40(4):261-269.

How to Perform a Salpingohysterectomy

Cheryl B. Greenacre, DVM, DABVP (Avian Practice),
DABVP (Exotic Companion Mammal Practice)

Session #3343

Affiliation: From College of Veterinary Medicine, University of Tennessee, Knoxville, TN 37996, USA (Emeritus). Current address: Office of Research, University Research Animal Resources, University of Georgia, 501 D.W. Brooks Drive, Room 3504, Athens, GA 30602, USA.

Abstract: The Association of Avian Veterinarians routinely offers a high-quality instructional hands-on laboratory on coelomic surgery of birds that includes instruction on how to perform a salpingohysterectomy. For various reasons, including cost, popularity of the laboratory with limited space, or timing, not everyone who has an interest in learning this topic can benefit from the laboratory. Although hands-on training is the best method to learn a surgical procedure, the following description provides step-by-step instruction of how to perform a salpingohysterectomy in various sizes and species of birds. A description of common clinical signs and diagnosis of reproductive disease is included as well as pathology. It is recommended that these skills be practiced on a cadaver and enroll in laboratories or consult a mentor for perfecting your technique before performing the surgery, including learning microsurgical techniques for delicate tissue handling.

Introduction

Salpingohysterectomy (SHE) refers to the surgical removal of the entire oviduct and uterus (i.e., shell gland). Female birds commonly present for reproductive disease necessitating medical or surgical intervention. If oviductal disease cannot be medically managed, then removal of the oviduct and uterus can be performed. A SHE can be performed to manage dystocia (i.e., egg binding), egg impaction, cystic right oviduct, chronic egg laying, chronic infection, and neoplasia. The following is a description of important anatomic landmarks, indication for the need for a salpingohysterectomy and step-by-step instruction on how to perform a salpingohysterectomy. There are many resources available to clinicians including published articles, webinars, websites, and a newly published book, *Surgery of Exotic Animals*.[1-7]

Anatomy

The gonads of birds are intracoelomic, therefore, to determine the gender of a bird, surgical sexing, blood testing, or observation of external or behavioral characteristics must be performed. Most parrots are not obviously sexually dimorphic. To determine the gender of a parrot, surgical sexing or blood sexing must be performed. An exception would be the eclectus parrot that at a few days of age the female will grow red feathers and the male will grow green feathers. Surgical sexing involves visualizing the gonads and reproductive tract via a rigid endoscope placed into the abdominal or caudal thoracic air sac. Blood sexing involves evaluating 0.2 ml of blood via PCR that tests for a heterogamete (female is ZW) or homogamete (male is ZZ).

The female reproductive tract consists of an infundibulum, magnum, isthmus, uterus (i.e., shell gland) and very short vagina.[6] The ovary is not typically surgically removed from birds due to its very close attachment via the mesovarium to the aorta and dorsal body wall making it prone to fatal hemorrhage. Therefore, a salpingohysterectomy is performed in birds. Usually this involves only removing the left oviduct since the right oviduct is rudimentary in most birds and does not develop. Occasionally, birds can develop a cystic right oviduct, which is most common in pet chickens but can occur in psittacine birds as well, and would also need to be removed.[1-7]

The oviduct is suspended within the dorsal part of the intestinal peritoneal cavity by the dorsal and ventral suspensory ligaments and extends from the cranial division of the left kidney to the cloaca. Blood supply to the oviduct consists of the cranial, accessory cranial, middle, and caudal, and oviductal arteries. These vessels are usually easily visible within the suspensory ligament.[4-6]

Before the Procedure

The bird should be in the best health possible for the surgical procedure. Correct any underlying problems (e.g., such as dehydration, infection, inflammation, negative energy balance). Decrease the size of the reproductive tissues several weeks prior to surgery, if possible, by administering leuprolide acetate or deslorelin. A chicken, which is a food animal, cannot be given hormones in the United States because they are a prohibited drug in any food-animal species (e.g., chicken or turkey), regardless of their use.

Once the patient is in the best possible health, consider premedication with an opiate pain reliever such as butorphanol about 15 minutes or more before anesthetic induction and midazolam or other injectable anesthetic or sedative to relax the patient and use less anesthetic gases.[2,3] Begin mask induction with isoflurane or sevoflurane. An uncuffed endotracheal tube is placed after all monitoring equipment is placed (e.g., Doppler, ECG, pulse oximeter, respiratory monitor, blood pressure cuff). Ideally, an intravenous or intraosseous catheter is placed and fluids are given. Thermal support is important. A heated surgical table can be used as well as a forced heated air blanket. For larger birds, a thermometer can be used in the cloaca.

Procedure

Preparation

The entire ventrum and/or left side of the coelom is plucked and the edges of the proposed sterile area are taped off with white tape to keep the feathers out of the surgical site. The area is aseptically prepared with chlorhexidine scrub and sterile saline (alcohol can drop the body temperature precipitously). For most psittacine birds, a left lateral incision is preferred that extends from 1 cm cranial to the left pubic (pin) bone cranially to about 1 cm dorsal to the uncinate process of the last two ribs. For large birds such as poultry a ventral midlne incision following the linea ventrally is preferred.[4,5] Some perform a T-shaped incision by making a ventral midline incision and then a 90-degree angle to the left. Castro-Viejo or eyelid retractors, or a Lone Star retractor can be used for better visualization. The author uses bipolar radiocautery with a Harrison tip to cut through the skin and the coelom wall musculature. Long, balanced, microsurgical instruments (ring-tip forceps, needle holders, scissors) are used for delicate manipulation of fine tissues, and good quality 5x magnifying loupes. Purchase good-quality tools for this job. The uterus is identified as a whitish-pink, longitudinally striated tissue that can be gently retracted with ring-tip forceps and followed cranially to the flimsy cranial end of the salpinx near the left ovary. Bipolar radiocautery is used to cut and coagulate through the fascia and use small to medium surgical clips to ligate vessels before cutting (usually two clips stay with the bird and one clip goes with the tissue being removed). Work your way from cranial to caudal ligating and cutting vessels until the juncture of the uterus to the cloaca is reached where it is ligated with 5-0 or appropriate sized polydioxanone suture depending on the size of the bird.[4,5] The same suture is used to close the coelomic musculature in a simple continuous pattern and the skin is also closed with the same suture in a continuous Ford-interlocking pattern. During recovery from anesthesia, the patient is given meloxicam, and then repeated butorphanol injections as needed approximately every 4 hours. There are no pain scales developed for birds. The patient usually eats within a few hours of surgery. Care is taken not to give fluids subcutaneously near the surgical site, so they are given in the right inguinal region post operatively. Sutures are removed in about 14 days.

Reproductive Diseases

Egg-related coelomitis or peritonitis

Egg-related coelomitis or peritonitis is acute or chronic inflammation or infection involving all or part of the reproductive structures. It can be septic or not. It is common for chickens to have some degree of egg-related coelomitis or peritonitis, and mild cases are commonly encountered at necropsy in production hens. Generally, chickens tolerate mild coelomitis or peritonitis better than parrots. Causes usually involve retrograde flow of shelled or shell-less eggs from the oviduct back into the coelomic cavity due to oviductal bacterial infection, oviductal impaction, or abnormal confirmation of the oviduct. Heavy production hens or those with inadequate calcium in their diet can have calcium depletion and uterine inertia leading to retrograde flow of egg material. Bacterial infections commonly involve *E. coli* migrating up the oviduct from the vent.[1]

Clinical signs of egg-related coelomitis or peritonitis can be obvious (e.g., a sudden drop or cessation in egg laying) or can be vague and seem unrelated to the reproductive tract (e.g., lethargy, partial anorexia, weight loss, and lameness). Physical examination findings can include a thin bird, increased respiratory and heart rates, crackles of lower respiratory tract, a large doughy abdominal region, fluctuant coelomic fluid, and a lameness not explained by other obvious reasons.

Diagnosis is based on typical clinical signs, ultrasound, and radiographs. Endoscopy is possible but may be hampered or risky due to coelomic fluid and limited air sac space. Radiographically, a large variety of findings can be encountered such as the ground glass appearance of coelomic fluid, multiple or single-shelled or shell-less eggs inside or outside the oviduct, hernia or expanded coelomic wall, thickened air sacs, and masses in the caudal thoracic and abdominal air sac area. An ultrasound is helpful in determining if the egg material is within or outside the oviduct. A CBC and chemistry profile including total and ionized calcium can be performed to determine the degree of infection or inflammation, amount of dehydration, if there is concurrent liver or kidney disease, or if calcium depletion is present.

Treatment can involve fluids, antibiotics, non-steroidal anti-inflammatory drugs (NSAID), butorphanol, and/or surgery. Sometimes it is difficult to ascertain whether surgery will provide a better outcome than antibiotics and pain relievers alone. Generally, the more severe the egg-related coelomitis or peritonitis then the more likely surgery will provide a better outcome. A few clients chose no surgery for their hens with what appeared to be a moderate egg-related coelomitis or peritonitis (i.e., lameness, large fluctuant coelomic cavity, cessation of egg laying, but no shelled eggs in the coelomic cavity) and it cleared up with just oral antibiotics and meloxicam. Oviductal impaction should be resolved with surgery to remove the egg material before it adheres to the oviduct. Parenteral calcium is given for oviductal inertia due to calcium depletion. There are minimal to no adverse effects from administering one calcium injection after drawing blood to perform an ionized calcium while you are waiting for the results.

Egg binding or dystocia

Egg binding describes a condition in which a fully formed egg in the uterus cannot be laid. There are many causes for this condition including too large of an egg or a misshapen egg, inability of uterus to contract sufficiently from low blood calcium, or other causes of uterine paralysis. Clinical signs usually include straining to lay the egg (which sometimes owners think is straining to defecate), dyspnea, fluffed appearance, anorexia and sitting at the bottom of the cage. Diagnosis can be made on history, palpation on physical examination (including digital cloacal examination), ultrasound, and radiographs showing an egg or eggs. Radiographs can be taken without anesthesia in a cardboard box because positioning is not necessary to determine the presence of a calcified egg. It cannot be radiographically determined if the egg is in the uterus or out loose in the coelomic cavity. In larger birds, an ultrasound may help to determine this. Total and ionized calcium are helpful, as is a CBC and chemistry profile, but egg binding is a true emergency and injectable calcium, subcutaneous fluids with or without 2.5% dextrose, and a warm incubator are needed immediately. The primary cause of egg binding is usually hypocalcemia from insufficient calcium in the diet, such as a seed diet. Oxytocin, or if available, vasotocin, or prostaglandin E2 alpha can be given, but only if the egg is deemed able to come out.[2,3] If the egg is too large or irregular to come out naturally, or if the egg is putting pressure on the kidneys and has progressed to shock, then the egg should be imploded via ovocentesis as soon as possible. Implosion involves creating a negative pressure within the egg to collapse the egg in on itself, not crushing the egg. Prevention of egg laying in the future involves removing all sexual stimuli such as mirrors, decreasing daylight (10 hours of light and 14 hours of dark daily), and removing the presence of a male. Leuprolide acetate, a GnRH agonist, can also be given monthly to non-food animal species, or a salpingohysterectomy can be performed.

References

1. Saif YM. *Diseases of Poultry.* 12th Ed. Ames, IA: Blackwell Publishing; 2008.

2. Crespo R, Petritz OA. Backyard poultry, gamebirds, and waterfowl. In: Carpenter JW, Harms CA. *Carpenter's Exotic Animal Formulary.* 6th Ed. St. Louis, MO: Elsevier; 2023:444-495.

3. Guzman DS, Beaufrere H, Welle KR, et al. Birds. In: Carpenter JW, Harms CA. *Carpenter's Exotic Animal Formulary.* 6th Ed. St. Louis, MO: Elsevier; 2023:222-443.

4. Mehler SJ, Bennett RA. Avian reproductive procedures. In: Bennett RA, Pye GW. *Surgery of Exotic Animals.* Hoboken, NJ: Wiley-Blackwell; 2022:163-174.

5. Bennett RA, Harrison GW. Soft tissue surgery. In: Ritchie BW, Harrison GW, Harrison LR. *Avian Medicine: Principles and Application.* Lake Worth, FL: Wingers Publishing; 1996:1096-1136.

6. Joyner KL. Theriogenology. In: Ritchie BW, Harrison GW, Harrison LR. *Avian Medicine: Principles and Application.* Lake Worth, FL: Wingers Publishing; 1996:748-804.

7. Espinosa RA. Cystic right oviduct in poultry. Merck Veterinary Manual. Available at: https://www.merckvetmanual.com/poultry/disorders-of-the-reproductive-system/cystic-right-oviduct-in-poultry?query=cystic%20right%20oviduct%20poultry. Accessed April 4, 2023.

Egg Binding in Small Psittacine Species: Tips, Tricks, Treatment and Outcomes

Ariana Finkelstein, DVM, and
Laurel Harris, DVM

Session #3455

Affiliation: From Sebastian Animal and Bird Hospital, 948 US 1, Sebastian, FL 32958, USA (Finkelstein), Wasatch Exotic Pet Care, 892 Fort Union Blvd, Cottonwood Heights, UT 84121, USA (Harris).

Abstract: This presentation covers different presentations of egg binding in small psittacine species (e.g., budgerigar, cockatiel). It addresses the complexity of the egg binding process and recommended treatments. Pathophysiology and nutrition are addressed. Outcomes and possible scenarios are discussed. Discussions with the pet owner on reasonable expectations also are addressed. This is more of a how-to or in-our-experience discussion than a scientific presentation. Questions are addressed during the lecture. Videos are provided.

Introduction

Egg binding can present differently in small psittacine birds. Egg binding is a complex process. Pathophysiology (e.g., hypothalamus-pituitary-ovarian axis [HPG] is upregulated to have increased estrogenic activity), nutrition (e.g., excessive food, especially high-fat food sources), and behavior play a large part in egg binding cases.

Egg Binding

New to a practice, small birds often present on an emergency basis. They are fluffed in the bottom of the cage, not eating or drinking and often not able to pass urates and feces. Often the owner is unaware of how long their pet has been in distress.

On presentation, these patients are fragile. They should be managed as quickly and handled as minimally as possible. Sometimes a quick weight, subcutaneous fluids, brief examination and placement in an incubator is the starting point. A good history from the owner and discussion of all potential outcomes and scenarios can be done while the bird rests. Once this is accomplished, pain management (e.g., premedication with meloxicam or gabapentin, especially if the bird is in shock) is an important next step. Opioids can drop blood pressure too. It is important to remember that unless there is oviductal pathology, the painfulness of this condition is temporary. Sedation (butorphanol ± midazolam), lube around the cloaca and potentially a calcium injection can be performed as the next step. After some time, gentle manipulation may be attempted if the egg has not passed. If at this stage there is still no relief, oxygenation, radiology, and/or ultrasound should be the next step to potentially guide you in the manipulation process. Oxytocin or prostaglandin E2 can be considered but they are not as effective in avian patients as in mammalian patients. If the egg is still not progressing, general anesthesia to attempt oviposition and manipulation from the vent is the next step. With a technician watching the patient's breathing closely, place one or two fingers above the egg and apply constant resistance while waiting for the cloaca to stretch. Active pushing or manipulation is not required, just firm constant pressure.

In the event that the cloaca is not still functionally able to oviposit normally or the bird has hormonally cycled out of the laying phase (meaning an old egg,), then consideration is the avian equivalent of a closed mammalian cervix and you have to collapse the egg. These eggs are sometimes also adhered to the oviduct and are always more problematic.

In the event that this fails consideration for ovocentesis, surgical removal may be the next step. The potential for complications from both of these different treatments would need to be addressed and discussed at this time.

Once the bird is stable and the egg has been removed, nutrition, proper light cycle, and behavior modification will all be necessary to help downregulate the HPG axis to decrease estrogenic activity and potentially prevent future occurrences.

Excessive egg producers and small psittacine birds with egg binding issues are among the scariest presentations that require triage on a seasonal basis. Management of these cases provides unique challenges.

The Avian Uropygial Gland in Health and Disease

Lauren V. Powers, DVM, DABVP (Avian Practice),
DABVP (Exotic Companion Mammal Practice)

Session #3437

Affiliation: From Avian and Exotic Pet Service, Carolina Veterinary Specialists, 12117 Statesville Road, Huntersville, NC 28078, USA.

Abstract: The uropygial (preen) gland is a unique structure to birds and is easily overlooked on the routine physical examination. Uropygial gland disease is relatively uncommon in psittacine birds and can be challenging to manage. The most common conditions of the uropygial gland in psittacine birds are impaction and neoplasia. Surgery is often curative for advanced or aggressive disease but can pose significant surgical challenges.

Introduction

The uropygial (preen) gland is a unique anatomic structure to birds. The gland is situated on the dorsal rump and produces an oily to waxy substance that is placed on the feathers by the bird's beak during autogrooming. The gland serves to maintain feather condition and may have additional purposes in birds.

Anatomy

The skin in birds is glandless apart from the uropygial gland, glands of the ear canal, and pericloacal glands. The uropygial gland is the only compact gland in the avian integument.[1,2]

Presence and size

The gland is generally larger in aquatic birds compared to terrestrial species.[2-4] Some birds lack a uropygial gland. The uropygial gland is absent in the following avian species.

Large flightless birds

The uropygial gland is absent in the ostrich (*Struthio camelus*), emu (*Dromaius novaehollandiae*), rhea (e.g., *Rhea americana*), and Southern cassowary (*Casuarius casuarius*).[2]

Columbiformes

The uropygial gland is absent or naked (lacking tuft feathers) in columbiform birds.[2,5] In the rock dove (*Columba livia*), the gland is present but indistinctly bilobed and the papilla is large and naked.[2] The gland is absent in crowned pigeons (e.g., *Goura victoria*).[2]

Psittaciformes

The uropygial gland is absent in all Amazon parrots and pionus parrots examined to date, non-*Ara* blue macaws (hyacinth macaw [*Anodorhynchus hyacinthinus*], Lear's macaw [*Anodorhynchus leari*], Spix's macaw [*Cyanopsitta spixii*]) and some *Brotegeris* species examined to date (e.g., grey cheeked parakeet [*Brotogeris pyrrhopterus*]).[2]

Topography and macroscopic anatomy

The uropygial gland is generally situated dorsal to the last sacral and first caudal vertebrae but is occasionally found more caudally such as over the pygostyle.[2,6] The gland lies dorsal and medial to the levator caudae and lateralis caudae muscles of the tail.[2,6] It is embedded in a cushion of fat and has deep fibrous attachments between the ventral capsule and the connective tissue covering the rectrice follicles and caudal vertebrae.[1] The gland is covered by skin that is often feathered.[2,7]

The uropygial gland is bilobed in nearly all avian species examined to date and has a characteristic heart or pear shape.[1,8] The two lobes are divided by the interlobular septum. In the chicken and all psittacine species evaluated in one report,[1] there is one duct for each of the two lobes with each excretory duct exiting through a nipplelike papilla to a single external orifice on each side; however, some avian species (generally those with no primary storage cavity) have up to 18 orifices.[1,2] The papilla itself is normally bare apart from a small tuft of feathers at the tip, termed the circulus uropygialis or uropygial wick.[1,2,8] These feathers are either modified down or modified semiplume feathers.[2] The number and appearance of these feathers varies widely across avian species.[2]

The blood supply to the uropygial gland arises from the median caudal (coccygeal) artery,[1,6] which gives rise to four or more paired caudal segmental arteries.[6] One of these caudal segmental arteries is a specialized uropygial gland artery that then divides into two or more branches that vascularize the gland. Branches from adjacent caudal segmental arteries may also feed the gland while other branches feed the lateral caudae and levator caudae muscles.[6] There may be anastomoses between the branches of the uropygial gland along the dorsal surface.[6] The gland lacks striated muscle of its own, and movement of the gland is through movement of the tail or rectrices.[1] The glandular uropygialis nerve arises from the caudalis dorsalis nerve that also innervates the skeletal muscles of the tail.[1]

Microscopic anatomy

The gland is covered by a dense external connective tissue capsule comprised of collagenous, non-elastic fibers and may contain some smooth muscle fibers.[1,6] The interlobar septum is comprised of similar connective tissue but may also possess elastic fibers.[1,6] The septum is perforated in some avian species (e.g., many passerine species).[1] The capsule can be deeply pigmented with melanin granules in some psittacine species.[1]

The inner aspect of the gland is comprised of glandular tubules where secretions are generated and various cavities that store the secretions.[1] The presence of storage cavities is highly variable, even in psittacine species.[1,8] For example, a larger central primary cavity is present in the sulfur-crested cockatoo (*Cacatua sulphurea*) but not in the cockatiel (*Nymphicus hollandicus*).[1] Although many avian species with a compact-type papilla possess smooth muscle layers in the papilla, the cockatiel was the only psittacine bird in which discrete muscle fibers were detected in the papilla in one report.[1]

Secretions

The uropygial gland is a compound tubuloalveolar holocrine gland with a lipoid sebaceous secretion containing cell fragments and secretory granules.[1,3,8] Chemical substances found in the secretions include lipids, fatty acids (both volatile and non-volatile), aliphatic monoester waxes, and long-chain monohydroxy wax-alcohols. Secretions also contain 7-dehydrocholesterol, a vitamin D_3 precursor.[1,9]

Flora

The normal flora of the uropygial gland has been investigated in several avian species but no psittacine species.[10] For example, Enterococcaceae, Clostridiales Family X1, Clostridiaceae 1, Comamonadaceae, Enterobacteriaceae, and Micrococcaceae were identified in the preen glands of free-living house sparrows (*Passer domesticus*) in one report.[10]

Function

It is widely accepted that the uropygial gland is involved with maintaining feather condition in those species that possess them; however, the understanding of the full functions of the gland is still in its infancy.[2,8,11,12] Postulated functions of the uropygial gland include those listed below.

Waterproofing and in-flight drag reduction

A waterproofing function of the uropygial gland is well-supported in the literature but the exact mechanism remains uncertain.[2,12,13] Postulated mechanisms include direct waterproofing by providing plumage with a covering layer of lipids and indirect waterproofing through the general maintenance of feather structure.[2]

Antimicrobiosis

Preen oil contains compounds that inhibit bacterial and fungal growth and has been shown to reduce growth of keratinase-containing, feather-degrading bacteria such as *Bacillus licheniformis*.[2,11,12,14]

Intraspecific communication and predator deterrence

Preen gland oil shows some level of hormonal control and exhibits individual, sex, and age-specific patterns. Support for the hypothesis that the uropygial gland produces pheromones is mixed.[2,12,15] An intact olfactory system appears to be critical to courtship and mating in mallard ducks (*Anas platyrhynchos*).[2]

Feather coloration

Preen gland oil likely does not significantly alter coloration or ultraviolet light reflectance in birds but the evidence is mixed.[2] For example, preen gland oil in the greater flamingo (*Phoenicopterus roseus*) contains carotenoids which make the pink plumage more intense.[2]

Vitamin D_3 synthesis

Preen gland oil contains 7-dehydrocholesterol, a vitamin D_3 precursor; however, evidence of a definitive role in vitamin D_3 synthesis through exposure to ultraviolet light is dated and mixed. The presence of the gland was not essential for calcium metabolism altered by rickets in birds.[9,12]

Clinical Evaluation

Physical examination

The uropygial gland should always be visually inspected and palpated during the physical examination in species that possess one.[7,16] A diseased preen gland is easy to miss during the exam. The gland is comparatively large in grey parrots (*Psittacus erithacus*) and is often larger in juvenile birds compared to adults.[17] This large size may be mistaken for impaction or infection.[17] Routine expression of the gland during the exam can be painful and injurious and is generally not advised.

Evidence for uropygial gland disease includes signs of localized pain or discomfort during autopreening or the physical examination; excessive enlargement or cystic protrusion of one or both lobes; asymmetry of the lobes; crust, ulcerated skin, or masses on or around the gland; feather loss or damage on or around the gland; automutilation of the gland and/or surrounding tissues; presence of blood on the surrounding feathers or beak, and abnormal (e.g., hemorrhagic) secretions from the duct(s).[18] Many birds with preen gland disease do not display overt signs of pain or discomfort and disease.

Clinical pathology and imaging

While routine diagnostic testing such as hematology and clinical biochemistry are generally not specifically diagnostic for diseases of the uropygial gland, clinical pathology evaluation is important in assessing the overall health of the bird, particularly in the presurgical patient.

The diagnostic strength of routine radiography is generally considered to be low for preen gland diseases although the uropygial gland can often be observed on the lateral-lateral radiographic projection. Computed tomography can be useful to evaluate the uropygial gland and evaluate for evidence of distant metastatic disease in cases where neoplasia is confirmed or suspected.[4]

Microbiologic evaluation

The microbiologic flora of secretions from the uropygial gland and excised tissue specimens can be evaluated by cytology, Gram stain and other biological stains, culture and sensitivity, PCR, and next generation sequencing.

Cytology and histopathology

Cytologic evaluation can be performed on uropygial gland secretions, impression smears from ulcerated masses, and fine needle aspirates of solid masses. Masses within or around the uropygial gland can be carefully biopsied and tissue samples submitted for histopathology. Caution is warranted when collecting biopsy samples that involve the papilla. Excised masses and in toto uropygial glands should also be evaluated by histopathology.[7,20] Masses can be biopsied prior to more aggressive surgery such as mass excision or uropygialectomy. A small dermal biopsy punch can be used for sample collection.[7]

Diseases

Disease of the uropygial gland is not commonly diagnosed in companion birds, possibly in part due to the difficulty in visualizing the gland without handling and restraint and failure to inspect the gland during the routine physical examination. The most recognized clinical conditions of the preen gland are impaction and neoplasia.

Impaction

Impaction is often a sequela to squamous metaplasia and hyperkeratosis associated hypovitaminosis A and can also be secondary to uropygial gland adenitis.[4,7,20] Impaction of the uropygial gland can be unilateral or bilateral. Clinical signs include excessive enlargement of one or both lobes, cystic protrusions, evidence of localized pain or discomfort, and inability to manually express the gland. Caution is warranted with attempts to express the gland as the gland may rupture internally or externally with excessive force. In severe cases of impaction, the gland may rupture spontaneously[7] but this author has found this to be uncommon.

In one report, a king penguin (*Aptenodytes patagonicus*) had uropygial impaction and abscessation and died 6 months later with bacterial septicemia and myiasis of the preen gland.[4] In the same report, 21 gentoo penguins (*Pygoscelis papua*) had uropygial gland disease (all impactions) with an annual incidence from 2.6 to 11.5%. No uropygial gland disease was seen in macaroni penguins (*Eudyptes chrysolophus*) or rockhopper penguins (*Eudyptes chrysocome*).[4]

Neoplasia

The most reported neoplasms of the uropygial gland are squamous cell carcinoma and adenocarcinoma,[16,18,20] although other neoplasms have been reported including adenoma,[20] epithelioma,[19] and melanoma.[21,22] Squamous cell carcinoma often appears as areas of crust or ulceration and may progress to solid masses.[18] Neoplastic lesions generally advance or enlarge over time. Deeper tissues are frequently involved with more aggressive neoplasms.[18] Histopathology is necessary for a definitive diagnosis. Distant metastasis has been reported.[21]

Infection

Viral (e.g., poxvirus[4]), bacterial, and mycotic (e.g., *Candida albicans*[23]) causes of uropygial gland adenitis have been reported. Infected glands are often enlarged or thickened and can be abscessed.[4,16] Thick, pasty, fetid, or hemorrhagic glandular secretions may be present.[23]

Treatment

Conservative management

In most cases of simple impaction, conservative management is the preferred option. For birds that are not displaying signs of localized pain or discomfort, benign neglect with periodic monitoring is a reasonable option. Nutritional deficiencies (e.g., hypovitaminosis A) should be corrected. Warm compresses can be applied periodically to the gland if this is possible.[4,7] Careful attempts at manually expressing the gland can be attempted but caution is warranted. Applying direct pressure to the gland can be painful and risks rupturing the gland. Recurrence is common after manual expression.

Non-surgical treatment of impaction and infection

Cannulation (catheterization)

Cannulation of the uropygial gland duct is possible in larger birds.[24] Cannulation should be performed using appropriate local and/or systemic analgesia and sedation or general anesthesia.[24] Flushing with warm sterile saline by catheterization was ineffective in all cases of impaction in gentoo penguins in which it was attempted.[4]

Antimicrobial therapy

In cases of uropygial gland infection or abscessation, antibiotics and/or antifungal agents can be used. The selection of suitable antibiotics ideally is based on results of culture and sensitivity testing. Antibiotics are unlikely to resolve impactions when used alone.

Analgesic therapy

Appropriate analgesic therapy should be used in cases with evidence of localized pain or birds with conditions of the preen gland likely to be painful.

Non-surgical treatment of neoplasia

Cryotherapy

Cryotherapy can be attempted but it can be difficult to freeze tumors completely and recurrence is likely.[18] Injury to surrounding tissues is challenging to avoid. Cryotherapy of the uropygial gland is not advised.

Radiation therapy

The successful use of Strontium-90 (S-90) β-ray plesiotherapy for the treatment of uropygial gland neoplasia has been reported in birds including cockatiels and budgerigars (*Melopsittacus undulatus*).[20,25] The depth of penetration is shallow (often cited as 5 millimeters or less)[18,25] and so larger masses should be debulked prior to treatment. The gland may need to be divided into sections for treatment in larger birds. Single and repeated fractions of 100 Gray (Gy) have been reported.[20,25]

Chemotherapy, electrochemotherapy

There are no reports of the use of systemic chemotherapy alone for the treatment of uropygial gland neoplasia in birds. The successful use of electrochemotherapy has been reported in birds.[26,27] Electrochemotherapy consists of intratumoral application of a chemotherapeutic agent such as cisplastin, carboplatin, or bleomycin followed by electroporation through localized treatment with electrical pulses with either plate or needle electrodes with the patient under general anesthesia. Two or more treatment sessions are often needed for a successful outcome.

Immunotherapy

Immunotherapy is a treatment for canine melanomas that utilizes the human tyrosinase enzyme to stimulate the host immune system against tyrosinase enzyme-producing cells.[23] Immunotherapy with a canine melanoma vaccine (Oncept, Merial, Duluth, GA, USA) was reported in a black-footed penguin (*Spheniscus demersus*) with melanoma of the uropygial gland.[23] The penguin was euthanized after the second dose and had extensive metastatic disease.[23]

Surgical treatment

Surgery should be considered for cases of uropygial gland neoplasia and for impaction, infection, or abscessation if the impaction recurs, if there is a chronic non-responsive infection, or if the gland has ruptured. Surgical options for impaction and infection include surgical debridement and partial or complete uropygialectomy. Smaller isolated neoplastic lesions can often be completely excised with primary closure or healing by second intention.[16,18] Larger, more aggressive masses and those involving the papilla may require a partial or complete uropygialectomy. Very large or aggressive masses may require amputation of the tail. Potential surgical complications include hemorrhage, incision dehiscence, infection, incomplete surgical margins, and vascular metastasis.[20,27] Birds should be monitored postoperatively for picking at the surgical site and a neck collar or brace should be considered if this occurs.

Surgical treatment options

Surgical debridement

An incision (or incisions) is made directly over the affected lobe(s) and the contents of the primary cavity are expressed or otherwise drained and the lumen is irrigated with warm sterile saline.[28] The incision(s) can be left open or closed routinely. Attempts can be made to flush and/or cannulate the duct(s) intraoperatively.[28] In gentoo penguins this approach resulted in fewer repeat episodes of impaction and was the recommended treatment for uropygial gland impactions in captive penguins.[4]

Complete uropygialectomy

Complete excision (extirpation) is indicated for advanced cases of impaction or infection that have not responded to medical management and with large and aggressive neoplasms.[19,27] Complete excision can be challenging and significant hemorrhage is possible. Radiosurgical equipment or a surgical (CO_2) laser,[30] should be considered for proper hemostasis. Magnification using surgical loupes or an operating microscope and proper illumination are strongly advised as the borders of the gland can be difficult to identify, particularly with chronic or advanced disease (Nemetz L, personal communication, March 2021). Local anesthesia with lidocaine or bupivacaine is recommended. The patient is placed in lateral or semi-ventral recumbency. A towel can be placed under the tail and the rectrices can be securely taped. A circumferential skin incision is made around the gland.[17,31] Or, alternatively, the skin over the dorsal aspect of the gland can be incised transversely and carefully dissected free from the underlying gland and reflected cranially. This technique may improve the likelihood of successful primary closure (Nemetz L, personal communication, April 2023). The soft tissues underlying the gland are then carefully undermined.[31] The gland has deep soft tissue facial attachments to the rectrice follicles and caudal vertebrae. Care must be taken not to damage the blood supply to the rectrice follicles.[17] Closure is routine.[17,31] If the remaining defect is too large to close by primary closure it can be allowed to heal by second intention. A hydroactive or hydrocolloid dressing can be used. Although a drain can be placed postoperatively,[17] drains are generally not necessary and may result in irritation or inflammation.[17]

Partial uropygialectomy

Partial uropygialectomy with physical removal or cauterization of remaining glandular tissue of one or both lobes can be considered for severe or persistent impactions or infections unresponsive to conservative treatment (Nemetz L, personal communication, March 2021). Partial excision avoids the risks of serious hemorrhage and damage to surrounding important tissues such as the rectrice follicles.[32,33] Partial excision is not recommended for large and aggressive neoplasms (Nemetz, personal communication, March 2021). As with complete excision, magnification and illumination are strongly advised and local anesthesia is recommended. For a unilateral approach, the dorsal wall and sides of the affected lobe are excised. For a bilateral approach, a circumferential skin incision is made that incorporates the papilla and the dorsal and lateral walls of both lobes. The remaining contents of the affected lobe(s) are evacuated. Deeper biopsy samples can be collected at this time. The remaining glandular walls are then manually debrided using a bone curette or similar device. The walls can be chemically cauterized using silver nitrate applicator sticks. Caution is warranted with the use of silver nitrate and the area should be copiously lavaged after application. Hemorrhage is controlled with direct pressure and/or chemical or thermal cauterization. The incision is left open and the area is packed with antibiotic ointment or cream and allowed to heal by second intention. Postoperative antibiotics and analgesics are recommended. Repeated treatment sessions are often required for complete resolution, particularly if removal or cauterization of remaining glandular tissue is unadequate. If repeated treatments are unsuccessful, complete excision can be then performed. The gland is often significantly smaller after partial excision and easier to fully excise (Nemetz L, personal communication, March 2021).

Tail amputation

Very large or aggressive neoplasms may require amputation of the tail. In one report of a 33-year-old male blue and gold macaw (*Ara ararauna*) with uropygial gland adenocarcinoma, a tail amputation consisting of transection of musculature to the vertebral column, disarticulation between the second and third caudal vertebrae, and routine closure was performed but the bird died of an undetermined cause approximately 4 months after surgery.[29]

Acknowledgments: The author extends deep appreciation and gratitude to Larry Nemetz, MS, DVM for advisement on clinical and surgical management of uropygial gland disease and for contributions to this manuscript.

Disclosure Statement: The author has no financial or commercial conflicts or interests to disclose.

References

1. Jacob J, Ziswiler V. The uropygial gland. In: Farner DS, King JR, Parkes KC, eds. *Avian Biology*. New York: Academic Press; 1982:199-324.

2. Johnston DW. Morphological atlas of the avian uropygial gland. *Bull Br Mus Nat Hist (Zool)*. 1988;54(5): 199-259.

3. Baumel JJ, Raikow RJ. Integumentum speciale. In: Baumel JJ, Kings AS, Breazile JE, et al, eds. *Handbook of Avian Anatomy: Nomina Anatomica Avium*. 2nd Ed. Cambridge, MA: Nuttall Ornithological Club; 1993:22-44.

4. Keeble EJ. The incidence and treatment of uropygial gland disease in a mixed species exhibit of captive penguins. *Proc Int Conf Dis Zoo Wild Anim*. 2007.

5. Chiale MC, Carril J, Montalti D, et al. The uropygial gland of the eared dove and its evolutionary history within the Columbiformes (Aves). *J Ornithol*. 2019;160:1171-1181.

6. Atalgin H, Kürtül İ. Arterial vascularization of the uropygial glands (gl. uropygialis) in the Japanese quail (*Coturnix coturnix japonica*) and silver polish (*Gallus gallus domesticus*). *Anat Histol Embryol*. 2008;37(3):177-180.

7. Abou-Zahr T. Avian dermatology. *Vet Clin North Am Exot Anim Pract*. 2023;26(2):327-346.

8. Carril J, Chiale MC, Barbeito CG. The uropygial gland of the monk parakeet *Myiopsitta monachus*: Histology, morphogenesis, and evolution within Psittaciformes (Aves). *Evol Dev*. 2020;22(3):269-282.

9. Salibian A, Montalti D. Physiological and biochemical aspects of the avian uropygial gland. *Braz J Biol*. 2009; 69: 437-446.

10. Grieves LA, Bottini CL, Gloor GB, et al. Uropygial gland microbiota differ between free-living and captive songbirds. *Sci Rep.* 2022;12(1):18283.

11. Fülöp A, Czirják GÁ, Pap PL, et al. Feather-degrading bacteria, uropygial gland size and feather quality in house sparrows *Passer domesticus*. *Ibis.* 2016;158(2):362-370.

12. Moreno-Rueda G. Preen oil and bird fitness: a critical review of the evidence. *Biol Rev.* 2017;92(4): 2131-2143.

13. Møller AP, Laursen K. Function of the uropygial gland in eiders (*Somateria mollissima*). *Avian Res.* 2019;10(1):1-6.

14. Bodawatta KH, Schierbech SK, Petersen NR, et al. Great tit (*Parus major*) uropygial gland microbiomes and their potential defensive roles. *Front Microbiol.* 2020;28;11:1735.

15. Galván I, Møller AP. Odor transmission and olfaction: the tuft of the uropygial gland and olfactory ability in birds. *The Condor.* 2013;115(4):693-699.

16. Howard L, Lamberski N, Newton AL. Uropygial gland inflammation, neoplasia and surgical resection in six northern carmine bee-eaters (*Merops nubicus nubicus*). *Proc Annu Conf Am Assoc Zoo Vet.* 2006.

17. Wissman MA. Diseases of the uropygial gland. Available at: http://www.exoticpetvet.net/avian/uropygial.html. Accessed April 9, 2023.

18. Quesenberry K. Neoplastic diseases in birds - treatment options. *Proc Conf Pacific Vet.* 2018.

19. Ambar N, Eshar D, Njaa BL. Intraluminal uropygial gland epithelioma in an African grey parrot (*Psittacus erithacus*). *J Avian Med Surg.* 2022;35(4):433-439.

20. Nemetz LP, Broome M. Strontium-90 therapy for uropygial neoplasia. *Proc Annu Conf Assoc Avian Vet.* 2004.

21. Reavill D. Pet bird oncology. *Proc Conf Assoc Avian Vet Australasian Comm Unusual Exot Pet.* 2011.

22. Steeil J, Schumacher J, Ramsay EC, et al. Application and limitation of advanced diagnostic and therapeutic modalities in zoologic species with malignant melanoma. *Proc Annu Conf Am Assoc Zoo Vet.* 2013.

23. Hines RS, Sharkey P, Friday RB. Itraconazole treatment of pulmonary, ocular, and uropygeal aspergillosis and candidiasis in birds – data from five clinical cases and controls. *Proc Annu Conf Am Assoc Zoo Vet.* 1990;364-370.

24. Faux CM, Logsdon ML. Impacted uropygial (preen) gland. In: Orsini JA, Grenager NS, de Lahunta A, eds. *Comparative Veterinary Anatomy: A Clinical Approach.* London, UK: Elsevier; 2022.

25. Pignon C, Azuma C, Mayer J, et al. Radiation therapy of uropygial gland carcinoma in psittacine species. *Proc Annu Conf Assoc Avian Vet.* 2011.

26. Racnik J, Svara T, Zadravec M, et al. Electrochemotherapy with bleomycin in the treatment of squamous cell carcinoma of the uropygial gland in a cockatiel (*Nymphicus hollandicus*). *J Exot Pet Med.* 2019;29: 217-221.

27. Hirata SB, Brunner CHM, Santos SV. Electrochemotherapy for the treatment of fibroma in red-billed curassow (*Crax blumenbachii*). *Proc Conf Am Assoc Zoo Vet, Eur Assoc Zoo Wildl Vet, Int Zoo Wildl Vet.* 2016.

28. Hochleithner M, Hochleithner C, Fuchs A. Surgical treatment of impacted uropygial glands. *ExoticDVM.* 1999;1(1):23-25.

29. Robertson J, Guzman DS, Sinnott D, et al. Modified tail amputation technique in a blue and gold macaw (*Ara ararauna*) with uropygial gland adenocarcinoma. *J Avian Med Surg.* 2020;34(1):57-64.

30. Van Sant F, Kureshi M. Successful use of a surgical (CO_2) laser in squamous cell carcinoma in companion psittacines. *Proc Symp Southwest Vet.* 2019.

31. Guzman DS. Avian soft tissue surgery. *Vet Clin North Am Exot Anim Pract.* 2016;19(1):133-157.

32. Jawad HS, Idris LH, Bakar ZB, et al. Partial ablation of uropygial gland effect on carcass characteristics of Akar Putra chicken. *Poult Sci.* 2016;95(8):1966-1971.

33. Hasan SAJ. Uropygialectomy method overview. *Dairy Vet Sci J.* 2017;2(4):555592.

Section 8

AAV Masterclasses Track 4

Brian Speer, DVM, DABVP (Avian Practice), DECZM (Avian); Anthony Pilny, DVM, DABVP (Avian Practice)
Session Chairs

Avian Bornavirus: A Global Perspective on Pathology, Current Diagnostics, Research and Treatment

Sharman Hoppes, DVM, DABVP (Avian Practice),
Michael Lierz, DZooMed, DipECZM (Wildlife Population Health), DipECPVS,
Helene Pendl, Dr med vet

Session #3173

Affiliation: From Texas Avian & Exotic Hospital, Grapevine, TX 76051, USA, Texas A&M University, College Station, TX 77843, USA (Hoppes), Clinic for Birds, Reptiles, Amphibians and Fish, Justus Liebig University, Giessen, Germany (Lierz), Pendl Lab, Zug, Switzerland (Pendl).

Abstract: Avian bornavirus (ABV) is a neurotropic virus affecting parrots and other species of birds worldwide including North America, South America, Europe, South Africa, Asia and Australia. ABV is the cause of proventricular dilation disease also known as avian ganglioneuritis. The diagnosis, treatment and management of this disease both in the individual pet bird and in flocks has been a source of frustration to veterinarians, aviculturists and owners for decades. This masterclass is an international collaboration discussing pathology, immunology, current research and treatment for birds exposed and infected with avian bornavirus. Current research on transmission and susceptivity is reviewed. Treatment options are discussed along with management and prevention protocols.

Introduction

Proventricular dilatation disease (PDD) or avian ganglioneuritis (AG) was first recognized in the late 1970s in macaws imported into North America and Europe.[1-12] Initially diagnosed in macaws (*Ara* spp.) it was later identified in other parrot species. It is also known as Macaw wasting disease, neuropathic ganglioneuritis, lymphoplasmacytic ganglioneuritis, infiltrative splanchnic neuropathy, and psittacine encephalomyelitis. The nomenclature of proventricular dilatation disease (PDD) and avian bornavirus (ABV) has evolved over the years. The recommended names now are psittacine bornavirus (PaBV) for infected psittacine birds and either avian ganglioneuritis or avian bornaviral ganglioneurits for the clinically diseased state or PDD bird. This presentation uses avian ganglioneuritis (AG). AG better defines the disease process and takes the focus off the proventriculus as other organ systems can be affected.

The disease has been identified in over 80 species of psittacine and non-psittacine birds. In parrots, African grey parrots (*Psittacus erithacus*), macaws (*Ara* species), Amazon parrots (*Amazona* species), and cockatoos (*Cacatua* species) are the most common species affected.[3] The disease is minimally represented in Quaker parrots (*Myiopsitta monachus*) and lovebird species (*Agapornis* species).[3] It has been reported in captive psittacine birds worldwide. AG/PDD has historically been considered an infectious disease based on its spread through aviaries. But the etiologic agent proved elusive for over thirty years. Then in 2008, two different research groups Honkavuori and colleagues and Kistler and colleagues detected a novel bornavirus in parrots with PDD.[13,14] During this same time, Ouyang and colleagues detected ABV Nucleoprotein (N-protein) in the brains of birds with AG/PDD.[15]

Bornaviruses

The bornaviruses are neurotropic, negative-encoded, single-stranded, nonsegmented RNA viruses and are members of the family Bornaviridae of the order Mononegavirales. The replication of and transcription of bornaviruses takes place in the cell nuclei of nerve cells. Currently, the family comprises the three genera Orthobornavirus, Carbovirus and Cultervirus. Due to the discovery of various new bornaviruses during the past 15 years, the taxonomy and nomenclature of the family has undergone several revisions.[16-18] Borna disease virus (BDV) causes encephalitis in horse, sheep, and other domestic mammals in Europe.[3] BDV has also been detected in wild mallards and corvids in Scandinavia, and ostriches in Israel.[4,5,19] Experimental infections of chickens showed that 1-day-old chicks inoculated intracerebrally with brain homogenates from rabbits with Borna disease virus developed paralysis of legs and wings.[20] There are currently two BDV genotypes. BoDV-1 infects mammals: horses, sheep and the white toothed shrew (its natural reservoir). BoVD-2 was isolated from a horse.[21] A new mammalian bornavirus, variegated squirrel bornavirus 1 (VSBV-1) has been associated with a fatal encephalitis in humans.[22]

Avian Bornavirus

Currently there are 15 genetically diverse Bornavirus genotypes comprising six distinct viral species identified within the genus Orthobornavirus, which can occur in avian species. Eight genotypes are found in parrot populations worldwide that share 91 to 100% of their nucleotide sequences within genotypes and 68 to 85% between genotypes.[17] The two species known to cause AG/PDD in psittacine birds are psittaciform bornavirus 1 and 2 and include PaBV genotypes 1-8. PaBV-2 and PaBV-4 are the most reported genotypes in parrots.[17]

It has been speculated that the diversity of bornavirus genotypes in captive parrots compared to free-living waterfowl is the result of the spread of bornaviruses across species during the unregulated housing and transport associated with wild captures decades ago.[23] Genotypes for bornaviruses in parrots do not appear to be clearly linked to geographic regions unlike bornaviral genotypes identified in waterfowl.[24] Avian bornavirus does not appear to be zoonotic.

Other known ABV's includes passeriform bornavirus 1 (CNBV-1,2,3 canary) and (MuBV-1 munia finch), passeriform bornavirus-2 (EsBV-1, estrildid finch bornavirus 1) and waterbird bornavirus 1 (ABBV-1, aquatic bird bornavirus-1).[17]

Epidemiology of Avian Bornavirus

Avian bornavirus infections have been described in a wide range of families within the order Psittaciformes. It is assumed that all psittacine species are susceptible to some degree of infection. The viruses can be found in captive psittacine populations worldwide.[1-12] This information on the distribution of ABV is based on sampling of birds in aviaries, veterinary clinics, and information gained by pathologists. Few actual prevalence studies have been published. In the most extensive study, Heffels-Redmann et al.[25] found 23% of 1442 captive psittacine birds from several European countries were positive for markers of avian bornavirus infection (viral RNA detection and/or bornavirus-reactive antibodies), suggesting a wide distribution of these viruses.[25] In captive psittacine birds, PaBV-2 and PaBV-4 are the most commonly reported genotypes.[3,16,21] In contrast, all other known psittacine bornaviruses (PaBV-1, PaBV-3 and PaBV-5 to PaBV-8), have been detected in less than ten birds each.[24]

Mixed infections with two psittacine bornaviruses, such as PaBV-2 and PaBV-4, have been described rarely. Very little data is available on avian bornaviruses in free-ranging psittacine populations, but it has been reported to be present in South America and Australia.[26,27]

Transmission

Transmission of avian bornaviruses has only been studied for parrot bornaviruses and canary bornavirus. Direct transmission between birds does not occur easily. The virus is shed by urine and feces, but infection via oral and nasal route were not successful in an experimental study compared to parenteral application routes which consistently induce infection. This is supported by proof of horizontal transmission via wounds.[28] Vertical transmission seems to play a major role in virus epidemiology, but the influencing factors are not clear. It seems that carrier birds infected at a young age transmit the virus more easily via vertical transmission compared to birds infected as adults.[29] In summary, introducing the virus into collections is more likely via carrier birds rather than by indirect introduction through contaminated accessories, feed or humans. Introduction of untested birds and especially the release of untested birds back into wild populations in a species conservation program should be avoided.

Pathology and Pathophysiology of Avian Borna Virus Related Disease

The non-cytopathic nature of Bornaviruses

Bornaviruses cause a persistent noncytolytic infection. Persistence is achieved by tight regulation of viral replication involving the X, N, and P proteins. A central role seems to be played by the X protein, which inhibits cellular apoptosis and neuronal degeneration via protective effects on mitochondria. This is also likely to be true for avian bornaviruses, as they are detected in many tissues without concurrent signs of tissue damage.[16]

Wounds as a possible entry point of transmission

Traumatic skin wound transmission has recently been proposed for PaBV-2,[30] PaBV-4[28] in parrots and variegated squirrel bornavirus-1 (VSBV-1) in humans. The presence of VSBV-1 in the oral cavity of infected variegated squirrels further supports the bite wound transmission hypothesis and prompts comparisons with rabies virus (RABV), another neurotropic negative-sense single-stranded RNA virus with a similar mode of spread to bornaviruses.[30]

Non-suppurative meningoencephalomyelitis and ganglioneuritis as the key lesion

Avian Borna virus related disease is based on non-suppurative inflammatory lesions in the central and peripheral nervous system. In contrast to bornavirus disease (BVD) in mammals and humans, the enteric nervous system is particularly affected. This is especially true in parrots, where the disease was first described, and has led to the misleading initial term of proventricular dilation disease (PDD). Although inflammatory lesions of nervous tissue have been demonstrated in correlation with canary and aquatic bornaviruses, psittacine borna viruses, namely PaBV-2 and PaBV-4, currently remain the most clinically relevant avian borna viruses.[16] Several successful experimental infections fulfilled Koch's postulates and confirmed PaBVs as the causative agent of PDD in psittacine birds.[31-38] Moreover, it appears that different parrot bornaviruses (PaBVs) cause differing clinical patterns with predominant neurologic or predominant gastrointestinal manifestations.[34] It remains a subject of intense investigation, how these non-cytopathic viruses induce clinical disease and which factors contribute to the high variability of clinical onset and outcome.

T-cell mediated immunopathogenesis

Based on the histopathologic findings of a mononuclear inflammation of nervous tissues, immune mediated mechanisms have long been suspected to contribute to the pathogenesis of AG/PDD. It seems to be mainly mediated by T-cells, as in BoDV-1 - onset of clinical neurologic signs is independent from virus load and distribution but is correlated to the presence of CD3+ T cells in the brain.[21]

For BoDV-1 infections, it is known that this cell-mediated response is driven mainly by CD4+ and CD8+ T cell subsets with CD8+ cells representing the effector cells targeting the viral nucleoprotein. Experimental studies in parrots suggest a similar pathogenesis for PaBV related disease, but parrot specific or cross-reactive tools for further immunophenotyping are lacking.[16,39,40]

Immunohistochemical investigations in brain sections of cockatiels experimentally infected with PaBV-2 confirm a predominance of T lymphocytes (CD3+) with lesser amounts of macrophages/microglia (IbA1+) and very few B-cells (PAX5+) in perivascular cuffs,[39] which is in alignment with immunophenotypic patterns seen BoDV-1 infected rodents/horses and VSBV-1 infected humans.[22,41,42]

In the gastrointestinal tract, inflammatory infiltrates were more frequent in the proventricular ganglia than in the small intestine, again consisting mainly of T lymphocytes, with lower levels of macrophages and B lymphocytes. Small intestinal ganglia showed a milder infiltration, consisting of a more uniform mixture of the three cell types.[39]

The route of PaBV-2 spread and inflammation

Recently, time-based studies in cockatiels (*Nymphicus hollandicus*) provided important insights into the spread of the virus and the development of lesions after intramuscular inoculation of parrot PaBV-2.[30,43] Based on the first time point of PaBV N-protein detection by IHC in histologic sections, viral spread in days post infection (dpi) could be characterized as an initial retrograde (centripetal) axonal migration from the injection site to the CNS (spinal cord 20 dpi, brain 25 dpi), followed by an anterograde (centrifugal) axonal migration to peripheral neural tissues (adrenal glands and ganglia in the gastrointestinal tract 30 dpi, epicardial ganglia, kidney 35 dpi) and, with a clear delay, to non-neural tissues (skin 80 dpi).

Mononuclear inflammatory lesions showed an identical time course and spread, appearing earliest in the brachial plexus at 20 dpi, in the thoracic and cervical spinal cord at 20-25 dpi (then spreading evenly to more caudal parts of the cord) and reaching the brain at 25 dpi, where the lesions remained confined to the central ventral parts of the brain until 40 dpi. The most severe lesions were found in the thalamic nuclei and midbrain, with the optic tectum involved at 35 dpi. The cerebrum, cerebellum, and the meninges showed no signs of inflammation before 40 dpi and were only mildly affected.

Consequently, defining the anatomic location of the inflammatory lesion in the central nervous system may allow conclusions to be drawn about the duration of the infection. Furthermore, there is a good chance that cerebrum and cerebellum (the most common samples taken from the brain) will be (falsely) negative, especially in the early stages of clinical disease, which can occur as early as 35 dpi. A similar timeline of virus spread has been discussed for embryonic brain tissue of cockatiels (*Nymphicus hollandicus*) experimentally infected with PaBV-4 and PaBV-2.[44]

Retrograde axonal transport to the brain has also been demonstrated for the spread of Borna disease virus 1 (BoDV-1) in experimentally infected rodents via nerve endings of the nasopharyngeal mucosa and the olfactory bulb and for Rabies virus (RABV) via motor neurons and spinal cord to the brainstem and limbic system.[30]

Clinical presentation from the perspective of pathology and pathophysiology

Factors affecting clinical presentation

Given that immunopathogenesis requires a functional immune system, it is reasonable to assume that any stressor that alters the host's immune competence will influence clinical onset, severity, duration and outcome. Examples of such stressors include age, genetic background, and immunosuppression due to co-infection with other pathogens or during certain life stages such as migration or breeding. In addition, viral dose and strain, route of infection, and degree of virus-host adaptation are likely to influence disease development. This explains, at least in part, the high variability in incubation period, clinical presentation, course, and outcome of parrot bornavirus infections.[16]

CNS lesions

Depending on the location in the CNS affected, different neurologic symptoms may be observed clinically. Cerebral perivascular cuffing and glial cell injury may cause seizures; optic lobe lesions result in cortical blindness; disruption of cerebellar cell layers results in fine motor deficits such as ataxia, intention tremor, incoordination, and proprioceptive abnormalities. In addition, cognitive deficits, and abnormalities in vocalization may be noted.[45]

Vagus nerve innervated organs as target tissues

Peripheral tissues mainly affected in histologic sections of AG (i.e., the anterior gut, the heart, the adrenal gland, and the kidneys) correlate with the innervation areas of the 10th cranial nerve (*N. vagus*) suggesting a pivotal role of this nerve for anterograde axonal migration of the virus from the brain to the periphery.[21]

Proventricular dilation

Comparison of the pathophysiology of gastroparetic disease events in humans and laboratory rodents, as well as proventricular dilations of other etiologies in birds, suggests that the dilation of the proventriculus (and sometimes other parts of the upper digestive tract) in AG is primarily due to a loss of nitrergic neurons in the autonomic enteric nervous system (ENS) and the vagus nerve plexus.[21] A simple causal relationship between the lymphocytic infiltrations and the destruction of nitrergic neurons is unlikely, because ganglioneuritis affects all neurons and not selectively nitrergic neurons. Rather, anatomic and neurophysiologic findings in laboratory rodents and poultry suggest that a complex interplay between direct inflammatory damage and indirect impairment of neuroimmunologic and neuroendocrine reflex arcs within the ENS and the gut-brain axis (communicating between the ENS and CNS) leads to the disruption of peristalsis and impairment of mucosal secretion and absorption. Together, these imbalances result in a net excess of cholinergic (stimulating, muscle-contracting) over nitrergic (inhibiting, muscle-relaxing) stimuli in the gastric muscle wall, leading to its permanent contraction and functional obstruction of the isthmus. The upstream proventriculus expands as a result of the downstream obstruction. This expansion is a reversible process in physiologic peristalsis, mediated by vagal pathways and mechanoreceptors in the gastric wall. With this reflex relaxation, the proventriculus regulates food flow and pressure in the downstream ventriculus and duodenum. If the gastro-duodenal contraction pattern is disturbed over a long period of time or if there is continuous contraction, a point of no return is passed and chronic (i.e., irreversible) dilation of the proventriculus with complete loss of function develops.[21] Clinical symptoms reflecting impaired proventricular or esophageal function include delayed crop emptying, regurgitation, anorexia, impaired GI transit, and dilatation and impaction of the upper GI tract. Passage of undigested food material in the feces and diarrhea are signs of impaired mucosal secretion and absorption and pancreatic activity (both of which depend on intact vagal innervation), leading to weight loss and starvation.[45]

Sudden death due to cardiac failure

In a retrospective study of the prevalence of cardiac lesions in naturally and experimentally ABV-infected birds with clinical signs of AG, epicardial ganglioneuritis was diagnosed as the most common finding in more than 66% of both groups, followed by myocarditis and endocarditis. PaBV-2 antigen was most frequently detected in the epicardial ganglia by IHC. These results suggest that heart failure caused by PaBV infection may be the underlying pathogenesis of sudden death in psittacine birds with AG.[46]

Blindness

Blindness is a common clinical feature of bornavirus disease in horses, sheep and rodents and has been associated with both central blindness[40] and degeneration of retinal neurons.[30] BoDV-1 spreads from the brain along the second cranial nerve (*N. opticus*) into the retina and causes immune-mediated retinopathy. A similar pathogenesis may be applicable for blindness and retinitis in avian bornavirus disease. Blindness has been reported in the course of bornaviral infections in parrots and canaries with occasional demonstration of mononuclear infiltrates within the retina. ABBV-2 has been detected within the retinas of infected ducks and ABBV-1 in the retinas of gulls.[21]

Feather destructive and auto-mutilative behavior

AG-associated peripheral neuritis has been suggested as a cause of feather picking and self-mutilation, possibly mediated by impaired sensory fiber signaling in the dorsal root ganglia of the spinal cord.[45] To date, this hypothesis lacks sufficient experimental evidence.

Autoimmunity as an alternative hypothesis for pathogenesis

A humoral autoimmune response to brain gangliosides has been proposed as an alternative pathogenesis for AG, based on the hypothesis that avian borna viruses induce the production of anti-ganglioside antibodies.[45,47-49] According to the original experiment, administration of purified gangliosides to cockatiels resulted in the development of clinical signs of AG. Based on these results it was claimed that AG was a suitable animal model for human Guillain-Barré syndrome (GBS) and that serologic testing for anti-ganglioside antibodies was a useful antemortem diagnostic for AG. Further statements suggested the presence of Campylobacter in the gut as a prerequisite for the development of AG and antibody-mediated phagocytosis of axons by intraganglionic and intracerebral macrophages recruited by CD4+ T cells as the cause of demyelination.[45] To date, despite attempts by other groups to substantiate the findings, these hypotheses still lack sufficient experimental evidence.[16,21,50,51]

Nevertheless, autoantibodies against brain antigens can be found in the sera of birds experimentally infected with PaBV-2. Similar to results from human VSBV-1 infection and human achalasia, these antibodies do not appear to play a central role in the pathogenesis and clinical course of AG and their significance in the disease process requires further investigation. In addition, future studies should also investigate the role of possible autoreactive T cells, a direction that has not yet been explored in AG.[51]

Clinical disease and influencing factors

Avian Bornavirus is a neurotropic virus affecting primarily psittacine birds. The symptoms include gastrointestinal signs such as dilatation of the proventriculus (PDD), regurgitation and maldigestion. Central nervous system signs include head shaking, ataxia and tremors, seizures as well as blindness and sudden death. Recently it was demonstrated that PaBV infection was more common in birds with neurologic disorders than in clinically healthy birds. The neurologic birds had higher antibody titers and higher levels of viral shedding.[52] The same study discussed that feather plucking was observed in birds with PaBV-infection at a higher prevalence than controls. Statistically the difference was not significant and feather plucking should be considered multi-factorial and studies determining that PaBV incites feather destructive behavior still need to be performed.

Virus strain appears to cause different disease patterns. and may influence the clinical outcome of infection.[34] Whether this is really linked to the genotype or the virus strain within the same genotype still needs to be evaluated. As described above the status of the immune system seems to play a major role in the clinical outcome of an infection. Therefore it is not surprising that under experimental conditions, cockatiels infected during their first week of life (immature immune system) developed into carriers without clinical symptoms, whereas cockatiels infected as adults developed severe clinical signs.[53]

In non- psittaciform species clinical symptoms are rarely described. ABBV-1 associated clinical signs in waterfowl included esophageal dilation and encephalitis.

Laboratory diagnosis

When testing for avian bornavirus, it needs to be determined if the diagnostic focus is detection of an infected (PaBV positive) bird, or if the goal is to determine if a clinical symptom is caused by avian bornavirus. Healthy infected birds are common. Therefore, PaBV positive birds may present with clinical signs similar to ABV infection that are not caused by ABV infection. Other potential differential diagnoses should be ruled out. Demonstration of the infection is important for introducing birds into a collection, releasing it to the wild, pairing, pre-purchase exam or as a potential diagnosis for a disease.

Ideally, testing a bird for avian bornavirus should include serology and PCR testing. Some birds shed the virus without seroconverting and in birds which seroconverted shedding may be irregular. So, performing both tests provides the highest accuracy. Detection of the virus is usually done by PCR using a pooled choanal and cloacal swab. In deceased birds, the brain or retina is the best material for viral detection. Feces, feather and blood are not ideal sample materials and are not preferable. Not all PCR-protocols detect all genotypes of avian bornaviruses, therefore it should be mentioned to the lab which genotype to focus on. In suspected bornaviral clinical cases with a negative PCR, additional protocols involving other bornavirus genotypes might be necessary. Sampling, storage and sample transport are very important as viral RNA degrades quickly leading to false negative laboratory results. Gold standard for the detection of anti-ABV-antibodies is the immunofluorescence assay.[54] Alternatively, ELISA or Western Blot test are described but seem to have a lower sensitivity, especially in cases with a low titer.

A bird should be considered infected if both PCR and serology provide a positive result. In cases where only the PCR is positive, the bird can be retested 4-6 weeks later, to confirm infection or exclude contamination of the sample. In addition, a seroconversion of the bird might be detected.[55] If the bird seroconverted or the PCR detects bornaviral-RNA repeatedly, the bird should be considered positive. Seropositive birds with a negative PCR should be considered infected if the titer is high. Low titers might become negative in following tests and clearance of an infection cannot be ruled out, although very unlikely.

Histopathology of tissues detects lesions typical for bornaviral disease. Tissue biopsies might be helpful in live birds although crop biopsies have been shown to have a low sensitivity. Therefore, demonstration of infection and ruling out other differential diagnoses is preferable in clinical affected cases. However, it has been shown that high antibody titers often correlate with clinical symptoms and therefore may be more useful for clinical diagnosis.

The detection of antiganglioside antibodies in clinically affected PaBV infected birds has been proposed. This theory may have some merit, but proof for the occurrence of antiganglioside- antibodies in all and only affected birds is lacking, as well as a validation of the tests. As a direct test for clinical disease, further investigations are necessary.

Treatment Protocols

Although currently, AG/PDD is considered an incurable disease, there are treatments to help diminish clinical signs and prolong life. Based on the observed histopathologic lesions, that are inflammatory in nature, nonsteroidal anti-inflammatory drugs (NSAID's) are widely utilized. Anecdotally this treatment appears beneficial. Effectiveness is difficult to measure, however, with no controlled studies. For example, the effectiveness of treatment may depend on the time of onset of treatment and the stage of the disease when initiated. Most clinicians believe the preferential and selective COX-2 inhibitors improve and extend the quality of life of AG/PDD birds. For birds manifesting neurologic deficits, Gabapentin may be helpful.

Drugs used to treat AG/PDD

The main stay of AG/PDD treatment has been the nonsteroidal anti-inflammatory drugs (NSAIDs) that inhibit cyclooxygenase (COX) isozymes. Celecoxib, meloxicam and robinacoxib are commonly utilized to treat AG. Celecoxib is believed to be more selective to COX-2 enzymes than meloxicam and therefore less likely to affect the GI tract and kidney and was the initial NSAID utilized for AG/PDD.[56] Dalhausen and colleagues administered celecoxib at 10 mg/kg PO q24h to treat birds with clinically diagnosed PDD and birds showed marked clinical improvement. Unfortunately, there were no control groups provided so the significance of this report is unclear. Currently, Celecoxib is administered at 20-30 mg/kg PO q12-24h. Anecdotally the COX-2 inhibitors appear to reduce clinical signs of disease and are utilized by veterinarians to treat AG worldwide. Anecdotally most practitioners also report beneficial effects in AG/PDD birds using oral and injectable Meloxicam. The dose for meloxicam for AG/PDD birds is 1-4 mg/kg IM or PO q12-24h.

Robinacoxib is also a COX-2 selective and COX-1 sparing NSAID. It is recommended at 1-2 mg/kg IM q3days or 10 mg/kg IM weekly. After 4 weeks of weekly treatment decrease to 2-10 mg/kg IM monthly.[45,57] Anecdotally, robinacoxib appears to be especially effective in parrots presenting with neurologic clinical signs.

Cyclosporine is a calcineurin inhibitor that selectively suppresses T-cell function that has anecdotally lessened symptoms in birds who presented with end stage AG/PDD who had not responded to either meloxicam or celecoxib (this was prior to the availability of robinacoxib injections). Cyclosporine at 10 mg/kg PO q12h along with meloxicam at 0.5 mg/kg PO q12h was given to severely affected birds who were unresponsive to NSAID alone. The addition of cyclosporine appeared to reduce the clinical signs of disease and prolonged their life. Since cyclosporine is an immunosuppressive drug, the birds were on antibiotics and antifungals prophylactically.[58]

Ribavirin is a ribonucleic analog that stops viral RNA synthesis and readily kills mammalian bornavirus and ABV in tissue culture.[59] Pharmacokinetic studies of ribavirin in birds show that it is excreted rapidly and that appropriate mammalian doses are insufficient to reach and maintain therapeutic levels in birds. Reuter and colleagues[60] demonstrated a synergistic effect with Ribavirin and interferon-a (IFN-a) against PaBV in avian cells, suggesting that a combination of ribavirin with type 1 IFNs may be more effective in vivo. More studies are needed on the use of ribavirin with and without IFNs.

Favipiravir (T-705): In Japan, researchers performed a study looking at favipiravir (T-705) an antiviral similar to ribavirin. Favipiravir was found to be a potent inhibitor of both BoDV-1 and PaBV-4 in vitro.[61] Further studies are needed to determine if T-705 would be effective in vivo.

Gabapentin is a gamma-Amino butyric acid (GABA) agonist. It is used to relieve neuropathic pain in mammals and birds. Although there have been no controlled studies in birds, anecdotally there have been positive clinical responses. The dose of 10-25 mg/kg PO q12h is recommended for central or peripheral nervous system signs such as seizures or ataxia. Gabapentin 50 mg/kg PO has been used in birds that self- mutilate.[45,57]

Adjunctive therapies for birds with PDD include drugs such as cisapride and metoclopramide to improve gastrointestinal tract transit. Antibiotics or antifungals may be necessary for treatment of secondary infections. Essential fatty acids such as flax seed oil (0.1 ml/kg PO daily or Vet Omega 0.22-0.44 ml/kg PO daily) are natural anti-inflammatories and help support kidney function. Reproductive stress has been reported to exacerbate clinical signs, so in some cases GNRH agonists such as Leuprolide acetate or the deslorelin implants have been used to reduce reproductive hormones. Anecdotally, stress has been deemed a factor in the development of clinical signs. Owners should avoid overcrowding of birds, practice good hygiene, and provide nutritionally adequate diets.

Vaccination

RNA viruses lack the proofreading enzymatic mechanisms that are present in DNA viruses, leading to increased mutation rates and increased changes to their genetic code, making a vaccine potentially more challenging. Several studies looking at vaccination options have been performed,[37,38,62] but to date, no ABV/PABV vaccines are commercially available. The studies have employed either recombinant bornavirus N protein or viral vector vaccines expressing bornavirus N and P proteins. The viral vector vaccines have shown promise, but more research and study in the area of vaccines is needed. In particular, the vaccination trials to date did not reach a sterile immunity and infectious virus was still detectable in vaccinated birds. As vaccinated birds seroconvert, PaBV-antibody test will not be able to differentiate between antibodies due to a field infection or due to vaccination.

Control

AG/PDD is an incurable and fatal disease. With limited treatment options and no vaccine available, control must be sought by other methods. Fortunately, ABV is not an environmentally stable virus. It is easily destroyed with soap or dilute bleach and is inactivated by ultra-violet light and desiccation. Bornaviruses can be inactivated with chlorine-containing disinfectants, and research suggests that allowing infected cages to dry thoroughly for 8 and 24 hours reduced the amount of infectious virus by 48% and 86%, respectively. Bornaviruses become rapidly less infective with heat treatment at 56 °C (or 132.8 °F).[63-65]

Controlling the spread of ABV in homes and in aviaries involves good hygiene, testing, and good quarantine procedures. In an aviary setting, the goal of such measures is the eradication of the virus from a flock and the prevention of its reintroduction. An eradication program is based on repeated rT-PCR testing of all birds combined with serology, followed by separation and ultimately removal of all positive birds from the flock. Newly introduced birds from flocks with a known positive or unknown status would be tested and kept in quarantine until deemed negative. With the challenges in testing and the unknown mechanism of transmission, this could be a long process (i.e., taking years). Along with practicing good hygiene, testing and quarantine measures, a nutritious diet should be provided.

The following are guidelines to control PaBV in aviaries and homes.

- Good hygiene and housing birds outdoors when temperature permits.
- For pets, reduce reproductive stress via diet and husbandry modifications and use of GNRH agonists. Avoid overcrowding and provide good nutrition.
- Isolation of all new, sick, or PaBV-positive birds, separation of birds from a flock when they are PaBV positive.
- Existing birds and newly presenting birds should be tested using both serology and PCR, and birds should be separated based on the results of this testing. There should be at least three negative tests at 3- to 6-week intervals before considering a bird negative.
- With the likelihood of vertical transmission, chicks from infected birds should be incubator hatched, hand-raised, kept separate from other non-infected birds and monitored for development of disease.
- Due to the intermittent shedding and inconclusive serology results, testing and separating birds may require years to obtain PaBV-negative aviaries, depending on the willingness of the owner.
- Healthy birds that test positive with either PCR or serology testing should not be euthanized.

References

1. Daoust PY, Julian RJ, Yason CV, et al. Proventricular impaction associated with nonsuppurative encephalomyelitis and ganglioneuritis in two Canada geese. *J Wildl Dis.* 1991;27(3):513-517.
2. Delnatte P, Ojkic D, Delay J, et al. Pathology and diagnosis of avian bornavirus infection in wild Canada geese (*Branta canadensis*), trumpeter swans (*Cygnus buccinator*) and mute swans (*Cygnus olor*) in Canada: a retrospective study. *Avian Pathol.* 2013;42(2):114-128.
3. Hoppes S, Gray PL, Payne S, et al. The isolation, pathogenesis, diagnosis, transmission, and control of avian bornavirus and proventricular dilatation disease. *Vet Clin North Am Exot Anim Pract.* 2010;13(3):495-508.
4. Payne SL, Delnatte P, Guo J, et al. Birds and bornaviruses. *Anim Health Res Rev.* 2012;13(2):145-156.
5. Perpiñán D, Fernández-Bellon H, López C, et al. Lymphoplasmacytic myenteric, subepicardial, and pulmonary ganglioneuritis in four nonpsittacine birds. *J Avian Med Surg.* 2007;21(3):210-214.
6. Weissenbock H, Sekulin K, Bakonyi T, et al. Novel avian bornavirus in a nonpsittacine species (Canary; *Serinus canaria*) with enteric ganglioneuritis and encephalitis. *J Virol.* 2009;83(21):11367-11371.
7. Avian Bornavirus Research Forum. *Proc Annu Conf Assoc Avian Vet.* 2014.
8. Berg M, Johansson M, Montell H, et al. Wild birds as a possible natural reservoir of Borna disease virus. *Epidemiol Infect.* 2001;127(1):173-178.

9. Payne S, Covaleda L, Jianhua G, et al. Detection and characterization of a distinct bornavirus lineage from healthy Canada geese (*Branta canadensis*). *J Virol.* 2011;85(22):12053-12056.

10. Shivaprasad H. Proventricular dilatation disease in a peregrine falcon (*Falco peregrinus*). *Proc Annu Conf Assoc Avian Vet.* 2005;107-108.

11. Payne S, Guo J, Tizard I. Bornaviruses. In: Hambrick J. Gammon LT, eds. *Ducks: Habitat, Behavior and Diseases.* New York, NY: Nova Scientific Publishers; 2013:1-20.

12. Smith D, Berkvens C, Kummrow M, et al. Identification of avian bornavirus in the brains of Canada geese (*Branta canadensis*) and trumpeter swans (*Cygnus buccinator*) with non-suppurative encephalitis. *Proc Wildl Dis Assoc.* 2010.

13. Honkavuori KS, Shivaprasad H, Williams BL, et al. Novel borna virus in psittacine birds with proventricular dilatation disease. *Emerg Infect Dis.* 2008;14(12):1883-1886.

14. Kistler AL, Gancz A, Clubb S, et al. Recovery of divergent avian bornaviruses from cases of proventricular dilatation disease: identification of a candidate etiologic agent. *Virol J.* 2008;5(1):1-15.

15. Ouyang N, Storts R, Tian Y, et al. Histopathology and the detection of avian bornavirus in the nervous system of birds diagnosed with proventricular dilatation disease. *Avian Pathol.* 2009;38(5):393-401.

16. Rubbenstroth D. Avian bornavirus research-a comprehensive review. *Viruses.* 2022;14(7):1513.

17. Kuhn JH, Dürrwald R, Bào Y, et al. Taxonomic reorganization of the family Bornaviridae. *Arch Virol.* 2015;160:621-632.

18. Rubbenstroth D, Briese T, Dürrwald R, et al. ICTV virus taxonomy profile: Bornaviridae. *J Gen Virol.* 2021;102(7):001613.

19. Malkinson M. Borna disease in ostriches. *Vet Rec.* 1993;132:304.

20. Ludwig H, Kraft W, Kao M, et al. Borna virus infection (Borna disease) in naturally and experimentally infected animals: its significance for research and practice. *Tieraerztl Prax.* 1985;13(4):421-453.

21. Tizard I, Shivaprasad H, Guo J, et al. The pathogenesis of proventricular dilatation disease. *Anim Health Res Rev.* 2016;17(2):110-126.

22. Hoffmann B, Tappe D, Höper D, et al. A variegated squirrel bornavirus associated with fatal human encephalitis. *New Engl J Med.* 2015;373(2):154-162.

23. Fluck A, Enderlein D, Piepenbring A, et al. Correlation of avian bornavirus-specific antibodies and viral ribonucleic acid shedding with neurological signs and feather-damaging behaviour in psittacine birds. *Vet Rec.* 2019;184(15):476-476.

24. Rubbenstroth D, Schmidt V, Rinder M, et al. Phylogenetic analysis supports horizontal transmission as a driving force of the spread of avian bornaviruses. *PLoS ONE.* 2016;11(8):e0160936.

25. Heffels-Redmann U, Enderlein D, Herzog S, et al. Occurrence of avian bornavirus infection in captive psittacines in various European countries and its association with proventricular dilatation disease. *Avian Pathol.* 2011;40(4):419-426.

26. Encinas-Nagel N, Enderlein D, Piepenbring A, et al. Avian bornavirus in free-ranging psittacine birds, Brazil. *Emerg Infect Dis.* 2014;20(12):2103.

27. Sutherland M, Phalen DN, Herzog S, et al. Detection of Immunoreactivity to Psittaciform Bornavirus in the Serum of a Wild Cacatuid in Victoria, Australia. *J Wildl Dis.* 2021;57(2):471-473.

28. Heckmann J, Enderlein D, Gartner AM, et al. Wounds as the portal of entrance for parrot bornavirus 4 (pabv-4) and retrograde axonal transport in experimentally infected cockatiels (*Nymphicus hollandicus*). *Avian Dis.* 2020;64(3):247-253.

29. Link J, Herzog S, Gartner AM, et al. Factors Influencing Vertical Transmission of Psittacine Bornavirus in Cockatiels (*Nymphicus hollandicus*). *Viruses.* 2022;14(12):2721.

30. Leal de Araujo J, Rech RR, Heatley JJ, et al. From nerves to brain to gastrointestinal tract: A time-based study of parrot bornavirus 2 (PaBV-2) pathogenesis in cockatiels (*Nymphicus hollandicus*). *PLoS ONE.* 2017;12(11):e0187797.

31. Piepenbring AK, Enderlein D, Herzog S, et al. Pathogenesis of avian bornavirus in experimentally infected cockatiels. *Emerg Infect Dis.* 2012;18(2):234-241.

32. Rubbenstroth D, Brosinski K, Rinder M, et al. No contact transmission of avian bornavirus in experimentally infected cockatiels (*Nymphicus hollandicus*) and domestic canaries (*Serinus canaria forma domestica*). *Vet Microbiol.* 2014;172(1-2):146-156.

33. Payne S, Shivaprasad HL, Mirhosseini N, et al. Unusual and severe lesions of proventricular dilatation disease in cockatiels (*Nymphicus hollandicus*) acting as healthy carriers of avian bornavirus (ABV) and subsequently infected with a virulent strain of ABV. *Avian Pathol.* 2011;40 (1):15-22.

34. Piepenbring AK, Enderlein D, Herzog S, et al. Parrot Bornavirus (PaBV)-2 isolate causes different disease patterns in cockatiels than PaBV-4. *Avian Pathol.* 2016;45(2):156-168.

35. Gray P, Hoppes S, Suchodolski P, et al. Use of avian bornavirus isolates to induce proventricular dilatation disease in conures. *Emerg Infect Dis.* 2010;16(3):473-479.

36. Mirhosseini N, Gray PL, Hoppes S, et al. Proventricular dilatation disease in cockatiels (*Nymphicus hollandicus*) after infection with a genotype 2 avian bornavirus. *J Avian Med Surg.* 2011;25(3):199-204.

37. Olbert M, Römer-Oberdörfer A, Herden C, et al. Viral vector vaccines expressing nucleoprotein and phosphoprotein genes of avian bornaviruses ameliorate homologous challenge infections in cockatiels and common canaries. *Sci Rep.* 2016;6(1):1-14.

38. Runge S, Olbert M, Herden C, et al. Viral vector vaccines protect cockatiels from inflammatory lesions after heterologous parrot bornavirus 2 challenge infection. *Vaccine.* 2017;35(4):557-563.

39. Leal de Araújo J, Rech RR, Rodrigues-Hoffmann A, et al. Immunophenotype of the inflammatory response in the central and enteric nervous systems of cockatiels (*Nymphicus hollandicus*) experimentally infected with parrot bornavirus 2. *Vet Pathol.* 2022;59(3):493-497.

40. Herden C, Briese T, Lipkin W, et al. Bornaviridae. In: Knipe D, Howley P, Griffin D, et al., eds. Fields Virology. Philadelphia, PA: Wolters Kluwer Health Lippincott Williams & Wilkins; 2013:1124-1150.

41. Tappe D, Schmidt-Chanasit J, Rauch J, et al. Immunopathology of fatal human variegated squirrel bornavirus 1 encephalitis, Germany, 2011-2013. *Emerg Infect Dis.* 2019;25(6):1058.

42. Caplazi P, Ehrensperger F. Spontaneous Borna disease in sheep and horses: immunophenotyping of inflammatory cells and detection of MHC-I and MHC-II antigen expression in Borna encephalitis lesions. *Vet Immunol Immunopathol.* 1998;61(2-4):203-220.

43. Leal de Araujo J, Rodrigues-Hoffmann A, Giaretta PR, et al. Distribution of viral antigen and inflammatory lesions in the central nervous system of cockatiels (*Nymphicus hollandicus*) experimentally infected with parrot bornavirus 2. *Vet Pathol.* 2019;56(1):106-117.

44. Wuest E, Malberg S, Petzold J, et al. Experimental infection of embryonic cells and embryonated eggs of cockatiels (*Nymphicus hollandicus*) with two parrot bornavirus isolates (PaBV-4 and PaBV-2). *Viruses.* 2022;14(9):1984.

45. Rossi G, Dahlhausen RD, Galosi L, et al. Avian ganglioneuritis in clinical practice. *Vet Clin North Am Exot Anim Pract.* 2018;21(1):33-67.

46. Leal de Araujo J, Hameed S, Tizard I, et al. Cardiac lesions of natural and experimental infection by parrot bornaviruses. *J Comp Pathol.* 2020;174:104-112.

47. Pesaro S, Crosta L, Bertoni P, et al. Anti-ganglioside antibodies production as a theory of PDD pathogenesis. *Proc Euro Assoc Avian Vet.* 2009;89.

48. Rossi G, Ceccherelli R, Crosta L, et al. Anti-ganglioside specific auto-antibodies in ganglia of PDD-affected parrots. *Proc Eur Assoc Avian Vet.* 2011;177-178.

49. Rossi G, Crosta L, Pesaro S. Parrot proventricular dilation disease. *Vet Rec.* 2008;163(10):310.

50. Nobach D, Müller J, Tappe D, et al. Update on immunopathology of bornavirus infections in humans and animals. *Adv Virus Res.* 2020;107:159-222.

51. Leal de Araujo J, Tizard I, Guo J, et al. Are anti-ganglioside antibodies associated with proventricular dilatation disease in birds? *PeerJ.* 2017;5:e3144.

52. Fluck A, Enderlein D, Piepenbring A, et al. Avian bornavirus infection in psittacine birds with neurological signs and feather plucking. *Proc ICARE.* 2013;388.

53. Gartner AM, Link J, Bücking B, et al. Age-dependent development and clinical characteristics of an experimental parrot bornavirus-4 (PaBV-4) infection in cockatiels (*Nymphicus hollandicus*). *Avian Pathol.* 2021;5(2):138-150.

54. Herzog S, Enderlein D, Heffels-Redmann U, et al. Indirect immunofluorescence assay for intra vitam diagnosis of avian bornavirus infection in psittacine birds. *J Clin Microbiol.* 2010;48(6):2282-2284.

55. Lierz M. Avian bornavirus and proventricular dilatation disease. In: Speer B, ed. *Current Veterinary Therapy in Avian Medicine and Surgery.* St. Louis, MO: Elsevier; 2016:28-46.

56. Dahlhausen RD, Aldred S, Colaizzi E. Resolution of clinical proventricular dilatation disease by cyclooxygenase 2 inhibition. *Proc Annu Conf Assoc Avian Vet.* 2002;9-12.

57. Clubb S, Meyer M. Clinical management of psittacine birds affected with proventricular dilatation disease. *Proc Annu Conf Assoc Avian Vet.* 2006;85-90.

58. Gancz AY, Elbaz D, Farnoushi Y, et al. Clinical recovery from proventricular dilatation disease following treatment with cyclosporine A in an African grey parrot (*Psittacus erithacus*). *Proc DVG Tagung über Vogelkrankheiten.* 2012;130-133.

59. Musser JM, Heatley JJ, Koinis AV, et al. Ribavirin inhibits parrot bornavirus 4 replication in cell culture. *PLoS ONE.* 2015;10(7):e0134080.

60. Reuter A, Horie M, Höper D, et al. Synergistic antiviral activity of ribavirin and interferon-α against parrot bornaviruses in avian cells. *J Gen Virol.* 2016;97(9):2096-2103.

61. Tokunaga T, Yamamoto Y, Sakai M, et al. Antiviral activity of favipiravir (T-705) against mammalian and avian bornaviruses. *Antiviral Res.* 2017;143:237-245.

62. Rall I, Amann R, Malberg S, et al. Recombinant modified vaccinia virus Ankara (MVA) vaccines efficiently protect cockatiels against parrot bornavirus infection and proventricular dilatation disease. *Viruses.* 2019;11(12):1130.

63. Boatright-Horowitz SL. Avian bornaviral ganglioneuritis: current debates and unanswered questions. *Vet Med Int.* 1-9; 2020. [Review article]; doi: https://doi.org/10.1155/2020/6563723.

64. Kranz JB. A study on the chemical and physical inactivation of avian bornavirus. Doctoral dissertation. College Station, TX: Texas A&M University; 2014.

65. Kranz JB, Escandon P, Musser JM. Environmental stability of avian Bornavirus: pH and drying. *Proc ExoticsCon.* 2015;89.

Practical Applications of Fluoroscopy in Avian Practice

Rachel Baden, DVM, DABVP (Avian Practice), and
Brian Speer, DVM, DABVP (Avian Practice), DipECZM (Avian)

Session #3217

Affiliation: From the Medical Center for Birds, 3807 Main Street, Oakley, CA 94561, USA.

Abstract: Fluoroscopy is an under-utilized imaging modality that has numerous clinical applications in avian practice. This masterclass will briefly review the physics and methods behind fluoroscopy and its use in avian clinical practice. Real clinical cases and video images will be used to illustrate its potential applicability in diagnostically shaping cases. Although the primary focus will be on diseases of the gastrointestinal tract, other systems and disease conditions will be included to best show the potential broad value and scope of this imaging modality as an addition to the diagnostic tools for clinical practice.

Physics of Radiology

While this masterclass does not spend much time on the physics involved in fluoroscopy, it can be useful to recall the basic principles and techniques, as these may inform interpretation of the resulting images. Both radiographs and fluoroscopic images are generated by x-rays, which are part of the electromagnetic spectrum. In particular, x-rays are given off by electron interactions outside of the atomic nucleus, and the energy of a given x-ray is dependent upon the energy of the original electron and its subsequent photonic energy. Since they are on the electromagnetic spectrum, x-rays (energetic photons) are simultaneously both a particle and a wave, like visible light.[1]

In fluoroscopy, x-rays are produced when high-speed electrons strike a metal x-ray tube. To generate electrons, electrical current is run through the x-ray tube filament, similar to that of a lightbulb, creating heat that causes electrons to leave the filament and form an electron cloud. The number of electrons in the cloud is directly related to the current running through the filament, which is controlled by the milliamperage (mA). Increasing mA is similar to increasing the wattage of a lightbulb. The milliampere-second (mAs) describes the total amount of radiation produced in a period of time.

Once produced, the electrons in the electron cloud are accelerated into metallic target, driven by a voltage differential between the filament and the target. The voltage difference is controlled by the kilovoltage peak (kVp); increasing kVp increases the voltage differential which more strongly attracts electrons to the target, producing higher velocity electrons and leading to higher-energy x-rays.

An important distinction to make is between the kiloelectron volt (keV, the unit of energy for electromagnetic radiation) and kVp, the kilovoltage peak. The first is a unit that describes the amount of energy of an electron or photon, and the second is the maximum voltage difference between the x-ray tube generating electrons and the target.

When the electrons strike the metallic anode (often tungsten, with current configurations), they are suddenly slowed down (Bremsstrahlung process) creating multiple photonic energy wavelengths. The creation of an x-ray image, ultimately, involves the x-rays (photons) that then pass through a patient (or don't) and strike a detector, the x-ray plate. One of the reasons that x-rays can produce an interpretable image is because of the difference between radiation exposure and absorption. For a given exposure to radiation, different tissues may absorb different amounts; for example, bone absorbs much more radiation than soft tissue, leading to a different appearance on radiography and fluoroscopy between the two tissues.

Fluoroscopy of Different Organ Systems

Gastrointestinal tract

Evaluation of the size, shape, and function of the gastrointestinal organs is likely the most common use of fluoroscopy in veterinary medicine, and can be done either with or without contrast. For contrast fluoroscopy, a barium sulfate suspension is anecdotally used most often, but iodine-based media (iohexol) has also been reported.[2-4] Some of the advantages of fluoroscopic assessment of the GI over standard radiography include the ability to monitor contractions of the crop and ventriculus, and evaluate intestinal filling and peristalsis.[5] With the addition of contrast, radiolucent intraluminal foreign objects, mural irregularities, or abnormal diverticula can be identified.[2,3,6] Additionally, contrast can be used to measure overall gastrointestinal transit time, as a way of evaluating overall motility, and has been assessed in several avian species to date.[4,7-11] Contrast-enhanced intestinal loops can also assist in locating specific segments of the tract as well as mass effects relative to other organs in the abdomen.[12]

One of the major advantages of fluoroscopy-based contrast GI study compared to radiography is the reduction in intrusiveness and risk. With radiographs, in order to get a properly positioned and interpretable film without causing undue stress to the patient, sedation and/or anesthesia is often required, and the patient is placed in lateral recumbency; both of these things can be dangerous with a crop full of liquid contrast medium, as regurgitation and subsequent aspiration are possible. Fluoroscopy avoids the need for chemical restraint and lateral recumbency both, as the patient can be standing on a perch or in a radiolucent container for the duration of the study.[13] In addition, sedation has been shown to artificially slow GI motility and prolong transit time.[4]

Cardiovascular system

Protocols for contrast-based fluoroscopy of the cardiovascular system have been reported in humans and other mammals, and are beginning to be reported in birds as well. There are two main methods, nonselective and selective contrast fluoroscopy, with the latter not (yet) published as being used in birds. The difference between the two is the selective fluoroscopy, also called digital subtraction angiography, involves digital processing of the images, where the non-contrast images are digitally subtracted from the views containing contrast, to enhance visualization of the important structures and reduce interference from other organs.[14]

Several reports of fluoroscopic angiography exist in birds, including the diagnosis of infundibular pulmonic stenosis in a Moluccan cockatoo (*Cacatua moluccensis*), a right coronary artery aneurysm in an umbrella cockatoo (*Cacatua alba*), and cardiac chamber dilation in a whooping swan (*Cygnus cygnus*).[15-17] One study has established a protocol with iodinated contrast medium for fluoroscopic angiography in Hispaniolan Amazon parrots (*Amazona ventralis*), but the literature on avian cardiovascular fluoroscopy remains sparse.[18]

Other organs

Reports on the use of fluoroscopy to evaluate other organ systems in birds are few, but include guidance for placement of a wire stent for tracheal stenosis in an eclectus (*Eclectus roratus*) and introduction of a tube into the ventriculus for foreign body removal in a honey buzzard (*Pernis apivorus*).[19,20] One study has investigated the kinematics of bipedal locomotion in common quail (*Coturnix coturnix*) using fluoroscopic video analysis.[21]

References

1. Thrall D, Widmer W. Radiation protection and physics of diagnostic radiology. In: Thrall D, ed. *Textbook of Veterinary Diagnostic Radiology.* Philadelphia, PA: Elsevier Saunders; 2018:2-22.

2. Degernes LA, Wolf KN, Zombeck, DJ, et al. Ventricular diverticula formation in captive parakeet auklets (*Aethia psittacula*) secondary to foreign body ingestion. *J Zoo Wildl Med.* 2012;43(4):889-97.

3. Cotton RJ III, Divers SJ. Endoscopic removal of gastrointestinal foreign bodies in two African grey parrots (*Psittacus erithacus*) and a hyacinth macaw (*Anodorhynchus hyacinthinus*). *J Avian Med Surg.* 2017;31(4):335-343.

4. Martel A, Mans C, Doss GA, Williams JM. Effects of midazolam and midazolam-butorphanol on gastrointestinal transit time and motility in cockatiels (*Nymphicus hollandicus*). *J Avian Med Surg.* 2018;32(4):286-293.

5. Keller D, Honkavuori KS, Briese T, et al. Proventricular dilatation disease associated with avian bornavirus in a scarlet macaw (*Ara macao*). *J Vet Diagn Invest.* 2010;22(6):961-965.

6. Laniesse D, Beaufrère H, Mackenzie S, et al. Perforating foreign body in the ventriculus of a pet pigeon (*Columba livia domestica*). *J Am Vet Med Assoc.* 2018;253(12):1610-1616.

7. Doss GA, Williams JM, Mans C. Determination of gastrointestinal transit times in barred owls (*Strix varia*) by contrast fluoroscopy. *J Avian Med Surg.* 2017;31(2):123-127.

8. Doss GA, Williams JM, Mans C. Contrast fluoroscopic evaluation of gastrointestinal transit times with and without the use of falconry hoods in red-tailed hawks (*Buteo jamaicensis*). *J Am Vet Med Assoc.* 2017;251(9):1064-1069.

9. Beaufrère H, Nevarez J, Taylor WM, et al. Fluoroscopic study of the normal gastrointestinal motility and measurements in the Hispaniolan Amazon parrot (*Amazona ventralis*). *Vet Radiol Ultrasound.* 2010;51(4):441-6.

10. Vink-Nooteboom M, Lumeij JT, Wolvekamp WTC. Radiography and image-intensified fluoroscopy of barium passage through the gastrointestinal tract in six healthy Amazon parrots (*Amazona aestiva*). *Vet Radiol Ultrasound.* 2003;44(1):43-48.

11. Kubiak M, Forbes NA. Fluoroscopic evaluation of gastrointestinal transit time in African grey parrots. *Vet Rec.* 2012;171(22):563.

12. Murphy B, Fitzgerald B, Olsen G, Speer B. Diagnosis and surgical repair of an acute abdominal wall hernia and partial cloacal strangulation in a yellow-naped Amazon parrot (*Amazona auropalliata*). *J Avian Med Surg*. 2018;32(4):336-341.

13. Dennison S. Common imaging modalities: selecting the right modality and getting the most out of your imaging. In: Speer B, Van Zeeland Y, eds. *Current Therapy in Avian Medicine and Surgery.* Vol II. St. Louis, MO: Elsevier; 2023.

14. Beaufrère H, Pariaut R, Rodriguez D, Tully TN. Avian vascular imaging: a review. *J Avian Med Surg*. 2010;24(3):174-84.

15. Vink-Nooteboom M, Schoemaker NJ, Kik MJ, et al. Clinical diagnosis of aneurysm of the right coronary artery in a white cockatoo (*Cacatua alba*). *J Small Anim Pract*. 1998;39:533-537.

16. Zoller G, Guzman DS-M, Summa N, et al. Infundibular pulmonic stenosis in a Moluccan cockatoo (*Cacatua moluccensis*). *J Avian Med Surg*. 2017;31(1):53-61.

17. Fischer I, Christen C, Scharf G, Hatt JM. Cardiomegaly in a whooper swan (*Cygnus cygnus*). *Vet Rec*. 2005;156:178-182.

18. Aller TL, Guzman DS-M, Stern JA, et al. Evaluation of a fluoroscopic angiography protocol in Hispaniolan Amazon parrots (*Amazona ventralis*). *J Avian Med Surg*. 2022;36(2):178-186.

19. Mejia-Fava J, Holmes SP, Radlinsky, MA, et al. Use of a nitinol wire stent for management of severe tracheal stenosis in an eclectus parrot (*Eclectus roratus*). *J Avian Med Surg*. 2015;29(3):238-249.

20. Lumeij JT, Wolvekamp WT, Bron-Dietz GM, Schotman AJ. An unusual case of lead poisoning in a honey buzzard (*Pernis apivorus*). *Vet Q*. 1985;7(2):165-8.

21. Ogihara N, Oku T, Andradaet E, et al. Planar covariation of limb elevation angles during bipedal locomotion in common quails (*Coturnix coturnix*). *J Exp Biol*. 2014;217(22):3968-3973.

Recognition, Assessment and Management of Pain Improves Avian Welfare

Michelle G. Hawkins VMD, DABVP (Avian Practice), and
Joanne Paul-Murphy DVM, Dipl ACZM, Dipl ACAW

Session #3166

Affiliation: From the Department of Medicine and Epidemiology, School of Veterinary Medicine, University of CA, Davis, Davis, CA 95616, USA.

Abstract: This masterclass provides an overview of current information regarding avian pain. The review includes recognition of avian pain, and the current evidence-based clinical analgesic use in birds. Recognition and assessment of behavioral changes associated with pain in birds are crucial tools towards pain management. The field of avian analgesia has expanded dramatically, identifying more options for alleviating acute and chronic pain for many species. Opioids, nonsteroidal anti-inflammatory drugs (NSAIDs), local anesthetics, and drugs including gabapentin, amantadine, and cannabinoids are being used for birds. Analgesic combinations acting at different points in the nociceptive system provide greater pain relief, reduce the risk of adverse effects and improve welfare of the patient.

Avian Analgesia and Welfare

There is still much to be determined to create criteria to recognize and assess pain in birds because this is a highly diverse class of species. Our team recently published two papers[1,2] that will be helpful. Both publications provide information on the physiology of avian pain, observed behavioral and physiologic alterations with pain, and how different sources and degrees of pain affect behavior. Species vary widely in their response to pain, and individual animals of the same species can have different responses to different types of pain. Layered on top of that, the individual animal's response to pain under different conditions, for example a bird under direct vs remote observation, or a bird under different environmental conditions such as home or the clinical setting. The World Organization for Animal Health (OIE) defines an animal as having good welfare if it is "healthy, comfortable, well nourished, safe, able to express innate behavior, and if it is not suffering from unpleasant states such as pain, fear and distress". At the heart of welfare definitions is the understanding that the welfare of a bird belongs to the bird.[3] The definition of welfare includes the bird's psychological well-being. Applying these concepts to recognition and management of pain often occurs without conscious effort when the clinical staff have experience and advanced training. Knowledge of the normal appearance and behavior of the species, and the individual bird involved, is essential for recognition of pain. When a bird experiences pain, one or more of the normal behavior typically shift or are absent. Given the overall reduction in ability to express pain via facial expression in birds, in contrast to mammals, evaluation of the avian patient should instead focus on the whole individual, and variations in positioning of the tail, wings, and crest (if present) may provide useful information.[1] In most circumstances, pain is assessed based on a bird's clinical appearance and overall behavior. This approach can be unreliable; however, it is usually effective in detecting severe pain in many species. It is also effective when pain is localized to one limb causing lameness or wing droop, or to a specific body area resulting in a marked behavioral response if that area is palpated.[4] Assessing pain is going to be highly subjective and based largely on preconceived ideas about the appearance and behavior of birds in response to pain.[4] A team approach can reduce subjectivity, and requires communication and cooperation among all involved; veterinarians, veterinary technicians, animal care staff and owners will advance the welfare of the bird in pain. Accurate assessment of avian pain can be confounded by the effect of the observer on the bird's behavior (especially in terms of wildlife species), effects of some analgesics or sedatives that can alter behavior, and concurrent physiologic conditions (anesthetic recovery, hypothermia, and others.) that could make a bird appear painful despite adequate analgesia.[2] The intensity of a bird's pain needs to be recognized to aid in the selection of the correct analgesic. Differences in the source (visceral, somatic, neuropathic), chronicity (acute vs. chronic), and degree (mild, moderate, severe) of pain can lead to differences in pain response in birds. If the wrong analgesic or the wrong dose is selected, it could result in suboptimal treatment.

The options to provide reduction of pain and analgesia are expanding for birds as the field of analgesia for all vertebrate species is advanced. Management of pain has numerous components with administration of analgesic drugs being an important tool. Published dosages of many analgesic drugs for birds are determined by pharmacokinetics (PK) studies using a limited number of animals and a restricted set of species. Pharmacodynamic studies to determine effectiveness is lacking for most analgesic drugs currently used to treat birds; however, the information provides a critically helpful reference point to extrapolate from, whether it be to another species or a different type of painful condition. Successive observations are critical to assess the effect of treatment and determine if the bird's condition is improving or declining. Having a structure for assessment, shared by the care team, is always helpful when monitoring throughout a period of time, especially when several people are involved. Additionally, as a clinical team shares observations and evaluations of birds' responses to analgesic treatments, the team's proficiency to make accurate assessments will improve. Published pain scales and assessment schemes are useful as a starting point to adopt the aspects that resonate with the observers. Generalized methods of quantifying pain such as visual analogue or numeric rating pain scales can be applied as a tool for evaluation of an avian patient's pain while awaiting development of more avian-specific pain scales.[2]

Specific Analgesia

Balanced analgesia

Combinations of drugs acting at different points in the nociceptive system are more effective and less toxic than either drug given alone. For example, opioids act centrally to limit nociceptive input whereas NSAIDS act peripherally to decrease inflammation, limiting nociceptive information that enters the CNS. Synergy has been demonstrated in laboratory animals and is now being tested in the clinical environment.

Opioids

Analgesia is species- and intra-species variable at equipotent dosages. Opioids are used for moderate to severe pain, such as traumatic or surgical pain. Opioids reversibly bind to specific receptors in the central and peripheral nervous system. These drugs are categorized as full agonists, partial agonists, mixed agonist/antagonists, or antagonists based on their ability to induce an analgesic response once bound to a specific receptor. The action of opioid drugs on these receptors activates G-proteins, leading to a reduction in transmission of nerve impulses and inhibition of neurotransmitter release.

Agonist drugs have a linear dose–response curve that may be titrated to reach the desired effect, whereas the agonist/antagonist drugs may reach a ceiling effect after which increasing the dose does not appear to provide additional analgesia. During anesthesia, opioids are used to provide perioperative analgesia that may reduce the concentrations of volatile anesthetics (i.e., gas anesthesia-sparing effects) while maximizing analgesia. The most common adverse effects reported with opioids are cardiac and/or respiratory depression. In many cases, these drugs may be reversed with antagonists, which will also terminate analgesia. The application and dosages of few opioid formulations have been scientifically evaluated and clinically applied in birds.

Few studies have been conducted regarding the distribution, quantity, and function of each opioid receptor type in birds, and although marked species variability is assumed, further studies are necessary.[5] There are small differences in receptor densities between cockatiels and pigeons but these may not be sufficient to make functional changes.[6] In conditioning experiments pigeons were able to differentiate μ from κ receptor agonists. Isoflurane-sparing effects of μ and κ agonists are quite variable in mammals and it is likely that similar variability will be evident in birds.

Morphine (MOR) is a μ-receptor agonist that has been infrequently used in avian medicine because early studies in domestic fowl yielded conflicting results and suggested very high dosages were required. The PK of MOR at 2 mg/kg IV in the domestic chickens found that they cleared MOR at a higher rate than humans or cats and plasma MOR analgesic concentrations were similar to humans for only approximately 2 hours. Following 2 mg/kg IV MOR, chickens had an increased time to finish the obstacle course and exhibited sedation, confounding assessment of its analgesic effects.

Hydromorphone HCl (HYDRO) is a μ-opioid receptor agonist with an approximate potency 5-7 times that of MOR. Hydromorphone showed a dose-responsive thermal antinociceptive effect when administered at 0.1, 0.3, and 0.6 mg/kg IM to American kestrel (Falco sparverius; AMKE) for up to 6 hours.[7] Severe sedation was detected in four birds after 0.6 mg/kg HYDRO, and when the sedated birds were handled, they appeared agitated for a few seconds before resuming sedated behaviors. Hydromorphone PK following IM and IV administration in AMKEs found high bioavailability (75%) and rapid elimination.[8] A similar thermal nociception study in cockatiels (*Nymphicus hollandicus*) administered the same dosages showed no significant effect on the thermal threshold, with only a mild sedative effect.[9] Thermal antinociceptive effects of HYDRO were evaluated following IM administration to orange-winged Amazon parrots (*Amazona amazonica*; OWAP).[10] Hydromorphone at 1 and 2 mg/kg significantly increased thermal foot withdrawal responses compared with the response after saline administration, and plasma concentrations were maintained above 1 ng/ml for 6 hours in OWAP receiving 1 mg/kg HYDRO IV and IM, respectively. Significant agitation was also observed for 1 and 2 mg/kg HYDRO doses. This kind of agitation in response to an opioid in normal birds may not be present in birds with pain. Other significant adverse effects after administration of 1 and 2 mg/kg HYDRO included nausea-like behavior, ataxia and pupillary constriction. The presence of adverse effects warrants consideration of lower doses for clinical use in this species.

Methadone (MET) is a μ-opioid receptor agonist that also is a non-competitive inhibitor of NMDA receptors. Methadone can reduce re-uptake of norepinephrine and serotonin, which may contribute to its analgesic effects. Due to its additional mechanisms of action, MET potentially may be more efficacious than other μ-agonists, particularly for neuropathic or chronic pain. In hens, MET at 6 mg/kg IM decreased the MAC_{iso} by 30% but it also produced mild respiratory acidosis, decreased heart rate, and increased systemic blood pressure, with occasional cardiac arrhythmias. The IM bioavailability was approximately 80%. [11] In OWAP, 6 mg/kg IM MET increased thermal thresholds significantly; lower doses (1 and 3 mg/kg) did not result in a significant effect. Methadone caused significant agitation and nausea-like behavior when compared to saline administration. The IM mean bioavailability was 85.7 %. While 6 mg/kg MET IM appears to have desired analgesic effects, the presence of agitation, nausea-like behavior and ataxia warrants consideration of lower doses for clinical use in OWAP.

Fentanyl (FEN) is a short-acting μ-opioid receptor agonist often administered as a continuous rate infusion (CRI) or as a transdermal patch. Early work evaluating FEN in psittacine birds was challenging to interpret. Fentanyl administered as an IV constant rate infusion (CRI) in red-tailed hawks (*Buteo jamaicensis;* RTHA) reduced the minimum anesthetic concentration of isoflurane 31-55% in a dose-related manner, without significant effects on heart rate, blood pressure, $PaCO_2$, or PaO_2. In Hispaniolan Amazon parrots (*Amazona ventralis*; HAPs), FEN administered as an IV CRI reduced $MAC_{isoflurane}$ similarly to the RTHAs; however, approximately 20 times higher dosing was necessary to achieve the same effects with significant effects detected in HR and BP suggesting that close monitoring is advised.[12] In several breeds of chickens, the plasma concentrations of FEN were evaluated following application of 25 μg/hour transdermal FEN patches.[13] There was significant inter-individual difference in maximum plasma FEN concentrations attributed to absorption variability, but all chickens achieved human target plasma concentrations within 2-4 hours and had a rapid decrease in concentrations following patch removal. The FEN transdermal patch resulted in plasma concentrations within the range considered therapeutic in people for 72 hours.[13]

Butorphanol (BUT) is a mixed κ-opioid receptor agonist and μ-opioid receptor antagonist opioid drug. This is one of the most commonly used opioid analgesics in birds, despite many studies suggesting it requires frequent dosing. There is evidence that BUT does not produce dose-related respiratory depression. Butorphanol (1 mg/kg) produced an isoflurane-sparing effect in cockatoos (*Cacatua* spp.) and grey parrots, *Psittacus erithacus*), but not in blue-fronted Amazons (*Amazona aestiva*), indicating species variability in response to the drug itself or to the dose administered. When 4 mg/kg IV of BUT was administered rapidly, there were significant arrhythmias in all birds, significant decreases in heart rates and arterial blood pressures, and one bird died.[14] Other studies have shown efficacy of BUT administered 2-6 mg/kg IM without significant anesthetic or cardiopulmonary changes. Oral administration had very limited bioavailability (5.9%) in HAPs and is not recommended. In contrast, hyperesthesia, hyperalgesia and agitation were noted in male AMKE receiving 6 mg/kg; thus BUT is not recommended in this species for pain management. Likewise, when a dose of 2 mg/kg IV BUT was administered to chickens, its terminal half-life was shorter than in humans, dogs, cattle, rabbits, and owls. Chickens finished an obstacle course faster after 2 mg/kg BUT injection, but concentrations only lasted 2 hours. All these data confirm frequent BUT dosing is necessary in birds. This may be impractical because of lack of personnel to provide frequent dosing and the stress of frequent handling on the patient. In common peafowl, the PK of butorphanol were assessed after administration via an osmotic pump, a novel delivery method.[15] Plasma concentration reached 60 ng/ml after 24 hours in most birds, and most birds stayed above this threshold for 7 days, then decreased rapidly after implant removal. This delivery method holds promise for pain management in larger birds with the limitation that it will likely require deep sedation or general anesthesia for implantation and removal.

Buprenorphine HCl (BUP) has a slow onset of action with longer duration than other opioid drugs and a complex pharmacological profile. It is thought to act as a partial μ agonist but its κ-receptor activities are less well defined. Several studies suggest that BUP demonstrates κ-receptor agonist/antagonist activity. Buprenorphine has unusual receptor binding characteristics as a result of slow drug dissociation from opioid receptors. Buprenorphine 0.1 mg/kg IM in grey parrots did not demonstrate any antinociceptive effect but pigeons administered 0.25 and 0.5 mg/kg IM BUP had increased withdrawal thresholds from a noxious electrical stimulus for 2 and 5 hours, respectively. In AMKEs, BUP did cause a significant dose-dependent thermal antinociceptive response for 6 hours compared with controls. At 0.6 mg/kg, a mild sedative effect was appreciated. The PK of 0.6 mg/kg IM and IV BUP was also evaluated in AMKEs and high bioavailability (94.8%) after IM dosing was identified. In a PK study in AMKEs, SC and IM administration of 1.8 mg/kg BUP SR-LAB™ (Wildlife Pharmaceuticals, Windsor, CO, USA) were both characterized by rapid absorption and mean plasma concentrations remained above the target concentrations for 48 hours after administration.[16] The elimination half-life was approximately 9 times longer than those reported in AMKEs receiving 0.6 mg/kg IM standard BUP. A follow-up study evaluating the thermal antinociceptive effects of this formulation in the same species concluded that 1.8 mg/kg IM requires administration at least every 24 hours to manage pain but might have a longer duration of action depending on the severity and type of pain, adjunctive therapy, and the individual response.[17] In RTHAs, the PK of two doses of the concentrated formulation of BUP (Simbadol®, Zoetis, Kalamazoo, MI, USA) were evaluated after SC administration.[18] Plasma concentrations were maintained above 1 ng/ml for at least 24 and 48 hours, respectively. Adverse effects were not noted in any birds. The use of this concentrated BUP formulation might help reduce stress related to repeated handling and administration of treatments while providing consistent pain management.

In contrast, no significant difference was identified in thermal withdrawal threshold using the same dosages of BUP in cockatiels and no sedative effects were observed at any dosage.[19] In OWAPs, the thermal antinociceptive effects of the concentrated formulation of BUP (Simbadol®) after 2 mg/kg SC administration was evaluated. Concentrated BUP resulted in a small but significant increase in thermal threshold, while lower doses did not result in any significant effect. These results provide evidence of weak efficacy at lower doses for pain management in this species and highlight the differences in efficacy. Further studies are needed to fully determine the analgesic and adverse effects of buprenorphine in psittacines and warrants evaluation in other species.

Opioid-like medications

Tramadol HCl (TRAM) is an analgesic drug often used due to its ability to be dosed orally. It is a synthetic μ-opioid receptor agonist that appears to have variable efficacy depending on the species. The efficacy is thought to be primarily related to the production of the O-desmethyl-tramadol metabolite (M1). This metabolite acts centrally over the opioid receptor and is thought to have almost 200 times higher affinity for this receptor than the parent drug. Tramadol also has effects on the k and δ- opioid receptors, as well as inhibiting reuptake of norepinephrine and serotonin. The conversion to the M1 metabolite is variable among species, but it is produced in the bird species studied thus far. Pharmacokinetic profiles have been reported for a multitude of avian species, and important differences exist with PK parameters between species. Oral bioavailability of 11 mg/kg TRAM PO was higher (98%) for American bald eagles achieved concentrations in the human analgesic range for 10 hours; M1 plasma concentrations reached human ranges only in two eagles at much earlier time points. In RTHAs administered 11 mg/kg PO, analgesic thresholds for humans were maintained for approximately 4hrs.[20] In HAPs, doses of 10 or 20 mg/kg PO did not result in significant antinociception to thermal stimuli and human analgesic thresholds were only reached after dosing with 30 mg/kg orally or 5 mg/kg IV.[21] Transient bradycardia was noted after IV administration. In adult peafowl gavaged with 7.5 mg/kg, M1 concentrations remained above the human therapeutic threshold for 12-24 hours.[22] This underscores the importance of assaying for pharmacologically active metabolites in these studies. In AMKEs, 5 mg/kg PO TRAM significantly increased thermal thresholds for 9 hours when compared with baseline thresholds, while 15 and 30 mg/kg PO resulted in less antinociceptive effect.[8] In Muscovy ducks (*Cairina moschata domestica*), 30 mg/kg PO TRAM was effective in improving gait in a temporary arthritis model and assessed by ground-reactive forces measured by a pressure-sensitive walkway (PSW) system.[23] In this study, Muscovy ducks readily produced the M1 metabolite and maintain it well above concentrations known to be therapeutic for humans and other avian species. In African penguins (*Spheniscus demersus*) following 10 mg/kg PO administration, the elimination half-life was 7.3 hours, significantly longer than other species reported.[24] Tramadol and M1 concentrations were consistent with therapeutic concentrations in humans for 12 hours for tramadol and 36 hours for M1.

Tapentadol (TAP) is an analgesic agent which acts as both a μ-opioid receptor agonist and a norepinephrine reuptake inhibitor. It is a common therapeutic agent in human medicine for management of acute and chronic pain and is beginning to be investigated for use in veterinary medicine. The PK of TAP were evaluated in HAPs following a single 30 mg/kg PO of a compounded 5 mg/ml TAP suspension.[25] Low oral bioavailability and short $t_{1/2}$ of this drug in HAPs suggest that this drug might be of limited clinical use in this species.

Non-steroidal anti-inflammatory drugs (NSAIDs)

NSAIDs are used to relieve musculoskeletal and visceral pain, acute pain and chronic pain such as in osteoarthritis. The pharmacologic activity of NSAIDs has been reviewed elsewhere and the mechanism of action is similar when administered to birds. The relative expression of COX-1 and COX-2 enzymes varies between species and both enzymes are important in pain and inflammation. The application and dosages of NSAID formulations continues to be evaluated and clinically applied in birds. The intention of recently developed NSAIDs has been to spare COX-1 and emphasize COX-2 inhibition with the goal of providing analgesia and suppressing inflammation without inhibiting physiologically important prostaglandins. The common NSAIDs used in avian medicine include meloxicam (MEL), carprofen, celecoxib, robenacoxib and piroxicam, each with a distinct COX-1/COX-2 ratio and differing reports of effectiveness and toxicity in birds.

The most common adverse effects of NSAIDs include effects on the gastrointestinal system, renal system, and coagulation. NSAIDs have been implicated in mammals with an increased risk of myocardial infarction and delays in bone healing, but these effects have not been substantiated in birds. These adverse effects are often dose dependent and associated with chronic administration. Originally it was hypothesized that the adverse renal effects of NSAIDs were linked primarily to COX-1 inhibition but COX-2 metabolites have been implicated in maintenance of renal blood flow, mediation of renin release, and regulation of sodium excretion. Therefore, in conditions of relative intravascular volume depletion and/or renal hypoperfusion such as dehydration, hemorrhage, hemodynamic compromise, heart failure, and renal disease, interference with COX-2 activity can have significant deleterious effects on renal blood flow and glomerular filtration rate. The PK of NSAIDs is typically measured from plasma samples but it is important to remember that tissue kinetics may differ significantly, with drugs that have short plasma half-lives being active in tissues long after the drug has disappeared from the plasma. In mammals this has been evaluated using tissue cages where it is possible to measure the effect and duration of the NSAID by quantifying the effect in tissue PGs. A single report describes the use of this model in turkeys for evaluation of sodium salicylate.[26]

Meloxicam (MEL) is the most widely used anti-inflammatory medication today. Meloxicam is a COX-2-selective NSAID that demonstrates COX-1 selectivity at higher doses. Single dose MEL PK studies have been evaluated in a plethora of avian species; references for these can be requested from the authors. Differences in dosage, bioavailability, clearance and elimination half-lives likely contribute to the clinical differences seen with the use of this NSAID. For example, the MEL elimination half-life in chickens and pigeons is three times as long as that of ducks, ostriches and turkeys.[27] Wyandotte hens had greater AUC, longer $t_{1/2}$ and MEL persisted longer in their egg yolks compared with white leghorn hens, suggesting their PO dosing intervals may be every 24 hours and every 12 hours in white leghorns. Based on these results, a 2-week withdrawal time should be adequate to avoid drug residues in eggs meant for consumption.[28,29] Meloxicam was also identified in 2/3 unfertilized eggs of captive bearded vultures after breeding failure, but the implication is unclear.[30] Following 1 mg/kg MEL PO the elimination half-life in grey parrots was twice that of HAPs with significantly lower oral bioavailability (38.1% ± 3.6%).[31] In cockatiels, only 11% oral bioavailability following 1 mg/kg meloxicam PO was found with and elimination half-life of less than 1 hour.[32] The highly variable bioavailability in avian species[33-35] is quite different than mammals where it is usually 85% or greater.[28,29,32] A possible explanation could be variability in proventricular pH among species. In a multiple dose PK study performed in grey parrots evaluating three different dosages, plasma concentrations were greater than effective concentrations described for other avian species. Evidence of moderate drug accumulation was found with both IM and PO routes of administration.[36] The large variability in PK between avian species is one of the main issues with analgesic therapy in birds, as cross-species allometric scaling is not reliable and therefore PK/PD analysis is required in every species.[27] A sustained-release formulation of MEL (MEL SR, Wildlife Pharmaceuticals, Windsor, CO, USA) has been evaluated in two species- HAPs[37] and American flamingos.[38] Its use cannot be recommended due to huge PK variability (parrots) and lack of difference from standard MEL (flamingos).

There are very few studies evaluating the PDs of MEL in birds. In lame chickens, MEL significantly increased the thermal threshold compared to saline.[39] Administration of MEL at 0.5 mg/kg IM or 2 mg/kg IM reduced clinical signs associated with experimental systemic inflammation induced by *E. coli* LPS in broiler chickens, but dexamethasone was ultimately preferred by the authors for its clinical effects of reducing the acute inflammation.[40]

Since higher doses of MEL can block COX-1 enzyme activity, this drug has the potential to induce the adverse effects of gastrointestinal, renal and cardiac toxicity seen in mammals, yet reports of intoxication at doses used in avian practice are minimal. Numerous studies evaluating the adverse effects of MEL in different species[41-45] with no significant adverse effects were reported at doses in therapeutic range; however, 2/9 AMKEs had gastric ulceration after a dose of 20 mg/kg.[44] Plasma and urine N-acetyl-β-d-glucosaminidase activities were significantly increased in grey parrots at 1.6 mg/kg MEL administered PO q24h for 7 days.[36] Elimination half-lives of MEL are highly variable and those species with rapid elimination are less likely to have adverse effects.[32] Anecdotal reports of visceral gout and gastrointestinal ulceration in brown pelicans, using dosages commonly used in other species, prompted a PK study using 0.2 mg/kg PO which found the MEL elimination $t_{1/2}$ was 36 hours, the highest reported to date in any species, heightening suspicion that adverse effects, including renal toxicity, may be attributed to this.[46] The renal effects of NSAIDs have been of particular concern in the Asian vultures, but MEL may be the exception because of its overall short elimination half-life (< 1 hour); however, recently MEL was detected in eggs of captive bearded vultures (*Gypaetus barbatus*).[30] Despite MEL's large therapeutic range and relative safety compared with other NSAIDs, these new data demonstrate that dose extrapolation is not a suitable method for MEL dose determination in avian species. Species-specific PK, efficacy and toxicity studies are necessary to determine appropriate dosages for different species.

The newer NSAID coxibs celecoxib, mavacoxib, robenacoxib are COX-2 selective as opposed to MEL which is COX-2 preferential, inhibiting both COX-1 and COX-2. Celecoxib in humans is used for acute pain, osteoarthritis, rheumatoid arthritis and spondylitis. Its use in veterinary medicine is off label in the United States; it is not approved for use in any veterinary species. Celecoxib 10 mg/kg PO q24h has been used to reduce inflammation associated with avian ganglioneuritis.[47] Anecdotally there are suggestions of very high doses (30-40 mg/kg) PO q12h, then long term 15-20 mg/kg PO q12h for GIT signs, and 60-80 mg/kg divided PO q12h for CNS involvement with ABV.[48] Mavacoxib administered orally has a long mean half-life of 15.5 days in dogs, allowing for monthly dosing after two loading doses administered 1 week apart. [49] Plasma concentrations of 0.4 µg/ml mavacoxib were associated with efficacy in reducing osteoarthritis in dogs.[50]

The single dose PK and oral bioavailability of celecoxib 10 mg/kg PO and mavacoxib 4 mg/kg PO were evaluated in the cockatiel.[32] Mavacoxib was completely absorbed (bioavailability = 111%) but celecoxib only had a mean PO bioavailability of 56%.[32] Bioavailability of the coxibs is often directly associated with feeding status; the presence of food has been shown to increase absorption of both NSAIDs.[51] The cockatiels in this study were fed ad libitum. A further study in cockatiels with ad libitum feeding and additional gavage feedings during the study showed that changes in body temperature, mentation, and posture were less pronounced with a single dose of mavacoxib 4 mg/kg PO compared to celecoxib 10 mg/kg at the tested dosages.[52] This is likely associated with the higher oral bioavailability of mavacoxib in cockatiels compared to celecoxib and MEL. Plasma concentrations of 0.4 µg/ml mavacoxib, effective in osteoarthritic dogs[53] were maintained for 1,000 hours in the cockatiels[32,52] and Caribbean flamingos maintained plasma concentrations > 0.4 µg/ml for 120 hours after a single dose of 6 mg/kg.[54] Elimination half-lives in cockatiels (135 hours)[32] and Caribbean flamingos (*Phoenicopterus ruber*, 74.5 hours)[54] were not as long as in dogs but are still much longer than most NSAID elimination half-lives in other bird species and prompts additional studies to determine dosing regimens.[54]

Piroxicam is a nonselective NSAID used for its anti-inflammatory properties and its value as a chemopreventive and antitumor agent. It has a much higher potency against COX-1 than COX-2. In chickens, intramuscular administration of 1-2 mg/kg of piroxicam twice daily for 4 days failed to result in gastrointestinal lesions consistent with toxicity.[55] Piroxicam has been used in cranes for more than 20 years to treat a variety of conditions including trauma, inflammation, and postoperative pain with minimal observed side effects.[56] A PK study in brolga cranes (*Grus rubicunda*) evaluated 0.5 mg/kg and 1 mg/kg PO, finding that daily administration might be appropriate for this species.[56]

Scavenging bird species such as vultures and eagles may be exposed to NSAIDs when they feed on livestock carcasses. The NSAID diclofenac played the major role in the catastrophic decline in vulture populations on the Indian sub-continent. Subsequently, other NSAIDs have also been identified as toxic to vultures. Ketoprofen and nimesulfide, a sulfonamide NSAID, caused visceral gout and death in vultures whereas MEL and tolfenamic acid (TA) were reported to be safe.[57-61]

Other NSAIDs must be identified to reduce the use of those toxic to vultures and other scavengers. Tolfenamic acid is an NSAID belonging to the fenamates group. This NSAID inhibits both COX-1 and COX-2 receptor activity and exerts potent anti-tumor activity by both COX-dependent and independent pathways. Tolfenamic acid was detected in 20% Eurasian griffon vulture (*Gyps fulvus*) nestling samples, [62] suggesting it is being used more frequently by veterinarians in the region. A study evaluating the safety of TA in three *Gyps* species, all of which are susceptible to diclofenac, was performed with 3.5 mg/kg TA PO via gavage and two birds died with elevated uric acid levels and severe visceral gout; the remainder showed no adverse clinical or biochemical signs at the same or even higher dosages, suggesting the deaths may have been an anomaly due to the high dose used and the high ambient temperatures at the time of the experiment.[58] Tolfenamic acid is likely to be safe to *Gyps* vultures at concentrations encountered by wild birds and could therefore be promoted as a safe alternative to toxic NSAIDs.[58] Otherwise the pharmacokinetics of TA in chukar partridge (*Alectoris chukar*),[63] Japanese quail (*Coturnix japonica*)[64] and geese (*Anser cygnoides*)[65] have been reported. The $t_{1/2\lambda z}$ of TA administered PO in chukar partridges, Japanese quail and geese was similar to MEL.[63,65] These short $t_{1/2\lambda z}$ are important in preventing drug accumulation and delaying toxic effects in repeated administrations in birds.[66] Bioavailability after a 2 mg/kg PO dose was 88%, 78%, and 76% for the IM, SC, and oral routes, respectively in geese, and 57% in Japanese quail suggesting relatively good absorption in these species. Currently there are no specific dosing recommendations for TA in birds but in mammals and reptiles 2-4 mg/kg PO is used.[67,68] While the few studies to date are promising, additional studies evaluating clinical efficacy, multi-dose pharmacokinetics, safety, and residue data of TA should be conducted before widespread use of TA in birds at mammalian doses.

Prostaglandin receptor antagonist

Grapiprant (GRAP) is a prostaglandin E2 (PGE_2) receptor antagonist approved for use in dogs in the United States. Grapiprant binds to the EP4 receptor, blocking the sensitization of sensory neurons and stimulation of inflammation, mediated by the PGE_2 receptor. It does not rely on inhibition of COX enzymes and preserves functions associated with COX enzymes pathways. Oral administration of GRAP at 30 mg/kg to fasted RTHAs achieved plasma concentrations approximately 100 times the canine therapeutic concentrations with no adverse effects identified. It is still unknown whether the dog concentrations are analgesic in birds or toxic in hawks after multi-dosing.[69] Grapiprant administered at the same dosage to the same RTHAs with food had significantly lower plasma concentrations compared to when fasted, with delayed absorption and similar half-lives (Hawkins MG, personal communication, April 1, 2023). Due to the concern that target plasma concentrations may not be maintained if this dosage was administered every 24 hours versus every 12 hours, multi-dose modeling simulations showed that the concentrations troughs are still above the target plasma concentrations in dogs for analgesia (Hawkins MG, personal communication, January 9, 2022). Further studies evaluating the PK/PD of multi-dose GRAP with food in RTHAs are underway.

Gabapentin

Gabapentin is an analog of the neurotransmitter gaba-aminobutyric acid (GABA) that binds to voltage-gated calcium channels acting presynaptically to decrease the release of excitatory neurotransmitters, especially glutamate. In humans and animals, it is best used as an adjunct to other analgesics and seems to be most effective for neuropathic pain. Some formulations of gabapentin contain xylitol and these should not be used in birds until further safety evidence is available. Oral bioavailability was found to be high (> 80%) in HAPs.[70] The terminal $t_{1/2}$ has been in the range of 3-5 hours and so gabapentin should be administered 2-3 times per day to maintain adequate plasma concentrations. In great horned owls (*Bubo virginianus*) following oral administration of 11 mg/kg PO, plasma gabapentin concentrations were maintained above 2 μg/ml for 9.6 hours.[71] In Caribbean flamingos (*Phoenicopterus ruber ruber*), gabapentin dosed at 25 mg/kg PO maintained plasma concentrations above the human therapeutic range for approximately 12 hours; however, dosages as high as 37 mg/kg PO q12h may be needed in some individuals.

Amantadine

Amantadine is an antiviral medication that increases dopamine in the central nervous system and weakly inhibits the NMDA receptors that contribute to chronic pain. Amantadine is synergistic with NSAIDs, opioids, and gabapentin/pregabalin. In humans, amantadine has been shown to decrease opioid use in post-operative patients and aid in treating neuropathic pain. Amantadine administered to dogs at 3-5 mg/kg PO q24h with meloxicam showed significant improvement in management of osteoarthritis pain refractory to meloxicam alone. In OWAPs, the single and multiple dose PK profile was evaluated following 10 mg/kg PO once and 5 mg/kg PO q24h for 7 days.[72] No obvious adverse effects have been observed. Based on these results, once-daily administration of amantadine at 5 mg/kg PO in OWAPs can maintain above target plasma concentrations with weak accumulation. For long term administration, it is recommended to check plasma concentrations over time. The use of amantadine in chickens, ducks, and turkeys is prohibited in the United States by the Federal Drug Administration.

Cannabidiol

Cannabidiol (CBD) oil is the primary non-psychoactive cannabinoid derived from the cannabis plant (*Cannabis sativa*). The pharmacokinetics of a CBD oral formulation (50 mg/ml; Canna Companion, Mill Creek, WA, USA) was evaluated in HAPs. The highly variable results and short half-life of the drug in HAPs, even at high doses, suggests that this formulation would need to be administered at least three to four times daily. Evaluation of other formulations and in other species is recommended to further characterize the analgesic potential of CBD in birds.

References

1. Mikoni NA, Guzman DSM, Fausak E, et al. Recognition and assessment of pain-related behaviors in avian species: an integrative review. *J Avian Med Surg.* 2022;6:153-172.

2. Mikoni NA, Sanchez-Migallon Guzman D, Paul-Murphy J. Pain recognition and assessment in birds. *Vet Clin North Am Exot Anim Pract.* 2023;26:65-81.

3. Paul-Murphy J. Foundations in avian welfare In: Speer BL, ed. *Current Therapy in Avian Medicine and Surgery.* St. Louis, MO: Elsevier; 2015;669-677.

4. National Research Council (US). *Recognition and Alleviation of Pain in Laboratory Animals.* National Academies Press; 2010.

5. Duhamelle A, Raiwet DL, Langlois I, et al. Preliminary findings of structure and expression of opioid receptor genes in a peregrine falcon (*Falco peregrinus*), a snowy owl (*Bubo scandiacus*), and a blue-fronted Amazon parrot (*Amazona aestiva*). *J Avian Med Surg.* 2018;32:173-184.

6. Fousse SL, Golsen BM, Sanchez-Migallon Guzman D, et al. Varying expression of mu and kappa opioid receptors in cockatiels (*Nymphicus hollandicus*) and domestic pigeons (*Columba livia domestica*). *Front Genet.* 2020;11.

7. Guzman DS, Drazenovich TL, Olsen GH, et al. Evaluation of thermal antinociceptive effects after intramuscular administration of hydromorphone hydrochloride to American kestrels (*Falco sparverius*). *Am J Vet Res.* 2013;74:817-822.

8. Guzman DS, Drazenovich TL, Olsen GH, et al. Evaluation of thermal antinociceptive effects after oral administration of tramadol hydrochloride to American kestrels (*Falco sparverius*). *Am J Vet Res.* 2014;75: 117-123.

9. Houck EL, Guzman DS-M, Beaufrère H, et al. Evaluation of the thermal antinociceptive effects and pharmacokinetics of hydromorphone hydrochloride after intramuscular administration to cockatiels (*Nymphicus hollandicus*). *Am J Vet Res.* 2018;79:820-827.

10. Sanchez-Migallon Guzman D, Douglas JM, Beaufrère H, et al. Evaluation of the thermal antinociceptive effects of hydromorphone hydrochloride after intramuscular administration to orange-winged Amazon parrots (*Amazona amazonica*). *Am J. Vet Res.* 2020;81:775-782.

11. Escobar A, Barletta M, Pypendop BH, et al. Pharmacokinetics and pharmacodynamics of methadone administered intravenously and intramuscularly to isoflurane-anesthetized chickens. *Am J. Vet Res.* 2021;82: 181-188.

12. Hawkins MG, Pascoe PJ, DiMaio Knych HK, et al. Effects of three fentanyl plasma concentrations on the minimum alveolar concentration of isoflurane in Hispaniolan Amazon parrots (*Amazona ventralis*). *Am J. Vet Res.* 2018;79:600-605.

13. Delaski KM, Gehring R, Heffron BT, et al. Plasma concentrations of fentanyl achieved with transdermal application in chickens. *J Avian Med Surg.* 2017;31:6-15.

14. Escobar A, Valadão CA, Brosnan RJ, et al. Cardiopulmonary effects of butorphanol in sevoflurane-anesthetized guineafowl (*Numida meleagris*). *Vet Anaesthes Analg.* 2014;41:284-289.

15. Clancy MM, KuKanich B, Sykes JM. Pharmacokinetics of butorphanol delivered with an osmotic pump during a seven-day period in common peafowl (*Pavo cristatus*). *Am J Vet Res.* 2015;76:1070-1076.

16. Guzman DS-M, Knych HK, Olsen GH, et al. Pharmacokinetics of a sustained release formulation of buprenorphine after intramuscular and subcutaneous administration to American kestrels (*Falco sparverius*). *J Avian Med Surg.* 2017;31:102-107.

17. Guzman DS, Ceulemans SM, Beaufrère H, et al. Evaluation of the thermal antinociceptive effects of a sustained-release buprenorphine formulation after intramuscular administration to American kestrels (*Falco sparverius*). *J Avian Med Surg.* 2018;32:1-7.

18. Gleeson MD, Guzman DSM, Knych HK, et al. Pharmacokinetics of a concentrated buprenorphine formulation in red-tailed hawks (*Buteo jamaicensis*). *Am J Vet Res.* 2018;79:13-20.

19. Guzman DS-M, Houck EL, Knych HKD, et al. Evaluation of the thermal antinociceptive effects and pharmacokinetics after intramuscular administration of buprenorphine hydrochloride to cockatiels (*Nymphicus hollandicus*). *Am J Vet Res.* 2018;79:1239-1245.

20. Souza MJ, Martin-Jimenez T, Jones MP, et al. Pharmacokinetics of oral tramadol in red-tailed hawks (*Buteo jamaicensis*). *J Vet Pharmacol Ther.* 2011;34:86-88.

21. Sanchez-Migallon Guzman D, Souza MJ, Braun JM, et al. Antinociceptive effects after oral administration of tramadol hydrochloride in Hispaniolan Amazon parrots (*Amazona ventralis*). *Am J Vet Res.* 2012;73: 1148-1152.

22. Black PA, Cox SK, Macek M, et al. Pharmacokinetics of tramadol hydrochloride and its metabolite O-desmethyltramadol in peafowl (*Pavo cristatus*). *J Zoo Wildl Med.* 2010;41:671-676.

23. Bailey RS, Sheldon JD, Allender MC, et al. Analgesic efficacy of tramadol compared with meloxicam in ducks (*Cairina moschata domestica*) evaluated by ground-reactive forces. *J Avian Med Surg.* 2019;33: 133-140.

24. Kilburn JJ, Cox SK, Kottyan J, et al. Pharmacokinetics of tramadol and its primary metabolite O-desmethyltramadol in African penguins (*Spheniscus demersus*). *J Zoo Wildl Med.* 2014;45:93-99.

25. Duvall A, Tully Jr TN, Carpenter JW, et al. Pilot study of a single dose of orally administered tapentadol suspension in Hispaniolan Amazon parrots (*Amazona ventralis*). *J Avian Med Surg.* 2021;35:45-50.

26. Cramer K, Schmidt V, Richter A, et al. Investigations on the acute, carrageenan-induced inflammatory reaction and pharmacology of orally administered sodium salicylate in turkeys. *Berl Munch Tierarztl Wochenschr.* 2015;128:240-251.

27. Baert K, De Backer P. Comparative pharmacokinetics of three non-steroidal anti-inflammatory drugs in five bird species. *Comp Biochem Physiol C Toxicol Pharmacol.* 2003;134:25-33.

28. Souza MJ, Bailey J, White M, et al. Pharmacokinetics and egg residues of meloxicam after multiple day oral dosing in domestic chickens. *J Avian Med Surg.* 2018;32:8-12.

29. Souza MJ, Gerhardt LE, Shannon L, et al. Breed differences in the pharmacokinetics of orally administered meloxicam in domestic chickens (*Gallus domesticus*). *J Am Vet Med Assoc.* 2021;259:84-87.

30. Zorrilla I, Richards NL, Benítez JR, et al. Case study: detection of two nonsteroidal anti-inflammatory drugs (NSAIDs) in the eggs of captive-reared bearded vultures at a breeding center in southern Spain. *J Wildl Rehabil.* 2018;38.

31. Montesinos A, Ardiaca M, Gilabert JA, et al. Pharmacokinetics of meloxicam after intravenous, intramuscular and oral administration of a single dose to African grey parrots (*Psittacus erithacus*). *J Vet Pharmacol Ther.* 2017;40:279-284.

32. Dhondt L, Devreese M, Croubels S, et al. Comparative population pharmacokinetics and absolute oral bioavailability of COX-2 selective inhibitors celecoxib, mavacoxib and meloxicam in cockatiels (*Nymphicus hollandicus*). *Sci Rep.* 2017;7:12043.

33. Montesinos A, Ardiaca M, Diez-Delgado I. Pharmacokinetics of parenteral and oral meloxicam in African grey parrots (*Psittacus erithacus erithacus*). *Proc Annu Conf Eur Assoc Avian Vet.* 2011;129-134.

34. Molter CM, Court MH, Cole GA, et al. Pharmacokinetics of meloxicam after intravenous, intramuscular, and oral administration of a single dose to Hispaniolan Amazon parrots (*Amazona ventralis*). *Am J Vet Res.* 2013;74:375-380.

35. Lacasse C, Gamble KC, Boothe DM. Pharmacokinetics of a single dose of intravenous and oral meloxicam in red-tailed hawks (*Buteo jamaicensis*) and great horned owls (*Bubo virginianus*). *J Avian Med Surg.* 2013;27:204-210.

36. Montesinos A, Encinas T, Ardiaca M, et al. Pharmacokinetics of meloxicam during multiple oral or intramuscular dose administration to African grey parrots (*Psittacus erithacus*). *Am J Vet Res.* 2019;80:201-207.

37. Guzman DS, Court MH, Zhu Z, et al. Pharmacokinetics of a sustained-release formulation of meloxicam after subcutaneous administration to Hispaniolan Amazon parrots (*Amazona ventralis*). *J Avian Med Surg.* 2017;31:219-224.

38. Sim RR, Cox SK. Pharmacokinetics of a sustained-release formulation of meloxicam after subcutaneous administration to American flamingos (*Phoenicopterus ruber*). *J Zoo Wildl Med.* 2018;49:839-843.

39. Hothersall B, Caplen G, Parker RM, et al. Thermal nociceptive threshold testing detects altered sensory processing in broiler chickens with spontaneous lameness. *PloS One.* 2014;9:e97883.

40. Nakhaee P, Mosleh N, Nazifi S, et al. Comparative effects of meloxicam and dexamethasone in chickens with experimental systemic inflammation: clinical outcome and cardiovascular parameters. *Comp Clin Pathol.* 2021;30:681-691.

41. Montesinos A, Ardiaca M, Juan-Sallés C, et al. Effects of meloxicam on hematologic and plasma biochemical analyte values and results of histologic examination of kidney biopsy specimens of African grey parrots (*Psittacus erithacus*). *J Avian Med Surg.* 2015;29:1-8.

42. Miller KA, Hill NJ, Carrasco SE, et al. Pharmacokinetics and safety of intramuscular meloxicam in zebra finches (*Taeniopygia guttata*). *J Am Assoc Lab Anim Sci.* 2019;58:589-593.

43. Desmarchelier M, Troncy E, Fitzgerald G, et al. Analgesic effects of meloxicam administration on postoperative orthopedic pain in domestic pigeons (*Columba livia*). *Am J Vet Res.* 2012;73:361-367.

44. Summa NM, Guzman DS-M, Larrat S, et al. Evaluation of high dosages of oral meloxicam in American kestrels (*Falco sparverius*). *J Avian Med Surg.* 2017;31:108-116, 109.

45. Dijkstra B, Guzman DS, Gustavsen K, et al. Renal, gastrointestinal, and hemostatic effects of oral administration of meloxicam to Hispaniolan Amazon parrots (*Amazona ventralis*). *Am J Vet Res.* 2015;76:308-317.

46. Horgan MD, Knych HK, Siksay SE, et al. Pharmacokinetics of a single dose of oral meloxicam in rehabilitated wild brown pelicans (*Pelecanus occidentalis*). *J Avian Med Surg.* 2020;34:329-337.

47. Hoppes SM, Tizard I, Shivaprasad HL. Avian bornavirus and proventricular dilatation disease. *Vet Clin North Am Exot Anim Pract.* 2013;339-355.

48. Rossi G, Dahlhausen RD, Galosi L, et al. Avian ganglioneuritis in clinical practice. *Vet Clin North Am Exot Anim Pract.* 2018;21:33-67.

49. Lees P, Pelligand L, Elliott J, et al. Pharmacokinetics, pharmacodynamics, toxicology and therapeutics of mavacoxib in the dog: a review. *J Vet Pharmacol Ther.* 2015;38:1-14.

50. Payne-Johnson M, Becskei C, Chaudhry Y, et al. Comparative efficacy and safety of mavacoxib and carprofen in the treatment of canine osteoarthritis. *Vet Rec.* 2014.

51. Cox SR, Lesman SP, Boucher JF, et al. The pharmacokinetics of mavacoxib, a long-acting COX-2 inhibitor, in young adult laboratory dogs. *J Vet Pharmacol Ther.* 2010;33:461-470.

52. Gasthuys E, Houben R, Haesendonck R, et al. Development of an in vivo lipopolysaccharide inflammation model to study the pharmacodynamics of COX-2 inhibitors celecoxib, mavacoxib, and meloxicam in cockatiels (*Nymphicus hollandicus*). *J Avian Med Surg.* 2019;33:349-360.

53. Cox SR, Liao S, Payne-Johnson M, et al. Population pharmacokinetics of mavacoxib in osteoarthritic dogs. *J Vet Pharmacol Ther.* 2011;34:1-11.

54. Huckins GL, Carpenter JW, Dias S, et al. Pharmacokinetics of oral mavacoxib in Caribbean flamingoes (*Phoenicopterus ruber ruber*) *J Zoo Wildl Med.* 2020;51:53-58.

55. Awan AF, Ashraf M, Malik A, et al. Toxicity study of piroxicam in broilers. *Lat Am J Pharm.* 2012;31.

56. Keiper NL, Cox SK, Doss GA, et al. Pharmacokinetics of piroxicam in cranes (Family Gruidae). *J Zoo Wildl Med.* 2017;48:886-890, 885.

57. Galligan TH, Mallord JW, Prakash VM, et al. Trends in the availability of the vulture-toxic drug, diclofenac, and other NSAIDs in South Asia, as revealed by covert pharmacy surveys. *Bird Conserv Intl.* 2021;31: 337-353.

58. Chandramohan S, Mallord JW, Mathesh K, et al. Experimental safety testing shows that the NSAID tolfenamic acid is not toxic to *Gyps* vultures in India at concentrations likely to be encountered in cattle carcasses. *Sci Total Environ.* 2022;809:152088.

59. Galligan TH, Green RE, Wolter K, et al. The non-steroidal anti-inflammatory drug nimesulide kills *Gyps* vultures at concentrations found in the muscle of treated cattle. *Sci Total Environ.* 2022;807:150788.

60. Nambirajan K, Muralidharan S, Ashimkumar AR, et al. Nimesulide poisoning in white-rumped vulture *Gyps bengalensis* in Gujarat, India. *Environ Sci Pollut Res.* 2021:1-7.

61. Stokstad E. Vultures face new toxic threat. *Science.* 2021;373:1187-1187.

62. Gómez-Ramírez P, Blanco G, García-Fernández AJ. Validation of multi-residue method for quantification of antibiotics and NSAIDs in avian scavengers by using small amounts of plasma in HPLC-MS-TOF. *Int J Environ Res Public Health.* 2020;17.

63. Cetin G, Corum O, Corum D, et al. Pharmacokinetics of intravenous meloxicam, ketoprofen and tolfenamic acid in chukar partridge (*Alectoris chukar*). *Br Poult Sci.* 2021:1-7.

64. Turk E, Tekeli IO, Corum O, et al. Pharmacokinetics of meloxicam, carprofen, and tolfenamic acid after intramuscular and oral administration in Japanese quails (*Coturnix coturnix japonica*). *J Vet Pharmacol Ther.* 2021;44:388-396.

65. Turk E, Tekeli IO, Durna Corum D, et al. Pharmacokinetics of tolfenamic acid after different administration routes in geese (*Anser cygnoides*). *J Vet Pharmacol Ther.* 2021;44(3):381-387.

66. Naidoo V, Wolter K, Cromarty AD, et al. The pharmacokinetics of meloxicam in vultures. *J Vet Pharmacol Ther.* 2008;31:128-134.

67. Corum O, Atik O, Durna Corum D, et al. Pharmacokinetics of tolfenamic acid in red-eared slider turtles (*Trachemys scripta elegans*). *Vet Anaesth Analg.* 2019;46:699-706.

68. Tekeli IO, Turk E, Durna Corum D, et al. Effect of dose on the intravenous pharmacokinetics of tolfenamic acid in goats. *J Vet Pharmacol Ther.* 2020;43:435-439.

69. Rodriguez P, Paul-Murphy JR, Knych HK, et al. Pharmacokinetics of grapiprant administered to red-tailed hawks (*Buteo jamaicensis*) after food was withheld for 24 hours. *Am J Vet Res.* 2021;82:912-919.

70. Baine K, Jones MP, Cox S, et al. Pharmacokinetics of compounded intravenous and oral gabapentin in Hispaniolan Amazon parrots (*Amazona ventralis*). *J Avian Med Surg.* 2015;29:165-173.

71. Yaw TJ, Zaffarano BA, Gall A, et al. Pharmacokinetic properties of a single administration of oral gabapentin in the great horned owl (*Bubo virginianus*). *J Zoo Wildl Med.* 2015;46:547-552.

72. Berg KJ, Sanchez-Migallon Guzman D, Knych HK, et al. Pharmacokinetics of amantadine after oral administration of single and multiple doses to orange-winged Amazon parrots (*Amazona amazonica*). *Am J Vet Res.* 2020;81:651-655.

Section 9

AAV Special Panel on Geriatrics

Shangzhe Xie, BSc/BVMS, MVS, PhD, DABVP (Avian Practice), DACCM
Session Chair

Avian Geriatric Diseases

Sharman Hoppes, DVM, DABVP (Avian Practice),
Scott Echols, DVM, DABVP (Avian Practice),
Yvonne van Zeeland, DVM, MVR, PhD, DipECZM (Avian),
DipECZM (Small Mammal), CPBC,
Susan Orosz, PhD, DVM, DABVP (Avian Practice), DipECZM (Avian)

Session #3282

Affiliation: From Texas Avian & Exotic Hospital, Grapevine, TX 76051, USA, (Hoppes), Medical Center for Birds, 3807 Main Street, Oakley, CA 94561, USA (Echols), Parrish Creek Veterinary Hospital and Diagnostic Center Centerville, UT 84014, USA (Echols), Echols Veterinary Services, Cottonwood Heights, UT 84093, USA (Echols), Scarlet Imaging, PO Box 57515, Murray, UT 84157, USA (Echols), Division of Zoological Medicine, Department of Clinical Sciences of Companion Animals, Faculty of Veterinary Medicine, Utrecht University, Utrecht, The Netherlands (van Zeeland), Bird and Exotic Pet Wellness Center, Toledo, OH 43623, USA (Orosz).

Abstract: As a result of improved psittacine husbandry and nutrition, and advances in the diagnosis and treatment of diseases, life expectancy of captive parrots has substantially increased over the years. This in turn has led to a growing number of elderly birds seen in veterinary practices and associated with it a higher incidence of geriatric-related diseases such as cataracts, cardiovascular disease, chronic lung, hepatic or renal disease, spinal and joint disease, cystic ovarian disease, and neoplasia. In this masterclass, four of the most common conditions seen in aging psittacine birds (i.e., atherosclerosis and cardiac disease, osteoarthritis, renal disease and neoplastic conditions of the reproductive system) are discussed including their clinical presentation, diagnostics, and the preventive and therapeutic management. A panelist team, consisting of specialists in avian medicine that work in private practice and academia present their approach to each of these conditions from their respective backgrounds to help provide a broad and international perspective on the approach to these conditions. This session provides practitioners with guidelines for immediate implementation in practice, as well as insight into some of the more advanced, novel, or future options to optimize care for the aging psittacine bird.

Introduction

Avian geriatric medicine comprises the care, treatment, and management of the elderly bird. This presentation focuses primarily on pet parrots. In the past, few pet birds lived to be geriatric due to poor husbandry, nutrition, and infectious diseases. Over the years, avian veterinarians, avian industry, and the pet bird owner have gained knowledge in husbandry, nutrition, medicine, and surgery, hence resulting in an increased life expectancy of companion birds. Nevertheless, despite their increased longevity, few parrots outlive their owners. Despite life spans of up to 60 to 80 years having been documented (dependent on the species), most parrots do not live that long as a result of a sedentary lifestyle, over abundant nutrition, and possible genetic factors.

Organ Changes Over Time

Physiologic changes that occur in geriatric pet birds have not been well studied. Similarly, data from poultry cannot be easily extrapolated because studies are based on reproductive capacity for production and not on a bird's actual potential life span.

In humans and small animals, there are a number of documented age-related issues. In a study looking at young versus older dogs (> 10 years), older dogs had significantly lower venous oxygen saturation and lower plasma glucose after exercise, and a slower hematopoietic response to acute anemia compared to young dogs.[1]

In humans over 80 years of age, a hypercoagulable shift can be noted, which is associated with a higher incidence of pulmonary thromboembolism. Similar changes occur in some species of small mammals. The thoracic cage becomes more rigid, the lungs less elastic, and respiratory muscle strength decreases by 25%. Respiratory diseases are generally more common and more difficult to treat in elderly patients of all species.[1] In addition, both older humans and older parrots are prone to atherosclerosis, which leads to hypertension and cardiac disease.[2]

In mammals, glomerular filtration rate, urine concentrating and diluting ability and creatinine clearance all diminish with age, resulting in decreased renal function.[2] Similarly, a decrease in liver function is common in geriatric mammals. Geriatric mammalian patients can have up to 50% age related reduction in liver mass. Less information is available on the effects of age on the avian kidneys and liver, but we do know that older avian patients have an increased incidence of renal and liver disease overall, of which many are chronic in nature.[1,2]

Osteoarthritis is common in geriatric patients of all species, including older companion birds. Older animals lose lean body muscle mass which is replaced by fat. Aging changes in the musculoskeletal system, most notably the cells and extracellular matrix of joint tissues, likely increase the susceptibility of older individuals to osteoarthritis, especially when other risk factors such as joint injury, obesity, sedentary lifestyle, or anatomic features that affect joint mechanics are present.

Aging also affects the reproductive system, leading to changes in gonadal tissues, decreased production of sperm, declining stores of primary oocytes, and overall reduced fertility as a result of changing hormone levels. Preliminary comparisons of reproductive aging of female hens, quail, and budgerigars with mammals of similar size show that birds age slightly to much slower than their mammalian counterparts.[3,4] Nevertheless, certain reproductive diseases are seen more frequently in older hens (e.g., cystic ovarian disease, neoplasms).

Cognitive changes have been documented in many mammalian mammals; however, effects of aging on avian cognition remain largely unknown. There are anecdotal accounts from owners and veterinarians that suggest senility may occur. For example, changes in behavior and cognition have been reported in birds with cardiac disease or atherosclerosis.

Geriatric Diseases

Defining geriatric age in parrots is challenging. The life span of many parrots has been documented to be 60 to 80 years in larger parrots and 30 to 40 years in smaller parrots; however, due to malnutrition or over abundant feeding and obesity, combined with the sedentary lifestyles of most pet birds, many parrots develop old-age diseases much earlier than expected. Since many owners still view their bird as a teenager at age 20, they are hardly prepared for their bird to be diagnosed with cardiac disease or atherosclerosis.

The list of common geriatric diseases in the bird is similar to the list for mammals, including humans. Osteoarthritis, cataracts, metabolic bone disease (osteoporosis), reproductive disease, renal disease, hepatic disease, atherosclerosis, cardiac disease, and neoplasia are amongst the diseases commonly seen in the elderly patient. In this masterclass osteoarthritis, metabolic bone disease, renal disease, atherosclerosis and cardiac disease, and reproductive disease, most notably neoplastic conditions of the reproductive system, are discussed.

Advanced Imaging

New imaging technology, most notably computed tomography (CT) and to a lesser degree magnetic resonance imaging (MRI), has been instrumental in identifying many geriatric diseases. That statement does not discount the benefits of high-resolution radiograph and ultrasound studies. This presentation focuses on the use of newer imaging technology to identify the diseases discussed in this masterclass.

Newer high-resolution flat-panel CT scanners and 3T and greater MRI often are required to identify fine details necessary for accurate diagnoses in small patients. While older CT and MRI technology may be acceptable for identifying large masses and gross changes that may otherwise be identified via physical examination and radiographs and/or ultrasound, high resolution and fine detail obtained from newer scanners are often required to make accurate diagnoses.

Contrast studies for both CT and MRI improve the diagnosis of normal versus abnormal tissue. Without contrast, CT (ionizing radiation) is only capable of differentiating five basic densities (same as with radiographs): air, fat, soft tissue, bone and metal. Properly performed contrast studies provide greater differentiation of soft tissue structures. This allows the reviewer to visualize blood vessels, different organs, masses and other structures that would otherwise appear as one density. Contrast studies in MRI (radiofrequency and magnetization) also help to better differentiate soft tissue structures and highlight pathology.

Osteoarthritis

The prevalent form of arthritis in birds is osteoarthritis (OA). Osteoarthritis is a common cause of lameness and chronic pain, particularly in psittacine birds over 15 years of age and in birds that have experienced trauma. It also occurs at a younger age in waterfowl and poultry kept as pets, especially those kept on packed earth, concrete or other hard surfaces. Bumblefoot, ulcerative dermatitis, and callus formation may be secondary effects of OA. Osteoarthritis is caused by damage to chondrocytes. This damage or stress on the cartilage may be related to trauma, obesity, normal wear on abnormal joints or abnormal wear on normal joints. Osteoarthritis is a progressive disease; unless the cycle can be interrupted, further degradation of the cartilage causes continued damage to the joint.

Joint pain is the essential problem associated with OA. The synovium and subchondral bone are the major sources of OA pain, but pain can also arise from periarticular muscles and ligaments. Clinical signs can vary, depending on the location and severity of the disease. Birds may present with lameness and a history of decreased activity, inability or reluctance to perch, falling off the perch, or altered preference for perches. In addition, a flighted bird may not want to fly or cannot fly as well as it used to fly. Some birds may demonstrate a poor grip response, which may mimic more commonly recognized disorders such as leg fractures, renal disease, nerve trauma, heavy metal poisoning and more. Other, less obvious signs of OA include swollen joints, decreased range of motion, self-injurious behavior such as feather destructive behavior or mutilation, and excessive vocalization or biting/aggression.

Diagnosis is based on clinical signs, and findings from the physical examination and radiographs or CT. Radiographic lesions include narrowing of the joint space, sclerosis of subchondral bone, misalignment of the joint, and osteophyte formation. Such changes are more readily recognized using a CT scan, as this is a more sensitive tool for diagnosing osteoarthritis than radiology. Commonly affected joints include the phalangeal joints, the tibiotarsus-tarsometatarsal joint and stifle joint. Joints of the thoracic limb appear less commonly affected than the pelvic limb, probably due to many companion birds not being flighted.[2] The shoulder appears to be the most commonly identified wing joint with OA based on CT evaluation (MSE). Another, possibly most underdiagnosed region is in the spine. While high-resolution CT is often required for diagnosis, one author (MSE) would argue that spinal arthritis is fairly common in aged birds.

Although not formally adopted in avian medicine, the Kellgren and Lawrence system for radiographic classification of osteoarthritis and the more recently published CT arthritis classification system used in humans may serve as a guide in birds.[5-7] The tenets of Kellgren and Lawrence system for radiographic OA are described in Table 1.

Table 1. Kellgren and Lawrence system for radiographic osteoarthritis.

Grade	Description
Grade 0 (none)	Absence of radiographic changes of osteoarthritis
Grade 1 (doubtful)	Questionable joint space narrowing and possible osteophytic lipping
Grade 2 (minimal)	Possible joint space narrowing and clearly defined osteophytes
Grade 3 (moderate)	Definite joint space narrowing with some sclerosis and possibly deformed bone ends and moderate multiple osteophytes
Grade 4 (severe)	Severely narrowed joint space, severe sclerosis and definite deformity of bone ends with large osteophytes

The CT scoring system used in humans describes osteophytes sizes that may be too large for most bird species. The basic concepts mostly involve evaluation of the joint space, presence of osteophytes and identifying cysts and can be applied to avian patients.[7]

A multimodal treatment plan is recommended to provide the best outcome for the patient. Additionally, use of multiple drugs may allow for decreased drug dosages, tapered to the individual drug to prevent toxicity (which may be especially important given the increased incidence of renal or liver disease at old age). Management of OA incorporates both medical and nonmedical modalities to achieve therapeutic goals. Medical options include the (combined) use of nonsteroidal anti-inflammatory drugs (NSAIDs), chondroprotectants, and possibly opioids. The most common NSAID utilized in avian medicine is meloxicam, a COX-2 inhibitor, which is commonly used at a dose of 1-1.6 mg/kg PO q12-24h . Celebrex can be used at 20 mg/kg q12-24h for up to several months. There are potential side effects of NSAIDs such as renal ischemia or gastric ulceration. NSAIDs should be used with caution, with the goal to use the lowest therapeutic dose possible to reduce the risk of adverse effects. In addition, gastric protectants (e.g., sucralfate) or antacids (e.g., omeprazole, ranitidine, cimetidine) may be added in patients in which long-term treatment with NSAIDs is indicated. Chondroprotectants, including glucosamine, glycosaminoglycan and chondroitin can also be useful for treatment of OA, as these have been shown useful to slow the progression of arthritis and pain in dogs. Glucosamine has been shown to promote production and repair of cartilage and has anecdotally been used in birds at dosages of 20-40 mg/kg q12h or 35 mg/kg PO q24-48h. Adequan, a polysulfated glycosaminoglycan, has also been used in birds with OA without apparent side effects when used in dosages of 5 mg/kg IM once weekly for 4 weeks, then once monthly. As a note of caution, fatal hemorrhage has been reported in two raptors, one psittacine and one coraciiforme after an intramuscular injection of polysulfated glycosaminoglycans.[8] For chondroitin sulfate, which promotes water retention, elasticity of cartilage and inhibits breakdown of cartilage, no known dosages exist in birds.

Opioids may be needed for acute exacerbations of pain due to OA, or for birds that initially do not respond adequately to NSAIDs. Depending on the species, tramadol at dosages of 11 mg/kg PO q6-12h up to 30 mg/kg PO q12h) or butorphanol (0.5-3 mg/kg IM q4h) may be used for a short period of time until the anti-inflammatory effects of the NSAIDs take effect. Gabapentin (10-25 mg/kg PO q8-12h) is also being used more in avian medicine for both neuropathic and chronic pain. Gabapentin is a lipophilic structural analogue of the inhibitory neurotransmitter GABA. This drug was initially developed as an anticonvulsant and is now indicated for the treatment of seizures and neuropathic pain. In mammals, gabapentin has been shown to be highly effective as an adjunctive therapy (i.e., combined with opioids or NSAIDs).

In addition to aforementioned medications, omega-3 fatty acids, ginger and turmeric have been used in birds with OA.[9] Omega-3 fatty acids (e.g., VetOmega; 0.22-0.44 mL/kg PO daily) have been shown in many species to be anti-inflammatory and renal protective (an added bonus when using NSAIDs which can be damaging to the kidneys in high doses or in renal compromised or dehydrated patients). Ginger and turmeric have similar anti-inflammatory effects, but not the side effects of NSAIDs. Ginger (gingerols), which inhibits platelet cyclo-oxygenase synthase, thereby leading to a decrease of thromboxanes and leukotrienes, is suggested to have greater COX-1 inhibition than aspirin as well as being a potent inhibitor of COX-2.[10] It appears to have hepatoprotective properties, which can be helpful for older birds that are suspected of or diagnosed with liver disease. Turmeric is considered to have anticarcinogenic properties.[11] White willow bark is the original source of aspirin; however, it does not inhibit platelet aggregation. The active ingredients are converted from salicylic acid to acetylsalicylic acid. As such, this herb has also been implemented in the treatment of osteoarthritis, muscle pain, and gout.[12]

Non-medical treatment modalities for OA include husbandry changes, weight management, an exercise plan, feeding a healthy diet rich in omega-3 fatty acids, and physical and/or laser therapy. An exercise plan in the captive avian patient can be challenging, particularly in non-flighted birds. In flighted birds, allowing them to fly or encouraging flight is the best form of exercise, provided that it can be done safely. If this is not possible, encouraging climbing and walking in birds, or even repeatedly stepping up and down for several times during the day can be exercise for parrots. Foraging for food can also be used to encourage exercise (e.g., by scatter feeding or placing foraging boxes on opposite sides of the cage or enclosure). Moreover, providing foraging opportunities can entertain and distract the bird from its discomfort, thereby diminishing feather destructive behavior and mutilation.

Physical therapy is important for any patient with OA as it encourages blood flow to the joints and cartilage and strengthens the muscles which helps to support the joints. Should the bird be overweight, weight loss is essential as studies have identified obesity as a risk factor for OA in many species. Weight management includes transitioning the bird slowly to a healthy pelleted diet with the addition of omega-3 fatty acids. Husbandry changes can dramatically affect the life of a bird with OA. Cages should be equipped with adequately sized perches and platforms for the bird to rest on. Perches may need to be lowered to prevent trauma if a patient is frequently falling and can be wrapped with self-adhesive bandage or felt to provide better grip. For smaller birds, hutches can be provided as an alternative means of support. In severe cases, the bird may need to be moved to a box or flat surface environment with soft bedding or towels to prevent bumblefoot. To prevent the bird from getting its nails caught in toys, floor or bars of the enclosure, nails should be trimmed and dull. The toenails should be trimmed and dull to prevent them from getting caught in toys or the bars of the cage.

Metabolic Bone Disease

Nutritional secondary hyperparathyroidism, or metabolic bone disease, occurs commonly in elderly birds that are on a poor diet and housed indoors. Diets deficient in calcium and vitamin D and high in phosphorus (i.e., all seed diets) are risk factors. The decreased bone density that results from the vitamin/mineral imbalance increases the risk for pathologic fractures during restraint or following a fall.[13]

Clinical signs include weakness, ataxia, inability to perch, and lameness as a result of (pathologic) fractures. Reproductively active hens may present with egg binding or cloacal prolapse.

Diagnosis is based on the bird's history, which typically reveals a sedentary bird on a poor diet, and findings during the physical exam. In addition, the bird may present with decreased blood calcium (plasma and ionized), increased phosphorus, and decreased vitamin D (25-hydroxycholecalciferol) plasma levels, while radiographs or CT may reveal decreased bone density and/or pathologic fractures.

While bone density evaluation using radiographs is quite subjective (pixel intensity), calibrated CT scans can be used to measure Hounsfield Units which is a measurement of voxel density. Most common DICOM viewers have measurement tools which allow the viewer to select a region to measure density. Numerous bone density studies have been published, mostly relative to poultry species.[14-16] At this time, there are no normal established bone density parameters for captive companion birds. Measurements can be taken between normal-appearing bone and abnormal (usually more radiolucent) bone in different birds of the same species. Bone health goes beyond the scope of this discussion and is a complex balance between developmental nutrition, long term nutrition, physical activity, sunlight exposure (importance varies highly between species), reproductive activity, sex and presence of certain diseases.

Some generalizations can be made when evaluating bone density. Birds that spend most of their time perched and do not climb around much will have poor relative bone density along the proximal femurs and at the stifle joint. Birds that are non or poor flyers tend to have low relative bone density at the proximal humeri and coracoids. The combination of limited flying and climbing may additionally result in poor relative density along the notarium and synsacrum. Focal or regional osteoporosis and can be associated with a variety of health concerns including disuse atrophy (as can be seen with certain orthopedic and neurologic diseases), localized bone cancer and infections, arthritis and more.

Reproductive activity and subsequent physiologic (and abnormal) hyperostosis can make bone density determination challenging. Hyperostosis is a physiologically normal phenomenon in hens that allows for rapid mobilization of calcium and other bone minerals to move from the cortex into the medullary cavity. By having these dense elements in the medullary space, minerals can be rapidly pulled from the medulla into the blood stream and used to make the eggshell. Short term, this physiologic process results in minor mineral depletion from the cortex that is (under normal circumstances and with a good diet) replenished once the hen is no longer reproductively active. Long term, this process results in cortex thinning and osteoporosis with increased risk of fracture development.

Radiographically, medullary hyperostosis makes the medulla appear more dense than normal. While the overall bone density (cortex and medulla combined) may be normal on a volume bases, the cortex thickness and density often suffer with chronic medullary hyperostosis. As has been reported in poultry, this may result in micro (common in the spine, costrochondral junctions and more) and macro fractures.[17] Micro CT is often needed to visualize small fractures; however, secondary features such as arthritis and poor relative density can be seen with high-resolution clinical CT scans. Specific to spinal osteoporosis, one author (MSE) has noticed a tendency towards increased incidences of intervertebral disk disease, spondylosis and other spinal column diseases as viewed on high-resolution clinical CT scans. This statement requires further investigation to prove a true correlation between osteoporosis and spinal disease.

Treatment includes dietary modifications (converting to a diet appropriate for the species), providing ultraviolet lighting (indoor) or, ideally time outdoors, increasing physical activity, reducing reproductive activity (especially if inappropriate or chronic), managing underlying disease (obesity, arthritis, etc), and providing supplemental omega-3 fatty acids. Omega-3 fatty acids added to the diet has been shown to increase bone mineral density and/or improve markers of bone health in multiple bird models.[18,19] Physical activity appears to be necessary for omega-3 fatty acids to have maximum benefit to bone health. Calcium supplementation and modifying the cage to prevent falling are also advised. In the case of an existing fracture, strict cage rest is recommended whereby climbing should be minimized and any potential for falling be avoided. Research has shown that birds may benefit from both oral and UV-B delivered vitamin D_3. Regardless, proper diet, exercise and managing the other factors described above are still required for proper bone health. Limited research also suggests African grey parrots (*Psittacus erithacus*) have a greater dependence on UV-B light for maintaining adequate serum calcium levels than do *Amazona* spp.[13] This suggests that species differences may account for unique UV light needs.

Renal Disease

Birds can develop renal disease at any age, but it is more common in the older bird. Often, there is a history of poor diet and inappropriate husbandry. Renal disease may also be due to infectious diseases, toxin exposure, poor genetics, neoplasia and more. Diets deficient in vitamin A, high in phosphorus, calcium or vitamin D, or just long-term generalized malnutrition from an all-seed diet can lead to renal disease. Numerous viruses can infect and affect the kidneys including polyoma virus, psittacine herpes virus, and possibly also avian borna virus (ABV), which has been found in the kidney, is shed in the birds' urine, and may induce polyuria/polydipsia (PU/PD), and lead to elevated blood uric acid levels. Chronic bacterial infections or septicemia can potentially lead to (chronic) nephritis, tubular dilatation, glomerulopathies and more, which in turn can lead to chronic renal insufficiency or renal failure.[20]

Clinical signs of renal disease include reduced or poor appetite, weight loss, PU/PD, depression, regurgitation, and (unilateral) lameness or paresis. Owners often report that the bird is less active and quieter. In some cases, articular or visceral gout may develop. Visceral gout is the accumulation of uric acid crystals on the internal organs (e.g., heart, kidney, spleen), and is often seen in advanced disease, and is usually fatal. Articular gout is the accumulation of uric acid crystals in the joints and tendons, particularly in the feet and hocks and is more commonly seen as a side effect of chronic renal disease. Uric acid crystals irritate surrounding tissues and cause severe inflammation and pain. The prognosis for gout is usually grave.

Diagnosis of renal disease includes blood work, which reveals persistent hyperuricemia that may be accompanied by hyperphosphatemia, hypocalcemia, and changes in electrolytes. A CBC may reveal leukocytosis and (non-regenerative) anemia. Due to the relative lack of sensitivity of uric acid plasma concentrations, diagnosing kidney disease is often difficult in avian patients. In well hydrated birds, persistently elevated blood uric acid and urinary GGT are moderate indicators of some forms of renal disease. Newer biomarkers that could prove useful include N-acetyl-b-d-glucosaminidase (NAG), which can be measured in the blood or urine to assess tubular (dys-)function, and symmetric dimethylarginine (SDMA), which can be measured in the blood to assess glomerular filtration rates.[21] However, further studies are needed to assess their value for diagnosing renal disease in birds. Radiographs or CT may reveal renal enlargement and mineralization and altered uptake of iodinated contrast media by the kidneys. Endoscopic examination can be performed to visualize the kidneys and collect biopsies for histopathologic examination, which may help determine the etiologic process and prognosis.

High-resolution CT with contrast is particularly helpful in identifying renal lesions, especially prior to the development of laboratory abnormalities. Kirk reported 47 avian cases with CT identifiable renal disease. Only six of those 47 cases (12.8%) had lab abnormalities (hyperuricemia) supporting renal disease. Eight of those cases had histology performed, all of which had histologically confirmed renal disease. Of those eight histology confirmed cases, only two had hyperuricemia.[22] CT with contrast not only provides shape, size, density and vasculature of the kidneys, a properly performed study will also show renal contrast clearance giving functional information. Areas of non or poor contrast uptake and distribution indicate abnormally functioning renal tissue. These areas of poor contrast processing may be due to infectious disease, cancer, degenerative processes, cysts and more and aid in targeting tissue selection for biopsy.

Treatment for renal disease includes fluid therapy and nutritional support. Allopurinol (10 mg/kg PO q24h), colchicine (0.01-0.2 mg/kg PO q12-24h), and omega-3 fatty acids (VetOmega 0.22-0.44 ml/kg PO q24h) are used as anti-inflammatory agents. Use caution as allopurinol has been shown to have a low margin of safety in red tailed hawks (*Buteo jamaicensis*) and is generally not recommended for *Buteo* species.[23] Butorphanol (0.5-3 mg/kg IM q4h) or tramadol (11 mg/kg PO q6-12h to 30 mg/kg PO q12h) are important if gout is present. In addition, birds may benefit from treatment with anti-emetics, antacids or gastric protectants to stimulate appetite and reduce uricemia-associated gastric upset. Changes in the diet and husbandry may be indicated in birds with renal disease.[20] In certain countries, diets formulated specifically for geriatric birds might be available. Older birds, especially cockatiels and budgerigars, on poor diets should be converted very slowly to a formulated diet. They may develop PU/PD, lethargy and elevated uric acid levels when placed suddenly on a formulated diet. This may be due to reduced renal and hepatic function from a life-long deficiency in vitamin A and certain amino acids.[20]

Rehmania is a nutraceutical that significantly improved renal function in rats with chronic nephritis,[24] and may therefore also be worthwhile to consider in the treatment of birds with chronic kidney disease. One of the authors (SEO) has used it for chronic renal disease in a number of species along with boosting omega fatty acids (balanced 3-6-9 PUFAs) and fluids, and found patients to benefit from this treatment.

Reproductive Disease and Neoplasia

Common reproductive diseases in pet birds include cystic ovaries, salpingitis, egg yolk peritonitis, egg binding, hernias and neoplasia. Many of these can occur in any reproductively active hen, but they tend to occur more commonly as the bird ages. Ovarian neoplasia is reported with increasing frequency in pet birds, and may be associated with egg binding, cystic ovarian disease, medullary hyperostosis, abdominal hernias, oviductal impaction, ascites, and general malaise. One study by the U.S. Department of Agriculture Animal and Plant Health Inspection Service of mature fowl noted that 38% of the birds were condemned because of neoplastic disease, most of which related to the genital tract. Granulosa cell tumors and ovarian adenocarcinomas are most frequently reported, but carcinomas, leiomyosarcomas/leiomyomas, adenomas, teratomas, dysgerminomas, fibrosarcomas, lipomas and lymphomatosis have all been reported in bird ovaries.[13] Oviductal tumors are less common than ovarian neoplasia and include adenocarcinomas/adenomas, adenomatous hyperplasia, carcinoma, and carcinomatosis.[25]

Testicular neoplasia has been reported in numerous bird species and includes Sertoli and interstitial cell tumors, seminomas, teratomas, and lymphoproliferative diseases. Sertoli cell tumors are most common, especially in budgerigars, and are typically benign neoplasms.[25] Nevertheless they can result in clinical signs as a result of their size, and production of hormones. Reported neoplasms of the epididymis and ductus deferens include leiomyosarcoma and carcinoma.[25]

Many reproductive tract neoplasms are functional and can cause increased plasma hormone levels. Granulosa cell tumors have been reported in older female budgerigars (*Melopsittacus undulatus*) and may be functional, resulting in hyperestrogenism. Similarly, Sertoli-cell tumors are functional and can lead to feminization, leading to a change in cere color from blue to brown in the budgerigar. Polyostotic (medullary) hyperostosis may occur as a paraneoplastic syndrome with functional ovarian, oviductal or testicular neoplasms.

Clinical signs include reduced appetite, depression, lethargy, ascites, dyspnea, straining, cloacal prolapse, paresis/lameness, and poor or altered reproductive performance. Birds may present with feather destructive behavior or nesting behavior. Ceres may become hypertrophied or change color in budgerigars.

Diagnostics include hematology, biochemistry profile, radiography, CT, magnetic resonance imaging (MRI), ultrasound, endoscopy and biopsy. Hematology is often non-specific, though it may reveal a leukocytosis and anemia. The biochemistry profile may reveal elevations in AST, creatine phosphokinase (CPK), total protein, (total) calcium, cholesterol, triglycerides, and lipemia. Imaging may reveal an enlarged ovary, testicle or oviduct, hepatomegaly, ascites, and/or mineralization of the medulla of the long bones. Exploratory celiotomy may confirm the diagnosis.

Advanced imaging, such as high-resolution contrast CT and MRI, can be especially beneficial when evaluating the reproductive tract. The ovary should appear as a grape cluster with each follicle taking up contrast in a similar manner. In reproductively active hens, active follicles (F series follicles) will appear enlarged. Enlarged 'F' follicles should all vary sequentially in size from F1 (largest and closest to being released into the infundibulum) to F2, F3, F4, F5 approaching the size of the small inactive ovarian follicles. Birds that lay only two eggs in a clutch, will only have F1 and F2 follicles visible at the time of egg production. Birds that are continuous or large clutch layers will often have visible F1-F5 follicles. All follicles should be smooth spheres with a relatively homogenous vascular pattern (that evenly transitions from the capsule to the center of the follicle). Some species will have two ovaries (and sometimes two oviducts).

Large follicles that are either too big to be a normal yolk for the species or are of similar size and do not appear to be sequentially getting smaller are abnormal. Ovarian cysts often have a thin rim of contrast uptake and no visible contrast in the remainder of the mass. Both abnormalities may be associated with cancer and chronic inflammation. Inflamed, infected and neoplastic follicles may have an irregular vascular pattern and uneven surface. Follicles or cysts found separate from the ovary and oviduct may indicate ectopic ovulation, coelomitis, or neoplastic spread (such as carcinomatosis). Ovarian cancer is sometimes associated with invasion into the cranial renal division (usually left, but may involve either side).

The oviduct can vary significantly from indistinguishable to massive and occupy the bulk of the caudal (usually left) coelom. As hens become reproductively active, the oviduct enlarges, folds upon itself, and becomes highly vascular. Eggs in various stages of development (follicles to shelled eggs) may be visualized within the oviduct.

A developing egg found outside the oviduct is abnormal and may represent reverse peristalsis, oviductal rupture or sometimes failure of the infundibulum to catch the follicle. If there is a fresh oviductal tear, contrast may be helpful to visualize the point of rupture. Oviductal impaction, hyperplasia and metritis often result in a thickened and vascular wall with a non-contrast enhancing lumen. Oviductal cysts, masses, abnormal anatomy, etc. should all be noted.

The testicles are also highly variable in size relative to the bird's reproductive status. Testicles are often slightly asymmetrical with the left testicle usually being larger. Testicular shapes also vary between species and should all have a smooth surface and homogenous vascular pattern. Any variations (e.g., cysts, avascular areas or mixed vascularity, irregular surfaces, shapes) should be noted and may indicate some form of testicular disease.

Treatment options include dietary modifications (if needed), antibiotics (in case of suspected bacterial involvement), anti-inflammatories or analgesics (e.g., NSAIDs and/or opioids), and GnRH agonists such as leuprolide acetate (700-800 μg/kg IM; Lupron Depot, AbbVie, Chicago, IL, USA) or deslorelin acetate implants (Suprelorin, Virbac AH, Inc. Fort Worth, TX, USA).[26] For severe reproductive disease that is unresponsive to medical management, a partial or complete ovariectomy, and in the case of oviductal neoplasia, salpingohysterectomy may be the only feasible option, despite this carrying a more guarded prognosis in an elderly, debilitated bird. Dependent on the neoplasia involved, adjunct chemotherapy or radiation therapy may be considered following consultation with a veterinary oncologist. For testicular tumors, surgical removal of the affected testicle is the treatment of choice and carries a good prognosis if metastasis is not present.

Atherosclerosis

Atherosclerosis is a chronic inflammatory and degenerative condition of the arterial wall. The mineralization and plaque formation that occurs results in narrowing of the lumen and decreased blood flow. In birds, changes associated with atherosclerosis are commonly seen in the cardiac outflow tract, aorta, brachiocephalic trunk, pulmonary arteries, and main cerebral vasculature. Lesions may also occur in the peripheral vasculature, most notably the coronary, carotid, celiac, cranial mesenteric, and ischiatic arteries. Atherosclerosis is likely the most common cardiovascular disease in middle aged to older pet birds. Reported incidence has been as high as 91% in some species. Risk factors include a high-fat diet, low omega-3 fatty acid diet, sedentary lifestyle, age, and there may be a genetic predisposition. Females are overrepresented, as are Amazon and African grey parrots.[27]

Clinical signs depend on the site of the lesion but may include sudden death, dyspnea, depression, lethargy, ascites, ataxia, tremors, seizures and paresis. Owners may report their birds act disoriented or are exhibiting extension and rigidity of a leg or having difficulty perching. Similarly, self-injurious behaviors and wing web dermatitis may be noted.

Antemortem diagnosis of atherosclerosis is challenging. Diagnostics include radiography, contrast-enhanced CT imaging, and measuring cholesterol and triglyceride levels. Radiographs or CT may reveal an increased density and size of the aortic arch and/or tortuous or mineralized vessels. Some birds will have elevated cholesterol and triglycerides levels and lipemic blood.[27] Direct correlation between cholesterol and triglyceride levels and atherosclerosis development in birds is weak.

High-resolution CT with contrast can be very beneficial in identifying atherosclerotic lesions and calculating blood flow in larger vessels. While vascular mineralization can often be seen with radiographs, high-resolution CT can also better define fatty changes within the vasculature. Fatty atheromas typically are less dense and have a mixed density compared to the arterial wall. These may be seen with or without contrast. The addition of contrast fills the arterial lumen allowing one to calculate rough percentage of patency through regions where atheromas are present. Such lesions may have no evidence of vascular tortuosity or mineralization as has been discussed with radiographs.

Treatment is challenging and includes dietary modifications (formulated diet with vegetables and fruit), weight loss, and omega-3 fatty acids. Omega-3 fatty acids (VetOmega 0.22-0.44 ml/kg PO q24h) have been advocated to reduce cholesterol and inflammation, and seem to be correlated with a lower incidence of atherosclerosis in multiple bird species, thereby suggesting supplementation may have benefit in managing this disease. Isoxsuprine (10 mg/kg PO q12-24h (Bi-Coastal Pharm International LLC, Red Bank, NJ, USA), a peripheral vasodilator that causes vascular smooth muscle relaxation, has previously been used anecdotally with some success in birds, but is currently no longer available. Other blood pressure lowering drugs such as ACE-inhibitors may therefore be considered. In addition, pentoxifylline, a vasoactive drug that improves blood flow by reducing blood viscosity, has been used with good success in many atherosclerotic patients by one of the authors (YvZ). Statins are a group of lipid-lowering drugs used in humans with atherosclerosis. Statins have been used in birds, but thus far there is little information to support their use. A study investigating rosuvastatin (Crestor, AstraZeneca Pharmaceuticals LP, Wilmington, DE, USA; 10 mg/kg and 25 mg/kg PO) did not yield detectable plasma concentrations.[28] Similarly, atorvastatin did not yield detectible changes in the lipid profile following daily dosing of 10 mg/kg PO for 30 days.[29] Further studies would be required to determine whether higher or more frequent dosing would be effective.

One of the authors (SEO) uses Chinese ginseng to help reduce cholesterol in the serum and liver. This is based on a study using hypercholesterolemic chickens that showed Chinese/Wisconsin ginseng reduces HMG-CoA reductase and cholesterol 7α-hydroxylase activities, which in turn resulted in lowering of cholesterol.[30] Anecdotally, a nutraceutical/vitamin combination developed by Dr. Cathy Johnson-Delaney has been suggested to be useful for parrots as a dietary supplement. Her formula consists of 1000 mg L-carnitine, 1000 mg taurine, 1000 IU vitamin E, and 400 mg CoQ10 that is placed in 30 ml of a balanced omega 3-6-9 fatty acid oil (such as VetOmega). For some patients, 10 mg of policosanol is added as a cholesterol reducer.[31]

Hawthorn leaf with flower is another nutraceutical that can be considered for patients with atherosclerotic disease. This herbal acts as a positive inotrope, increasing coronary and myocardial perfusion, and decreasing vascular resistance, and is also hypocholesterolemic as it upregulates LDL receptors in the liver.[32]

Cardiac Disease

Cardiac disease is being diagnosed more frequently in geriatric birds. In the past, diagnosis was commonly only made following postmortem examination. One study found heart disease in 11% of avian necropsies.[33] With advancing diagnostic capabilities like digital imaging, CT scan and ultrasound, cardiovascular disease is increasingly diagnosed antemortem. Major predisposing factors for cardiac disease in birds include poor diet, obesity and a sedentary lifestyle. Common diseases of the aging cardiovascular system include cardiomyopathy, valvular disease and atherosclerosis. Vegetative lesions of the aortic and mitral valves may occur secondary to chronic infections such as hepatitis, reproductive infections and pododermatitis.[33] Congestive heart failure is associated with fluid retention resulting from sustained inadequate cardiac output. Pulmonary edema predominates with left sided heart disease and hepatomegaly and ascites predominate in right sided heart disease.

Clinical signs include periodic or sustained weakness, syncope, lethargy, increased respiratory effort, hepatomegaly and ascites. The physical exam may reveal brady-/tachycardia, cardiac arrhythmias, a heart murmur, ascites and jugular distention, weakened pulse, tachypnea/dyspnea or slow respiratory recovery following restraint (3-5 minutes is normal). There may also be muffled heart sounds from pericardial effusion, or hepatomegaly.[33]

Diagnostics include CT imaging, radiography, electrocardiography, and cardiac ultrasound. Whole body radiographs can reveal an enlarged cardiac silhouette, the presence of tortuous vessels, hepatomegaly and/or ascites. Radiographic detection of an enlarged cardiac silhouette with muffled heart sounds is suggestive of pericardial effusion. Ultrasound is the standard in non-invasive cardiac evaluation. The probe can be directed cranially through the liver to get a four-chamber view of the heart. Chamber size, wall thickness, blood flow and valvular defects may be seen. CT allows visualization of the heart and the blood vessels and is more sensitive in evaluating vasculature than radiology. Additionally, CT with contrast can be used to measure chamber and vasculature size and calculate chamber volumes. The electrocardiogram (ECG) characterizes cardiac arrhythmias and conduction disturbances, but results may be affected by sedation and anesthesia. Indirect blood pressure can be measured using a Doppler flow detector placed over the superficial ulnar artery or cranial tibial artery and a cuff and sphygmomanometer.[34] Studies have shown consistently that this method does not agree with direct methods and may be of limited clinical use, especially in small avian patients.[34] Results suggested that indirect monitoring of a single bird during a single cuff placement such as during an anesthetic procedure or an emergency may be useful to detect changes within an event.[34] Direct blood pressure may also be performed in the anesthetized patient.

Management of a patient with congestive heart failure first aims at stabilization of the patient through provision of supportive care and oxygen therapy. Furosemide (1-10 mg/kg PO q4-24h) or coelomocentesis can be used to alleviate signs of congestion by promoting diuresis and eliminating excess fluids. Following stabilization, an ACE-inhibitor such as enalapril (1.25-5 mg/kg PO q12-24h) and positive inotrope such as pimobendan (0.1-10 mg/kg PO q12-24h) can be used to improve cardiac function by reducing afterload and contractility, respectively.[33] Consultation with a veterinary cardiologist is recommended, when possible, to optimize the medical treatment.

Omega-3 fatty acids (VetOmega at 0.22-0.44 ml/kg PO daily) have been advocated for cardiac patients and used to reduce cholesterol and inflammation. Dandelion may be used as a nutraceutical, in either the root or leaf form, or both. In rats, the diuretic effects of the leaves, often described as the herb, appeared comparable to furosemide, while the root was far less effective. Unlike furosemide, dandelion does not lead to hypokalemia, as the high potassium content of the leaf replaces the loss during diuresis.[35]

Long-term care may include exercise restriction, cage rest with appropriate housing, and conversion to a healthier diet, along with weight loss, if the bird is obese. Because many cardiac patients will need to be on long-term medication, positive reinforcement training for targeting and syringe training may aid in administration of medications. Medications should be formulated or compounded to be as palatable and concentrated as possible to provide for better acceptance and reduced-volume dosing. Nevertheless, the long-term prognosis of heart failure in birds is guarded to grave.

References

1. McMichael TM. Emergency and care of the geriatric patient. *Proc Conf Int Vet Emerg Crit Care.* 2012.

2. Lightfoot TL. Geriatric psittacine medicine. *Vet Clin North Am Exot Anim Prac*. 2010;13:27-45.

3. Holmes DJ, Thomson SL, Wu J, et al. Reproductive aging in female birds. *Exp Gerontol.* 2003;38:751-756.

4. Abdul-Rahman I, Jeffcoate I, Obese FY. Age-related changes in the gross anatomy of the reproductive organs, and associated steroid hormone profiles in male and female guinea fowls (*Numida meleagris*). *Vet Anim Sci.* 2018;6:41-49.

5. Kellgren J, Lawrence J. Radiological assessment of osteo-arthrosis. *Ann Rheum Dis.* 1957;16(4):494-502.

6. Kohn M, Sassoon A, Fernando N. Classifications in brief: Kellgren-Lawrence classification of osteoarthritis. *Clin Orthop Relat Res.* 2016;474(8):1886-1893.

7. Gielis WP, Weinans H, Nap FJ, et al. Scoring osteoarthritis reliably in large joints and the spine using whole-body CT: osteoarthritis computed tomography-score (OACT-Score) *J Pers Med.* 2021;11:5.

8. Anderson K, Garner MM, Reed HH, et al. Hemorrhagic diathesis in avian species following intramuscular administration of polysulfated glycosaminoglycan. *J Zoo Wildl Med.* 2013;44(1):93-99.

9. Frech TM, Clegg DO. The utility of nutraceuticals in the treatment of osteoarthritis. *Curr Rheumatol Rep.* 2007;9:25-30.

10. Ginger: *Zingiber officinale* Roscoe. In: Blumenthal M, ed. *The ABC Clinical Guide to Herbs.* Austin, TX: American Botanical Council; 2003:176.

11. Rao CV, Rivenson A, Simi B, et al. Chemoprevention of colon carcinogenesis by dietary curcumin, a naturally occurring plant phenolic compound. *Cancer Res.* 1995; 55:259266.

12. Beer AM, Wegener T. Willow bark extract (*Salicis cortex*) for gonarthrosis and coxarthrosis-results of a cohort study with a control group. *Phytomed.* 2008;15:907-913.

13. Stanford M. Calcium metabolism. In: Harrison GJ, Lightfoot TL, eds. *Clinical Avian Medicine.* Lake Worth, FL: Spix Publishing; 2006:141-152.

14. Aguado E, Pascaretti-Grizon F, Goyenvalle E, et al. Bone mass and bone quality are altered by hypoactivity in the chicken. *PLoS ONE.* 2015;10(1):e0116763.

15. Oviedo-Rondon EO, Mente PL, Arellano C, et al. Influence of gait on bone strength in turkeys with leg defects. *Poult Sci.* 2018;97(8):2621-2637.

16. Martinez-Cummer MA, Heck R. Leeson S. Use of axial X-ray microcomputed tomography to assess three-dimensional trabecular microarchitecture and bone mineral density in single comb white leghorn hens. *Poult Sci.* 2006;85(4):706-711.

17. Webster AB. Welfare implications of avian osteoporosis. *Poult Sci.* 2004;83(2):184-192.

18. Tompkins YH, Chen C, Sweeney KM, et al. The effects of maternal fish oil supplementation rich in n-3 PUFA on offspring-broiler growth performance, body composition and bone microstructure. *PLoS ONE.* 2022;17(8):e0273025.

19. Tarlton JF, Wilkins LJ, Toscano MJ, et al. Reduced bone breakage and increased bone strength in free range laying hens fed omega-3 polyunsaturated fatty acid supplemented diets. *Bone.* 2013;52(2):578-586.

20. Echols S. Evaluating and treating the kidneys. In: Harrison GJ, Lightfoot TL, eds. *Clinical Avian Medicine.* Lake Worth, FL: Spix Publishing; 2006:451-492.

21. Alberton S, Vergneau-Grosset C, Summa N. Advances in exotic animal clinical pathology. *Vet Clin North Am Exot Anim Pract.* 2019;22:367-385.

22. Kirk N, Echols MS. Comparison of contrast CT renal lesions and uric acid values in avian cases. *Proc ExoticsCon.* 2022;21.

23. Poffers J, Lumeih JT, Timmermans-Sprang EPM, et al. Further studies on the use of allopurinol to reduce plasma uric acid concentrations in the red-tailed hawk (*Buteo jamaicensis*) hyperuricaemic model. *Avian Pathol.* 2002;31(6):567-572.

24. Su ZZ, He YY, Chen G. Clinical and experimental study on effects of man-shen-ling oral liquid in the treatment of 100 cases of chronic nephritis. *Zhongguo Zhong Xi Yi Jie He Za Zhi.* 1993;13:259-272.

25. Lightfoot TL. Clinical avian neoplasia and oncology. In: Harrison GJ, Lightfoot TL, eds. *Clinical Avian Medicine.* Lake Worth, FL: Spix Publishing; 2006:560-566.

26. Ottinger MA. Neuroendocrine regulation of reproduction in birds and clinical applications of GnRH analogues in birds and mammals. *Semin Avian Exotic Pet Med.* 2002;11(2):71-79.

27. Beaufrere H. Avian atherosclerosis: parrots and beyond. *J Exotic Pet Med.* 2014;22(4):336-347.

28. Beaufrere H, Papich MG, Nevarez J, et al. Pharmacokinetics of orally administered Rosuvastatin in Hispaniolan Amazon parrots (*Amazon ventralis*). *Proc Annu Conf Assoc Avian Vet.* 2014:25.

29. Robertson JA, Guzman DS, Graham JL, et al. Evaluation of orally administered atorvastatin on plasma lipid and biochemistry profiles in hypercholesterolemic Hispaniolan Amazon parrots (*Amazona ventralis*). *J Avian Med Surg.* 2020;34:32-40.

30. Qureshi AA, Abuirmeileh N, Din ZZ, et al. Suppression of cholesterogenesis and reduction of LDL cholesterol by dietary ginseng and its fractions in chicken liver. *Atherosclerosis.* 1983;48:81-94.

31. Orosz SE, Johnson-Delaney CA. Application of nutritional supplements and herbs in avian practice. *Proc Annu Conf Assoc Avian Vet.* 2007.

32. Upton R. *Hawthorn Leaf with Flower: Crataegus spp.* Santa Cruz, CA: American Herbal Pharmacopoeia and Therapeutic Compendium; 1999.

33. Fitzgerald BC, Beaufrere H. Cardiology. In: Speer B, ed. *Current Therapy in Avian Medicine and Surgery.* St. Louis, MO: Elsevier; 2016;252-328.

34. Johnston MS, Davidowski LA, Rao S, et al. Precision of repeated, Doppler-derived indirect blood pressure measurements in conscience psittacine birds. *J Avian Med Surg.* 2011;25:83-90.

35. Leung Ay, Foster S. *Encyclopedia of Common Natural Ingredients Used in Food, Drugs, and Cosmetics.* 2nd Ed. New York, NY: John Wiley & Sons; 1996.

Section 10

AAV Potpourri Session

Jessica Magnotti, DVM
Session Chair

Cryptosporidiosis in Cockatiels and Other Psittacine Birds

Susan L. Clubb, DVM

Session #3155

Affiliation: From Rainforest Clinic for Birds and Exotics, Inc., 3319 E Rd, Loxahatchee, FL 33470, USA.

Abstract: Cryptosporidium, sometimes informally called crypto, is a genus of apicomplexan parasitic alveolates that can cause a respiratory gastrointestinal illness (cryptosporidiosis) that primarily involves watery diarrhea (intestinal cryptosporidiosis), sometimes with a persistent cough (respiratory cryptosporidiosis) in humans.[1,2] Cryptosporidiosis is also widespread in many species of mammals, birds, reptiles and fish. This paper addresses intestinal cryptosporidiosis in psittacine birds.

Introduction

Cryptosporidiosis, caused by parasite species of the genus *Cryptosporidium,* is a major diarrheal disease in both people and animals globally, with *C. hominis* and *C. parvum* the main species infecting humans. Environmentally robust oocysts which are shed in high numbers in the feces of infected individuals are resistant to disinfectants, including levels of chlorine normally used in drinking water. As a result, *Cryptosporidium* is a major cause of waterborne and foodborne outbreaks.[1,2]

Intestinal cryptosporidiosis is typically an acute, short-term infection, can be recurrent through reinfection in immunocompetent hosts, and become severe or life-threatening in immunocompromised individuals. In humans, it remains in the lower intestine and may remain for up to 5 weeks. The parasite is transmitted by environmentally hardy cysts (oocysts) that, once ingested, exist in the small intestine and result in an infection of intestinal epithelium. Impaired intestinal epithelial barrier function and increased permeability are commonly associated with diarrheal diseases caused by enteric infections.[3,4]

Species delimitation in *Cryptosporidium* is largely based on four criteria: a description of oocyst morphology, a description of natural host specificity (where possible), genetic characterizations, and compliance with the International Code of Zoological Nomenclature (ICZN). There are now at least 44 validated species of *Cryptosporidium* (many of which were originally identified as genotypes and later described as species), including six avian *Cryptosporidium* spp. (*Cryptosporidium ornithophilus, Cryptosporidium proventriculi, Cryptosporidium avium, Cryptosporidium galli, Cryptosporidium baileyi* and *C. meleagridis*).[1,5]

Zoonotic potential of *C. meleagridis* or *C. baileyi* has been reported in domestic birds. *Cryptosporidium* has been detected in cockatiels worldwide, but the species or genotype of isolates usually remain unclear.[1,6,7] One study documents the first discovery of *C. meleagridis* and *C. baileyi* in cockatiels and suggests that pet birds may play an important role in the epidemiology of cryptosporidiosis.[6] Therefore, accurately identifying isolates is important for understanding the epizootiology of *Cryptosporidium* infection in birds and for the control of human cryptosporidiosis.

Studies in numerous countries find *Cryptosporidium* as a prevalent parasite in psittacine species. A study of a Japanese pet shop found *Cryptosporidium* in 27 of 98 (14.8%) of psittacine birds.[5]

A new species of *Cryptosporidium*, *Cryptosporidium proventriculi*, has been reported in cockatiels. Transmission studies failed to infect chickens, budgerigars and mice. Experimentally, cockatiels shed 4000 to 60,000 oocysts per gram feces for 30 days post inoculation; however, no infected cockatiels showed clinical signs.[7]

No vaccine and few treatments are currently available, which has greatly limited control of this disease to date. Livestock are both an important reservoir and source of human infections, and some authors have suggested pet birds may represent a source of human infections.

An outbreak of *Cryptosporidium* was reported in a psittacine nursery in 1996.[8] In this study it was demonstrated that *Cryptosporidium* often is missed on necropsy and histopathology if the tissues are not collected very shortly after death of affected birds. Postmortem autolysis in which the intestinal epithelium can preclude detection by histopathology. Diagnostic techniques were contrasted, but at that time PCR did not exist. In the author's opinion PCR is the most reliable diagnostic test. In the 1996 outbreak, oral paromomycin 100 mg/kg q12h for 7 days provided rapid and effective therapy and was used (by the author) subsequently with predictably excellent results.[8]

An Outbreak of *Cryptosporidium* in Handfed Cockatiels (2019-2020)

Because of low wholesale and retail prices, many breeders in the late 2000s reduced or ceased breeding of cockatiels resulting in a nation-wide shortage. In an attempt to gear up production, some breeders turned to utilizing imported cockatiels for breeding stock. In this case the new stock was batch tested for *Cryptosporidium* upon arrival at the aviary quarantine, and it was not detected; however, an outbreak occurred when these birds started to produce.

Chicks were removed from the nest at approximately 3-4 weeks of age. Young cockatiels exhibited a disease syndrome showing poor growth, diarrhea, crop stasis, high mortality, and secondary or concurrent infections with gram-negative bacteria and yeast. Tests for viral pathogens were negative. Cryptosporidiosis was diagnosed by pathology and PCR testing of feces.

To determine how widespread *Cryptosporidium*, and later *Helicobacter*, were in breeding stock and chicks, 281 clutches of cockatiels were swabbed (pooled fecal swabs) for PCR when removed from the nest. Results are provided in Table 1.

Table 1. Clutches of cockatiels tested by PCR of pooled feces for *Cryptosporidium* and *Helicobacter*.[a]

Clutches tested	Organism	Number positive	Percent positive
281	*Cryptosporidium*[b]	99/281	35%
64	Also tested for *Helicobacter*	20/64	31%

[a]*Cryptosporidium* were identified by sequencing as *C. parvum* and *C. meleagridis*.
[b]217 clutches only tested for *Cryptosporidium*.

Because *Cryptosporidium* is typically considered a waterborne organism, the well, concentrated filtered water, and the reverse osmosis system membranes were tested by PCR with negative results. It was assumed that the organism was introduced by the imported birds, and either missed on initial screening, or potentiated by breeding activity.

At the time of the 2019-2020 outbreak paromomycin was not available commercially in the United States. It is currently available from compounding pharmacies. The only drug available for human use in the United States, Alinia (nitazoxanide), was cost prohibitive, so an alternative was necessary.

Many husbandry interventions, medications, even herbal and alternative treatments were tried. Finally, a feed additive, an acidified clay product, was added to the handrearing formula at the rate of 1 level teaspoon in 5 pounds of dry handrearing formula. The morbidity and mortality rates declined rapidly and the population stabilized. Unfortunately post treatment testing was not performed; however, when the supplement ran out, health issues returned. When the supplement was available again, mortality and morbidity and mortality rates again returned to normal levels.

Clinical *Cryptosporidium* Cases in Adult Psittacine Birds

While cryptosporidiosis is typically considered self limiting, it appears that it can cause disease, and possibly chronic illness, in adult psittacine birds.

Cryptosporidium was diagnosed by PCR of feces in 14 of 39 (35.8%), in clinical cases in psittacine birds. Most of the cases were in adult pet birds in home environments. Cases were selected based on chronic gastrointestinal disease, intermittent diarrhea, mal-odor of feces, or recurrent clostridium infections. In many cases birds were tested simultaneously for *Cryptosporidium* and *Helicobacter*. Most birds were swabbed (choanal and cloacal) and the swab also introduced into the feces. Swabs were stored under refrigeration until shipped overnight to the laboratory for PCR testing.

In the author's opinion, *Helicobacter* infections can show clinical presentation similar to *Cryptosporidium*, so in most cases testing for both was elected. The case positive rate for *Helicobacter* was five positive of 35 total cases or 14.2% (Table 2).

Birds that tested positive were treated with compounded paromomycin, 100 mg/kg, q12h for 7 days. Most owners reported clinical improvement. In five cases, post therapeutic testing was negative for *Cryptosporidium*.

Table 2. Incidence and signalment for *Cryptosporidium* and *Helicobacter* in psittacine birds tested by PCR.

Species	Age (years)	Presentation	*Cryptosporidium*	*Helicobacter*	
African grey parrot	0.5	Diarrhea	Negative	Negative	
Baby blue throat macaw	0.5	Delayed weaning	Negative	Not tested	
Blue-headed pionus	1	New birds, diarrhea	Negative	Not tested	
Cockatiel	2	Exposure	Negative	Negative	
Cockatiel	2	Exposure	Negative	Negative	
Eclectus parrot	2	Proventricular dilatation disease confirmed	Negative	Negative	
Black-headed caique	3	Chronic illness, second opinion	Negative	Negative	
Cockatiel	5	Chronic intestinal issues	Positive	Negative	*
White belly caique	6	Export exam	Negative	Negative	
Eclectus parrot	7	Export exam	Negative	Negative	
Blue and gold macaw	12	Weight loss, lethargy	Negative	Positive	
Senegal parrot	12	Anemia	Positive	Negative	
Blue and gold macaw	12	Chronic intestinal issues	Positive	Negative	
Blue and gold macaw	12	Bloody stools	Positive	Negative	
Cockatiel	17	Diarrhea	Negative	Negative	
Cockatiel	17	Abnormal stools	Positive	Negative	
Yellow-naped Amazon	18	Rehoming, health screen	Negative	Negative	
Senegal	19	Chronic intestinal disorder	Negative	Negative	
Yellow-crowned Amazon	19	Recurrent intestinal infections	Positive	Negative	*
Catalina hybrid macaw	19	Weight loss, lethargy	Positive	Not tested	
African grey parrot	20	Falling, anemic	Negative	Negative	
Blue and gold macaw	21	Smelly stools, *Clostridium*	Positive	Not tested	*
Eclectus parrot	22	Chronic liver disease	Negative	Negative	
Palm cockatoo	22	Chronic regurgitation	Positive	Negative	
Hyacinth macaw	22	Breeder, diarrhea	Positive	Negative	
Double yellow-headed Amazon	24	Chronic crop problems	Negative	Negative	
Cockatiel	26	Old bird, chronic gastrointestinal issues	Positive	Negative	
Scarlet macaw	27	Proventricular dilatation disease suspect	Negative	Positive	
Severe	28	Undigested seeds in stool	Positive	Negative	*
Harlequin hybrid macaw	29	Chronic gastrointestinal disorder, clostridium	Negative	Positive	
Blue-fronted Amazon	32	Sick, anemic	Negative	Negative	
Yellow-naped Amazon	37	Noise when defecates, cloacitis	Negative	Positive	
Yellow-naped Amazon	38	Chronic gastrointestinal issues, recurring infections	Negative	Positive	
Orange-winged Amazon	44	Papillomas, fecal accumulation around vent	Negative	Negative	
Timneh grey parrot	NA	Abnormal stools	Negative	Negative	
Red-lored Amazon	NA	Proventricular dilatation disease suspect, regurgitation	Negative	Negative	
Alexandrine parakeet	NA	Old bird, chronic illness	Negative	Negative	
Yellow-naped Amazon	NA	Old pair, purchased for breeders	Positive	Negative	
Number of positive tests		14/39	5/35		
Percent positive tests		35.8	14.2		

*Retest after paromomycin treatment: negative.

Discussion

Cryptosporidium has world-wide distribution in surface water and water systems, and is the most common waterborne disease in humans and many other species.[3,4] Likewise reports of cryptosporidiosis shows a world-wide distribution in many avian species, including cockatiels and other psittacine birds. While it is primarily self-limiting in immunocompetent humans, moderate to severe watery diarrhea, crampy abdominal pain, low grade fever, dehydration, weight loss and fatigue typically last 2-4 weeks. As few as 2-10 oocysts can initiate infection. Immunocompromised humans can be chronically infected. Re-infection can occur typically with reduced virulence.[3,4] While not documented in this study, it appears that cryptosporidiosis may not be self-limiting in many psittacine birds.

Cryptosporidium oocysts are extremely resistant in the environment. They are highly resistant to chlorine disinfection levels used in water systems. The required chlorine levels generally preclude the use of chlorine as a reliable control method. With high enough concentrations and contact time, oocysts will be inactivated by chlorine dioxide and ozone treatment. Ultraviolet light at relatively low doses will inactivate *Cryptosporidium.* Boiling at least 1 minute (3 minutes at 200 m altitude) or filtration with a 1 micrometer filter, will also decontaminate drinking water. Ammonia can inactivate *Cryptosporidium*, but requires an 18-hour exposure time. Hydrogen peroxide is more effective than chlorine. Alcohol-based hand sanitizers are ineffective. Outside the body, oocysts can remain infectious for 6-8 months.

In clinical cases, owners should be advised that *Cryptosporidium* can be a zoonotic disease and aggressive disinfection is required. Porous items such as perches and toys should be discarded or baked in the oven. Cages should be scrubbed to remove all organic material. UV light may be the most practical means of disinfection of cages, potential fomites such as equipment, and rooms.

Testing for *Cryptosporidium* and *Helicobacter* by PCR of feces, should be considered. This testing is especially warranted in cases of gastrointestinal disease where conventional therapy is ineffective.

References

1. Ryan UM, Feng Y, Fever R, Xiao L. Taxonomy and molecular epidemiology of *Cryptosporidium* and *Giardia* – a 50 year perspective (1971-2021). *Intl J Parasitol.* 2021;51(13-14):1099-1119.

2. Zahedi A, Ryan U. *Cryptosporidium* – an update with an emphasis on foodborne and waterborne transmission. *Res Vet Sci.* 2020;132:500-512.

3. Wikimedia Foundation. *Cryptosporidium.* In: Wikipedia. 2023. https://en.wikipedia.org/wiki/Cryptosporidium.

4. Centers for Disease Control and Prevention. Enteric Diseases Epidemiology Branch. 2022. https://www.cdc.gov/ncezid/dfwed/edeb/index.html.

5. Akira A, Marcelo N, Meireles V. *Cryptosporidium* infections in birds - a review. Infecção por *Cryptosporidium* em aves - uma revisão. *Parasitol Res.* 2010;106(6):1491-1497. doi: 10.1007/s00436-010-1810-5.

6. Abe N, Iseki M. Identification of *Cryptosporidium* isolated from cockatiels by direct sequencing of the PCR-amplified small subunit ribosomal RNA gene. *Parasitol Res.* 2004;92(6):523-526. doi: 10.1007/s00436-004-1082-z.

7. Holubova N, Zikmundova V, Limpouchova Z, et al. *Cryptosporidium proventriculi* sp. n. (Apicomplexa: Cryptosporidiidae) in Psittaciformes birds. *European J Protistol.* 2019;69:70-87.

8. Clubb S, Cray C, Greiner E, Latimer K. Cryptosporidiosis in a psittacine nursery. *Proc Annu Conf Assoc Avian Vet.* 1996.

Captive Monk Parakeets (*Myiopsitta monachus*) Allowed to Choose Sodium Intake Consume Less Sodium than Their Counterparts

EV Voltura, PhD,
Christopher A. Bailey, PhD,
J. Jill Heatley, DVM, MS, DABVP (Avian Practice), Dipl ACZM

Session #3401

Affiliation: From Schubot Center for Avian Health (Voltura, Heatley), the Department of Entomology (Voltura), the Department of Poultry Science, College of Agrilife and Life Sciences (Bailey), Clinical Small Animal Sciences Department, School of Veterinary Medicine and Biomedical Sciences (Heatley), Texas A&M University, 400 Bizzell St, College Station, TX 77843, USA.

Note: Abstract format selected by author(s). See Front Matter for description of conference proceedings manuscripts submitted as scientific abstracts.

Abstract: Many of the nutrient requirements of Psittaciformes, including sodium (Na), have been derived from poultry nutrition studies. This derivation is likely inappropriate because species in the Psittaciformes and Galliformes differ vastly from one another. This difference is supported by the fact that some free-living parrots appear to thrive on diets containing much lower sodium (Na) than what Psittaciformes are currently thought to require as a minimum level. We sought to determine whether captive monk parakeets (*Myiopsitta monachus*) fed a low Na diet would remain as healthy, reproduce effectively, and produce offspring similarly to those fed a high Na diet. Sixteen adult pairs were fed a custom-formulated diet containing either 280 or 1500 ppm [Na] and offered an aqueous Na solution containing either 500 or 1200 ppm [Na]. Feed, water, Na solution intake, and reproductive output were measured. Total Na intake was calculated. Health checks were conducted pre- and post-treatment (e.g., body weight, body condition score, venous biochemistries). Low Na diet pairs consumed less Na (i.e., 400 ppm [Na]) than high Na diet pairs (i.e., 1500+ ppm [Na]) as well as less water and total aqueous solution. Although venous [Na^+] of blood decreased over the study, the change was small and deemed clinically irrelevant. No other significant differences were detected in measures of health or reproductive output between treatments. Monk parakeets' Na requirements appear to be significantly lower than 1500 ppm [Na]. Additional research is needed to determine whether feeding parrots high Na diets over long periods of time negatively impacts their health.

Figure-of-8? More Like, Figure-I-Hate!

Renée Schott, DVM, CWR

Session #3443

Affiliation: From Wildlife Rehabilitation Center of Minnesota, 2530 Dale St. N, Roseville, MN 55113, USA.

Note: Abstract format selected by author(s). See Front Matter for description of conference proceedings manuscripts submitted as scientific abstracts.

Abstract: Figure-of-8 wraps are commonly used for various fractures of the avian wing; however, this author and others, anecdotally, have noted it is one of the most incorrectly applied bandages.[1] In addition, many peer-reviewed sources show improperly applied figure-of-8 wraps.[2-5] Complications of improperly placed figure-of-8 wraps are common due to the difficulty of placing this looping and winding wrap, which may take several tries to achieve correct placement, along with immense species anatomy variation.[2,6] While it has occasionally been recommended as a temporary fixation in addition to a body wrap for a humeral fracture, it is no longer recommended because it tends to distract the humeral fracture further.[6] Other complications of this wrap include over-tightening effect when using a self-adhesive bandage,[6] placing wrap caudal to elbow joint,[2,6] hyper-flexion of the carpus and/or elbow,[6,7] wrap placed too loose,[7] and patagial damage.[2,7-9] More studies are needed comparing the figure-of-8 wrap to a wing-to-body wrap when dealing with distal wing fractures.

References

1. Tully T. Simple avian surgical procedures. DVM360. Available at: https://www.dvm360.com/view/simple-avian-surgical-procedures-proceedings. Accessed June 1, 2023.

2. Darrow B, Bennett RA. Avian orthopedics. In: Bennett RA, Pye GW, eds. *Surgery of Exotic Animals*. John Wiley & Sons; 2021:112-153.

3. Kayikci C, Kuşcu Y, Durmuş A, Aslan L. Fractures and treatment methods in wild avians. *Van Vet J.* 2019;30(2):115-119.

4. Pollock C. Wing wrap placement in birds. LafeberVet. Available at: https://lafeber.com/vet/wing-wrap-placement-in-birds/. Accessed June 1, 2023.

5. Department of Conservation; New Zealand Government. Wildlife Health: Avian Bandaging Techniques (Figure-of-8 Module). Available at: https://www.doc.govt.nz/globalassets/system/training-courses/wildlife-health-7/index.html#/. Accessed June 1, 2023.

6. Graham JE, Doss GA, Beaufrère H. *Exotic Animal Emergency and Critical Care Medicine*. Hoboken, NJ: Wiley-Blackwell; 2021.

7. Wade L. Treating angel wing deformity: a sling for the wing. *Today's Vet Pract.* 2022;68-74.

8. Gull JM, Saveraid TC, Szabo D, Hatt J-M. Evaluation of three miniplate systems for fracture stabilization in pigeons. *J Avian Med Surg*. 2012;26(4):203-212. doi:10.1647/2010-062R1.1

9. Duerr RS, Purdin GJ. *Topics in Wildlife Medicine: Orthopedics*. St Cloud, MN: National Wildlife Rehabilitators Association; 2017.

Novel 3D Imaging of the Psittacine Beak and a Note on Papillae and Beak Trimming Procedures

Emily J. Lessner, PhD,
Scott Echols, DVM, DABVP (Avian Practice),
Joanne R. Paul-Murphy, DVM, Dipl ACZM, Dipl ACAW,
Brian L Speer, DVM, DABVP (Avian Practice), DipECZM (Avian),
Casey M. Holliday, PhD

Session #3165

Affiliation: From Denver Museum of Nature and Science, 2001 Colorado Blvd., Denver, CO 80205, USA (Lessner), Medical Center for Birds, 3807 Main Street, Oakley, CA 94561, USA (Echols, Speer), Scarlet Imaging, PO Box 57515, Murray, UT 84157, USA (Echols), Department of Medicine and Epidemiology, School of Veterinary Medicine, University of California-Davis, Davis, CA 95616, USA (Paul-Murphy), University of Missouri School of Medicine, 1 Hospital Drive, Columbia, MO 65212, USA (Holliday).

Note: Abstract format selected by author(s). See Front Matter for description of conference proceedings manuscripts submitted as scientific abstracts.

Abstract: The avian beak is a complex organ complete with bone, neurovascular tissue and a keratinized covering called the rhampotheca. Beak trimming is a common procedure described in avian species and may be used for corrective, cosmetic and sometimes behavioral modification purposes- often with little regard to the underlying neurovascular structures. Previous publications describe papillae extending from the premaxillary bone into the upper beak rhampotheca providing sensory input;[17,28] however, no recommendations as to how beak trimming can be safely performed without damaging these sensory papillae have been accurately defined. In this study, we evaluated the upper and lower beaks of two deceased grey parrots (*Psittacus erithacus*). In one bird, we used a novel stain (Scarlet Tissue Stain, Scarlet Imaging, Murray, UT, USA) and micro computed tomography (CT) to visualize the papillae in the upper and lower beaks. In a second grey parrot, we hand isolated the upper and lower beak dermal papillae and used high resolution photography plus traditional paraffin histology, and staining using a modified Masson's trichrome. The utilized processes clearly identified the papillae three-dimensionally (CT) and two-dimensionally (photography and histology) and in both cases demonstrated associated nerves. This allowed us to determine the approximate lengths of the papillae at the rostral tomia for the upper and lower beaks. Based on these findings, the authors recommend lateral radiographs of the bird's head and beak prior to considering performing beak trims to identify the location of the underlying bone relative to the overlying rhampotheca. Based on this preliminary study and specifically in grey parrots, the authors recommend the rostral tomia (upper and lower beaks) should not be trimmed closer than 8-10 mm from the underlying bone. Further work will be needed to support these recommendations and possibly provide guidelines for other species.

References

1. Bailleul AM, Witmer LM, Holliday CM. Cranial joint histology in the mallard duck (*Anas platyrhynchos*): new insights on avian cranial kinesis. *J Anat.* 2017;230(3):444-460.

2. Baumel JJ. Handbook of avian anatomy: *nomina anatomica avium. Publications of the Nuttall Ornithological Club.* 1993;23.

3. Berkhoudt H. Taste buds in the bill of the mallard (*Anas platyrhynchos L.*). *Neth J Zool.* 1976;27(3): 310-331.

4. Berkhoudt H. Touch and taste in the mallard (*Anas platyrhynchos L.*). *Rijksuniversiteit te Leiden.* 1980.

5. Bolze GV. Anordnung und bau der herbstschen korperchen in limicolenschnabeln im zusammenhang mit nahrungsfindung. *Zool Anz.* 1968;181:313-355.

6. Crole MR, Soley JT. Comparative distribution and arrangement of Herbst corpuscles in the oropharynx of the ostrich (*Struthio camelus*) and emu (*Dromaius novaehollandiae*). *Anat Rec.* 2014;297(7):1338-1348.

7. Cunningham SJ, Alley MR, Castro I, et al. Bill morphology of ibises suggests a remote-tactile sensory system for prey detection. *The Auk.* 2010;127(2):308-316.

8. Cunningham SJ, Corfield JR, Iwaniuk AN, et al. The anatomy of the bill tip of kiwi and associated somatosensory regions of the brain: comparisons with shorebirds. *PLoS One.* 2013;8(11): e80036.

9. Demery ZP, Chappell J, Martin GR. Vision, touch and object manipulation in Senegal parrots *Poicephalus senegalus*. *Proc Royal Soc B: Biol Sci.* 2011;278(1725):3687-3693.

10. Dubbeldam JL. The sensory trigeminal system in birds: input, organization and effects of peripheral damage. A review. *Arch Physiol Biochem.* 1998;106(5):338-345.

11. Dubbeldam JL, De Bakker MA, Bout RG. The composition of trigeminal nerve branches in normal adult chickens and after debeaking at different ages. *J Anat.* 1995;186(3):619.

12. Genbrugge A, Adriaens D, De Kegel B, et al. Structural tissue organization in the beak of Java and Darwin's finches. *J Anat.* 2012;221(5):383-393.

13. Gentle MJ, Breward J. The bill tip organ of the chicken (*Gallus gallus* var. *domesticus*). *J Anat.* 1986;145:79.

14. Gignac PM, Kley NJ, Clarke JA, et al. Diffusible iodine-based contrast-enhanced computed tomography (diceCT): an emerging tool for rapid, high-resolution, 3-D imaging of metazoan soft tissues. *J Anat.* 2016;228(6):889-909.

15. Gottschaldt KM. Structure and function of avian somatosensory receptors. In: King AS, McLelland J, eds. *Form and Function in Birds.* 1985;375-461.

16. Gottschaldt KM, Lausmann S. The peripheral morphological basis of tactile sensibility in the beak of geese. *Cell Tissue Res.* 1974;153(4):477-496.

17. Goujon DE. An apparatus of tactile corpuscles situated in the beaks of parrots. *J Anat Physiol Norm Pathol.* 1869;6:449-455.

18. Gustafson LA, Cheng HW, Garner JP, et al. The effects of different bill-trimming methods on the well-being of Pekin ducks. *Poult Sci.* 2007;86(9):1831-1839.

19. Halata Z, Grim M. Sensory nerve endings in the beak skin of Japanese quail. *Anat Embryol.* 1993;187(2): 131-138.

20. Heppleston PB. Anatomical observations on the bill of the oystercatcher (*Haematopus ostralegus occidentalis*) in relation to feeding behaviour. *J Zool.* 1970;161(2):519-524.

21. Hoerschelmann H. Strukturen der Schnabelkammer bei Schnepfenvögeln (Charadriidae und Scolopacidae). *Zeitschrift für wissenchaftliche Zool J.* 1972;185:105-121.

22. Homberger DG, Brush AH. Functional-morphological and biochemical correlations of the keratinized structures in the African grey parrot, *Psittacus erithacus* (Aves). *Zoomorphol.* 1986;106(2):103-114.

23. Kuenzel WJ. Neurobiological basis of sensory perception: welfare implications of beak trimming. *Poult Sci.* 2007;86(6):1273-1282.

24. Lederer RJ. The role of avian rictal bristles. *The Wilson Bulletin*, 1972;193-197.

25. Lucas AM, Stettenheim PR. Avian anatomy: integument. Washington, DC: US Department of Agriculture; *Agricult. Handbook 362.* 1972.

26. Lüdicke M. Aufbau und Abnutzung der Hörnzähne und Hornwülste des Vogelschnabels. *Zeitschrift für Morphologie und Ökologie der Tiere*, 1940;37(2):155-201.

27. Mench J, Paul-Murphy J, Klasing K, Cussen V. True parrots (Psittacidae). In: Yeates J, ed. *Companion Animal Care and Welfare: The UFAW Companion Animal Handbook*; 2018;338-354.

28. Menzel R, Lüdicke M. Funktionell-anatomische und autoradiographische Untersuchungen am Schnabelhorn von Papageien (Psittaci). *Zool Jahrb Anat.* 1974;93:175-218.

29. Nebel S, Jackson DL, Elner RW. Functional association of bill morphology and foraging behaviour in calidrid sandpipers. *Anim Biol.* 2005;55(3):235-243.

30. Orosz SE, Bradshaw GA. Avian neuroanatomy revisited: from clinical principles to avian cognition. *Vet Clin North Am Exot Anim Pract.* 2007;10(3):775-802.

31. Schneider ER, Mastrotto M, Laursen WJ, et al. Neuronal mechanism for acute mechanosensitivity in tactile-foraging waterfowl. *Proc Natl Acad Sci.* 2014;111(41):14941-14946.

32. Soliman SA, Madkour FA. A comparative analysis of the organization of the sensory units in the beak of duck and quail. *Histol Cytol Embryol.* 2017;1(4):1-16.

33. Soons J, Herrel A, Genbrugge A, et al. Multi-layered bird beaks: a finite-element approach towards the role of keratin in stress dissipation. *J R Soc Interface.* 2012;9(73):1787-1796.

34. Speer B, Echols S. Beak deformities: form, function and methodology for medical treatment. *Assoc Latinoamericana Vet Fauna Silv.* 2016.

35. Speer B, Powers LV. Anatomy and disorders of the beak and oral cavity of birds. *Vet Clin North Am Exot Anim Pract.* 2016;19(3):707-736.

36. Speer BL, Fecchio R, Holliday CM. The beak: form, function, and problems. In: Speer B, van Zeeland Y, eds. *Current Therapy in Avian Medicine and Surgery,* Vol 2. St. Louis, MO: Elsevier; 2023.

37. Stettenheim PR. The integumentary morphology of modern birds—an overview. *Am Zool.* 2000;40(4): 461-477.

38. Struthers S, Classen HL, Gomis S, Schwean-Lardner K. The effect of beak tissue sloughing and post-treatment beak shape on the productivity of infrared beak-treated layer pullets and hens. *Poult Sci.* 2019;98(9): 3637-3646.

39. Van Hemert C, Handel CM, O'brien DM. Stable isotopes identify dietary changes associated with beak deformities in black-capped chickadees (*Poecile atricapillus*). *The Auk.* 2012;129(3):460-466.

40. Wild JM. The avian somatosensory system: a comparative view. In: Scanes CG, ed. *Sturkie's Avian Physiology*, 6th Ed. New York NY: Academic Press; 2015:55-69.

Novel Use of Antifungal Nail Lacquer to Treat *Candida* Bill Infections in Ducks

Glenn H. Olsen, DVM, MS, PhD

Session #3290

Affiliation: From the US Geological Survey, Eastern Ecological Science Center (formerly Patuxent Wildlife Research Center), 12302 Beech Forest Road, Laurel, MD 20708, USA.

Abstract: Captive wild ducks in the research colony at the U.S. Geological Survey's Eastern Ecological Science Center in Laurel, Maryland frequently contract *Candida* infections on the upper and lower bills. Various standard treatments were attempted with mixed success. Several of the surf scoters suffered extreme erosions of the bill leading to euthanasia prior to using this new treatment. For the past 3 years a human medical product, ciclopirox 8% nail lacquer, has been used to treat the *Candida* successfully. The treatment requires careful trimming and debriding of the dead tissue before applying the ciclopirox, then keeping the ducks caged away from water for 15 minutes to allow the product to dry. Standard treatment is twice weekly. For persistent cases oral fluconazole is added to the treatment regime. Common eider, surf scoter, white-winged scoter, black scoter, long-tailed ducks, and harlequin ducks have been treated successfully with ciclopirox and no adverse reactions were observed.

Introduction

The bill or beak of birds can be affected by fungal pathogens, namely *Candida albicans*, *C. parapsilosis*, *C. krusei*, *C. tropicalis*, *C. glabrata*, and *Aspergillus*.[1-3] Presenting signs are described as white caseous exudate or plaques on the bill or oropharynx.[1,3] *Candida* species is a common cause of fungal infections in birds including flamingoes, raptors, parrots, pigeons, and poultry.[5-7] These *Candida* infections often are associated with antibiotic therapy, malnutrition, poor sanitation, other concurrent illnesses, or immunocompromised birds.[1,5]

Treatments have involved the use of topical and systemic antifungal medications. Topical applications of clotrimazole cream (Lotrim, Schering, Kenilworth, NJ, USA), amphotericin B ointment (Fungizone, Bristol-Myers Squibb Co, Princeton, NJ, USA), nystatin ointment (Mycostatin topical, Bristol-Myer Squibb Co, Princeton, NJ, USA), miconazole gel (Daktarin Oral Gel, Jansen-Cilag Ltd, Wycombe, Buckinghamshire, UK), or chlorhexidine (Nolvasan Solution, Zoetis, Inc, Kalamazoo, MI, USA) have all been used to treat *Candida* infections.[1,2,5] Systemic antifungal therapy used against *Candida* have included itraconazole (Sporonox, Janssen Pharmaceutical Products LP, Titusville, NJ, USA), nystatin (Mycostatin oral, Bristol-Myer Squibb Co, Princeton, NJ, USA), or fluconazole (Diflucan, Pfizer, Inc, New York, NY, USA).[1,5]

Clinically Relevant Data

The sea duck colony at the U.S. Geological Survey, Eastern Ecological Science Center (EESC) in Laurel, Maryland houses approximately 125 ducks of six species used for research and education. The sea ducks at the EESC include common eider (*Somateria mollissima*), black scoter (*Melanitta nigra*), surf scoter (*M. perspicillata*), white-winged scoter (*M. fusca*), long-tailed ducks (*Clangula hyemalis*), and harlequin ducks (*Histrionicus histrionicus*). In addition, two non-sea duck species are maintained, lesser scaup (*Aythya affinis*), also used for research, and wood ducks (*Aix sponsa*), used primarily for education by our visitor center. The EESC holds the first captive breeding record for surf scoters. The research with the captive sea ducks revolves around some physiologic studies such as hearing acuity[9] and vision,[10] and also, avian influenza.[11] The EESC Institutional Animal Care and Use Committee (IACUC) annually approves the standard operating procedures for the duck colony husbandry and the veterinary medical care of the ducks. The information reported in this paper resulted from several years of attempting to and successfully treating *Candida* infections of the bill and oropharynx of the captive ducks in the colony.

Animal husbandry staff often first find signs of *Candida* as an elongated bill or misshaped bill. Sometimes the lesions appear more in the oropharynx as white plaques or flaky exudate. Diagnosis has been made by fungal cultures submitted to Antech Laboratories. Cultures consistently yield *Candida* species, but not every duck was tested. Often the diagnosis was made on the signs of the lesions alone. Early treatments were with nystatin, clotrimazlole, or chlorhexidine topically (Table 1). Sometimes itraconazole was used orally (10 mg/kg twice daily), but often there was no resolution of the bill lesions with any of these treatments. Lesions could result in severe crusting and deformity of the upper bill in the surf scoters especially. No birds were euthanized because of these lesions, but a group of surf scoters were chosen for a terminal avian influenza study partly due to the bill deformities. Even with the bill deformities the birds otherwise were in good body condition.

Ciclopirox (Ciclopirox Topical Solution 8% Nail Lacquer, Perrigo, Bronx, NY, USA) was at first just used alone, with some initial success (Table 1), but some cases did not resolve, and one surf scoter male continued to get worse. In an effort to help this bird, fluconazole at 20 mg/kg orally twice weekly (Fluconazole Tablets USP 50 mg, Rising Health LLC, Saddle Brook, NJ, USA) was added to the treatment regime. This bird responded and improved markedly within 1 month. The current treatment involves handling ducks twice weekly to debride the tissues infected on the bill and oropharynx to remove any white plaques and debris, then to give fluconazole tablet (15-25 mg/kg), and finally to paint the affected areas with the ciclopirox nail lacquer. Birds are held out of the water for 15 minutes to allow the ciclopirox time to dry. Most ducks resolve in 2-3 months, mean days from onset of treatment to resolution is 108 (range 46-283 days).

Table 1. Select results from several years of treating *Candida* species bill and oropharynx infections in sea ducks at the U.S. Geological Survey Eastern Ecological Science Center, Laurel, Maryland USA. Clotrimazole, chlorhexidine, nystatin, and ciclopirox all applied topically to lesions. Fluconazole given orally as a tablet.

Date	Species[a]	Band #	Sex	Age	Treatment used[b]	Date resolved	Length of treatment	Comments
16 Sep 2016	SUSC	635	F	13	Clotrimazole Chlorhexidine	Not resolved		Discontinued 11 Jan 2017
30 Oct 2017	COEI	196	M	4 mo	Nystatin	20 Dec 2017	51	Treatment 1X or 2X day
6 Oct 2017	COEI	169	M	4 mo	Nystatin	Not resolved		Treatment 1X or 2X day
14 Nov 2018	COEI	164	M	1	Nystatin	Not resolved		Treatment 2X day
7 Sep 2018	SUSC	F-461	M	2	Ciclopirox	Not resolved		Treatment 1X week
12 Feb 2021	SUSC	F-461	M	5	Ciclopirox + Fluconazole	22 Nov 2021	283	Severe case Trt 2X week
24 Sep 2020	WWSC	157	M	6	Ciclopirox	Not resolved		Treatment 1X week
1 Nov 2021	WWSC	157	M	7	Ciclopirox + Fluconazole	28 Oct 2021	211	Treatment 2X week
7 Dec 2018	WWSC	181	F	1	Ciclopirox	Not resolved		Treatment 3X month
1 Nov 2021	WWSC	181	F	4	Ciclopirox + Fluconazole	2 Aug 2021	123	Treatment 2X week
20 Jun 2019	COEI	166	F	2	Ciclopirox	10 Oct 2019	112	Treatment 1X week
29 Apr 2021	HARD	285	M	5	Ciclopirox + Fluconazole	21 Jun 2021	53	Treatment 2X week
17 Mar 2022	HARD	548	M	16	Ciclopirox + Fluconazole	12 May 2022	46	Treatment 2X week
28 Sep 2022	BLSC	137	M	16+	Ciclopirox + Fluconazole	6 Feb 2023	125	Treatment 2X week
29 Sep 2022	LTDU	754	F	5	Ciclopirox + Fluconazole	12 Dec 2022	54	Treatment 2X week
26 Dec 2022	SUSC	176	F	10	Ciclopirox + Fluconazole	16 Mar 2023	80	Treatment 2X week

[a]Species: BLSC = black scoter (*Melanitta nigra*), COEI = common eider (*Somateria mollissima*), LTDU = long-tailed duck (*Clangula hyemalis*), HARD = harlequin duck (*Histrionicus histrionicus*), SUSC = surf scoter (*M. perspicillata*), WWSC = white-winged scoter (*M. fusca*).

[b]Treatment used: Chlorhexidine (Nolvasan Solution, Zoetis Inc, Kalamazoo, MI, USA), Ciclopirox (Ciclopirox Topical Solution 8% Nail Lacquer, Perrigo, Bronx, NY, USA), Clotrimazole (Lotrim Cream, Schering, Kenilworth, NJ, USA), Fluconazole (15-25 mg/kg, PO, twice weekly, Fluconazole Tablets USP 50 mg, Rising Health LLC, Saddle Brook, NJ, USA), Nystatin (Mycostatin topical, Bristol-Myer Squibb Co, Princeton, NJ, USA).

Conclusions

Candida infections, though not common, occur on the bill and oropharynx of many avian species. The author of this paper has diagnosed and treated the condition in several psittacine species. Generally, the available topical products cited in the introduction worked well in these other species; however, in ducks, especially sea ducks, the birds are almost constantly on water. Even restraining the birds in cages for up to 30 minutes after treatment with one of the other topical products failed to resolve the problem in ducks. The advantage of the ciclopirox is that after 15 minutes it is dry to the touch and persists, not being easily washed away in water.

Many ducks with mild cases resolved with just the ciclopirox (one example is listed in Table 1); however, adding fluconazole orally to the treatment regime seems to improve outcome, from a clinical perspective. The treatment frequency of twice weekly is dictated by the species. The sea ducks are all captive reared, except the black scoter, which was wild caught. Even so, they do not tolerate frequent handling. For long-term treatments such as required to control *Candida* infections, twice weekly has worked well, with one surf scoter receiving treatments over a 9-month period. The addition of fluconazole at 15-25 mg/kg twice weekly is similar to one published recommendation for psittacine birds of 20 mg/kg PO q48h.[12]

Acknowledgments: The author thanks veterinary technician, Carlyn Caldwell, for assistance provided and the leader of our animal husbandry unit, Sharon Peregoy, for support in preparing this manuscript.

Disclaimer: Any use of trade, product, or firm names is for descriptive purposes only and does not imply endorsement by the U.S. Government.

References

1. Olsen GH. Oral biology and beak disorders of birds. *Vet Clin North Am Exot Anim Pract.* 2005:281-291.

2. Samour JH, Naldo JL. Diagnosis and therapeutic management of candidiasis in falcons in Saudi Arabia. *J Avian Med Surg.* 2002;16(2):129-132.

3. Donnelly KA, Wellehan JFX, Quesenberry K. Gastrointestinal disease associated with non-*alibicans Candida* species in six birds. *J Avian Med Surg.* 2019;33(4):413-418.

4. Hoefer HL. Diseases of the gastrointestinal tract. In: Altman RB, Clubb SL, Dorrestein GM, et al, eds, *Avian Medicine and Surgery.* Philadelphia, PA: WB Saunders; 1997:419-453.

5. Rayment KM, Garner MM, Drees R, et al. Atypical candidiasis of the gnathotheca in a lesser flamingo (*Phoeniconaias minor*). *J Avian Med Surg.* 2021;35(4):457-463.

6. Silvanose C. Infectious diseases: candidiasis. In: Samour JH, ed. *Avian Medicine.* 3rd Ed. St. Louis, MO: Elsevier; 2016:472-473.

7. Crespo R, Franca MS, Fenton H, Shivaprasad HL. Galliformes and Columbiformes. In: Terio KA, McAloose D, St Leger J, eds. *Pathology of Wildlife and Zoo Animals.* Cambridge, MA: Academic Press; 2018: 747-773.

8. Ciclopirox Topical Solution, 8%. Bronx, NY; Perrigo. Drug insert.

9. Crowell SE, Wells-Berlin AM, Carr CE, et al. A comparison of auditory brainstem responses across diving bird species. *J Comp Physiol A.* 2015;201:803-815.

10. Cantlay JC, Bond AL, Wells-Berlin AM, et al. Ineffectiveness of light emitting diodes as underwater deterrents for long-tailed ducks *Clangula hyemalis*. *Global Ecol Conserv.* 2020: doi.org/10.1016/j.gecco.2020.e01102.

11. Prosser DJ, Schley H, Simmons N, et al. A lesser scaup (*Aythya affinis*) naturally infected with Eurasian 2.3.4.4 highly pathogenic H5N1 avian influenza virus – movement ecology and host factors. *Transbound Emerg Dis.* 2022;69(5):e2653-e2660. doi.org/10.1111/tbed.14614.

12. Hawkins MG, Guzman DSM, Beaufrere H, et al. Birds. In: Carpenter JW, ed. *Exotic Animal Formulary.* St. Louis, MO: Elsevier; 2018:167-375.

Assessment of Novel Serologic Assays in Avian Species

Carolyn Cray, PhD,
Adriana Callico, MS,
A. Leclerc, DVM, DipECZM (Zoo Health Management),
Ellen Bronson, Dr med vet, Dipl ACZM

Session #3227

Affiliation: From Division of Comparative Pathology, Department of Pathology & Laboratory Medicine, University of Miami Miller School of Medicine, P.O. Box 016960 (R-46), Miami, FL 33101, USA (Cray, Callico), Zoo De Beauval, 41110 Saint-Aignan-sur-Cher, France (Leclerc), The Maryland Zoo in Baltimore, 1 Safari Place, Baltimore, MD 21217, USA (Bronson).

Note: Abstract format selected by author(s). See Front Matter for description of conference proceedings manuscripts submitted as scientific abstracts.

Abstract: While antemortem serodiagnostics of avian aspergillosis remains problematic, recent advances in gliotoxin detection as well as quantitation of hydroxybutyrate and the use of protein electrophoresis have been documented for use especially in penguin species.[1-3] While gliotoxin detection is highly specific, the sensitivity as a routine element of a screening panel is questionable. In a review of 345 samples submitted for analysis, less than 5% were positive for this marker. With the goal of improving antibody detection, additional studies were undertaken to examine options for ELISA and Western blot technologies. For the ELISA, Afmp1p, a recombinant antigen was used as previously described in falcons.[4] More than 260 penguin samples were examined and 29% were positive. With limited case follow up, these were inclusive of penguins with confirmed and suspect disease and those with known exposure. In some cases, antibody levels were followed through repeated measures and limited results suggest that titers may have prognostic value. Lastly, a novel implementation of a commercially available Western blot for antibody detection[5] was examined using samples from African penguins from an institution in France and from several parks in the United States. In this method, the presence of antibody is scored by the reactivity to four different *Aspergillus* antigens. While greater reactivity was observed in penguins with confirmed infection (over presumed normal penguins), differences were found between France and U.S. samples suggesting that institutional differences may be present. Additional studies need to be conducted to address the application of these methods in other avian species.

Acknowledgments: Thank you to the clinicians who shared clinical information related to submitted samples.

References

1. Cray C, Watson T, Arheart KL. Serosurvey and diagnostic application of antibody titers to *Aspergillus* in avian species. *Avian Dis.* 2009;53:491-494.

2. Desoubeaux G, Rodriguez M, Bronson E, et al. Application of 3-hydroxybutyrate measurement and plasma protein electrophoresis in the diagnosis of aspergillosis in African penguins (*Spheniscus demersus*). *J Zoo Wildl Med.* 2018;49(3):696-703.

3. Reidy LR, Desoubeaux G, Cardenas J, et al. Detection of gliotoxin but not bis(methyl) gliotoxin in plasma from birds with confirmed and probable aspergillosis. *J Zoo Wildl Med.* 2022;53(1):60-69.

4. Wernery U, Tsang C, Hebel C, et al. Serodiagnosis of aspergillosis in falcons (*Falco* spp.) by an Afmp1p-based enzyme-linked immunosorbent assay. *Mycoses.* 2018;61:600-609.

5. Oliva A, Flori P, Hennequin C, et al. Evaluation of the *Aspergillus* Western blot IgG kit for the diagnosis of chronic aspergillosis. *J Clin Microbiol.* 2015;53:248-254.

Effects of a Single Dose of Gabapentin in Adult Domestic Pigeons (*Columba livia domestica*)

Laura Burns, DVM,
Kelsey Trumpp, DVM,
Connor Jacobson, DVM,
Stephen Divers, BVetMed, DZooMed, DipECZM, Dipl ACZM, FRCVS

Session #3375

Affiliation: From the Department of Small Animal Medicine and Surgery (Zoological Medicine), College of Veterinary Medicine, University of Georgia, Athens, GA 30602, USA.

Note: Abstract format selected by author(s). See Front Matter for description of conference proceedings manuscripts submitted as scientific abstracts.

Abstract: Avian species are commonly presented for routine wellness evaluations and clinical disease. Handling for physical examinations can lead to increased stress that worsens clinical signs. Gabapentin has been used in other species as a safe and effective treatment to reduce stress for transportation and physical examination.[1-3] This study aimed to evaluate the effects of gabapentin on physiologic parameters (heart rate and respiratory rate) and stress-associated behaviors of pigeons. The hypothesis was that gabapentin administered once would improve physiologic parameters, improve tractability, and reduce stress-associated behaviors. As part of a blinded, randomized, crossover study, twenty domestic pigeons received 50 mg/kg of oral gabapentin and an equal volume oral saline control. All animals were subjected to an 8-day washout period between treatments. Each pigeon was assessed prior to any handling and then assessed at 1 hour and 2 hours post-treatment. Individual behaviors and physiologic parameters were evaluated before and after handling at each of these time points. A modified human intruder test, sedation scale (1-5) and resistance score (0-3) were also developed and compared between time points.[1,4,5] Average heart rate was decreased by 30 beats per minute 2 hours post treatment. Resistance scores were 0.35 points lower than control 2 hours post treatment. Attempts to capture for examination were lower by 0.49 attempts 2 hours following treatment compared to the control. Mild sedation was noted in 20% of pigeons receiving gabapentin. Results from this study indicate that gabapentin may be beneficial when handling pigeons and future studies should be considered in other avian species.

References

1. Van Haaften KA, Eichstadt Forsythe LR, Stelow EA, Bain MJ. Effects of a single preappointment dose of gabapentin on signs of stress in cats during transportation and veterinary examination. *J Am Vet Med Assoc.* 2017;251:1175-1181.

2. Bleuer-Elsner S, Medam T, Masson S. Effects of a single oral dose of gabapentin on storm phobia in dogs: A double-blind, placebo-controlled crossover trial. *Vet Rec.* 2021;e453.

3. Conway R, Burton M, Mama K et al. Behavioral and physiologic effects of a single dose of oral gabapentin in rabbits (*Oryctolagus cuniculus*). *Proc ExoticsCon.* 2022;183-184.

4. Heatley JJ, Cary J, Kingsley L, et al. Midazolam sedates passeriformes for field sampling but affects multiple venous blood analytes. *Vet Med (Auckl)*. 2015;6:61-69.

5. Bethell EJ, Cassidy LC, Brockhausen RR, Pfefferle D. Toward a standardized test of fearful temperament in primates: a sensitive alternative to the human intruder task for laboratory-housed rhesus macaques (*Macaca mulatta*). *Front Psychol*. 2019;10.

Effect of Capromorelin (Entyce®) on Appetite and Weight Gain of Domestic Pigeons (*Columba livia domestica*)

Kelsey Trumpp, DVM,
Laura Burns, DVM,
Connor Jacobson, DVM,
Stephen J. Divers, BVetMed, DZooMed, DipECZM (Herpetology), DipECZM (Zoo Health Management), Dipl ACZM, FRCVS

Session #3179

AAV House Officer Manuscript Competition Winner

Affiliation: From the Department of Small Animal Medicine and Surgery (Zoological Medicine), University of Georgia College of Veterinary Medicine, 501 D. W. Brooks Drive, Athens, GA 30602, USA.

Note: Abstract format selected by author(s). See Front Matter for description of conference proceedings manuscripts submitted as scientific abstracts.

Abstract: Weight loss and decreased appetite are commonly encountered sequela of disease and stress in avian patients. There is currently minimal information in the literature regarding appetite stimulation in birds.[1,2] Capromorelin (Entyce®, Elanco Animal Health, Greenfield, IN 46140, USA) is a potent agonist of the growth hormone secretagogue receptor (GHS-R), which increases food consumption via direct stimulation of the hunger centers of the hypothalamus. This drug is approved by the U.S. Food and Drug Administration for use as an appetite stimulant in dogs and has been shown to also increase food consumption in New Zealand white rabbits and broiler chickens.[3-9] Twenty adult domestic pigeons, housed in groups of five, were involved in a randomized, placebo-controlled study to investigate the effect of capromorelin on appetite and weight gain in birds. Each pigeon was randomly assigned to the control or treatment group. The birds were individually weighed and given either oral water (control) or capromorelin at 12 mg/kg once daily for the duration of the 6-day study period. Total food consumed was recorded in grams per cage each day. Pigeons given Entyce® consumed a greater percentage of food daily and gained more weight over the course of the study period than those that were given the same volume of water. No adverse side effects were noted in any birds in this study. Entyce® shows promise as an appetite stimulant in pigeons and further investigation into its use in additional avian taxa is warranted.

References

1. Scagnelli A, Titel C, Doss G, Mans C. Effects of midazolam and lorazepam on food intake in budgerigars (*Melopsittacus undulatus*). *J Exot Pet Med.* 2022;41:42-45.

2. Martel A, Berg C, Doss G, Mans C. Effects of midazolam on food intake in budgerigars (*Melopsittacus undulatus*). *J Avian Med Surg.* 2022;36(1):53-57.

3. Rhodes L, Zollers B, Wofford JA, Heinen E. Capromorelin: a ghrelin receptor agonist and novel therapy for stimulation of appetite in dogs. *Vet Med Sci.* 2018;4(1):3-16.

4. Zollers B, Rhodes L, Heinen E. Capromorelin oral solution (Entyce®) increases food consumption and body weight when administered for 4 consecutive days to healthy adult beagle dogs in a randomized, masked, placebo controlled study. *BMC Vet Res*. 2016;13(1):1-5.

5. Zollers B, Wofford JA, Heinen E, et al. A prospective, randomized, masked, placebo-controlled clinical study of capromorelin in dogs with reduced appetite. *J Vet Intern Med*. 2016;30(6):1851-1857.

6. Zollers B, Rhodes L, Smith RG. Capromorelin increases food consumption, body weight, growth hormone, and sustained insulin-like growth factor 1 concentrations when administered to healthy adult beagle dogs. *J Vet Pharmacol Ther*. 2017;40(2):140-147.

7. Zollers B, Huebner M, Armintrout G, et al. Evaluation of the safety in dogs of long-term, daily oral administration of capromorelin, a novel drug for stimulation of appetite. *J Vet Pharmacol Ther*. 2017;40(3): 248-255.

8. Draper JM, Savson DJ, Lavin ES, et al. Comparison of effects of capromorelin and mirtazapine on appetite in New Zealand white rabbits (*Oryctolagus cuniculus*). *J Am Assoc Lab Anim Sci*. 2022;61(5):495-505.

9. Ceron-Romero N, Taofeek N, Thomas A, et al. Capromorelin, a ghrelin receptor agonist, increases feed intake and body weight gain in broiler chickens (*Gallus gallus domesticus*). *Poult Sci*. 2021;100(8):101204.

Successful Conservative Management of a Neuro-Compressive Unstable Synsacral Fracture in a Yellow-Headed Amazon Parrot (*Amazona oratri*)

Lauren Thielen, DVM, DABVP (Avian Practice),
Ian Porter, DVM, Dipl ACVR,
Laura St Clair, DVM,
Nicola Di Girolamo, DVM, PhD, DipECZM (Herpetology), Dipl ACZM, Dipl ACVPM

Session #3473

Affiliation: From the Department of Clinical Sciences, College of Veterinary Medicine, Cornell University, 930 Campus Road, Ithaca, NY 14853, USA.

Note: Abstract format selected by author(s). See Front Matter for description of conference proceedings manuscripts submitted as scientific abstracts.

Abstract: Although conservative management for synsacral fractures in birds is suggested in the literature, case reports describing the conservative management and long-term outcomes have not been reported.[1] This case report describes successful conservative management of an acute, traumatic, neuro-compressive synsacral fracture in a parrot. A 32-year-old female Amazon parrot (*Amazona oratri)* was presented with increased respiratory effort and rate, depressed mentation, and an inability to stand after an acute fall. On initial assessment, cloacal tone was absent and proprioceptive deficits were noted in both hind legs. The patient had superficial sensation, grip of the feet, and movement of the legs. A contrast-enhanced computed tomography (CT) scan was performed and found an unstable complete linear circumferential fracture of the mid-caudal aspect of the synsacrum. The fracture resulted in ventral displacement and a step deformity attenuating approximately 40% of the vertebral canal diameter. Moderate pulmonary consolidation secondary to trauma was noted. Conservative management was elected due to the size of the synsacrum and the risk for surgical complications. The patient was hospitalized to provide analgesia, supportive care and multiple time daily cloacal stimulation to defecate. After 8 days in hospital, the patient began to develop cloacal tone, and was able to perch on a rolled towel. Strict cage rest was performed for 6 weeks, in a confined space devoid of perches. In the first week following discharge the patient returned to normal defecation and function of both hind limbs. A repeat CT scan performed 8 months after the trauma showed adequate healing with a moderate amount of smooth, bridging bone crossing the fracture site. Conservative management could be considered as a treatment option for similar synsacral fractures.

References

1. Powers LV, Bergman RL. Surgical spinal stabilization in an umbrella cockatoo (*Cacatua alba*). *Proc Annu Conf Assoc Avian Vet.* 2005.

Effects of Benzodiazepines on Food Intake in Budgerigars (*Melopsittacus undulatus*)

Christoph Mans, Dr med vet, Dipl ACZM,
Caroline E. Titel, BS,
Grayson A. Doss, DVM, Dipl ACZM

Session #3454

Affiliation: From the School of Veterinary Medicine, University of Wisconsin-Madison, 2015 Linden Drive, Madison, WI 53706 USA.

Note: Abstract format selected by author(s). See Front Matter for description of conference proceedings manuscripts submitted as scientific abstracts.

Abstract: Reduced food intake is a common concern in sick and hospitalized birds, and insufficient caloric intake can lead to a variety of detrimental effects and prolong hospitalization. Currently, insufficient food intake in pet birds is predominately managed by veterinarians by providing nutritional support via crop or proventricular gavage, depending on the species. Several medications have been used to stimulate appetite in dogs and cats, including benzodiazepines. In psittacine birds, our knowledge about the effects of benzodiazepines is currently limited to single-dose studies (1 mg/kg) of midazolam and lorazepam.[1-2] The purpose of this study was to evaluate two different doses of midazolam (1 mg/kg and 0.5 mg/kg IM once) and diazepam (1 mg/kg IM once) in a randomized, controlled, blinded, complete crossover study. Food intake was measured in hourly intervals for the first 8 hours and then one measurement for the 8- to 12-hour period. The birds were monitored by video recording for potential adverse effects. Both midazolam and diazepam at 1 mg/kg resulted in a significant increase in food intake compared to the control treatment. At 0.5 mg/kg, midazolam-induced food intake was less, and sedation behavior was of shorter duration compared to midazolam at 1 mg/kg. Results of this study indicate that diazepam has similar positive effects on food intake in budgerigars compared to midazolam, and lowering the dose of midazolam will result in a decrease in its appetite-stimulating as well as sedative effects. Either benzodiazepine can be used clinically as a short-acting appetite stimulant in budgerigars.

References

1. Martel A, Doss G, Berg C, Mans C. Effects of midazolam on food intake in budgerigars. *J Avian Med Surg.* 2022;36:53-57.

2. Scagnelli A, Titel C, Doss G, Mans C. Effects of midazolam and lorazepam on food intake in budgerigars. *J Exot Pet Med.* 2022;41:42-45.

Section 11

AAV "I'll Take Topics with the Acronym AI for 500 Please, Alex" Session

Nick Kirk, DVM, DABVP (Avian Practice)
Session Chair

Highly Pathogenic Avian Influenza in Raptors

Dana Franzen-Klein, DVM, MS,
Annette Ahlmann, DVM, MS, CWR,
Victoria Hall, DVM, MS, Dipl ACVPM

Session #3340

Affiliation: From The Raptor Center at the University of Minnesota, St. Paul, MN 55108, USA.

Note: Abstract format selected by author(s). See Front Matter for description of conference proceedings manuscripts submitted as scientific abstracts.

Abstract: The strain of highly pathogenic avian influenza (HPAI, strain EA/AM 2.3.4.4. H5N1) that arrived in the United States in early 2022 is behaving very differently from previous HPAI outbreaks. The outbreak is persisting through more than one migration season, and there is a high rate of transmission and clinical disease among wildlife in all four of the migratory flyways. From March 28-December 31, 2022, The Raptor Center at the University of Minnesota screened 1051 individuals for HPAI via polymerase chain reaction (PCR) testing. Of those tested, 215 individuals representing 12 species were confirmed HPAI positive. Great horned owls (*Bubo virginianus*, n = 92) were the most affected species, followed by bald eagles (*Haliaeetus leucoceophalus*, n = 48) and red-tailed hawks (*Buteo jamaicensis*, n = 39). Clinical signs were characterized and video examples of the various presentations of HPAI in raptors will be presented. Most birds admitted with the virus were exhibiting severe neurologic signs. One great horned owl recovered from the infection and was released to the wild. In addition, the utility of a point-of-care avian influenza test (Zoetis FluDETECT®) was assessed by comparing the results to PCR. Paired FluDETECT® and PCR tests have been run on 101 individuals representing 10 species to date, and sampling is ongoing. Based on preliminary data analysis, the FluDETECT® test has a high false negative rate compared to PCR results, making this diagnostic of limited value in a raptor rehabilitation setting.

Acknowledgments: The authors thank Dr. Irene Bueno and Dr. Carol Cardona for their contributions to this work, and Dr. Arno Wuenschmann for pathology results.

References

1. Bertran K, Busquets N, Abad FX, et al. Highly (H5N1) and low (H7N2) pathogenic avian influenza virus infection in falcons via nasochoanal route and ingestion of experimentally infected prey. *PLoS ONE*. 2012;7(3):e32107.

2. Caliendo V, Leijten L, van de Bildt MWG, et al. Pathology and virology of natural highly pathogenic avian influenza H5N8 infection in wild common buzzards (*Buteo buteo*). *Sci Rep*. 2022;12:920.

3. Hall JS, Ip HS, Franson C, et al. Experimental infection of a North American raptor, American kestrel (*Falco sparverius*), with highly pathogenic avian influenza virus (H5N1). *PLoS ONE*. 2009;4(10):e7555.

4. Nemeth NM, Ruder MG, Poulson RL, et al. Bald eagle mortality and nest failure due to clade 2.3.4.4 highly pathogenic H5N1 influenza a virus. *Sci Rep*. 2023;13:191.

5. Shearn-Bochsler VI, Knowles S, Ip H. Lethal infection of wild raptors with highly pathogenic avian influenza H5N8 and H5N2 viruses in the USA, 2014-15. *J Wildl Dis*. 2019;55(1):164-168.

6. Uno Y, Soda K, Tomioka Y, et al. Pathogenicity of clade 2.3.2.1 H5N1 highly pathogenic avian influenza virus in American kestrel (*Falco sparverius*). *Avian Pathol*. 2020;49(5):515-525.

7. U.S. Department of Agriculture. Final Report for the 2014-2015 Outbreak of Highly Pathogenic Avian Influenza (HPAI) in the United States. Revised August 11, 2016. Washington, DC: Animal and Plant Health Inspection Service. https://www.aphis.usda.gov/animal_health/emergency_management/downloads/hpai/2015-hpai-final-report.pdf.

8. van den Brand JMA, Krone O, Wolf PU, et al. Host-specific exposure and fatal neurologic disease in wild raptors from highly pathogenic avian influenza virus H5N1 during the 2006 outbreak in Germany. *Vet Res*. 2015;46:24.

Quarantined Emu (*Dromaius novaehollandiae*) Infected with Highly Pathogenic Avian Influenza H5 Survives, Despite Previous Fulminant Encephalitis Vaccine-Induced Rhabdomyolysis

April Romagnano, PhD, DVM, DABVP (Avian Practice)

Session #3383

Affiliation: From the Avian & Exotic Clinic of Palm City, Palm City, FL 34990, USA.

Note: Abstract format selected by author(s). See Front Matter for description of conference proceedings manuscripts submitted as scientific abstracts.

Abstract: This presentation describes a male emus' recovery from vaccine-induced rhabdomyolysis despite contracting highly pathogenic avian influenza (HPAI), as confirmed by the U.S. Department of Agriculture's (USDA) National Veterinary Services Laboratory, from a positive feral Egyptian goose. Avian influenza (AI) is being disseminated throughout Florida by migratory and feral birds, particularly Egyptian geese. The farm that maintained this emu reported two dead birds, a feral Egyptian goose and a rooster, to Florida's Department of Agriculture and Consumer Services (FDACS). The birds tested positive for AI. The farm was quarantined and over 60 ducks, poultry and guinea fowl were euthanized. FDACS sagely choose not to euthanize one black swan, two Sebastopol geese and two adult emus, including the male reported here. Thanks to FDACS, the farm experienced no further AI deaths and all bordering neighbors were unaffected. AI, which is a major threat to the poultry industry and animal health officials, has strains that vary in severity and pathogenicity. Those most likely strains to kill poultry are labeled highly pathogenic. When HPAI flocks are identified, FDACS collaborates with USDA to eliminate, control and contain the virus. The emu reported here, previously cared for by the farm's large animal veterinarians, was referred weeks after encephalitis vaccination and AI exposure. Down and moribund on examination, its swollen right leg and thigh was severely bruised and leathery at the vaccination site. Bloodwork confirmed severe rhabdomyolysis. The bird was treated diligently by its owners with parenteral fluids and medications. Daily handfeeding addressed anorexia. Intense daily physical therapy was used to get the emu up, including manual and ceiling-built-in support-track sling walking, which continued for many months. The emu recovered safely in a blanket-covered hay-bale enclosure, where it eventually ate, drank, defecated, urinated, stood, walked and finally ran on its own.

Artificial Intelligence Assisted Cell Morphology Assessment in Avian Blood Smear

Kaori Davis, DVM, PhD, DACVP (Anatomic and Clinical Pathology),
Liam Cattell, MEng, DPhil,
Bianca Murphy, DVM, DAVBP (Avian Practice),
Elyssa Dimaano,
Dan Huang,
Thanh Le, MS, PhD

Session #3415

Affiliation: From Moichor, 555 De Haro St Suite 220, San Francisco, CA 94107, USA.

Note: Abstract format selected by author(s). See Front Matter for description of conference proceedings manuscripts submitted as scientific abstracts.

Abstract: The complete blood count (CBC) is a foundational assay to ascertain the overall health status and detect various conditions, including inflammation, infection and neoplasia. The alterations in numerical parameters of the CBC, such as shifts in leukocyte absolute and differential counts are informative, although they can be non-specific. Morphologic assessment of leukocyte populations further aids in interpreting the leukogram. Morphologic assessment is typically performed manually; however, manual morphologic assessment has limitations as it is time and labor-intensive, making it only available in selected patients. In this study, we evaluated the effectiveness of cutting-edge artificial intelligence (AI) and computer vision techniques for analyzing leukocyte morphology in avian hematology samples, specifically in 500 Congo African grey parrots (*Psittacus erithacus*). Among them, 238 were clinically healthy, 196 had various diseases, and the health status and clinical history of 66 animals were unknown. AI-generated counts of morphologic abnormalities, including band heterophils and toxic heterophils, were compared to manual microscopic counts conducted by blinded clinical pathologists. Within-run, between-run, and between-observer coefficients of variation were significantly lower when counted by AI when compared to the manual microscopic count. There was a good agreement between the automated and manual methods. This study demonstrated automated techniques to screen CBC morphologic parameters in a high throughput manner.

Biosecurity and Quarantine Procedures During a Highly Pathogenic Avian Influenza Outbreak

Dana Franzen-Klein, DVM, MS,
Annette Ahlmann, DVM, MS, CWR,
Victoria Hall, DVM, MS, Dipl ACVPM

Session #3341

Affiliation: From The Raptor Center at the University of Minnesota, St. Paul, MN 55108, USA.

Note: Abstract format selected by author(s). See Front Matter for description of conference proceedings manuscripts submitted as scientific abstracts.

Abstract: The 2022 highly pathogenic avian influenza (HPAI) outbreak presented an extraordinary challenge to wildlife rehabilitators - how do they complete their work while preventing the spread of a highly contagious foreign animal disease? From March 28-December 31, 2022, The Raptor Center at the University of Minnesota (TRC) admitted 215 HPAI positive raptors into our wildlife hospital. Despite the high volume of positive cases, TRC was able to use biosecurity principles and quarantine procedures combined with polymerase chain reaction (PCR) testing to effectively prevent disease transmission in the wildlife hospital and protect the center's educational ambassadors from this deadly virus. PCR testing was conducted at admission and at each stage of the quarantine process to safely move over 175 individuals from quarantine into the main patient population. This data allows us to better understand incubation periods and associated disease transmission risk in a raptor rehabilitation setting. Affordable PCR testing with timely results is not readily available to the majority of wildlife rehabilitators and veterinarians. TRC has collected detailed clinical signs data and paired that with PCR results to help guide the use of clinical signs in place of PCR testing as an indicator of possible HPAI infection in a raptor patient. TRC has used the PCR and clinical data collected during the spring of 2022 to adjust our biosecurity and quarantine protocols to ensure sustainability as this outbreak continues. Biosecurity logistics, quarantine time frames, triage parameters, and the data that support our practices and recommendations are discussed.

Acknowledgments: The authors thank Dr. Irene Bueno and Dr. Carol Cardona for their contributions to the development of our biosecurity methods.

References

1. Bertran K, Busquets N, Abad FX, et al. Highly (H5N1) and low (H7N2) pathogenic avian influenza virus infection in falcons via nasochoanal route and ingestion of experimentally infected prey. *PLoS ONE*. 2012;7(3):e32107.

2. Hall JS, Ip HS, Franson C, et al. Experimental infection of a North American raptor, American kestrel (*Falco sparverius*), with highly pathogenic avian influenza virus (H5N1). *PLoS ONE*. 2009;4(10):e7555.

3. Stull JW, Bjorvik E, Bub J, et al. 2018 AAHA infection control, prevention and biosecurity guidelines. *J Am Anim Hosp Assoc*. 2018;54:6.

Avian Influenza: Practical Ways to Protect the Backyard Flock

Brynn McCleery, DVM, DABVP (Avian Practice)

Session #3134

Affiliation: From Red Bank Veterinary Hospital, 197 Hance Avenue, Tinton Falls, NJ 07722, USA.

Abstract: This session briefly reviews the origin of the current highly pathogenic avian influenza (HPAI) outbreak and its significance to wildlife, poultry, and humans. Clinical signs of disease and testing are reviewed. The focus of the session is on precautions to take for backyard poultry to help prevent infection with HPAI.

Introduction

The deadliest outbreak of avian influenza virus (AIV) in United States and European history began in 2020 and is ongoing. This outbreak of H5N1 AIV has been associated with enormous loss of poultry, as well as increased outbreaks and death of wild birds and sporadic spillover into mammals, including a handful of human cases. In the United States, >490 backyard poultry flocks have been affected.

Avian Influenza Viruses

The avian influenza viruses (AIVs) belong to the family Orthomyxoviridae.[1,2] They are all influenza type A viruses, which are single stranded, negative-sense ribonucleic acid (RNA) viruses with segmented genomes that code for 10 viral proteins.[2,3] The 10 viral proteins are encoded by 8 segmented genomic strands.[3] Compared to the other 2 types of influenza viruses, influenza type A has the most genetic variation and the broadest host range.[3] Viral naming is based on the influenza type, place of isolation, strain, designation, and year of isolation. Further classification into subtypes is based on the hemagglutinin (HA) and neuraminidase (NA) surface glycoproteins.[2] HA mediates viral attachment and entry into the host cell by binding to sialic acid receptors on the host cell surface and is responsible for neutralizing host antibodies. NA catalyzes the cleavage of virus particles from the sialic acid sites on the host cell, thereby allowing for spread of the virus within the host.[2,3] Many combinations of HA and NA proteins are possible to create different subtypes of AIV. There are 18 known HA subtypes and 11 NA subtypes.[3,4] Related viruses within a subtype are referred to as a lineage.

Influenza viruses have a high capacity to change due to both genetic drift (e.g., mutations) and genetic shift (e.g., reassortment through exchange of gene segments). Antigenic alterations occur frequently in these viruses, particularly in the HA and NA sites, and are a mechanism through which the virus adapts to the host and survives.[2] Antigenic drift is characterized by smaller variations secondary to accumulated amino acid changes. Mutations that avoid antibody neutralization are selected for over time. Antigenic drift tends to occur every year to few years and is the underlying reason why the annual flu vaccine is constantly changing for humans.[2] Reassortment of recombination of the 8 gene segments can lead to larger variations (i.e., antigenic shift) which may result in new viruses for which a host population has no immunity. Antigenic shift is what typically leads to influenza pandemics.[2]

Pathogenicity of the AIVs is classified as high (HPAI) or low (LPAI) based on the ability of the strain to produce disease and mortality in chickens.[1,3] AVIs have been recognized as a cause of disease in birds for almost 150 years. HPAI was first described in 1878 and referred to as "fowl plague." A viral cause was identified in 1901, but the link to influenza was not made until 1955.[1]

Natural infections of LPAI typically cause no clinical signs or tissue lesions in wild birds. LPAI infections can be asymptomatic in poultry or cause mild symptoms.[1,3,4] H5 and H7 subtypes have the potential to mutate or evolve into HPAI viruses. However, not all viruses of these subtypes cause HPAI.[1] HPAI are rarely isolated from wild birds, and when they do occur in wild birds, the infections typically occur in the vicinity of outbreaks of HPAI in poultry.[1] Instead, the proposed mechanism for emergence of HPAI viruses is that these strains occur only after the viruses have crossed from wild birds to poultry, where they mutate and adapt.[3,4]

Natural Hosts

AIVs infect over 100 different bird species globally and are endemic in some wild birds. Natural reservoirs of the viruses include ducks, swans, geese, terns, shorebirds (including storks, plovers, and sandpipers), and gulls.[3,4] Wild birds, especially dabbling ducks (mallards, pintails, and widgeons), may be infected in the gastrointestinal and respiratory tracts with no symptoms. Waterfowl congregate in huge flocks, often on lakes, annually prior to migration, at which time AIVs spread rapidly through the population. In one study evaluating infection rate of ducks in Alberta, Canada prior to southern migration, > 60% of juvenile ducks were estimated to be infected.[1] Previous surveillance studies reported prevalence rates of LPAI in wild birds of around 15% in ducks and geese and 2% in other species.[1]

Aberrant Hosts

Since LPAI naturally circulates in wild birds, spillover can occur into aberrant hosts in contact with these birds or their feces. Aberrant hosts include poultry (chickens, turkeys, pheasants, quail, domestic ducks, geese, and guinea fowl), pigs, wild mammals, and humans. After transmission to poultry, LPAI may lose its adaptation to wild birds preventing reinfection. Further mutations can occur leading to new strains of HPAI due to exchange of genetic material with other AIVs. Unfortunately, as with the current outbreak, H5 and H7 infections in poultry can spill back into wild birds resulting in further geographic spread as the birds migrate.[4]

Even in areas where AIV outbreaks occur with regularity, such as Minnesota, variation in virus subtype, number of outbreaks seen each year, and seasonal relationship of outbreaks suggest that the epizootics are a result of new primary introductions from wild birds and that AIVs are not enzootic in poultry species.[1]

Transmission

Shedding occurs primarily in the feces, as well as through oral and nasal secretions. Transmission is through the oral-fecal route and can occur primarily from direct contact with infected birds or shared environments (including contaminated water) or secondarily through mechanical transfer of infective feces to susceptible birds.[1,2]

In the USA, spread to poultry typically occurs indirectly through contaminated feed, clothing, and equipment, and less frequently from direct contact with wild birds as poultry are more commonly reared in bird-proof confinement. In other countries, birds may be reared in ways that allow contact with wild birds such as ducks reared on ponds and turkeys reared on range. These birds are at increased risk of infection, especially if situated on a migratory waterfowl route.[1] HPAI are extremely infectious with rapid flock-to-flock spread reported.

Clinical Symptoms

LPAI infections can be asymptomatic in poultry or cause mild symptoms and are most commonly associated with a mild primary respiratory disease, fluffed feathers, reduced egg production, and low mortality.[1,3,4] HPAI, on the other hand, can cause acute high mortality (up to 90-100% in chickens, often within 48 hours).[1] Symptoms of HPAI include acute death, severe depression, anorexia, neurologic disease (twisted neck, swimming in circles, tremors), lethargy, reduced appetite, decreased egg production, irregular eggs (e.g., soft, thin shelled, misshapen), dyspnea, nasal discharge, coughing, sneezing, diarrhea, and swelling of the head, comb, eyelids, wattles, and legs with purple discoloration.[3]

Impact

HPAI is associated with enormous economic losses in poultry and high numbers of bird loss, through death and culling ("stamping out") as part of control policies.[1,5] HPAI is considered a foreign animal disease in the United States. Internationally, HPAI is also reportable to the OIE. Although reportable in poultry, backyard flocks with no sales are considered non-poultry on an international level.

Previous Outbreaks

In 1996, H5N1 HPAI (A/goose/Guangdong/1/1996) reinfected wild birds after spilling over into poultry. This panzootic was unprecedented. The outbreak was contained through culling all the poultry in Hong Kong, but the AIVs continued to circulate in healthy duck populations in the surrounding areas. This strain of HPAI is now considered endemic in 6 countries worldwide, including Bangladesh, China, Egypt, Indonesia, India, and Vietnam.[3]

In 2014-2015, an outbreak of HPAI subtype H5Nx occurred in poultry in the USA and Canada. HPAI H5N8 spread through the flyways initially leading to infection in North America from migratory birds originating in Asia. First identified over the winter and last detected in poultry in June 2015, the outbreak affected 211 commercial flocks and 21 backyard flocks, leading to $3.3 billion in economic losses and the loss of 50 million birds. The outbreak affected 21 states. Fortunately, no human cases were detected. Three mortality events were also reported in wild birds, affecting snow geese and ringed-necked ducks. Multiple captive raptor deaths were reported after being fed meat from infected wild waterfowl. Subtypes involved included H5N8, H5N1, and H5N2.[6,7]

Current Outbreak

The current 2021 H5N1 outbreak is a different strain than the 2014 outbreak. The virus has spun off several variants within clade 2.3.4.4b making it even more successful. The first case was identified in an American wigeon duck in South Carolina. The infection travelled from western Europe to eastern Canada and has been detected in all four flyways. Most early infections centered around the intersection of the Central and Mississippi migratory flyways. There has been wide geographic spread in wild birds and poultry worldwide with sporadic spillover into a variety of mammalian species.[4]

Disease has been variable among species. The current outbreak affects layers more than broilers and turkeys.[4] In wild birds, swans, geese, waterbirds (e.g., pelicans, mergansers), raptors, and scavengers are all considered highly susceptible. The current outbreak has been associated with a higher wild bird mortality than that of 2014, as well as with a wider diversity of species affected. In addition to wild birds, HPAI has been detected in a variety of wild mammals (e.g., harbor seals, grey seals, foxes, skunks, bobcats, lynx, Eurasian otters, Eurasian polecats, European badger, domestic pigs, ferret, common raccoon dog, opossums, bears, farmed mink, sea lions).[4,5] A small number of human infections have been reported, all of which were associated with recent poultry exposure. Overall, symptoms in human cases have been mild. No cases of human-to-human transmission have occurred.[4]

Statistics (at the time of writing)

Global

- > 11,300 animal outbreaks of H5N1 reported by 73 member states to the World Organization of Animal Health since January 2022[4]
- Largest AIV epidemic thus far in European history (2398 outbreaks in poultry, 46 million birds culled, 168 detections in captive birds, 2733 die off events in wild birds, 36 European countries affected)[5]
- Humans: Jan 2022-Mar 2023, 10 cases in humans from 7 countries (1 in the USA)[4]

USA (as of March 2023)

- 815 confirmed poultry flocks: 323 commercial, 492 backyard (in 47 states)[4,8]
- Almost 6,500 wild birds (in 49 states)[4]
- 58.65 million birds affected[4,8]
- Iowa: two massive egg laying operations culled > 5 million birds in single incidents[8]
- Mammals: 144 mammals (in 23 states)[4]

Summary

The above statistics make this the deadliest AIV outbreak in USA and European history. While most previous outbreaks were halted by change in season, the current H5N1 outbreak made it through the summer and some authorities are suggesting that the observed persistence of the virus in wild birds since the 2020-2021 epidemic wave may indicate that this strain has become endemic in wild birds.[5]

Prevention

In some ways, preventing infection in smaller, backyard flocks is easier than in large, commercial flocks due to reduced farm-to-farm contact, equipment, and transport of birds; however, in other ways, prevention can be more difficult due to increased free range flocks and a reduced understanding of proper biosecurity practices in backyard flock owners.

A line of separation should be placed around the flock and may include the walls of a coop/barn, fence around the run, or fence around the yard. Consider the risks of anything crossing the line—such as other birds, other animals, and humans—and evaluate ways to reduce those risks. Consider both structural biosecurity (e.g., the enclosure the birds are housed in) and operational biosecurity (e.g., practices and procedures followed by the people caring for the birds).

Basic biosecurity principles

Basic biosecurity principles should be followed and include:

- Isolate newly acquired birds
- Become familiar with signs of illness; quarantine sick birds; seek veterinary care and testing
- Clean and disinfect enclosures and equipment regularly
- Limit visitors: only allow people caring for the flock (family, friends) interact with the birds, make sure all participants follow biosecurity principles, avoid having people with separate flocks interact with the birds
- Wash hands before and after contacting birds
- Consider a foot bath (remove debris first) or dedicated pair of shoes (rubber boot covers work well)
- Change clothes or wear protective outer garments when working with poultry

Specific prevention for AIVs

The biggest risk to the backyard flock is from waterfowl, which are less likely to encounter backyard flocks unless there is a pond or large amount of acreage. Songbirds and other visitors to the backyard (corvids, pigeons/doves) are less likely to be infected with AIVs including HPAI; however, due to the variety of species that can be infected, including wild mammals, care should be taken to avoid contact of all wild animals with the flock. Farm-to-farm transmission is even more common, and owners of backyard flocks should do their best to eliminate contact with other flocks during the outbreak (e.g., follow all in/all out procedures and not buy/adopt new birds from various sources, not share equipment).

Prevent cross-contamination with other flocks:

- Avoid having people with separate flocks interact with the backyard flock
- Do not share equipment
- Avoid use of soiled egg cases, do not reuse cardboard egg cases between different flocks

Prevent contact with free-ranging wild birds/animals:

- Prevent nesting/shelter seeking
  - Build structures with shorter eaves or no overhangs/ledges
  - Fill small burrows, tunnels, or holes with gravel

- Repair any holes or tears in barn/cool walls
- Exclusionary screens/netting: utilize small gauge (1-inch opening excludes most bird species)

- Reduce water access
  - Remove standing water (event small pools) through proper grading or filling areas where water stands < 24-48 hours after rain
  - Ponds: If present, consider using a boat, bird deterrent balls, or predator decoys to discourage wild birds.
  - Do not offer bird baths around the coop or run

- Reduce food sources
  - Mow grass frequently (< 3 inches)
  - Keep fruit trees and other edible plants father away from the coop or run
  - Remove fallen fruit
  - Avoid grass, weeds, and shrubbery next to the coop/run to prevent wildlife seeking shelter
  - Do not feed wild birds/wildlife near poultry (i.e., no bird seed, suet, corn, etc.)
  - Prevent access of wildlife to stored feed (lidded container, sweep up spills)
  - Keep compost and garbage away from poultry and prevent wildlife access

- Reduce fecal contamination
  - Shade cloth over the top of the run can prevent wild bird droppings from falling into the run area
  - If covering the entire area is not desirable, place food/waterers under cover to prevent fecal contamination from wild birds
  - Do not store extra equipment outside. Cleaned equipment should be stored in a garage/barn/basement until needed.
  - Large troughs/swimming pools: Cover with a tarp/lid when not in use. Drain and refill frequently to keep clean.

APHIS is an excellent resource for owners (www. aphis.usda.gov/animalhealth/defendtheflock).

Vaccination is not allowed in the United States at this time due to impacts on international trade as vaccinated birds cannot be distinguished from infected birds through testing. Vaccination is also complicated by timing for commercial poultry: birds would require two doses of vaccine with a withdrawal period after the vaccine prior to slaughter, which is not feasible.

Testing

In poultry with suspected HPAI, testing should be performed by a USDA APHIS accredited veterinarian. RT-PCR is recommended. Appropriate samples include an oropharyngeal swab in gallinaceous poultry or a cloacal swab in waterfowl.[2] Possible results include not detected or non-negative. A non-negative sample with clinical signs consistent with HPAI is considered a presumptive positive. If detected on PCR, virus isolation and growth of the virus in an embryonated chicken egg may be performed.[2] Serology is not recommended as mortality occurs acutely before antibody production. After reporting a presumptive positive flock, the flock is placed under quarantine with control and surveillance zones established. Depopulation and a fallow period (where repopulation is not allowed) will follow.

Testing of wild bird populations is performed for surveillance purposes only. Surveillance of AIVs in wild birds began in the mid 1970s.[1] The USDA began monitoring migratory birds in different flyways between 1998-2005.[2] Surveillance efforts have increased due to the current outbreak. Eradication of HPAI from wild birds is not being attempted. Instead, attempts to limit spillover to domestic species through preventive measures are recommended.

References

1. Alexander DJ. A review of avian influenza in different bird species. *Vet Microbiol.* 2000;74:3-13.

2. Thomas JK, Noppenberger J. Avian influenza: a review. *Am J Health Syst Pharm.* 2007;64:149-165.

3. Nunez IV, Ross TM. A review of H5Nx avian influenza viruses. *Ther Adv Vaccines Immunother.* 2019;7:1-15.

4. Centers for Disease Control and Prevention. Technical report: highly pathogenic avian influenza A (H5N1) viruses. Available at: https://www.cdc.gov/flu/avianflu/spotlights/2022-2023/h5n1-technical-report.htm#print. Accessed March 25, 2023.

5. Adlhoch C, Fusaro A, Gonzalez JL, et al. Avian influenza overview March-June 2022. *Eur Food Safety Author J.* 2022;20(8):7415.

6. U.S. Department of Agriculture. Final report for the 2014-2015 outbreak of highly pathogenic avian influenza (HPAI) in the United States. Available at: https://www.aphis.usda.gov/animal_health/emergency_management/downloads/hpai/2015-hpai-final-report.pdf. Accessed March 25, 2023.

7. Greene JL. Update on the highly-pathogenic avian influenza outbreak of 2014-2015. Available at: https://nationalaglawcenter.org/wp-content/uploads/assets/crs/R44114.pdf. Accessed March 25, 2023.

8. Animal and Plant Health Inspection Service. 2022-2023 confirmations of highly pathogenic avian influenza in commercial and backyard flocks. Available at: https://www.aphis.usda.gov/aphis/ourfocus/animalhealth/animal-disease-information/avian/avian-influenza/hpai-2022/2022-hpai-commercial-backyard-flocks Accessed March 25, 2023.

Measuring Disease Prevalence in Barred Owls (*Strix varia*) Presented to a Louisiana Wildlife Hospital (2014-2021)

Aspen La-Riene Settle, BSc, and
Mark A. Mitchell, DVM, MS, PhD, DipECZM (Herpetology)

Session #3487

Affiliation: From Louisiana State University School of Veterinary Medicine, Baton Rouge, LA 70803, USA.

Note: Abstract format selected by author(s). See Front Matter for description of conference proceedings manuscripts submitted as scientific abstracts.

Abstract: The Wildlife Hospital of Louisiana (WHL) sees an annual average of about 1,600 patients. Of that caseload, approximately 15% are raptors. This retrospective study catalogued the presentation of barred owls (*Strix varia*), the most numerous raptor patients presented to the WHL. Electronic records were categorized according to patient outcome (6/431 dead on arrival, 220/431 euthanized within 24 hours, 22/431 died within 24 hours, 22/431 died after 24 hours, and 161/431 released). Primary diagnoses of orphaning, trauma, neurologic disease, or unknown were recorded for each outcome, as well as classification of trauma type and existence of comorbidities (e.g., ballistic trauma, vehicular trauma, comminuted fractures, open fractures, ocular damage, luxation, and parasitism). Three-hundred thirty-two (77%) of all owls presented with evidence of trauma. Metal opacities outside of the gastrointestinal tract, entry or exit wounds, or at least one identified puncture wound combined with a fracture warranted classification of ballistic trauma, which was the case for six birds. Vehicular trauma was the most frequent suspected cause of trauma, which affected 119 (35%) birds. Of those presenting with trauma, comminuted fractures were identified 24 times, open fractures 82 times, ocular damage 93 times, luxation 21 times, and parasitism 28 times. Further results, including analysis of comorbidity impact on release rate, will be discussed during the lecture. The authors use this data to analyze the efficiency of the WHL triage assessment protocols, the success of the team's rehabilitation and release efforts, and the danger to these animals of anthropomorphic activity and encroachment.

Section 12

AAV Potpourri II Session

Rachel Baden, DVM, DABVP (Avian Practice)
Session Chair

How to Maximize Foraging Opportunities for Parrots: A Case Study

Mandy Beekmans, MSc,
Claudia Vinke, PhD,
Nico Schoemaker, DVM, PhD, DipECZM (Small Mammal), DipECZM (Avian),
Yvonne van Zeeland, DVM, MVR, PhD, DipECZM (Avian), DipECZM (Small Mammal), CPBC

Session #3275

Affiliation: From the Division of Zoological Medicine, Department of Clinical Sciences (Beekmans, Schoemaker, van Zeeland), Division of Animals in Science and Society, Department of Animal Population Health, Faculty of Veterinary Medicine, Utrecht University, Yalelaan 108, 3584 CM, Utrecht, The Netherlands (Vinke).

Note: Abstract format selected by author(s). See Front Matter for description of conference proceedings manuscripts submitted as scientific abstracts.

Abstract: Foraging enrichment is considered an effective tool to stimulate species-typical foraging behavior and to reduce abnormal (repetitive) behavior;[1-3] however, for parrots in captivity, even when using foraging enrichment it remains difficult to approximate foraging times of wild parrots.[4,5] As such current strategies may be insufficient to meet the parrots' foraging requirements. Moreover, currently available enrichment often stimulates a single aspect of foraging, while the behavior in fact comprises of a multitude of different activities including flying, searching, selecting and procuring, opening, extracting, processing, and finally the actual consumption of food.[2-4] We therefore designed a foraging enrichment for parrots, which consisted of two components that respectively stimulated the various appetitive (flying, searching, selecting, procuring) and consumptive behaviors (opening, extracting, processing), and studied its effect on foraging times in 12 grey parrots (*Psittacus erithacus*). The two-component enrichment was (tentatively) able to increase daily foraging times from 2 h/day (control condition) to 4 h/day, while the separate components increased daily foraging times to 3 h/day. Hence, through stimulation of both appetitive and consumptive behaviors, the two-component enrichment was able to increase daily foraging times to produce natural time budgets in captivity. While further studies are needed to investigate the importance of individual foraging activities for parrots, and evaluate whether naturalized time budgets can more effectively eliminate abnormal behaviors, findings of this study emphasize that efficacy of foraging enrichment can be increased by incorporating and combining multiple foraging activities typical for the species.

References

1. Meehan CL, Millam JR, Mench JA. Foraging opportunity and increased physical complexity both prevent and reduce psychogenic feather picking by young Amazon parrots. *Appl Anim Behav Sci.* 2003;80(1):71-85.

2. Lumeij JT, Hommers CJ. Foraging 'enrichment'as treatment for pterotillomania. *Appl Anim Behav Sci.* 2008;111(1-2):85-94.

3. Fangmeier ML, Burns AL, Melfi, VA, et al. Foraging enrichment alleviates oral repetitive behaviors in captive red-tailed black cockatoos (*Calyptorhynchus banksii*). *Zoo Biol.* 2020;39(1):3-12.

4. Rozek JC, Danner LM, Stucky PA, et al. Over-sized pellets naturalize foraging time of captive orange-winged Amazon parrots (*Amazona amazonica*). *Appl Anim Behav Sci.* 2010;125(1-2):80-87.

5. van Zeeland YRA, Schoemaker NJ, Ravesteijn MM, et al. Efficacy of foraging enrichments to increase foraging time in grey parrots (*Psittacus erithacus erithacus*). *Appl Anim Behav Sci.* 2013;149(1-4):87-102.

Desert Rose (*Adenium obesum*) Acute Toxicity and Treatment in 33-Year-Old Blue and Gold Macaw

April Romagnano, PhD, DVM, DABVP (Avian Practice),
Kian Martin,
Elliott Leonard,
Stephanie Monteith, CNA,
Lisa Smith, CVT,
Scott Martin, DVM, MS

Session #3380

Affiliation: From the Avian & Exotic Clinic of Palm City, Palm City, FL 34990, USA (Romagnano, Martin, Leonard, Monteith, Smith, Martin), Martin Animal Clinic Palm City, FL 34990, USA (Romagnano, Martin, Leonard, Monteith, Martin).

Note: Abstract format selected by author(s). See Front Matter for description of conference proceedings manuscripts submitted as scientific abstracts.

Abstract: This presentation reports the first case of desert rose *(Adenium obesum)* acute toxicity and successful treatment in an avian species. Desert rose *(Adenium obesum)*, of the family Apocynaceae, is a cardioactive steroid and cardiac glycoside plant. The entire plant produces a sap that is a cardiac toxin from its roots to its flowers. Historically, its sap was used to create poison arrows to hunt large game and fish in Africa. Desert rose poisoning has been reported in humans, dogs, cats and horses.[1] A 33-year-old male blue and gold macaw ate an entire desert rose flower and base from an indoor houseplant. After ingestion, the bird fell to the floor, ran to its owners who witnessed the bird foaming at the mouth, collapse, bluing of the cheeks and seizing, which rendered it moribund. The owners placed a call to the veterinary emergency phone and drove 2 hours to the clinic with the car's air-conditioner on high directed towards the bird's face. The macaw presented moribund, with dark-blue cheeks, rigid limbs and a severe bradycardia. It was weighed and then placed in a low heat and high oxygen incubator. Treatment included atropine, parenteral fluids, dextrose, calcium-gluconate, vitamins, anti-inflammatories, antibiotics and activated charcoal gavage. Initial bloodwork revealed a severe hyperkalemia > 10 mmol/L which was treated with fluids and furosemide. Diagnostics and supportive treatment continued for 3 weeks, including tube-feeding, sucralfate and hyperbaric oxygen. Constant incubator oxygen coupled with multiple hyperbaric oxygen treatments significantly improved perfusion, decreased inflammation and facilitated recovery. Presently the bird is bright and active with normal bloodwork.

References

1. Botelho AFM, Pierezan F, Soto-Blanco B, Martins Melo M. A review of cardiac glycosides: structure, toxicokinetics, clinical signs, diagnosis and antineoplastic potential. *Toxicon.* 2019;158:63-68.

Patient-Patient Owner Relationships and Their Influence on Budgerigar Husbandry Systems

Ruediger T. Korbel, Prof Dr med vet habil, DipECZM (Avian),
Cécile Danner, DVM,
Katrin Drasch, Dr phil,
Johannes Kallenbach,
Monika Rinder Prof Dr med vet, DVM,
Nicole J. Saam, Prof Dr phil

Session #3327

Affiliation: From the Center for Clinical Veterinary Medicine, Clinic for Birds, Small Mammals, Reptiles and Ornamental Fish, LMU University of Munich, Oberschleissheim, Germany (Korbel, Danner, Rinder), Institute for Sociology, Chair for Methods of Empirical Sociological Studies, Friedrich-Alexander-University of Erlangen-Nuremberg, Germany (Drasch, Kallenbach, Saam).

Note: Abstract format selected by author(s). See Front Matter for description of conference proceedings manuscripts submitted as scientific abstracts.

Abstract: According to the 1966 Veterinary Surgeons Act (an act of the Parliament of the United Kingdom), central tasks within the profession of veterinary medicine are: (1) diagnosis of disorders and injuries including diagnostics, (2) advisory (for the patient owner) based on diagnostics, (3) medical and (4) surgical treatment of animals. Mullan and Fawcett defined veterinary treatment as "interventions to prevent disease and suffering, diagnose and manage or cure disease in animals."[1] According to this definition, including prevention of suffering, everything must be undertaken to assure maximum animal welfare. Within the complex set of relations that may be described as a clinician-companion animal-client triangle (CCC-triangle), in routine daily clinical work, not only in pet bird medicine but also for state authority veterinary, commercially kept poultry veterinary work methods to date are based more on subjective approaches (i.e., relying on general unspecified experience within clinical daily routine) without validated standardized protocols.

Introduction

The housing, care, feeding and environmental demands of birds can vary greatly depending on the species and are often not easy to meet while in human care. In the present study, a husbandry index was developed to assess budgerigar husbandry and to examine the relevance of the human-bird relationship to budgerigar husbandry in captivity.

Husbandry Index

As a first step, a husbandry index for noncommercially kept budgerigars from private households (Budgerigar Husbandry Index BHIg-14) was developed, taking into account current valid husbandry recommendations and guidelines. The additive weighted index consists of 14 housing dimensions: housing environment, grating, perches, floor covering/floor substrate, nesting material, environmental enrichment, light, temperature, food, cleaning/hygiene, free flight, possibility of separation, socialization and bathing. It measures, as a preliminary stage of an animal welfare index, the quality of the existing husbandry conditions. The index was queried together with the owner-bird-relationship-scale (OBRS),[2] a psychometric scale that makes the human-bird relationship measurable in four factors in a standardized online questionnaire for private budgie owners in Germany. The evaluation was based on 1536 online questionnaires and was carried out using multivariate analysis methods.

Results

Overall study results reveal good housing conditions could be determined and confirmed objectively in the survey. Marked deficits were found in owners who kept exactly two or only one budgerigar, in sociologic groups with a low monthly net household income and in owners with a commercial background. Within the framework of the multivariate linear regression method, all four factors of the human-bird relationship described, namely the bird as human; the bird as meaning of life; empathy, mindfulness and respect; and the relationship of the bird-bird owner showed a highly significant positive correlation with the quality of budgerigar husbandry in human care. In addition, connections in relation to the behavior of the owners to gain knowledge, the number of budgerigars kept and other socio-demographic characteristics could be identified.

References

1. Mullan S, Fawcett A. *Veterinary Ethics: Navigating Tough Cases.* Sheffield, UK: 5m Publishing; 2017.

2. Burmeister A-K, Drasch K, Rinder M, et al. Development and application of the owner-bird relationship scale (OBRS) to assess the relation of humans to their pet birds. *Front Vet Sci.* 2020;7. https://doi.org/10.3389/fvets.2020.575221.

Clinical Signs Associated with Elevated Copper Levels in Mixed Species Captive Birds: A Case Series

Dawn Daugherty, DVM, MPH, CPH, and
M. Scott Echols, DVM, DABVP (Avian Practice)

Session #3248

Affiliation: From Parrish Creek Veterinary Hospital and Diagnostic Center, Centerville, UT 84014, USA (Daugherty, Echols), Echols Veterinary Services, Cottonwood Heights, UT 84093, USA (Echols), Scarlet Imaging, Murray, UT 84157, USA (Echols).

Note: Abstract format selected by author(s). See Front Matter for description of conference proceedings manuscripts submitted as scientific abstracts.

Abstract: While normal values of copper have been established at 0.07-0.19 ppm in Hispaniolan Amazon parrots (*Amazona ventralis*), there is little information on the effects of copper toxicity in captive birds.[1] Gastrointestinal distress, hemolytic anemia, liver disease and death are reported in mammals. The poultry literature describes cases of severe acute copper toxicity resulting in anorexia and death with ulcerations and congestion in the ventriculus and high levels of copper in various tissues.[2] Regardless, there is very little information on low level or chronic copper toxicity in birds. This case series examines six birds at the time of writing, (one cockatoo, one macaw, one Pionus parrot, one African gray parrot, two Amazon parrots) with variable clinical signs, recognizable metal in the proventriculus and/or ventriculus on standard projection radiographs and/or high-resolution computed tomography and elevated blood copper levels between 0.34-1.52 ppm. All birds were treated conservatively to remove the radiographically detected gastrointestinal metal and all birds returned to normal upon successful treatment. The case series describes clinical signs, diagnostics used, treatments and outcomes of birds entered in the study. The purpose of the study is to find associations between elevated copper levels and clinical signs in captive birds.

References

1. Osofsky A, Jowett PLH, Hosgood G, Tully TN. Determination of normal blood concentrations of lead, zinc, copper, and iron in Hispaniolan Amazon parrots (*Amazona ventralis*). *J Avian Med Surg.* 2001;15:31-36.

2. Giergiel M, Durkalec MM, Nawrocka A, et al. Ingestion of bedding material as a cause of acute copper sulfate poisoning in turkey poults. *Poult Sci.* 2019;98:707-711.

Determining Fecal Microbiome Using Next-Generation Genomic Sequencing in Cockatiels (*Nymphicus hollandicus*) Fed Different Diets

Nicole Sands, DVM,
Shachar Malka, DVM, DABVP (Avian Practice), DACEPM,
Gina Vecere, DVM, DACEPM,
Margaret Lee, DVM,
Jonathan Stockman, DVM, DACVIM (Nutrition),
Janina Krumbeck, PhD

Session #3425

Affiliation: From Long Island Bird and Exotics Veterinary Clinic, 333 Great Neck Road, Great Neck, NY 11021, USA (Sands, Malka, Vecere, Lee), Department of Clinical Veterinary Sciences, LIU College of Veterinary Medicine, Long Island University, 720 Northern Boulevard, Brookville, NY 11548, USA (Stockman), MiDOG LLC, 17062 Murphy Ave., Irvine, CA 92614, USA (Krumbeck).

Note: Abstract format selected by author(s). See Front Matter for description of conference proceedings manuscripts submitted as scientific abstracts.

Abstract: Formulated diets provide balanced nutrition with essential vitamins and minerals, and therefore are often recommended as the majority of a bird's diet by veterinarians.[1-4] The goal of this study is to characterize the normal fecal microbiome of client-owned cockatiels (*Nymphicus hollandicus*) fed a commercial seed mix as compared to cockatiels fed commercial brands of formulated pellets. Fecal samples from 34 healthy client-owned cockatiels, 15 eating a generic commercially available seed diet and 19 eating a formulated pellet diet, were collected and analyzed by next-generation DNA sequencing. samples were collected according to the manufacturer's instructions using a swab collection kit provided by the NGS technology diagnostic center and processed as previously described.[5] Next-generation DNA sequencing has been shown to provide practical applications for characterizing microbiomes and providing better identification of microbes as compared to traditional culture methods.[5-8] A total of 179 different bacterial species representing 94 genera were detected. The most abundant bacterial species in the pellet-fed group were *Lactobacillus alvi*, *Ureaplasma* spp., and *Lactobacillus salivarius* while *Ureaplasma* spp., *Lactobacillus alvi*, and *Lactobacillus crispatus* were the most abundant species in the seed-fed group. For fungi, 244 fungal species representing 156 genera were detected. *Cladosporium* spp., *Fusarium* spp., and *Alternaria* spp. were the most abundant in the pellet-fed group, and *Cladosporium* spp., *Cladosporium lignicola/sphaerospermum*, and *Meyerozyma carpophila/guilliermondii* in the seed-fed group. These results illustrate how various diets affect the fecal microbiome of cockatiels and should be taken into consideration by clinicians when identifying a disease state. Elaboration of how these results compare to each other are provided.

References

1. Brightsmith D. Nutritional levels of diets fed to captive Amazon parrots: does mixing seed, produce, and pellets provide a healthy diet? *J Avian Med Surg.* 2012;26(3):149-160.

2. Ullrey D, Allen ME, Baer DJ. Formulated diets versus seed mixtures for psittacines. *J Nutr.* 1991;121(11): 193-205.

3. Harrison G, Lightfoot T. *Clinical Avian Medicine.* Palm Beach, FL: Spix Publishing; 2006:63-94.

4. Samour J. *Avian Medicine.* St. Louis, MO: Elsevier Saunders; 2016:25-30.

5. Tang S, Prem A, Tjokrosurjo J, et al. The canine skin and ear microbiome: a comprehensive survey of pathogens implicated in canine skin and ear infections using a novel next-generation-sequencing-based assay. *Vet Microbiol.* 2020;247:108764.

6. Krumbeck J, Turner DD, Diesel A, et al. Skin microbiota of quaker parrots (*Myiopsitta monachus*) with normal feathering or feather loss via next-generation sequencing technology. *J Exot Pet Med.* 2022;42: 26-34.

7. Vecere G, Malka S, Holden N, et al. Comparison of ear canal microbiome in rabbits with and without otitis externa using next generation DNA sequencing. *J Exot Pet Med.* 2022;42:35-41.

8. Melgarejo T, Oakley BB, Krumbeck JA, et al. Assessment of bacterial and fungal populations in urine from clinically healthy dogs using next-generation sequencing. *J Vet Intern Med.* 2021;35(3):1416-1426.

Military Macaw with Chronic Active *Klebsiella* Stomatitis Diagnosed by Next-Generation Sequencing

April Romagnano, PhD, DVM, DABVP (Avian Practice), and Kian Martin

Session #3468

Affiliation: From the Avian & Exotic Clinic of Palm City, Palm City, FL 34990, USA (Romagnano, Martin), Martin Animal Clinic, Palm City, FL 34990, USA (Romagnano, Martin).

Note: Abstract format selected by author(s). See Front Matter for description of conference proceedings manuscripts submitted as scientific abstracts.

Abstract: This presentation reports a chronic oral proliferative stomatitis case in a 27-year-old male military macaw that progressed and evolved over a 6-year period. The oral growths were pale tan to dark brown, raised, and ranged from 0.2-1.3 centimeters. They were smooth to roughened, firm and irregularly shaped, often with a head and pedunculated stalk seated in an inflamed circular base surrounded by normal oral mucosa. These growths caused chronic-active stomatitis, complicated by rhinitis that initially started in 2017 as a proliferative ear growth referral case. Despite extensive diagnostics, various treatments-including antimicrobials based on culture + sensitivity, nebulization and hyperbaric-oxygen, and repeated surgical and cautery removal, the growths recurred every 4-6 weeks impeding mastication. Histopathology had repeatedly shown locally extensive, chronic active, proliferative heterophilic stomatitis with marked hyperkeratosis which evolved into chronic mixed mononuclear and heterophilic inflammation. Early histopathologic examination suggested a bacterial component. Numerous cultures revealed gram-negative and gram-positive bacteria with *Klebsiella* sp., *Pseudomonas* sp., and *Staphylococcus* sp. predominating. PCRs from multiple labs were negative for herpesvirus, papillomavirus, poxvirus, circovirus, chlamydia and *Aspergillosis*, *Candida* and *Malassezia*. Bloodwork had repeatedly shown an elevated WBC with a heterophilia, low bile acid, elevated beta-globulins, and negative *Aspergillosis* AB and galactomannan. MiSEQ next-generation sequencing of an oral mass from late in the disease process identified no viruses, but a predominance of *Klebsiella* sp. and particularly *K. pneumoniae*. Based on the compilation of all the findings and diagnostics in this case, it is believed that *Klebsiella* played a primary role in the disease process throughout this still-thriving bird's history.

Fourteen Cases of Naturally Infected Birds with Avian Bornavirus: Usefulness of Histologic Examination of Skin Lesions and Virus Localization of Feather Damaging Behavior for Early Diagnosis

Reiko Soga, DVM,
Souichi Yoshino,
Naoyuki Aihara,
Junichi Kamiie,
Akiko Makino,
Keizo Tomonaga

Session #3200

Affiliation: From Azabu University Veterinary Medicine and Pathology (Yoshino, Aihara, Kamiie), Kyoto University Medical and Biological Research Institute (Makino, Tomonaga), Grow-Wing Animal Hospital (Soga).

Abstract: An outbreak of avian bornavirus (ABV) disease was observed in a domestic breeding farm for pet Psittaciformes. Some individuals infected with ABV are known to exhibit feather destructive behavior. In this study, we report the results of a pathologic study of the defeathered skin area in ABV-positive individuals found to be defeathered.

Materials and Methods

Two live (i.e., surviving) birds and 14 dead birds with feather shedding and confirmed bornavirus infection were studied. Surviving birds had a clinical follow-up by multiple fecal PaBV PCR tests, and skin biopsies were pathologically examined. Dead birds were submitted for pathologic autopsy, and sampling of brain tissue, digestive tract, and defeathered skin area. Subsequently, histopathologic examination and PaBV immunostaining was performed.

Discussion and Conclusions

In infected birds that exhibited defeathering, infection and inflammation of the peripheral nerves surrounding the feather follicle was observed, as well as infection of the feather pulp. It was suggested that viral infection of the cutaneous peripheral nerves or feather pulp may be involved in defeathering. PaBV antigen was detected in skin peripheral nerves by immunostaining two live biopsied birds. All 14 dead birds were PaBV-positive in the brain and 14 were also PaBV-positive in the skin. Four of the 10 birds whose intestines could be examined had no bowel lesions.

Despite a negative fecal PCR test, ABV infected birds present with feather destructive behavior. Early PCR fecal testing is often negative. Therefore, if the diagnosis of ABV and feather destructive behavior can be confirmed early by anti-ABV antibody immunostaining, treatment will be successful and life may be prolonged.

References

1. Horie M, Veda K, Veda A, et al. Detection of avian bornavirus 5 RNA in *Eclectus roratus* with feather picking disorder. *Microbiol Immunol.* 2012;56(5):346-349 .

2. Kolb SJ. Feather Destruction Disorder. Available at: https://www.vin.com/ce/TECH281-0120_VSPN.htm. Accessed May 2, 2023.

3. Antinoff N. Feather damaging…for a reason. *Pacific Vet Conf.* 2019.

4. Sassa Y, Horie M, Fujino K, et al. Molecular epidemiology of avian bornavirus from pet birds in Japan. *Virus Genes.* 2013;47(1):173-177.

5. Philadelpho NA, Rubbenstroth D, Guimarães MB, et al. Survey of bornaviruses in pet psittacines in Brazil reveals a novel parrot bornavirus. *Vet Microbiol.* 2014;174(3-4):584-590.

6. Horie M. Parrot bornavirus infection: correlation with neurological signs and feather picking. *Vet Rec.* 2019;184(15):473-475.

Antemortem Diagnosis and Successful Long-Term Management of Disseminated Xanthogranulomatous Disease in an Eclectus Parrot (*Eclectus roratus*)

Christoph Mans, Dr med vet, Dipl ACZM,
Neta Ambar, DVM,
David Gasper, DVM, PhD, DACVP

Session #3476

Affiliation: From the School of Veterinary Medicine, University of Wisconsin-Madison, 2015 Linden Drive, Madison, WI 53706, USA.

Note: Abstract format selected by author(s). See Front Matter for description of conference proceedings manuscripts submitted as scientific abstracts.

Abstract: A 12-year-old male eclectus parrot was referred for evaluation of coelomic distention. A computed tomography scan and blood work revealed coelomic effusion with free coelomic mineral attenuating material and severe elevation of plasma bile acids. Coelomic effusion was consistent with macrophagic inflammation with abundant intracellular lipids. Initial treatment with meloxicam showed minimal improvement. Disseminated xanthogranulomatous inflammation was suspected based on imaging and laboratory results, which were consistent with those previously reported.[1-2] Biopsies of the liver and intracoelomic masses confirmed this diagnosis. Treatment with prednisolone at 1 mg/kg/day for 6 months, followed by 0.5 mg/kg/day for 3 months, was initiated. Clinical improvement was assessed based on owner evaluation, plasma bile acid levels, and diagnostic imaging. After 2 months of prednisolone treatment, the bird's behavior and appetite improved, which persisted throughout treatment and at follow-up 17 months after the cessation of steroid therapy. Bile acid levels were normal 10 months after the cessation of therapy. Diagnostic imaging showed minimal coelomic effusion at 10 months post-cessation of therapy, with progressive expansion of the air sacs and static to improved dystrophic mineral foci. This is the first report describing an antemortem diagnosis and treatment of disseminated coelomic xanthogranulomatous disease in a psittacine species, with an objectively documented response to prednisolone therapy.

References

1. Hanson ME, Donovan TA, Quesenberry K, et al. Imaging features of disseminated xanthogranulomatous inflammation in eclectus parrots (*Eclectus roratus*). *Vet Radiol Ultrasound.* 2020;61:409-416.

2. Donovan TA, Garner MM, Phalen D, Reavill D, Monette S, et al. Disseminated coelomic xanthogranulomatosis in eclectus parrots (*Eclectus roratus*) and budgerigars (*Melopsittacus undulatus*). Vet Pathol. 2022;59(1):143-151.

Prevalence, Distribution, and Risk Factor Correlates of Adipocytic Tumors in Psittaciformes: 648 Cases (1998-2018)

Lucyanne Megan, DVM,
David Sanchez-Migallon Guzman, LV, MS, DipECZM (Avian), DipECZM (Small Mammal), Dipl ACZM,
Kevin Keel, DVM, PhD, DACVP,
Hugues Beaufrère, Dr med vet, PhD, DABVP (Avian Practice), DipECZM (Avian), Dipl ACZM

Session #3361

Affiliation: From the William R. Pritchard Veterinary Medical Teaching Hospital, University California-Davis, Davis, CA 95616, USA (Megan), Department of Medicine and Epidemiology, School of Veterinary Medicine, University of California-Davis, Davis, CA 95616, USA (Guzman, Beaufrère), Department of Pathology, Microbiology, and Immunology, School of Veterinary Medicine, University of California-Davis, Davis, CA 95616, USA (Keel).

Note: Abstract format selected by author(s). See Front Matter for description of conference proceedings manuscripts submitted as scientific abstracts.

Abstract: Adipocytic tumors are mesenchymal tumors that are commonly reported in psittacine birds; however, large-scale studies evaluating their prevalence and associated risk factors are lacking. A retrospective study of adipocytic tumors in psittacine birds was performed by reviewing pathology submissions from the UC Davis Drury Reavill Pathology Database, containing 26,013 submissions from psittacine birds. Age, sex, genus, anatomic distribution, and pathologic diagnosis were collected for each case when available. Four-hundred fifty cases of lipoma, 129 cases of myelolipoma, 35 cases of hemangiolipoma, and 31 cases of liposarcoma were identified. Adipocytic tumors were identified in 27 genera. Lipomas were most commonly identified in *Amazona* (130/450), *Nymphicus* (65/450), and *Melopsittacus* (41/450) spp., while liposarcomas were most commonly identified in *Nymphicus* (6/31), *Myiopsitta* (5/31), and *Melopsittacus* (4/31) spp. The mean ages at diagnosis of lipoma and liposarcoma were 16.7 years (range 0.3-46 years) and 15.7 years (range 1-41 years), respectively. Of the psittacine birds diagnosed with lipoma, 35.3% (159/450) were female, 34% (153/450) were male, and 30.7% (138/450) were of unknown sex. Of those diagnosed with liposarcoma, 45.2% (14/31) were female, 25.8% (8/31) were male, and 29% (9/31) were of unknown sex. Anatomic location was specified in 87.7% (568/648) of cases of adipocytic tumors. The most common sites of lipoma formation were the wing and chest, and the most common site of liposarcoma formation was the wing. The prevalence of adipocytic tumors was determined as a function of age, sex, and genus. Risk factors and association with other comorbidities, such as atherosclerosis, hepatic lipidosis, and xanthoma, also were evaluated.

Radiation Therapy for the Treatment of Squamous Cell Carcinoma in a Barn Owl (*Tyto alba*)

Tracey K. Ritzman, DVM, DABVP (Avian Practice), DABVP (Exotic Companion Mammal), and
Tara M. Harrison, DVM, MPVM, Dipl ACZM, DACVPM, DipECZM (Zoo Health Management), CVA

Session #3130

Affiliation: From Carolina Raptor Center, 6000 Sample Road, Huntersville, NC 28078, USA (*Ritzman*), Carolinas Veterinary Care Clinic, 10110 Northcross Center, Ste 100, Huntersville, NC 28078, USA (*Ritzman*), North Carolina State University College of Veterinary Medicine, 1010 William Moore Drive, Raleigh, NC 27607, USA (*Harrison*).

Note: Abstract format selected by author(s). See Front Matter for description of conference proceedings manuscripts submitted as scientific abstracts.

Abstract: This presentation describes the collaborative medical and surgical treatment of an 18-year-old female captive barn owl with squamous cell carcinoma neoplasia of the thoracic limb (wing). Surgical, medical therapy, and radiation treatments, were utilized to treat the bird, and preserve the function of the wing. The tumor originated from the soft tissues of the right medial ulna and was initially treated with surgical debulking. Biopsy samples collected at the time of the surgery were submitted for histopathology which confirmed the diagnosis of squamous cell carcinoma. Medical therapy after surgery consisted of analgesia, systemic antibiotic therapy, bandaging, when necessary, physical therapy for the wing, and monitoring of the affected site. The tumor recurred 6 months after the initial debulking surgery. Palliative radiation therapy was utilized to treat the tumor and preserve wing function. Radiotherapy was effective at reducing the tumor size and delayed progression of disease, thus extending the life expectancy of this bird with a good quality of life. Radiotherapy for this patient was performed manually with single 12e-electron beam with 6 X 6 cone, prescribed to 90% IDL. A total of six (6) weekly fractions of 6 Gy was delivered for six consecutive weeks at the North Carolina State University College of Veterinary Medicine. The owl tolerated the radiation therapy very well and there were minimal side effects in this patient. For this patient there was delayed wound healing and delayed feather growth at the site. Other potential side effects of radiotherapy in general can include thermal injury at the treatment site and the inherent risk of the repeated general anesthesia sessions required for the treatments. This case report illustrates the positive benefits and feasibility of radiotherapy for avian patients with neoplasia.

References

1. Adair JE, Riggs GL. Treatment of recurrent multifocal cutaneous squamous cell carcinoma in an African grey parrot (*Psittacus erithacus*) with intralesional carboplatin in Poloxamer 407. *J Avian Med Surg.* 2022;36(2):200-205.

2. Alexander AB, Griffin L, Johnston MS. Radiation therapy of periorbital lymphoma in a blue- and -gold macaw (*Ara ararauna*). *J Avian Med Surg.* 2017;31(1):39-46.

3. Cutler DC, Shiomitsu K, Liu CC, Nevarez JG. Comparison of calculated radiation therapy versus actual radiation delivery in military macaws (*Ara militaris*). *J Avian Med Surg.* 2016;30(1):1-7.

4. Ferrell ST, Marlar AB, Garner M, et al. Intralesional cisplatin chemotherapy and topical cryotherapy for the control of choanal squamous cell carcinoma in an African penguin (*Spheniscus demersus*). *J Zoo Wildl Med.* 2006;37(4):539-540.

5. Guzman DS. Avian soft tissue surgery. *Vet Clin North Am Exot Anim Pract.* 2016;19(1):133-157.

6. Guthrie AL, Gonzalez-Angulo C, Wigle WL, et al. Radiation therapy of a malignant melanoma in a thick-billed parrot (*Rhynchopsittica pachyrhyncha*). *J Avian Med Surg.* 2010;24(4):299-307.

7. Hinkle D, Eschar D, Ambar N, et al. Successful radiation treatment of undertail of fibrosarcoma in a major Mitchell's cockatoo (*Lophochroa leadbeateri*). *J Am Vet Med Assoc.* 2022;260(12):1-4.

8. Kent MS. Principals and application of radiation therapy in exotic animals. *Vet Clin North Am Exot Anim Pract.* 2017;20(1):255-270.

9. Murtaugh RJ, Ringler DJ, Petrak ML. Squamous cell carcinoma in the esophagus of an Amazon parrot. *J Am Vet Med Assoc.* 1986;188(8):872-873.

10. Ramis A, Gibert X, Majó N, Grifols J. Metastatic oral squamous cell carcinoma in a Montagu's harrier (*Circus pigargus*). *J Vet Diagn Invest.* 1999;11(2):191-194.

11. Rettenmund CL, Newton AL, Calle PP. Uropygial gland squamous cell carcinoma in chinstrap (*Pygoscelis antarcticus*) and gentoo (*Pygoscelis papua*) penguins at the Wildlife Conservation Society's Central Park Zoo. *J Zoo Wildl Med.* 2015;46(1):113-119.

12. Rocha CMDS, Wilson TM, Sousa DER, et al. Squamous cell carcinoma of the footpad with systemic metastasis in a captive crowned solitary eagle (*Buteogallus coronatus*). *J Vet Med Sci.* 2020;82(11):1627-1630.

13. Staudenmaier AM, Simone-Freilicher EA, Noonan BP, Ewing PJ. Antemortem diagnosis of esophageal squamous cell carcinoma in a blue-fronted Amazon parrot (*Amazona aestiva*) and a mealy Amazon parrot (*Amazona farinosa*). *J Am Vet Med Assoc.* 2019;254(11):1324-1328.

14. Suedmeyer WK, Henry C, McCaw D, Boucher M. Attempted photodynamic therapy against patagial squamous cell carcinoma in an African rose-ringed parakeet (*Psittacula krameri*). *J Zoo Wildl Med.* 2007;38(4):597-600.

15. Swisher SD, Phillips KL, Tobias JR, et al. External beam radiation therapy of squamous cell carcinoma in the beak of an African grey parrot (*Psittacus timneh*). *J Avian Med Surg.* 2016;30(3):250-256.

16. Weinstock D, Correa MT, Rives DV, et al. Histopathology and epidemiology of condemnations in broiler chickens due to squamous cell carcinoma in North Carolina. *Avian Dis.* 1995;39(4):676-686.

Analyzing Baseline and Post-Restraint Plasma Corticosterone Levels in Hispaniolan Amazon Parrots (*Amazona ventralis*)

Shelby N. Parks, BS,
Thomas N. Tully, Jr. DVM, MS, DABVP (Avian), DipECZM (Avian),
Aspen L. Settle, BS,
Christine R. Lattin, PhD

Session #3171

Affiliation: From the Department of Veterinary Clinical Sciences, School of Veterinary Medicine, Louisiana State University, 1909 Skip Bertman Dr., Baton Rouge, LA 70803, USA (Parks, Tully, Settle), Department of Biological Sciences, Louisiana State University, 202 Life Science Building, Baton Rouge, LA 70803, USA (Lattin).

Note: Abstract format selected by author(s). See Front Matter for description of conference proceedings manuscripts submitted as scientific abstracts.

Abstract: The purpose of this study was to measure baseline plasma corticosterone levels in 22 Hispaniolan Amazon parrots (*Amazona ventralis*) and assess the effects of handling and restraint on corticosterone levels over 1 hour. To do this, each parrot was removed from its cage and wrapped in a towel for restraint, similar to that performed in a clinical setting. Once restrained, an initial baseline blood sample was collected in < 3 minutes,[1] after which blood samples were taken every 15 minutes for 1 hour (a total of five blood samples). To determine concentrations of plasma corticosterone, an enzyme-linked immunoassay was used. This assay was validated for Hispaniolan Amazon parrots by assessing parallelism of diluted parrot plasma samples with a corticosterone standard curve. On average, parrots showed a significant increase in corticosterone between baseline samples and all subsequent post-restraint time points. Females, on average, displayed significantly higher corticosterone levels than males after 30, 45, and 60 minutes of restraint. The plasma corticosterone concentrations measured in parrots with feather destructive behavior were compared to those without the behavior, and no significant differences were found between the groups at baseline or any post-restraint time points. Understanding the physiologic stress response in companion psittacine birds during routine handling will allow clinicians to better evaluate how this may affect the patient's condition and diagnostic test results.

References

1. Romero LM, Reed JM. Collecting baseline corticosterone samples in the field: is under 3 min good enough? *Comp Biochem Physiol A Mol Integr Physiol.* 2005;140(1):73-79.

Gap in pagination due to formatting issues.

Pages 272-729

Section 13

Hybrid Session 1

Andrea Winkel, DVM, MPH
Session Chair

How to Incorporate Exotic Animals into General Practice

Jesse A. Fallon, DVM, PhD

Session #3316

Affiliation: From Cheat Lake Animal Hospital, Morgantown, WV 26508, USA.

Abstract: Veterinarians love their exotic animal patients. Whether furred, feathered, or scaly, these patients are what get them excited about going to work; however, exotic animal medicine poses challenges when trying to incorporate this much-needed care into general practice. From scheduling problems to adequate staff training, to financial productivity, to appropriate medical and surgical equipment, there are many hurdles to successfully building an exotic case load. This presentation establishes a clear path for the general practitioner to establish themselves as an exotic animal veterinarian for the public, their staff, and their administrators.

Introduction

The practice of exotic animal medicine has improved markedly over the past several decades, and each year, after this ExoticsCon conference, veterinarians feel energized, ready to take on new knowledge and apply it to patients. Then, as they get back to clinical practice, they encounter numerous hurdles that prevent them from practicing at the level of efficiency and confidence desired. There have been many presentations on what equipment to purchase to make care of exotic animal patients easier and more effective. More common than equipment inadequacies are challenges of scheduling, productivity, and staff training. This presentation focuses on overcoming these more nuanced obstacles.

Scheduling

One of the biggest challenges that veterinarians face occurs well before they meet exotic animal patients. What strategies can you employ to help fill your schedule with exotic animal patients? How do you deal with no-shows and late appointments? How do you address a wellness appointment with a patient that really has significant deficits in husbandry or has a serious illness? These questions must be addressed with a structured approach directed to relatively uneducated front desk staff that want to make the right decisions about when and how to schedule, but need the right training and tools to succeed.

The first step when training receptionists is to help them understand that the initial approach to exotic animal patients is very similar to dog and cat patients. Get the patient's name, species, sex, status, age, and presenting complaint. Teach them to be specific (e.g., 17-year-old, male, blue and gold macaw presenting for regurgitation and decreased appetite rather than parrot: ain't doin' right). Staff are capable of this, and are typically eager to learn. This training can occur one-on-one in a small practice or may occur at routine reception meetings in larger practices. If it is not already a routine, encourage the practice manager or front desk manager to have routine receptionist meetings. Ask to attend one of these meetings, or ideally attend them semi-annually, as a reminder for experienced staff or to engage with new staff members.

Remember, developing a full but not overloaded schedule starts with the receptionist. The front desk needs to know which patients can wait until the next available appointment (e.g., chronic feather destructive behavior) and which need to be addressed quickly (e.g., a rabbit that isn't eating or passing stool). These descriptions can be outlined in a simple, bulleted document or dichotomous key that is laminated and displayed next to each reception workstation. Categorize common presenting complaints into emergencies, priority cases, and cases that can wait until the first available appointment. Unless staff are well trained and there is low turnover (unlikely), these documents should be in language that can be easily understood by the layperson. This will help prevent downtime and inappropriately scheduled sick patients. To help provide optimum care for their patients, many clinicians choose to have most of their appointments open for the receptionist to schedule any animal but keep one or two reserved each day for urgent care cases that remained blocked until a predetermined time before the appointment (e.g., 24 hours). This will help avoid a really complicated, sick patient from being squeezed into an inadequate time slot.

No-shows

Appointments that do not show up can be extremely frustrating. The downtime associated with these no-shows is detrimental to staff morale and reduces your production. To help reduce the number of no-shows, call ahead, text, and email reminders can be helpful. It is important not to rely on a single form of communication. Some clients are very responsive to text messages, while others need a phone call reminder. Most digital practice software can automatically text or email clients at a predetermined time. If a client does not show up for an appointment, have a staff member call them when they are 15 minutes late. Phrase this conversation in a way that is not accusatory, but gets the message across with a phone call saying something like "We noticed that you are late for your 3:30 pm appointment with Polly; we just wanted to check and make sure everything was okay."

Even when everything had been done to remind the client of their upcoming appointment, no-shows are still inevitable. As a clinician, there are essentially only two different strategies to deal with no-show exotic appointments, which seem to be more common than no-show dog and cat appointments. The first is to expect that there will be the occasional no-show, and structure your appointments with this in mind. This might mean scheduling more cases than would be comfortable during each day, with the expectation that one or two will not show up. The benefit to this approach is that it is relatively easy to accomplish – just book more appointments. The downside of this approach is the feeling of being overloaded if everyone shows up. The second approach is to preempt the no-show appointment by: 1. Have the client pre-pay for the exam, which will serve both to increase incidence of clients showing up for the appointment and if they don't show up for the scheduled appointment some financial compensation for the prescheduled time is attained. 2. Charge a no-show fee, which is a typical response for other professions (e.g., dentists). If this concept is new to the practice, it may generate resistance from the practice manager or owner. These strategies are not as negatively received by clients as some might think, particularly a pre-pay policy.

With any approach chosen, a full appointment calendar is the goal. While at work, it makes sense to be seeing patients!

Productivity

Let's face it, it can be hard to be a high-producing associate when a large percentage of exotic animal patients is being seen. Owners and practice managers that value exotic pets recognize this, and hopefully want to support efforts to see more exotic patients. How can you both see exotics routinely and be productive?

First, you must charge for your time. Exotic animal exams, even wellness exams, often take more time than dog and cat patients. These patients are harder to restrain and often require more lengthy husbandry and diet discussions. Pre-empt these challenges by having your exotic animal client owners fill out a simple diet and husbandry form. With the acceptance of digital communication, this can be done before the client shows up at the office. Take the initiative to make a husbandry and diet form for the species or classes of animals seen (e.g., pet birds, backyard poultry, reptiles, amphibians, mammals) and make it a routine for receptionists to email these forms the week of the client's appointment. Then, train technicians or assistants to review these forms prior to presenting the case to the veterinarian so they already know what areas need further discussion and which are already adequately addressed. Congratulate them in the areas where they are succeeding early during the appointment. This will help strengthen relationships and will make diagnostic and treatment recommendations more palatable. Remember, veterinarians are working with clients as a team to help care for their pet. Finally, charge more for exotic animal examinations than for dog and cat visits. Again, these patients require more time and expertise than domestic species, so it is perfectly reasonable to charge more.

After (or before if it is better for a particular patient) the examination and initial discussion, don't make judgements on what the client will or won't be willing to do diagnostically for their pet. Recommend the appropriate diagnostic, offer an estimate if needed, then give them time to decide how they want to proceed. If sedation would benefit the patient, explain to the owner why this is the safest, most efficient, and thorough method for obtaining the diagnosis or performing a complete examination. Again, charge appropriately for these services. They require expertise that is not likely captured on a per volume of drug basis.

Finally, schedule rechecks as appropriate for the patient and condition, and do it before the client leaves the exam room. Let the client know that their pet should be seen again in 3-4 weeks and have the technician schedule them while in the exam room. This will ensure that your patients are well cared for and that the appointment calendar remains full.

Staffing

One of the areas that most veterinarians can do better is training and utilizing staff appropriately. This is particularly true for exotic animal patients. Find a technician or assistant that can appropriately discuss husbandry and diet, safely restrain, monitor sedation, and perform diagnostics on patients. Consider how a dog or cat is treated; the technician collects the blood, takes x-rays, and educates the client. There is no reason that the same can't be done for exotic animal patients; however, to get to this level of efficiency and competency requires training.

Once an individual with aptitude and ability has been identified, encourage them to improve their knowledge and skills. Take videos of proper restraint, blood collection, and client discussions and have staff review them. Ask the practice manager if they will support sending the technician to ExoticsCon or another conference where they can learn from other veterinarians and technicians. Many times, these staff members are excited to learn how to help care for these patients. Encourage them, provide verbal reinforcement, and show enthusiasm! If exotic-animal-capable staff are only available 2 days per week, use those days as your exotic-animal-heavy appointment days.

Equipment and Medications

This topic has been thoroughly covered in other lectures. There are some necessities that are worth mentioning. Examination and treatment equipment that you cannot do without include:

- Gram scale. At least two, with appropriately sized perches for birds.
- Magnification and light. These are critical in small patients and can be as simple as an otoscope light with magnification or hobby loupes.
- Drugs for species-appropriate sedation (e.g., midazolam, alfaxalone, butorphanol).
- Liquid antibiotics, NSAIDs, gabapentin, and parasiticides.
- Small climate-controlled oxygen cage.

Conclusions

Exotic animal medicine is exciting, and staff will respond to a veterinarian's enthusiasm for it. Following the suggestions outlined here will make the time seeing exotic animal patients more efficient, more productive, and more enjoyable!

How to Convince Your Local Emergency Practice to See Avian and Exotic Animal Patients

Megan Turner, DVM, and
Bruce Nixon, DVM

Session #3389

Affiliation: From Animal Emergency Hospital of North Texas, Grapevine, TX 76051, USA (Turner), Texas Avian and Exotic Hospital, Grapevine, TX 76051, USA (Nixon).

Abstract: A challenge for any practice that treats avian and exotic animal patients is how to provide after-hours and emergency care. Traditionally, this responsibility has fallen on primary care veterinarians meaning they are forced to take emergency calls at night and on the weekends. This leads to a poor work-life balance and burnout over time as these veterinarians never get a break from work. Ultimately, this results in poorer outcomes for these patients when those tasked with providing them with care are exhausted and overwhelmed. The first part of this presentation addresses the best strategy that primary care veterinarians can employ to convince their local emergency practice to start seeing avian and exotic patients in addition to dogs and cats. Tips like identifying and persuading the key decision maker of the emergency practice and encouraging buy-in from the entire emergency practice staff—not just a couple of veterinarians that might be willing to see these patients when they are on shift—are discussed in detail. The second part of the lecture provides ideas for how to support and train emergency practice staff once they have agreed to start treating avian and exotic patients.

How to Convince Your Local Emergency Practice to See Avian and Exotic Patients

Below is an outline of information covered in this lecture.

Introduction

How many of you see emergencies yourself?

How many use intern(s)?

- Foreshadow a unique way to use interns later in the lecture.

Just don't see them?

- State ethics requirements.

Other methods?

Identify Neo- s/he is "The One"

Every emergency practice has a person, usually singular, who has the ability to make decisions.

Positions.

- Medical or Hospital Director (especially corporate practices)
- Chief of Staff (especially shareholder-owned practices)
- Owner and privately owned practices
- Area or Regional Administrator (corporate practices)
- Highly motivated associate

The Pitch

Determine what motivates "The One."

- Money
  - Show actual financial data.
  - New income source.
  - Avian and exotic animal owners are also dog and cat owners.
- Expanding their referral base
  - If you are not already exclusively referring to a particular emergency practice, then promise to start doing so.
- Institution building
  - This is especially valuable for 24/7 emergency practices associated with a referral center.

Close the Deal

Train and support their staff.

If at all possible, send them personnel.

- ERs everywhere are desperate for staff: doctors, technicians and receptionists
- Obviously a part-time only arrangement
  - Enter agreement about poaching staff in either direction.
  - This involves weekends mostly but also day and night weekday work for 24/7 facilities.
- This is good for your staff too
  - Emergency practices tend to pay extremely well.
  - It also helps them sharpen their emergency skills for those patients who present to you in critical shape.

- Offer for their staff to do part-time shifts at your practice
    - Having several staff members who work both practices strongly bonds their commitment to seeing avian and exotic animal patients.

Unique solutions

- Sharing an internship with the ER

How to Support and Train Your Local Emergency Practice to See Avian and Exotic Animal Patients

Megan Turner, DVM, and
Bruce Nixon, DVM

Session #3391

Affiliation: From Animal Emergency Hospital of North Texas, Grapevine, TX 76051, USA (Turner), Texas Avian and Exotic Hospital, Grapevine, TX 76051, USA (Nixon).

Abstract: In the first session, ideas for how to convince your local emergency practice to start treating avian and exotic patients were addressed. This session dives into how to support and train the emergency practice staff. Many veterinarians are intimidated by avian and exotic animal patients. They received minimal, if any, training with them during veterinary school and they do not know where to start. Furthermore, they do not feel qualified to treat these unique and special patients. It is important to provide support to the emergency practice staff as they gain confidence. Simple strategies like providing onsite training, recommending textbooks and other resources they should have on hand, providing constructive feedback on cases, and making yourself available for phone consultations can be an extremely efficient way at bringing emergency practice staff up to speed. Remember, all levels of staff will need training including receptionists, technicians and veterinarians. You are the best person to teach the staff how to care for your patients. By putting in the time up front and creating open lines of communication between your hospital and the emergency practice, you will ensure that your clients and patients are receiving the same quality of care they would from you. In turn, you will get your free time back to enjoy life. Finally, you will need to have a strategic plan to maintain your emergency practice's commitment to these patients as these facilities typically have high staff turnover and are a frequent target of buy-outs and mergers.

How to Support and Train Your Local Emergency Practice to See Avian and Exotic Animal Patients

Below is an outline of information covered in this lecture.

Now you have them committed, so it is time for you to commit to them

Training

Reference materials

Phone access

Follow-up

Training

Explicitly state the goal of the emergency practice

- We are not avian and exotic animal experts, but we will try to help your pet to the best of our ability.
- Stress that the emergency practice doctors are not specialists.
- Provide phone scripts for receptionists and others.
- Help them create a list of patients they are not willing to see as client owned patients (e.g., primates, state protected wildlife, venomous snakes).

Focus on stabilization and support, not definitive diagnosis and treatment

- Help them understand that all species are approached the same way during evaluation
- Based on history and PE, create a problem list and then figure out where in the DAMNIT scheme the patient falls and develop your plan from there.
- Stress that it is not important that emergency practice veterinarians know every single illness that can befall every single species

If there is interest, then you can provide them with pathways for diagnostics and more intensive treatment

Stress that there is no reason to be intimidated

- Remind them that most owners are grateful that somebody is willing to help their pet

Handling labs

- Physical examinations for veterinarians and handling of the patients for technicians
- Can do on the same day, but might be best to split into morning for veterinarians and afternoon for technicians

Species

- At least two birds: one large and one small
- Exotic mammals
- Rabbit
- Ferret
- GP (possibly a rat as well)
- Hedgehogs

- Reptiles
- Lizard
- Snake
- Tortoise

Triage and supportive care

- Emphasize that these unique patients are even better at hiding illness than dogs and cats when they get stressed and have an adrenaline dump
    - Do not underestimate their degree of illness because they look pretty good at admission.
    - Listen to the history. If owners say the animal is lethargic and hardly moving at home, believe them.
- All birds should immediately receive heat and oxygen; it will never hurt them can help a lot

Top five problems for each species group

Reference materials

Your own personal, laminated cheat sheet

Books

The top five recommended texts are a starting place. They are listed below.

- Carpenter JW, ed. *Exotic Animal Formulary.* 5th Ed. St. Louis, MO: Elsevier; 2018. (Have more than one copy and always have the latest edition on hand.)
- Graham J, Doss G, Beaufrere H. *Exotic Animal Emergency and Critical Care Medicine.* Hoboken, NJ: Wiley-Blackwell; 2021.
- Quesenberry KE, Orcutt CJ, Mans C, Carpenter JW, eds. *Ferrets, Rabbits, and Rodents: Clinical Medicine and Surgery.* 4th Ed. St. Louis, MO: Elsevier; 2021.
- Morishita TY, Greenacre CB, eds. *Backyard Poultry Medicine and Surgery: A Guide for Veterinary Practitioners.* 2nd Ed. Hoboken, NJ: Wiley-Blackwell; 2021.
- Speer B, ed. *Current Therapy in Avian Medicine and Surgery*. St. Louis, MO: Elsevier; 2016.

Online support

- Veterinary Information Network (vin.com)

Phone support

- Phone support is essential. Someone on your staff must always be available for a phone consultation. Failure to do this two or three times will undermine all your work. Failure to do this means you will likely have to start over with another emergency practice.

Follow-up

- Quarterly personal visits (with food)
- Feedback on individual patients
- Positive and constructive criticism; nobody only wants to hear what they did wrong

Compounding Considerations and Making Modified Pharmaceuticals Work for Patients

Julia K. Whittington, DVM, and
Lauren R. Forsythe, PharmD, DICVP

Session #3411

Affiliation: From University of Illinois College of Veterinary Medicine, 1008 W. Hazelwood Dr., Urbana, IL 61802, USA.

Abstract: Commercial pharmaceutical products can feel like one-size-fits-no-one as professionals try to figure out how to treat patients, especially when trying to get an accurate dosage and struggling to the dosage into the patient reliably. Compounded medications are options that are made specifically to suit a particular patient's needs. Compounded medications are not without risk, which requires consideration of the risk-benefit ratio of utilizing these medications. This session identifies pros and cons of commonly compounded dosage forms, as well as the risks to be aware of when opting for compounded medications. The session covers the points that should be considered when reaching for a compounded product as well as considerations to ensure the product is the best option to meet the specific needs of the individual patient.

Veterinary Pharmaceutical Regulations

The U.S. Food and Drug Administration (FDA) oversees the development and approval of pharmaceutical agents in animals. The FDA approval process is where safety and efficacy are proven before a drug can be marketed. Two pieces of legislation provide the framework for this oversight: The Federal Food, Drug, and Cosmetic Act (21 CFR 9) and the Animal Medicinal Drug Use Clarification Act (AMDUCA) (21 CFR 530). The Federal Food, Drug, and Cosmetic Act defines what a drug is and gives authority to the FDA to approve and regulate drugs for humans and animals including safety standards, proper labeling, and purity of foods, drugs, vaccines, devises, and cosmetics. AMDUCA was passed into law in 1994 and permits veterinarians to prescribe or use certain approved animal and human medications in a manner that is outside of the conditions described on the product's label. This off-label or extra-label use is defined as the actual or intended use of a drug in an animal in a manner that is not in accordance with the approved labeling and includes such situations as use in a non-target species, use for a condition for which the drug was not intended, and deviations in dosing, frequency or route of administration, as well as subsequent withdrawal times. There are few drugs that are approved for use in zoological species, making extra-label drug use common. AMDUCA requires that any drug used in this manner must be administered under the oversight of a veterinarian, and that a valid veterinarian-client-patient relationship (VCPR) exist. Furthermore, extra-label drug use is limited to cases where the health of the animal is threatened, and significant suffering or death of the animal will result if it is not treated. If a drug is available that is labeled for the species or condition, the extra-label use of an alternate drug is not allowed. AMDUCA also regulates the compounding of approved animal and human drugs. Specifically, veterinarians and pharmacists may compound drugs for use in animals but only if there is not an FDA-approved product that can be used as labeled or in an extra-label manner.

Compounding Medication

Compounding medication has vastly expanded the availability of some medications that were previously impractical due to dose and/or dosage form for use in veterinary patients. Compounding is the process of mixing, combining, or altering ingredients to create a medication that meets the unique needs of a specific patient. Compounding includes the combining of two or more drugs. Compounded medications are not FDA approved.[1] Diluting or reconstituting medication in accordance with the approved labeling is not considered compounding. The United States Pharmacopeia (USP) is an independent, scientific nonprofit organization that collaborates with the FDA to set standards for safe, quality medications. The USP compounding standards are USP chapter <795> for compounding nonsterile preparations, USP chapter <797> for compounding sterile preparations, and USP chapter <800> for compounding hazardous drugs.[2] With the wide variety of species and the size variety within each species, it is not possible to have approved products available to meet the needs of every, or even most, patients. Compounded medications are relied on to fill the gaps where manufactured products fail to meet the needs. In veterinary medicine, compounded medications are used to treat animals when an FDA-approved drug is not available in the required dosage form, strength, or flavor. FDA-approved drugs can be used for compounding as long as all criteria for extra-label drug use are met. Compounding using drugs or materials such as bulk powder that are not FDA-approved is only allowed in certain circumstances.[3] Compounded products can improve patient care in several ways, but they also have inherent risks. Therefore, it is important to limit compounded medication prescribing to instances where manufactured products are not a feasible option and consider the risk-benefit analysis of using a compounded product.

Benefits Of Compounded Medications

There are five main benefits of compounding medication for veterinary species: dosage form, flavoring, strength, availability, and elimination of toxic ingredients. Modifying the dosage form may facilitate medication administration. Veterinary patients, and specifically zoological species, aren't always willing to take medications and forced administration may be challenging. Therefore, compounding to modify the dosage form to facilitate administration is warranted. Some strategies that are facilitated by compounding include a more desirable dosage form such as a liquid instead of a tablet or a transdermal instead of an oral medication, a medication flavored to the patient's preference, or a treat formulation that encourages consumption of the drug.

In addition to dosage form, medications are compounded because of the need for customized dosage strength. For example, enrofloxacin is commonly used in rabbits but is not commercially available as an oral liquid and available tablets are not practical. Compounding medications is also indicated when a medication is needed and is unavailable commercially or no longer being manufactured. Cisapride is a commonly used medication in veterinary patients; however, it was pulled from the human market in 2000 due to its cardiac side effect profile in humans. Therefore, compounding is the only way to obtain this medication for use in veterinary patients. Similarly, with the ever-present supply chain issues, medications that are unavailable for long stretches of time may need to be compounded. Compounding provides a mechanism to continue therapy until the manufactured product is available.

Risk of Compounded Medications

While there are benefits and indications for compounding medications, it is not without risk. One risk is that the compound contains either more or less active ingredient than stated on the label. The compounding process has several points where ingredient loss can occur to the point that even though the correct amount is weighed or measured, the final product does not contain enough active ingredient. Errors during the compounding can also occur, with a decimal point transposed or a scale read incorrectly, leading to too much active ingredient. Errors also include the wrong medication being put into the compound. Many compounding ingredients are white powders. A stringent verification process is necessary to prevent an incorrect ingredient or an incorrect strength of an active ingredient from accidentally being used.

Compounded medications may be unstable. If medication compounding is done in practice, the best way to minimize the risk of an adverse event is to take the simplistic approach.[4] Minimal changes to the product, such as diluting a solution or adding a flavor, are less likely to have a negative impact on the final product. Trissel's Stability of Compounded Formulations[5] is a resource that compiles all currently available stability information on drugs in compounded oral, enteral, topical, ophthalmic, and other specialized preparations. The information contained is based on human medicine, but it has useful data on compound stability for the anticipated duration of use and proper storage of compounds. If a compounded medication is unstable, the active ingredient degrades before the beyond-use date (BUD) is reached. The USP standards provide default BUDs that are generally short enough that drug degradation and microbial contamination will not occur during the use period of the product; however, that is not guaranteed. There is also concern about whether extended dates are based on true stability-indicating assays or if they are based on potency over time. The latter is not sufficient for determining extended BUDs.

Efficacy and safety are two concerns one must consider when weighing the risks of compounding medication. The compound may be prepared correctly but not produce the intended effect in the patient. One cause may be that the bioavailability (amount absorbed) is too low. Likewise, the compound may be prepared correctly but there is something that makes it unsafe for the patient. This could be an inactive ingredient, dose or route that has a high potential of causing harm to a patient.[6]

Non-sterile compounds are exactly as the name implies; they are not sterile. They still need to have adequate preservatives or short-enough dating that excessive microbial growth does not occur while the medication is in use. Sterile products, on the other hand, must be sterile. While a sterile environment that complies with USP chapter <797> greatly increases the probability that the final product will be sterile, there is still the potential for contamination to occur resulting in a final product that is not actually sterile. Even if batch testing is done and a certain percentage of the batch is determined to be sterile, there is still the potential that there is contamination in an item that was not tested as contamination is not uniformly distributed.

Controlling the risks of medication compounding

AMDUCA stipulates that the first choice of pharmaceutical for any patient be an FDA-approved drug proven to be safe and effective and used as indicated on the label. In veterinary medicine, the first choice is often not feasible. The second choice is to use an approved drug, but in an extra-label manner. When the use of an approved drug is not an option, that is when AMDUCA allows for compounding medication using a commercial product. AMDUCA does not address the use of compounding using bulk drug. By legally requiring the use of approved products if possible, AMDUCA prevents unnecessary compounding while protecting the integrity of the drug approval process. Compounding may be performed by a licensed pharmacist in a state-licensed pharmacy, federal facility, or by a physician or veterinarian. In 2013, the Drug Quality and Security Act established the provision that compounding can be performed in outsourcing facilities if under the direct supervision of a licensed pharmacist and the facility is registered with the FDA.[1]

While AMDUCA legalizes compounding by manipulating approved products, it does not address compounding from bulk chemical. Bulk chemical is necessary in some cases such as when the approved product is on backorder, or when the active ingredient does not exist as an approved product. There are also cases where a high-quality formulation cannot be prepared from a commercial product, such as when the only approved product is a chewable tablet. Therefore, although bulk chemical compounding is technically illegal, it is necessary under certain circumstances. Because of this, the FDA has released GFI #256 to describe its expectations and when it plans to use regulatory discretion in the enforcement of bulk chemical compounding. While a full discussion of GFI #256 is beyond the scope of this session, Figure 1 provides a flow chart of what to expect regarding clinical impact.[7]

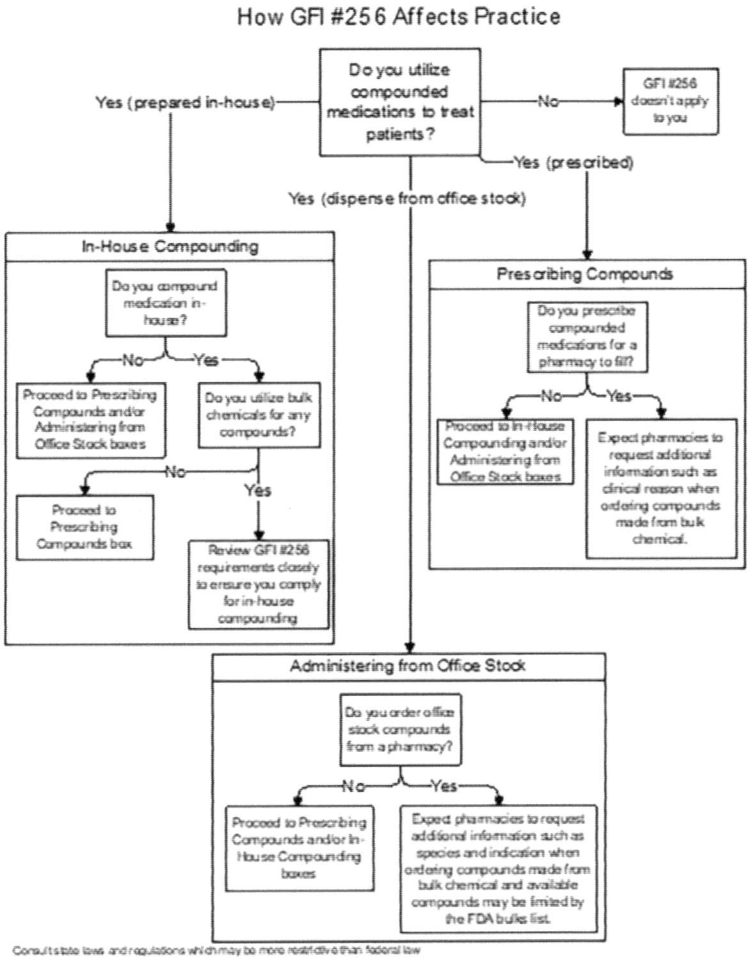

Figure 1. Expected impacts of GFI #256 on compounding in practice.
Source: DVM360. Forsythe L. What is GFI #256 and how does it impact my practice? Available at: https://www.dvm360.com/view/what-is-gfi-256-and-how-does-it-impact-my-practice-. Accessed April 3, 2023.

The risks associated with compounded medications are partially dependent on whether the compound is prepared in a 503a (traditional pharmacy) or 503b (outsourcing) facility. Traditional pharmacies should follow USP standards when compounding medications which are designed for small scale preparation of compounded products. These pharmacies are not held to the current good manufacturing practice (CGMP) requirements and are subject to less USP standards as they are enforced by the state(s) regulating them. This results in a higher risk of an adverse event when using compounded medication from these facilities. Conversely, compounds prepared in an outsourcing facility are held to CGMP requirements which are the same requirements that drug manufacturing facilities must adhere to. The final testing that is required by CGMP requires testing a percentage of each batch of compounded drug prior to release. This testing verifies the correct identity and potency of the active ingredient, as well as sterility if the product is marketed as sterile. There are also requirements that the outsourcing facility has proven the preparation process yields a consistently uniform batch, as well as that the formulation has sufficient stability for the expiration date given. For sterility, there is potential that the tested products are sterile but other products in the batch are contaminated. The large-scale manufacturing process designed to ensure consistency and repeatability between batches decreases the risks of an adverse event when using compounded medications from these facilities; however, because 503b outsourcing facilities are manufacturing compounds and not FDA-approved drugs, there is still a risk of lack of safety and/or efficacy. Safety and efficacy proof is required for the FDA-approval process, and not required for 503b compounds.

Compounding in Practice

Despite the risks, the benefits of compounding medications to meet the needs of veterinary patients often outweigh the risks. Knowledge of both state and federal compounding laws is essential for any veterinarian intending to compound medications. Most veterinarians receive little or no training in medication compounding as part of their veterinary training,[3] and veterinary practitioners report interacting with dispensing and compounding pharmacists more than any other professional collaboration.[8] A survey of zoo veterinarians found that only 25% were comfortable compounding medications for their patients. The majority of respondents stated that the reasons for not compounding medications included a lack of training, lack of access to ingredients, and the availability of commercial products. In situations where a compounded medication was needed for patient care, these veterinarians sought a pharmacist or compounding pharmacy to fill this need.[9] Another survey of over 1500 veterinary practitioners indicated that most used compounded medications in the treatment of their patients. Over 60% of the respondents indicated that they compounded medications in their practice, with the remainder utilizing compounding pharmacies. Less than 25% of the medications dispensed in primarily dog/cat practices were compounds, as compared to 50-95% of the medications used by practitioners treating companion zoologic species. Despite the significant number of practitioners engaged in medication compounding, relatively few understood their state laws regarding this practice.[3] State veterinary boards have regulatory oversight of veterinary practitioners through enforcement of their state practice act(s). Many states do not have specific compounding regulations for veterinarians.

Unlike in human health care, veterinarians can dispense medications directly to a client for a patient without a pharmacy prescription. Therefore, most veterinary prescriptions that are written are for compounded medications or for products that the veterinarian chooses not to stock. Any pharmacist can dispense a prescription medicine at the direction of a veterinarian if there is a valid prescription. The prescription becomes the legal documentation of the medical order. If the drug proves to be unsafe, or is not efficacious, both the pharmacist and veterinarian share the responsibility when it comes to compounded medications.[4] Therefore, it is important for the veterinarian to maintain documentation regarding the clinical need for a compounded medication and that the client was made aware of the risks and benefits of the compounded medication.

A veterinary pharmacist focuses on animal health in collaboration with the veterinarian, and veterinary pharmacists have a deeper understanding of animal health which translates to better patient care.[9] In many instances, utilizing a veterinary pharmacist to compound medications for veterinary patients is preferred because of their access to cost-effective drugs and advanced training in compounding. This has the added advantage of reducing preparation errors, contamination and having access to formulations that require specialized equipment to make.[3]

When partnering with a compounding pharmacy, take the time to investigate the facility and speak to the pharmacist. Ensure that they are experienced with veterinary compounds, that they understand animal toxicities, and that they utilize veterinary specific formulary information.[10] Compounding pharmacies must be licensed or registered to practice in the resident state and the state of dispensing. In addition, the pharmacy must follow USP standards as outlined in applicable state pharmacy regulations and adhere to criteria for compounding set by AMDUCA. Active pharmaceutical ingredients must be used in accordance with GFI #256, should be obtained from FDA-registered and DEA-registered suppliers, should be substances of USP grade, and should be accompanied by a valid certificate of analysis. Ideally, the pharmacy should be accredited by the Pharmacy Compounding Accreditation Board.[11] The key is to consider all options and have an open discussion with the client so they are aware of the benefits and risks of compounded medications and can participate in considering if a compounded medication is the best option in that particular case.

Resources described in Table 1 can be used for additional information on compounded medications.

Table 1. References for information on medication compounding in practice.

Reference	Web link	Information available
USP website	www.usp.org	Background information on USP. Current expert committee members. Updates on chapter updates in progress and key dates for comment submission. USP publications available for purchase.
USP Compounding Compendia	Purchase at https://www.usp.org/products/usp-compounding-compendium	Compounding chapters. Compounded preparation monographs.
State Pharmacy Boards	List of all websites found at: https://nabp.pharmacy/about/boards-of-pharmacy/	State regulations.
State Veterinary Boards	List of all websites found at: https://www.aavsb.org/public-resources/find-regulatory-board-information	State regulations.
AVMA	www.avma.org	News and position statements on compounding legislation. Summary of state office use laws. AMDUCA ELDU requirements.
FDA	www.fda.gov	GFIs and CPGs. Warning letters issued to compounding pharmacies. Current and resolved drug shortages. New drug approvals. Approved animal drugs and marketing status. Approved human drugs and marketing status. Appropriate drug disposal. Indexed drugs for Minor Use/Minor Species. Adverse event reporting. Compounding risk alerts.
DailyMed	https://dailymed.nlm.nih.gov/dailymed/	Package inserts for FDA-approved human and veterinary medications. Note: Package inserts different depending on the manufacturer.
AMDUCA	https://www.ecfr.gov/cgi-bin/text-idx?SID=054808d261de27898e02fb175b7c9ff9&node=21:6.0.1.1.16&rgn=div5	ELDU requirements.
DQSA	https://www.fda.gov/drugs/human-drug-compounding/text-compounding-quality-act	503a versus 503b pharmacies.
cGMP	https://www.accessdata.fda.gov/scripts/cdrh/cfdocs/cfcfr/CFRSearch.cfm?CFRPart=211	Requirements for medications prepared by 503b pharmacies and for manufacturing approved medications.
GFI #256	https://www.fda.gov/regulatory-information/search-fda-guidance-documents/cvm-gfi-256-compounding-animal-drugs-bulk-drug-substances	Current FDA guidance on when compounding from bulk chemicals is appropriate.
Non-sterile and sterile compounding for the veterinary practitioner	Textbook only; currently available for preorder through Barnes and Noble	Guide to all aspects of prescribing compounds as well as considerations for preparing simple compounding in house.

References

1. U.S. Food and Drug Administration. Compounding and the FDA: Questions and Answers. Available at: https://www.fda.gov/drugs/human-drug-compounding/compounding-and-fda-questions-and-answers. Accessed March 18, 2023.

2. United States Pharmacopeia. USP Compounding Compendium. Available at: https://www.usp.org/compounding. Accessed March 19, 2023.

3. Gochenauer AE, Rishniw M, Holmes ER, Forsythe LR. The current landscape of veterinary compounding in the veterinary setting. *Int J Pharm Compd.* 2019;23(5):428-433.

4. Boothe DM. Veterinary compounding in small animals: a clinical pharmacologist's perspective. *Vet Clin North Am Small Anim Pract.* 2006;36:1129-1173.

5. Trissel LA, Ashworth LD, Ashworth J. *Trissel's Stability of Compounded Formulations.* 6th Ed. Washington, DC: The American Pharmacists Association; 2018.

6. Davidson G. Veterinary compounding: regulation, challenges, and resources. *Pharmaceutics.* 2017;9(1):5.

7. DVM360. Forsythe L. What is GFI #256 and how does it impact my practice? Available at: https://www.dvm360.com/view/what-is-gfi-256-and-how-does-it-impact-my-practice-. Accessed April 3, 2023.

8. Kustritz MVR, Molgaard LK, Tegzes JH. Frequency of interactions between veterinarians and other professionals to guide interprofessional education. *J Vet Med Educ.* 2013;40(4):370-377.

9. Besley N, Browne P, Park M, et al. Pharmacists in zoos? A qualitative study investigating the potential for pharmacist involvement in wildlife health care in Aotearoa New Zealand. *J Am Pharm Assoc.* 2023; 63(3):825-831. doi:10.1016/j.japh.2023.01.008.

10. Anderson M. Compounding: myths and facts. *VMX.* 2021;38-40.

11. Powers LV, Davidson G. Compounding and extralabel use of drugs in exotic animal medicine. *Vet Clin North Am Exot Anim Pract.* 2018;21:195-214.

Recession Proof Your Exotic Animals Practice: For Owners and Associates

Len Donato, VMD, DABVP (Avian Practice)

Session #3294

Affiliation: From Radnor Veterinary Hospital Wayne, PA 19087, USA, the Department of Clinical Sciences and Advanced Medicine, The University of Pennsylvania School of Veterinary Medicine, Philadelphia, PA 19104, USA.

Abstract: During the COVID-19 pandemic, clinical veterinary medicine did very well financially. As time moves on from the pandemic, veterinary medicine is seeing that client visits are beginning to drop to pre-pandemic levels. When the number of visits decreases, a decrease in gross revenue is typically observed. Veterinary medicine has not experienced that yet because inflation caused fees to increase significantly, thereby increasing the average client transaction (ACT). Raising fees alone will not sustain meaningful growth, and in fact contributes to practice stagnation and regression. Understanding a variety of strategies to promote more visits and increase ACT will benefit the financial health of the hospital overall. Exotic animal practitioners have to be savvier than dog and cat counterparts as a majority of the patients do not get vaccines, flea and tick medication, and heartworm prevention. Those products and services help gross revenue by increasing transactions and ACT. Strategies affecting the number of transactions and the ACT at the same time will benefit the practice owner, the non-owner veterinarians, and staff. Increasing the number of transactions involves veterinarians maximizing time management, proper follow up and reminders, along with a comprehensive internal and external marketing campaign. ACT is addressed by appropriate fee increases, managing lost fees, and expanding services and skills. Knowing these strategies and how they affect gross revenue, veterinarian salary, and the health of the clinical environment will minimize the effects of a recession and will likely result in growth financially and professionally.

Introduction

Recessions are defined by the National Bureau of Economic Research (NBER) as a significant decline in economic activity that is spread across the economy and last for more than a few months.[1] Historical data demonstrates the most recent prolonged recessions were The Gulf War Recession (7/90-3/91), The Dot Com Recession (3/01-11/01), The Great Recession (12/07-6/09), and The COVID Recession, which is on-going.[2] When recessions are prolonged they are defined as depressions, but there is debate as to definition.[1,2]

It is widely thought that veterinary medicine is recession-proof, particularly with the most recent experience. In reality there have been industry-wide effects from the beginning of the COVID-19 pandemic. Food-animal supply chains were dramatically affected,[3] zoo income was reduced by lack of visitors and donations,[4] and a large number of veterinarians retiring from the profession or going part time has even made the everyday news.[5] The Great Recession in 2008 resulted in decreased gross revenue for veterinary practices and a significant decrease in net that was previously 35-45% and fell to 10% or below for some practices. The effects of that are still felt today as net revenue for a non-specialty practice reached 15%.[6] Non-owner veterinarians also feel recessions as numbers of clients typically decrease as well as their number of visits (N) and how much they spend at each visit (average client transaction, ACT). Pro-sal is a common method to pay associates which is based on revenue generated by the associate. Since N and ACT decrease, that affects additional pay over base salary. This makes Pro-sal a bit more risky for the non-owner veterinarian during a recession. Also, if a practice has a decrease in net revenue, there are fewer resources that could go to additional staff, equipment, or capital improvements for the hospital building, all of which make for a better-quality hospital and work environment.

Everything feels good financially right now because of the COVID/post COVID bump, but data show the number of visits decreased by 4% in 2022.[7] As seen from the recessions listed above, it is not if a recession occurs. The questions are: how many recessions will be experienced in a career, and how severe will they be? What can be done to maintain income during those times? There are strategies to make sure a similar drop-off is not seen as has been seen with other recessions. It is better to implement and practice the strategies now, rather than when the economic environment is not healthy.

Understanding Profitability

Veterinary practices should be managed to be profitable. This is a function of gross revenue and expenses to run the practice. While this simplifies what a practice needs to achieve, executing is a big challenge. Focusing on gross revenue, but not ignoring the expenses, will make for a profitable practice. Expenses need to be monitored as most will be within a percentage range of gross revenue. As an example, as the number of staff increases, so does payroll. The goal behind the increased staff is to accommodate more patients, which increases gross revenue. In this example, as the expense of staff increases so does revenue. There are two exceptions to expenses where a difference can be made; these are discussed later. The focus is on revenue generation.

Managing gross revenue has the greatest impact on profitability. To understand income there is a simple formula:

# of transactions (N) × average client transaction (ACT) = gross revenue[8]

It is important to understand the components that make up both factors in the equation, and the best ways they can be affected to produce higher income or gross revenue. Increasing ACT has a greater impact on gross revenue than N. Both are important to manage. This is why specialists typically generate more than general practitioners as one transaction could be in the thousands regularly.

One of the easiest ways to increase revenue through ACT is to raise prices. This has occurred in the last few years because of the extreme increase in N during the pandemic. It was meant to handle overall increases in payroll, increases due to inflation, and to become more profitable as a practice. The continued price increases during a recession eventually catch up and surpass what owners are willing to pay.[7] A precipitous drop in N occurs because the high prices keep clients from scheduling. This was observed in the Great Recession as the years before were quite prosperous. This isn't to say prices should not be raised. It should be a thoughtful process understanding the value of services, inventory costs, lab fees, and other expenses. Raising fees to weed out clients will result in having a hard time during a recession.

How to properly raise ACT

Yes, raise fees

As outlined above, make this a thoughtful process with good intentions.

- Exams and surgery, generally value based.
- Lab fees, generally 2 × cost ± a technician fee. Monitor monthly as fees from reference lab or materials for in-house equipment may increase suddenly.
- Inventory, generally 2 × cost ± a pharmacy fee. Monitor invoices as they arrive for fee increases.

Practice quality medicine

This sounds obvious. This is the greatest part of our careers, as not only does that satisfy our desire to be an exotics veterinarian, but it also increases ACT. It allows us to offer a variety of services. Consider adding a new service this coming year.

- Preventive care.
- In-hospital diagnostics.
- Advanced imaging, radiology, ultrasound, CT, and endoscopy.
- Soft tissue surgery and orthopedics.
- Behavior consultations.
- Senior wellness programs.
- Many others; you know what you like!

Address missed fees

This refers to fees not charged by the veterinary practice for services that were performed. This is a long-standing problem that can lose a practice 5-10% of gross revenue a year.[9] A practice that generates $2,000,000 in gross revenue can be missing up to $200,000/year.

- Identify areas where fees may be missed and address.
- Institute a checks-and-balance system that works for the doctors and staff.

How to properly raise N

Time management

Managing time sounds like it is obvious and intuitive. It is not. It must be practiced like sports or playing an instrument. Having more time allows you to see more cases or at least offer more services, which can affect ACT too. So how should time be managed?

- Delegate. This is an art and a skill. The art is you must be able to trust others. The skill is that you have to follow up. Even though you delegate a task, does not mean you are not to check on said task and provide feedback. It is the most forgotten step of delegating. As a veterinarian one should perform exams and surgery and follow up with the clients on patient care. No one else can do those tasks. Everything else can be delegated, but you still must check. Take a class on delegating if needed.

- Have the right people. In the practice have proper staffing type, numbers, and skills. Your external team is just as important. These are the payroll company, accountant, attorneys, and financial planner. Performing these financial and legal tasks on your own will end just as badly as if you tried to be the only one handling all the patients.

- Evaluate scheduling. This becomes a topic of debate. There are many scheduling strategies ranging from straight booking of timed appointments to high density schedule with catch-up slots. There are even schedules with overlapping appointments considering support staff contact and doctor contact for each patient. Scheduling considers the veterinarians and staff as a whole entity. Evaluate your schedule to determine if it is efficient and fits the practice philosophy.

- Be paperless or paper light. Time savings looking for files alone is worth every penny spent on an electronic medical record system. This also aids in addressing missed fees.

Client reminders

These are not for just annual examinations and vaccines. Set reminders for blood work when monitoring chronic patients, behavior visits for young birds to make sure clients get into good habits early, examination and husbandry reviews for reptiles before winter, and anything important to the specific patient.

Recall system

Call clients far more than if they were in for a sick visit. Call clients or have staff call for weight checks, diet conversions, recheck reminders, blood work monitoring, setting spay/neuter, post surgery check in, and whatever you think is remotely important. It helps with owner compliance, checks on the pet, increases visits, and builds trust, which results in return visits.

Marketing

Managing a successful veterinary practice involves strategic planning within and between several functional areas of the business; just a few of those areas include finance, human resources, inventory, operations, and marketing. The hospital's marketing strategy should provide a clear depiction of the specific elements that set your practice apart from the crowd. The marketing strategy's key elements ideally include defined internal and external marketing components, brand messaging, print and digital-based marketing tools, word-of-mouth communications, and social media. All these marketing mediums work together to reach and engage different types of consumers that would benefit from your services. Having a general understanding of the different types of marketing tools and their costs will help a practice owner determine specific marketing mediums to focus on, relative to their overall return on investment.

Expenses

As previously mentioned, expenses are not the focus when evaluating profitability. They cannot be ignored and must be managed, but your profitability will not dramatically change by continually switching companies for electricity. An office manager or hospital administrator will make sure those expenses are kept in check, but not a lot of resources should be placed on expenses with a couple of exceptions.

Payroll

The cost of labor should be calculated as a percentage of gross revenue. It will be close to 20% for veterinary salaries and 20% for non-veterinary salaries, totaling 40% of gross revenue.[10] The office manager or hospital administrator should monitor overtime. That is the one area payroll can get out of hand. If overtime is managed and payroll is above 40%, review the amount and type of staff to find where costs are going.

Inventory management

Inventory is an expense that can get way out of control. There are a few ways this can happen. The first is that there are no controls over how much is ordered or fear buying. There can be a lot of anxiety of running out of a product. Use ordering history to determine how much is needed over a given period. While taking longer than it has in the past, products still arrive in a fairly timely manner. Learn the patterns of delivery from your distributors. The second is that an owner of a practice may just want what they want. This can be challenging to overcome as everything and anything gets ordered. Cost of goods should be about 17.5% of revenue.[11] If it is above that, look for excess inventory, duplication of medications, lack of use of an online pharmacy, and even theft.

Conclusions

These outlined concepts are not challenging to understand. Implementation is the most challenging part. The strategies and tactics above will prevent having a challenging time during a recession. They require practice while growth and gross revenue is high so when it is not, you are prepared.

References

1. National Bureau of Economic Research. Frequently Asked Questions. Available at: https://www-nber-org.proxy.library.upenn.edu/research/business-cycle-dating/business-cycle-dating-procedure-frequently-asked-questions. Accessed March 18,2023.

2. Rodeck D. What is a recession? *Forbes.* 2022;7.

3. Hayes DJ, Schultz LL, Hart CE, KL Jacobs. A descriptive analysis of the COVID-19 impacts on U.S. pork, turkey, and egg markets. *Agribusiness.* 2021;37(1):122-141.

4. Fine L, Barnes C, Niedbalski A, Deem S. Staff perceptions of COVID-19 impacts on wildlife conservation at a zoological institution. *Zoo Biol.* 2022;41(3):234-243.

5. Columbia Broadcasting System (CBS). Available at: https://www.news9.com/story/62900baa45556307273f185c/future-of-veterinarian-care-uncertain-as-many-begin-to-retire. Accessed March 18, 2023.

6. Felsted KE. What is your veterinary clinic worth? *DVM360.* 2021;53(3):54.

7. Zirkle K. Reading the signs of healthy veterinary practices. *Trends.* 2022;15(1):26-28.

8. Donato L. Avian practice management: the art of delegating. *Proc Annu Conf Assoc Avian Vet.* 2008.

9. Fuhrman J. Don't misfire on missed charges. *Today's Vet Bus.* 2021;5(1):46-47.

10. O'Neil T. 5 Ways to better manage veterinary hospital labor. *Today's Vet Bus. 2019;2(6):12-16.*

11. American Veterinary Medical Association. The Economic State of the Veterinary Profession 2019. Available at: https://ebusiness-avma-org.proxy.library.upenn.edu/files/productdownloads/HR-Secured-ECO-StateofProfessionReport-20190927.pdf. Accessed March 18, 2023.

The Opioid Epidemic and Why a Human Health Crisis Matters in Veterinary Medicine

Julia K. Whittington, DVM, and
Lauren R. Forsythe, PharmD, DICVP

Session #3410

Affiliation: From University of Illinois College of Veterinary Medicine, 1008 W. Hazelwood Dr., Urbana, IL 61802, USA.

Abstract: The complexity of controlled substance regulations as well as a limited understanding about the prevalence of drug diversion can make compliance difficult. This session covers background information about the opioid epidemic to provide perspective on why the regulations are in place. It also includes information about prevalence and signs of in-hospital drug diversion. Human statistics are tied to veterinary medicine with specific case examples, and considerations when utilizing controlled substances to treat animal patients. Appropriate precautions to avoid supplying controlled substances to a drug-seeking client are also covered.

The Opioid Crisis

The opioid crisis in the United States is not a new problem. In fact, it has been brewing since the introduction of morphine in the late 1800s. Throughout history, there are examples of addiction resulting from the popularity of opioids, sedatives, and stimulants, and this popularity has been largely driven by pharmaceutical companies that embrace the popularity to enhance their profits. Along with addiction and the associated problems, an evolution of regulatory oversight culminating in our current tiered structure of controls and oversight has occurred. While many prescribers were content to decrease their prescribing of these addictive medications, some resisted the restrictions and continued as they always had. As a result, history has shown that regulations still leave gaps and are only as effective as their enforcement. One example is the case of Dr C. who was a physician with an illustrious career practicing in Madison, WI, near the University of Wisconsin from the 1960s until his retirement in 1979. He became known for readily prescribing Quaalude (methaqualone), to tense and restless college students during a time when other practitioners had largely ceased prescribing this drug due to the addiction concerns. Every college student that wanted a sedative or a stimulant (or both) learned to go to Dr. C and say that they had problems sleeping. They would then receive prescriptions for whatever they wanted with directions to one of the three pharmacies in town that would readily supply the medications. While this case was dramatic, it is only one example in the history of addictive drugs where a respected doctor served as a legal source of controlled substances.[1]

From a more recent perspective, the current opioid crisis has been progressively worsening since the 1990s (Fig 1). While veterinarians were not the original source for obtaining controlled substances, as controlled substance restrictions have continued to tighten, those addicted to a prescription medication have had to seek alternative options. Veterinary medicine presented a possible alternative. There are a variety of reasons for this; however, it is important to understand the significance of the current opioid crisis first.

Concerning statistics

Between 1999 and 2020, more than 932,000 Americans died of a drug overdose. While this is concerning on its own, it is also important to note that the problem is getting worse each year (Fig 1). Overdose deaths in the United States were six times higher in 2017 compared to 1999, with two-thirds of the 2017 deaths attributed to opioids. In 2020, overdose deaths increased 38% over 2019 deaths, with two-thirds again contributed to opioids. While this problem is not specific to the United States, the United States does account for 80% of the world's opioid consumption and 99% of the world's hydrocodone consumption despite only being 4.4% of the world's population. The result is 187 Americans dying each day from opioid overdoses (Fig 2).[3]

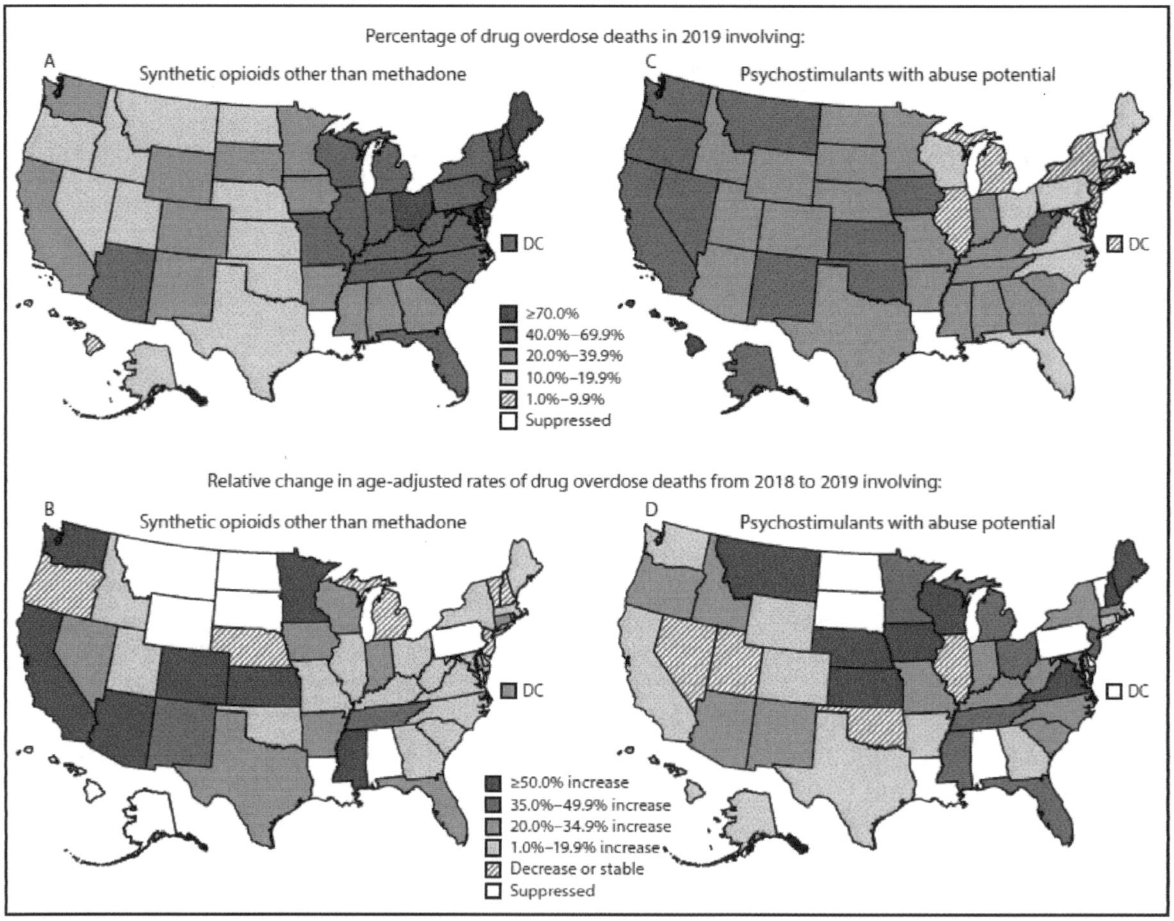

Figure 1. Drug overdose deaths by state and drug class in 2019.
Source: Mattson CL, Tanz LJ, Quinn K, et al. Trends and geographic patterns in drug and synthetic opioid overdose deaths-United States, 2013-2019. *Morb Mortal Wkly Rep.* 2021;70(6):202-207.

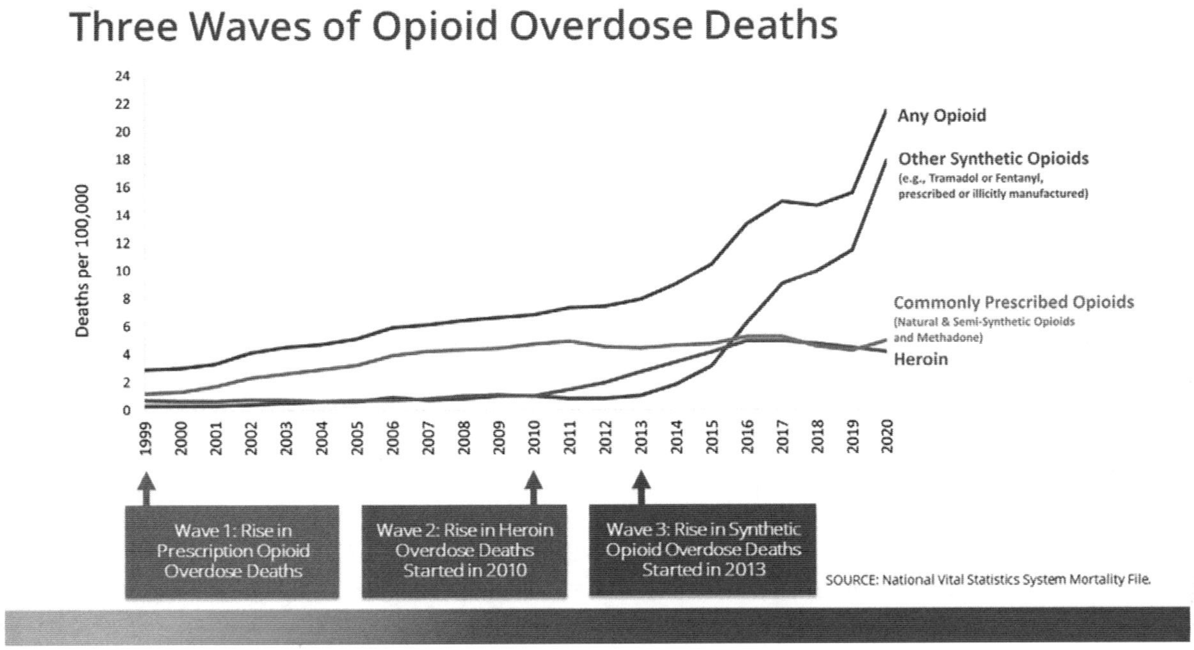

Figure 2. The three waves of the opioid crisis.
Source: Centers for Disease Control and Prevention. Understanding the Opioid Overdose Epidemic. Opioids. Available at: https://www.cdc.gov/opioids/basics/epidemic.html. Accessed March 26, 2023.

There are several studies looking at the prevalence of drug diversion in human healthcare. Based on the knowledge that the incidence of drug diversion in healthcare reflects that of the general population but with increased access, the authors assert that 10% of healthcare workers are abusing drugs at any given time. During their career, 15% of pharmacists, 10% of nurses and 8% of physicians will abuse drugs at some point. Despite knowing this is a problem and having stringent regulations designed to address it, 80-93% of diversion goes undetected.[4]

A 2021 study in human healthcare found that 96% of providers are concerned about hospital drug diversion, with 73% indicating that they believe most diversion goes undetected. Of the executives surveyed, 82% indicated that they personally know someone who has diverted drugs; however, 94% are confident or very confident in the effectiveness of the drug diversion program at their facility.[5]

While it is common knowledge in human healthcare that drug diversion in healthcare professionals is a significant problem, there is a disconnect between healthcare professionals believing it's a problem in general and believing that it could occur at their facility. In a 2019 study of over 650 hospital executives and providers, 85% of providers were concerned about hospital drug diversion, and 50% reported observing suspicious activity; however, despite the level of concern and observation of suspicious activity, only 20% believed that drug diversion was a problem at their facility.[6]

Xylazine, a new player

Recently, a new threat emerged that is worsening the opioid crisis. The presence of xylazine in illicit fentanyl is quickly becoming more prevalent and poses significant problems in humans. Xylazine is being added to fentanyl by drug dealers to boost the effects of fentanyl and extend the associated high. Since the effects of fentanyl are short-lived, the ability to extend the effects is desirable in the illicit drugs market; however, xylazine is not safe for humans. In the 1960s, xylazine was evaluated for potential use in humans, but proved inappropriate due to severe CNS depression.[7] Naloxone is not an effective reversal agent for xylazine since it is not an opioid, and the veterinary-approved reversal agents are not available for use in humans. In addition, xylazine can potentiate the effects of opioids such as sedation and respiratory depression, making naloxone reversal of opioid overdoses less effective. This occurs through the CNS depression from xylazine blunting the response to hypoxia associated with an opioid overdose. In addition to the systemic effects of xylazine, eschars commonly occur and can lead to amputation.[8] The result is xylazine making the opioid epidemic worse through increased mortality and addiction to xylazine in addition to opioids.[7]

Given the immediate public health crisis resulting from xylazine, the federal government is moving forward with scheduling this drug. Permanent scheduling is a long process where the Attorney General, Drug Enforcement Administration (DEA), Department of Health and Human Services (DHHS) and the Food and Drug Administration (FDA) place substances into the controlled substance schedules based on eight factors, which are collectively termed the 8-Factor Analysis. These eight factors are the following:[8]

1. Actual or relative potential for abuse;
2. Known scientific evidence of pharmacologic effects;
3. Current scientific knowledge of the substance;
4. History and current pattern of abuse;
5. Scope, duration, and significance of abuse;
6. Risk to public health;
7. Psychic or physiologic dependence liability; and
8. Whether the substance is an immediate precursor to an already-scheduled substance.

There is also the option for emergency scheduling of a substance into Schedule I; however, since this path is only available for Schedule I, and xylazine has accepted use in veterinary medicine, it is unlikely that this route to scheduling will be taken.[8] As of February 2023, it was reported that the DEA has conducted the 8-Factor Analysis described above for xylazine and that the scheduling process has been in the works since October 2021, with supplemental information provided to HHS in September 2022 to accelerate the scheduling process.[9] On February 28, 2023, the FDA announced that it is taking action to restrict unlawful entry of xylazine in both finished dosage form and as bulk chemical into the United States. This includes reviewing all xylazine shipments to ensure that they are headed to appropriate places such as veterinarians.[10] While this will not solve the problem, it is the first step giving the FDA the authority to seize shipments that are not directed to an appropriate channel. In addition, legislation can be passed through Congress and signed by the President to schedule a substance. At the time of writing, the American Veterinary Medical Association (AVMA) has endorsed the Combating Illicit Xylazine Act, which was introduced into Congress on March 28, 2023. If passed, this act would make xylazine a Schedule III controlled substance.[11]

The introduction of xylazine into the illicit drug supply likely has origins in Puerto Rico, with the earliest report of xylazine contaminated drugs being from the early 2000s and a rapid increase in 2019. A recent report issued by the DEA and Department of Justice noted that identification of xylazine as an illicit drug is increasing throughout the country with 193% and 112% increases in a 2-year period in the South and West respectively. While the xylazine found initially may have been raw chemical illegally imported, the xylazine on the streets now is not purely illicit and has likely been obtained through diversion. Empty bottles of pharmaceutical grade xylazine with VetOne (distributed by MWI Animal Heath) labeling are being found in areas where dealers are bagging drugs.[7,9] Diversion has likely been going undetected due to the uncontrolled nature of xylazine and limited inventory systems in veterinary medicine; however, a 2020 break-in at Peace River Veterinary Clinic located in Alberta, Canada involved theft of three sedatives which included xylazine, ketamine, and detomidine.[12]

Applying the human healthcare data to veterinary medicine

The information up to this point has focused on the data from the perspective of the impact to human healthcare. Afterall, in veterinary medicine, our patients are not the ones we are worried about diverting drugs. It is the humans attached to them that regulations are designed to protect. Compared to human healthcare, the data on drug diversion in veterinary medicine is limited; however, there is evidence that this issue is equally prevalent in veterinary medicine based on reports of DEA fines and court cases against veterinarians. The following are a few examples of these instances of drug diversion occurring in veterinary practices.

- In Wyoming, a veterinary receptionist was accused of stealing tramadol and fraudulently calling in prescriptions.[13]
- In Alabama, a veterinary technician was arrested on 20 counts of controlled substance possession in connection with theft from Auburn.[13]
- In Florida, a veterinary technician was arrested for stealing controlled drugs from two veterinary clinics where they were employed.[13]
- In Massachusetts, a veterinary technician was charged with stealing 7,800 opioid pills and changing clinic records to cover.[13]
- In Virginia, a veterinarian was charged with diverting hydromorphone by injecting himself and replacing the volume with saline or butorphanol.[14]

A 2018 survey of 189 Colorado veterinarians, with an average time in practice of 22 years, found that 44% were conscious of opioid misuse by either a client or staff member, 12% were aware of opioid misuse and diversion, and 13% were aware of a pet owner intentionally making their pet appear ill or injured to obtain opioids.[15]

History has shown that as controlled substance sources tighten up, diverters will find alternatives, and veterinary medicine presents an alternative to human healthcare. Some possible reasons include lack of inventory systems, need to transport controlled substances, large volumes required for some species, and historically less emphasis on diversion prevention in veterinary medicine compared to human healthcare. There is also limited use of Prescription Drug Monitoring Programs (PDMPs) due to challenges in applying these reporting systems to veterinary prescribing.

Who is harmed by drug diversion in healthcare?

While it can be tempting to assume that drug diversion affects the person diverting the drugs more than anyone else, there are significant risks to non-diverting healthcare workers, the organization, and the patients. The following is a partial list of potential risks to each group (Table 1).

Table 1. Summary of risks associated with drug diversion.

Risks to the person diverting drugs	Risks to other healthcare workers
Medication side effects or overdose. Infection from non-sterile drugs or supplies. Loss of license. Criminal prosecution.	Shared patient care responsibilities may lead to shared medical-legal liability. Potential victim of manipulations by the diverter. Risk of disciplinary action for innocent colleagues. Poor-quality work environment.
Risks to the organization	**Risks to patients**
Loss of revenue. Civil liability and investigations. Reduced employee morale. Reduced trust of clients.	Not receiving prescribed drug or dosage. Poor-quality care. Decreased quality of life. Increased morbidity and mortality.

Understanding Drug Diversion in Healthcare

The Joint Commission defines drug diversion as "the transfer of any legally prescribed substance from the individual for whom it was prescribed to another person."[16] This definition includes both controlled and non-controlled drugs. Although the focus here is on controlled substances, xylazine provides a prime example of a non-controlled drug that may be diverted, thus considered for controlled substance scheduling after being marketed as uncontrolled for many years.

Reasons for diversion vary and include an individual with a substance use disorder diverting drugs for their own use, a desire to sell the medications for a profit in the illicit drug market, or to relieve physical pain that is being inadequately managed. The most diverted drug class is opioids with fentanyl being the most common drug.[17] Other commonly diverted drugs include benzodiazepines, ketamine, propofol, pregabalin, and gabapentin.

It is important to consider that drug diversion can happen at any point during the life of a drug. Opportunities for diversion in practice include the following:

- Procurement: Ordering or inventorying incorrectly and then diverting.
- Preparation and dispensing: Replacing the drug with a product similar in appearance.
- Prescribing: Altering prescriptions and/or prescribing unnecessary drugs for personal pets.
- Administration: Documented as given to a patient but not actually given.
- Wasting: Logged as wasted but taken by healthcare worker.

Identifying diversion in practice

Controlled substance regulations state "A healthcare organization and its employees are mandated to report any incident of drug diversion within the organization."[18] To report drug diversion, it must first be identified. Signs of drug diversion are generally non-specific and can look like healthcare professionals that are overworked, short on time, stressed, tired, or generally careless. While those are problems on their own, they are not illegal. Therefore, to determine if drug diversion suspicions are justified, it's important to consider the combination of signs and trends. Below are some potential signs that may indicate a colleague is diverting controlled substances.[17]

- Chronic absenteeism
- Frequent breaks or disappearances from the workplace
- Excessive amount of time near the drug supply
- Unreliability in keeping appointments and deadlines
- Variable work performance
- Confusion, memory loss, difficulty concentrating
- Sloppy recordkeeping
- Excessive drug wastage
- Personality changes
- Removing controlled substances without an order
- Frequent association with compromised product containers
- Frequent association with volume missing from premixed infusion
- Missing multi-dose vial overfill
- Late documentation of certain medications
- Lack of effective patient analgesia despite what should be adequate pain control
- Frequent/unusual efforts to help administer pain medication to patients
- Controlled substance waste tampering

Diversion Prevention in Practice

The approach throughout history to address the problems associated with potentially addictive drugs has been to restrict access without complete prohibition so that those that need the drugs have access and those that do not are protected from their addictive potential. This is based on the 1800s mentality that white market (i.e., physician involvement) use of addictive drugs such as opioids, sedatives, and stimulants is necessary in many cases, but that all illicit use is unmoral and therefore should be punished; however, this creates a situation where white market addiction is viewed as uninformed physicians and patients accidentally being harmed, thus necessitating additional protection and education. This is in direct contrast with those having less access to healthcare being viewed as criminals for obtaining more dangerous versions of the drugs through unregulated supply chains. Regardless of your personal opinions on the merits of this structure, this mentality has led us to our current situation where regulations are used to protect the potentially naive white market consumers while punishing those obtaining drugs through other channels. The result is increasingly stringent regulations with increasing oversight of compliance.

When regulations are applied correctly and completely, they are designed to make it difficult to divert drugs and quickly identify any diversion that does get through the system. When regulations are not fully implemented, they can result in a false sense of security as if there is a safety net when an effective one does not actually exist. DEA fines are more than $15,000 per violation[19] and the DEA does not schedule visits at your convenience. It is critical for veterinarians to take controlled substance security and recordkeeping seriously. Following the regulations is a solid start, and there are practice-specific considerations that need to be addressed to find the most effective and feasible way to stay in compliance.

To make sense of the complicated landscape of controlled substance regulations, a global view is needed. The federal- and state-controlled substance acts provide minimum requirements for handling controlled substances; however, these are minimums, and additional precautions may be warranted in some situations. The DEA monitors and tracks all controlled substances from synthesis or import through the time you purchase it. They expect that this level of tracking will continue through the time the drug is administered or disposed of. To track these drugs effectively and report diversion, there needs to be a certain level of security and documentation. Therefore, as a DEA license holder, you are responsible for ensuring that you maintain your controlled substances with adequate security and documentation. The problem is that what meets that requirement is a matter of professional judgement and what a DEA agent thinks is acceptable on inspection will likely vary from one agent to the next.

While the ambiguity of adequate controlled substance security cannot be completely simplified, the following steps will help to critically evaluate current processes to identify potential gaps that need to be addressed.

- Own the process. If you are the DEA registrant, you are ultimately responsible for all controlled substances obtained under your license. You are allowed to delegate management of this process as you deem appropriate, but when the DEA comes walking through the door, it will be you that is responsible for whatever has been occurring under your license.

- Review the Federal Controlled Substance Act as well as your state Controlled Substance Act to ensure that you are compliant with all stated requirements. Remember, in the case that federal and state laws differ, you must follow whichever law is most strict. You can find the Federal Controlled Substance Act at https://www.ecfr.gov/current/title-21/chapter-II.

- Think like a diverter. Evaluate all of the steps that occur in the lifecycle of a controlled drug in your practice and look for opportunities where drug could be pocketed without anyone likely to notice. As stated earlier, the Controlled Substance Acts provide the minimum requirements, but you will likely need additional checks and balances that fit your unique practice situation. Keep in mind that the possible opportunities are likely different for injectable drugs only administered in house versus medications that may be dispensed for a client to administer at home.

- Rank the potential diversion opportunities identified in order of likelihood of diversion not being noticed and address the most problematic areas first.

- Review the 2022 American Society of Hospital Pharmacists (ASHP) Guidelines on Preventing Diversion of Controlled Substances. These are written for human healthcare and in some cases refer to technology that isn't readily available in veterinary medicine. They still provide valuable insight into areas of high concern and a lot of the recommendations are adaptable to veterinary medicine.

- Consider reaching out to the local DEA field agents with questions. Each inspector will be looking for different things and will often interpret compliance differently. They want you to be successful, so they are willing to answer questions proactively. Your successful controlled substance management contributes to their goal of keeping the controlled drug supply secure. You can also contact your state veterinary medical association. Many associations have a DEA resource and/or are willing to contact the DEA for you in the event you are hesitant to ask a question.

Human healthcare is ahead of veterinary medicine in drug diversion prevention. That means there is a lot of valuable data and literature on problems, diversion potential, and compliance that will help guide actions and policies in veterinary medicine. In veterinary medicine, we need to do our part to help curb the opioid crisis, and effectively managing our access to controlled substances is the way to accomplish that goal.

References

1. Herzberg D. *White Market Drugs: Big Pharma and the Hidden History of Addiction in America.* Chicago, IL: The University of Chicago Press; 2020.

2. Mattson CL, Tanz LJ, Quinn K, et al. Trends and geographic patterns in drug and synthetic opioid overdose deaths-United States, 2013-2019. *Morb Mortal Wkly Rep.* 2021;70(6):202-207.

3. Centers for Disease Control and Prevention. Understanding the Opioid Overdose Epidemic. Opioids. Available at: https://www.cdc.gov/opioids/basics/epidemic.html. Accessed March 26, 2023.

4. Baldisseri MR. Impaired healthcare professional. *Crit Care Med.* 2007;35(Suppl):S106-S116.

5. Porter Research Survey 2021: Drug Diversion in Healthcare. Available at: https://invistics.com/2021-porter-survey/. Accessed March 26, 2023.

6. BD Institute for Medication Management Excellence. Health Care's Hidden Epidemic. Available at: https://go.bd.com/BD-Institute-for-Medication-Management-Excellence-Drug-Diversion.html. Accessed March 26, 2023.

7. MedPage Today. What Doctors Should Know About Xylazine in Fentanyl. Available at: https://www.medpagetoday.com/special-reports/features/102915. Accessed March 26, 2023.

8. Congressional Research Service. Xylazine: considerations for federal control. *Insight.* 2023;IN12086.

9. MedPage Today. Will Xylazine Become a Controlled Substance? Available at: https://www.medpagetoday.com/special-reports/exclusives/103244. Accessed March 26, 2023.

10. Food and Drug Administration. FDA Takes Action to Restrict Unlawful Import of Xylazine. Available at: https://www.fda.gov/news-events/press-announcements/fda-takes-action-restrict-unlawful-import-xylazine. Accessed March 26, 2023.

11. American Veterinary Medical Association. AVMA endorses federal legislation to address illicit xylazine. Available at: https://www.avma.org/blog/avma-endorses-federal-legislation-address-illicit-xylazine. Accessed March 30, 2023.

12. Global News Canada. Veterinary Drugs, Equipment Stolen from Peace River Clinic - Edmonton. Available at: https://globalnews.ca/news/7387857/peace-river-veterinary-clinic-theft/. Accessed March 26, 2023.

13. Today's Veterinary Business. Dirty little secrets. Available at: https://todaysveterinarybusiness.com/dirty-little-secrets/. Accessed March 26, 2023.

14. U.S. Department of Justice. Amherst County Veterinarian Sentenced for Opioid Theft. Western District of Virginia. Available at: https://www.justice.gov/usao-wdva/pr/amherst-county-veterinarian-sentenced-opioid-theft. Accessed March 26, 2023.

15. Mason DS, Tenney L, Hellyer PW, et al. Prescription opioid epidemic: do veterinarians have a dog in the fight? *Am J Public Health.* 2018;108(9):1162-1163.

16. The Bulletin. Joint Commission focuses on strategies to detect, prevent drug diversion. Available at: https://bulletin.facs.org/2019/06/joint-commission-focuses-on-strategies-to-detect-prevent-drug-diversion/. Accessed March 26, 2023.

17. American Nurse. Drug diversion in healthcare - prevention and detection for nurses. Available at: https://www.myamericannurse.com/drug-diversion-in-healthcare/. Accessed March 26, 2023.

18. Code of Federal Regulations. 21 CFR 1301.91. Employee responsibility to report drug diversion. Available at: https://www.govinfo.gov/app/details/CFR-2011-title21-vol9/CFR-2011-title21-vol9-sec1301-91. Accessed April 2, 2023.

19. DVM360. Commonly ignored controlled substance regulations that can increase your risk and cost you money when audited by the DEA. Available at: https://www.dvm360.com/view/commonly-ignored-controlled-substance-regulations-that-can-increase-your-risk-and-cost-you-money-when-audited-by-the-dea. Accessed March 26, 2023.

The Unkind Truth About Conflict Avoidance and Managing Difficult Conversations

Julia K. Whittington, DVM

Session #3407

Affiliation: From University of Illinois College of Veterinary Medicine, 1008 W. Hazelwood Dr., Urbana, IL 61802, USA.

Abstract: Conflicts among personnel and with clients can be incredibly disruptive to effective functioning of the veterinary care team. Disagreements are compounded by the stress and challenges facing veterinary medicine leading to client dissatisfaction and staff turnover. Most people are hard-wired to avoid conflict, which tends to cause these individuals to harbor on-going resentment and feelings of disgruntlement because the root cause of the conflict is not addressed. Whether you are the practice manager, a staff member, or the client, addressing conflict through implementation of effective communication and de-escalation techniques will help in managing difficult conversations. Strategies for managing difficult conversations will be discussed.

Communication

Communication is essential for effective interaction between individuals of any species. As communities and societies developed in human culture, unique language and forms of communication developed to meet the needs of those communities. People communicate for several different reasons and the form of communication varies from spoken word to images, body language and actions, writings, pictures, symbols, and music. The way that humans communicate build relationships, which is essential to our humanness.

Components of communication

Written and spoken words have the capacity to be direct and have clear meaning; however, the way that written or spoken words are delivered to the recipient may impact the message that is received. An example is the use of an exclamation point at the end of a sentence when trying to express excitement. Another example is changing your tone of voice to ask a question. Non-verbal communication involves body language which includes facial expressions, body movements, posture, and gestures that have specific meanings. Non-verbal communication can be used alone or with verbal communication. When used alone, non-verbal communication is often used to relay how the person is feeling, but this carries the risk of being misunderstood. When used with verbal communication, non-verbal communication can add further clarity to the meaning of the words and has less risk of being misunderstood because there is more information being provided to the receiver. Verbal communication is rule-bound since it is through established words and all the associated grammar. Non-verbal communication is rule-bound by social norms and a general understanding of meaning; however, communication is constantly evolving because society and cultures change.

Humans tend to privilege nonverbal communication to verbal communication when our nonverbal messages conflict with our verbal messages. This means that nonverbal communication has a greater impact on our emotions and feelings when compared to verbal communication. The reason for this may be because a primitive survival trait is to react to feelings and sensory stimulation. It is the fight or flight response when the body just takes over for the survival of the individual. In times like that, people are not really thinking clearly. So, especially during times of high stress or emotion, visual and sensory cues are more impactful than words that require thinking to understand. It is difficult to choose if there is a reason that humans would privilege nonverbal communication given our culture. Both types of communication are used every day. Verbal communication, delivered in a clear and polite way, is very effective. Nonverbal communication is open to interpretation under the influence of the receiver's perception; however, people are emotional beings and social interactions are very important. Sometimes, words take on more meaning when combined with body language. The two together make a bigger impact and are most effective. Indirect routes of communication, such as emails, memos, and social media posts, lose the benefit of non-verbal communication.

Communication stereotypes

In general, women are more indirect, use more words, and are more emotional in their communication than are men. Women historically are the caregivers in the family unit. Historically, women are nurturers and the 'weaker' sex. Men are the breadwinners of the family. They are the disciplinarian and set the rules. Obviously, this is not true in modern day and many families are breaking stereotypes. Communication style is individual, even though stereotypes exist in modern day culture. There are men who are emotional and freely show their feelings. Unfortunately, sometimes these individuals are viewed as being weak and are dismissed as being unmanly. There are also women that are very direct and don't display their feelings. These individuals are sometimes viewed as heartless and uncaring. Neither of these assumptions are true. Sometimes people are not honest in their communication to hide their true self or to fit in. Breaking stereotypes and avoiding actions based on unintended biases requires that people stop and think before they act and speak. Being open-minded and viewing an individual as a person, not as a male or female, will allow communication to be more meaningful. Breaking stereotypes requires an open mind and empathy. Empathy is the ability to share or understand what another person may be feeling in a given situation. It requires an understanding of the reality a person is facing and putting yourself into their shoes. If that is done, then communication will become less stereotypical and more effective.

Communication filters

Perception is how an individual becomes aware or views things by using their senses. A person's perception can change their experiences and experiences can change a person's perception. The adage "a glass is either half full or half empty" is an example of perception. If a person views a situation as being negative, then they will likely have a negative experience. The same is true for the opposite, with a person's positive view resulting in a positive experience. Conversely, a person may have a negative experience that then causes their perception to be negative about future similar experiences.

Perception relates to identity and self-esteem. Identity is how a person views themselves in comparison to social standards. These comparisons result in a person having the feeling of accomplishment or failure, inclusion or being-marginalized, joy or sadness, fear or confidence, etc., which are all based on perception. Self-esteem is the value a person places on themselves and their usefulness in society. It is impacted by the community and the individual's place in it. Perception of the community by the individual, as well as the community's perception of the individual and their role, directly impact a person's sense of self-esteem. Perception has the capacity to induce a wide spectrum of feelings in the recipient of the communication. These feelings are unique to an individual, so can't be assumed. These feelings may be the result of some unmet need which is also highly variable between individuals because it is based on their unique experiences and perception. For example, three people may hear the same story and one will feel happy, one will feel anger, and one will feel sadness.

Managing perceptions that are based on fact and actual experiences are easier to manage that those that are based on preconceived ideas and those that are developed using the attribution theory. Preconceived ideas are those that are based on opinions made without facts. These are usually emotional and can be biased. Sometimes, perceptions that are preconceived cause an individual to think or do something in a way that reinforces their perception. This results in a self-fulfilling prophecy which increases the chance that the perception will remain. Another perception pitfall is described by the Attribution Theory,[1] and occurs when a cause-and-effect relationship is ascribed to a person's behavior or a random event when no such relationship exists. For instance, a difficult client may be a "bad person," or an untoward event may have occurred because of "carelessness" on the part of an individual. Human nature induces us to try to make sense of the social world, and as scientists, veterinarians are trained to look for causes and effects. The danger in having this perception is that people will be incorrectly judged, especially during first impressions. One way to combat this is with the Rule of 6.[2] For every story that we tell ourselves, we need to come up with five other possible stories. Maybe our initial story is true, but maybe it's a variety of other things. This creates a conscious effort to be curious and not judgmental. Everyone delivers and receives information through their filter of perceptions and their own sense of self. In modern society, we are taught from a very young age to think in terms of right and wrong, to judge and demand, and to compete. This results in inevitable conflict.[3] Being self-aware and understanding how the filters of another might impact communication is important to effective communication.

Communication in Healthcare

Providing quality patient care and client service in veterinary medicine requires positive and collaborative working relationships among team members. Efficient and effective communication among personal, and with clients, is paramount to relationship building. This is becoming more important in present day veterinary medicine in which personnel are experiencing high levels of stress and burnout. Veterinarians are often trained to focus on problems and to identify solutions. Indeed, advances in veterinary medical knowledge and biotechnology makes it impractical to didactically deliver to veterinary students the sheer expanse of information available in veterinary medicine. Because of this, many curricula have adopted problem-based learning to teach students to be life-long learners, considering basic sciences and clinical knowledge to problem solve.[4] There is relatively little content in veterinary education regarding communication or managing conflict. Therefore, veterinary professionals tend to approach conversations, and often conflict, with a find-it-and-fix-it attitude like their approach to the diagnosis and treatment of their patients;[5] however, not everyone wants their problems to be fixed. Sometimes what a person is really seeking is an empathetic statement. In some situations, it may be worth asking the person if they want empathy or a solution. That sets you up to successfully address their need.

Communication within the veterinary healthcare team

Promoting effective and efficient communication within the veterinary healthcare team has been shown to increase team satisfaction, increase client satisfaction, reduce errors and complaints, and improve patient care.[5] Communication is most effective when team members trust each other; trust is developed through vulnerability, shared experiences, time, and a belief in mutual good will. Often, trust is promoted through shared goals and a commitment of individuals to their respective roles and responsibilities. Though each person's roles and responsibilities differ, each contributes to the delivery of patient care. Respectful engagement to hear opinions and provide feedback enhances the performance of all members of the team.

Conflict Management

Conflicts happen and can occur within an individual, an organization, or between persons or groups. A conflict is a disagreement that either has the potential to cause harm or causes harm; however, not all disagreements result in conflict. Multiple factors contribute to conflict development including differences in perspectives, priorities, goals, values, and even organizational hierarchy.[6] Not all conflict is disruptive. Conflicts that are managed well can result in sharing of ideas, problem-solving, and improved efficiency of the team. However, conflicts that go unresolved are often disruptive resulting in reduced efficiency, and decreased confidence, self-esteem, self-reliance, and morale in team members. The result is usually increased staff turnover and increased incidence of adverse patient care.[7] There are four main reasons underlying most conflicts: 1) differences in values that are real or imagined, 2) competing individual goals, 3) poor communication, and 4) personalizing a generic or organizational issue.

The anatomy of a conflict

Conflicts evolve through four phases.[6,7] A disagreement occurs, and the individuals or groups involved become frustrated. This is the first phase of a conflict. Frustration is a strong emotion that can be overwhelming and broadly directed. When frustration is experienced, the natural inclination is finding the cause. This is the second phase, and it allows us to rapidly identify a plan to make this unpleasant emotion go away. Unfortunately, the identified cause is often inaccurate or more complicated than we can appreciate while feeling frustrated. Regardless, we start implementing our plan of action directed at addressing the "cause" in the third phase. This leads to a series of destructive outcomes in phase four as the conflict becomes ingrained or escalates. Conflicts may escalate for a variety of reasons including differences in roles and responsibilities, with individuals feeling unsupported or unheard, and also personal feelings of insecurity and discomfort in the individual (e.g., introvert, imposter syndrome).

Managing conflicts

Conflict is common in healthcare teams, and is often unavoidable, however, the impact of the conflict can be minimized through effective management. Management of conflict must not diminish the importance of the existing interactions. There is an inherent power struggle in a conflict with competing interests: assertiveness in the extent the individual (or group) will go to meet their goals, and cooperativeness in the extent the individual (or group) will go to meet the goals of the other individual (or group) with whom they have the conflict. Five conflict management strategies are defined. These are Avoiding, Accommodating, Compromising, Competing, and Collaborating.[6,7] In each strategy, the balance between self-interest and the interests of others defines the outcome (Fig 1). Collaboration is the most time consuming and can be costly in emotional energy as well as in resources. It is also the most likely to result in sustainable change for the betterment of all. When there is long-standing conflict or mistrust among team members, it may be difficult to get to the point where collaboration is possible.

Unfortunately, avoiding is often the conflict management strategy selected by individuals. In this case, neither party "wins" and the conflict becomes entrenched in culture, and possibly escalates, with individuals or groups divided over negativity, and team productivity and satisfaction being diminished.

The goal of conflict management is resolution, and ideally that resolution will include unification of the parties involved and a recommitment to the common goals of the team. It is important to consider those goals and understand that goals immediately post-conflict need to be short-term and flexible. Try to understand the 'root cause' of the conflict and realize that it may not even be well understood by the conflicting parties. The environment for conflict discussions needs to be safe and neutral for all parties. Whenever possible, conflict management conversations should not be relegated to a third party (e.g., an arbitrator or mediator) as this is a form of conflict avoidance. Allow individuals to express themselves openly but focus the conversation on the problem. Personal attacks and disrespectful interjections that interrupt another person who is talking are disruptive to the process. Identify the common ground and expand on it to build consensus. Most importantly, stay calm. To move toward a productive solution, both parties need to identify their unmet needs that are central to the conflict. Once the unmet needs are identified, then both sides can work together to identify potential solutions to meet these needs.

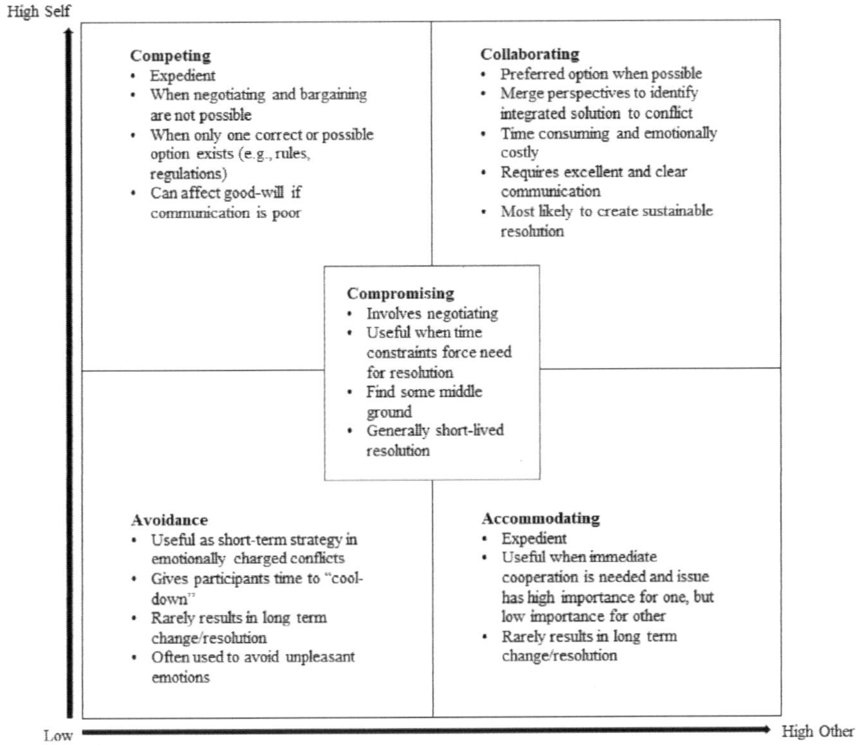

Figure 1. Conflict management strategies that address needs of self and other to varying degrees.

Communication Toolbox

Whether managing conflict, providing feedback, or having a difficult conversation, good communication skills are the foundation of a positive and successful interaction. The key to a positive interaction includes establishing a connection with the individual. Assume they have positive intent and thank them for their participation. Be curious and not judgmental. Be accountable for your role in any conflict, conceding or apologizing when appropriate. Be prepared to deliver facts clearly. When describing a specific situation, explain only the facts and avoid applying judgment through your filter. For example, say "You raised your voice and slammed a book when talking to your colleague. Tell me what was going on in that situation." instead of "Why did you get angry when talking to your colleague?" Be specific and avoid vague generalizations. Clear is kind, especially when delivering bad news or providing corrective feedback.[8] Communicate any pre-determined goals and the subsequent action plan that results from the interaction. When possible, make the action plan a collaborative approach by asking how you can support the other person.

Core skills

Communication skills are like any other skill in veterinary medicine and require practice to achieve and retain proficiency.[9] Communication in healthcare settings historically have the veterinarian, or physician in the case of human medicine, in the 'guardian' role. The veterinarian is viewed as the expert who provides the recommendations that the client is expected to follow with little shared decision making. With advances in veterinary medicine, more and more options have become available for our patients and veterinarians are needing to engage clients in decision making. In this manner, the veterinarian and client become collaborators and the result is relationship-centered care.[10] As in conflict resolution, effective communication and collaboration requires skill.

Making a connection

The first communication goal in any engagement is making a connection so that ideas may be shared clearly and respectfully. This goal may be hampered by factors including the nature of the interaction (e.g., emergency patient visit, providing negative feedback), but minimally the participants should acknowledge that each person has a valuable perspective and important information to share. A personal statement of introduction is a good first step. Using the person's name helps to make the individual feel seen.

Open-ended questions are those that cannot be answered with an affirmative or negative. They begin with when, what, how, or where, and allow the individual to share information openly. People want to share their ideas and information and often provide more detail when answering an open-ended question than if asked as series of yes-or-no questions. It is tempting to interrupt for clarification or to ask a related question when an individual is answering an open-ended question but avoid doing so as it may limit shared information at best, or negatively impact the collaborative interaction. Remember to pause before asking another question so that individuals have time to organize their thoughts.

Once information has been gathered, confirm that you have a clear and correct understanding of it by using reflective listening. By summarizing your understanding of the information relayed, you can correct misconceptions while making the individual feel heard.

Demonstrating empathy

To be empathetic is to be concerned about the feelings and values that another person has. Situational empathy is being able to feel and appreciate what another person is going through. Both are useful communication skills for promoting positive interactions. Empathy statements acknowledge, validate, and normalize the client's perceptions and emotions.[11] Fortunately, even persons who do not have innate feelings of empathy are able to make empathy statements which can help an individual feel supported and understood. People don't want to be judged and giving them the gift of empathy can go a long way towards collaborative agreements and conflict resolution. Beware of non-verbal cues that contradict your words.

Ensuring understanding

Through the sharing of information, a plan emerges. During the planning and ultimate execution of the plan, it is important to ensure that all people are on-board and understand what their responsibility will be. This can be done through a process called chunk-n-check.[11] This is especially helpful when working with clients who may feel overwhelmed with information, options, or simply the situation. By sharing information in manageable pieces, the receiver has time to absorb the news and the communicator can check-in to ensure understanding before moving on to the next piece of information. Another skill is the use of a signpost transition when moving between topics. In this way, the current information is summarized and then a verbal cue is given to indicate that the conversation is transitioning to a new topic.

Setting the stage

Anyone who has visited the Seattle, Washington, Pike's Place Market has likely enjoyed the energy exhibited by the employees of the World-Famous Pike Place Fish Market. This business is the focus of Fish! A Remarkable Way to Boost Morale and Improve Results, a book by Stephen C. Lundin, Harry Paul, and John Christensen.[12] In telling the story, the authors promote four key lessons to improve workplace morale. They are simple, common-sense lessons that can have a profound impact on promoting effective communication and conflict resolution.

Choose your attitude

You may not have a choice about what work you do, but you have a choice about the way you do it. If you have a good attitude, you can have a great day. If something bad happens to you or if someone makes you mad, you can choose to let it get to you or you can choose to let it go. If you have a grouchy attitude, you can have a bad day and make everyone else's day bad as well. By choosing your attitude you are being proactive, and you show personal accountability.

Play

You can be serious about your work, but you can have fun in the way you do it. Having fun is energizing.

Make their day

This is done by including others in your work and giving customers and co-workers a good experience, which leads to positive energy.

Be present

Look for opportunities to be fully engaged in your work and focus on other people. Be aware of customer needs so that you serve them well. Give people your full attention when engaging them, and really listen. Serving others brings personal satisfaction.

Conclusions

Communication style is as individualized as the individuals communicating, and every person approaches communication filtered through their personality, their perceptions, their experiences, their biases, and their sense of self. Communication skills are not a one-size-fits-all and not every tool is used in every interaction. Finding the tools that work best for you and practicing is the key to being an effective communicator. Take time to reflect on your own communication style and conflict resolution preferences. While not the only options, assessments such as Meyer-Briggs for personality type and the Thomas-Kilmann Instrument (TKI) for conflict resolution preferences can provide keys insights into your preferred processes. By knowing what you prefer, you can build off that to recognize your biases and hone your skills to be an effective communicator.

References

1. Mcleod S. Attribution Theory-Situational vs Dispositional. 2023. Available at: https://simplypsychology.org/attribution-theory.html. Accessed April 2, 2023.

2. Brown J. The rule of six. In: *The Art and Spirit of Leadership.* Bloomington, IN: Trafford Publishing; 2012. Available at: https://www.judysorumbrown.com/rule-of-six. Accessed April 2, 2023.

3. Bayer communication project-clear the air: tools for managing team conflict and enhancing team collaboration, module 15 syllabus. *Inst Healthcare Comm.* 2015;1-75.

4. Lane EA. Problem-based learning in veterinary education. *J Vet Med Educ.* 2008;35(4):631-636.

5. Bayer communication project-easy for you to say: communication within the veterinary healthcare team, module 12 syllabus. *Inst Healthcare Comm.* 2015;1-53.

6. Saltman DC, O'Dea NA, Kidd MR. Conflict management: a primer for doctors in training. *Postgrad Med J.* 2006;82:9-12.

7. Piryani RM, Piryani S. Conflict management in healthcare. *J Nepal Health Res Counc.* 2018;16(4):481-482.

8. Brown B. The call to courage. In: *Dare to Lead: Brave Work. Tough Conversations. Whole Hearts.* New York: Random House; 2018:45-69.

9. Coleman NL, Berlin A, Fischkoff K, et al. Annual structured communication skills training for surgery residents. *J Surg Res.* 2023;281:314-320.

10. Cornell KK, Kopcha M. Client-veterinarian communication: skills for client centered dialogue and shared decision making. *Vet Clin North Am Small Anim Pract.* 2007;37(1):37-47.

11. Shaw J. Guide to FRANK communication skills. *Frank Communication Workshop.* Ft. Collins, CO; 2014.

12. Lundin SC, Paul H, Christensen J. *Fish! A Remarkable Way to Boost Morale and Improve Results.* New York, NY: Hyperion; 2000.

Ten Tips for Supervising and Strengthening Teams When You Are Out of Bandwidth

Renée Schott, DVM, CWR

Session #3446

Affiliation: From Wildlife Rehabilitation Center of Minnesota, 2530 Dale St. N, Roseville, MN 55113, USA.

Abstract: Fortunately, or unfortunately, many veterinarians fall into formal or informal supervisory roles. Even if not a formal manager, veterinarians lead teams. Building a strong, cohesive team is not something taught well in many veterinary schools. The top 10 tips to help manage people and teams are provided in 15 minutes.

Tip 1

Show interest in each person's personal life. This is an easy way to foster connection.

Tip 2

Show vulnerability. If you cannot be vulnerable with your staff, they will feel they cannot be vulnerable with you, and will bottle up more emotions.

Tip 3

Let others know how you are feeling. It sets expectations and demonstrates vulnerability.

Tip 4

Ditch the formal reviews and instead schedule regular one-on-one meetings with your direct reports. It may seem laborious, but having a dedicated time and space for a regularly scheduled one-on-one meeting will naturally create the space for you and your report to connect on a human level. Meet weekly or a few times per month. The agenda could look something like: How are you? How is the work? How are we?

Tip 5

Know yourself. You need to dive into your own self-development to be able to be self-aware. Your inner world affects your outer world. Sometimes simple self-inquiry can make a difference in how you show up to work. Some self-inquiry questions may include: What's going on? What am I feeling? What is the dialogue inside my head? What are my feelings in my body? What is another way to see the situation?

Tip 6

Avoid blame (of yourself and others). Encourage accountability.

Tip 7

Ditch the golden rule; switch to the platinum rule. Focus on fairness (not equal) based in individual needs.

Tip 8

Recognition goes much further than praise!

Tip 9

Do not ask why. It implies blame. Instead ask what and how questions because they invite accountability.

Tip 10

Address conflict as soon as both parties are back to a grounded state. Discuss conflict at meetings and have staff create rules around how conflict should be repaired.

Bonus Tip

A great way to deliver feedback is to use the COIN method.

- C: Connect and/or give context.
- O: Observation. Describe the observable behavior.
- I: Impact. Describe how it impacted the team negatively.
- N: Next steps. Make sure to agree on the next steps.

There are multiple resources available to help build strong teams.[1-10]

References

1. Brown B. *Daring Greatly.* 2013. New York, NY: Random House; 2013.

2. Rock D. SCARF: A brain-based model for collaborating with and influencing others. *NeuroLeadership J.* 2018;1.

3. Brown B. *Dare to Lead: Brave Work. Tough Conversations. Whole Hearts.* New York, NY: Random House; 2018.

4. Wilson J, Gislason M. *Coaching Skills for Nonprofit Managers and Leaders: Developing People to Achieve Your Mission.* San Francisco, CA: Jossey-Bass; 2010.

5. Scott K. *Radical Candor: Be a Kick-Ass Boss Without Losing Your Humanity.* New York, NY: St. Martin's Press; 2017.

6. Rø, A, Fridjhon M. *Creating Intelligent Teams: Leading with Relationship Systems Intelligence.* Bryanston, JHB, South Africa: KR Publishing; 2016.

7. Stanfield RB. *The Art of Focused Conversation: 100 Ways to Access Group Wisdom in the Workplace.* BC, Canada: New Society Publishers; 2000.

8. Patterson K, Grenny J, McMillan R, Switzler A. *Crucial Conversations Tools for Talking When Stakes Are High.* New York, NY: McGraw-Hill; 2011.

9. Lencioni P. *5 Dysfunctions of a Team.* San Francisco, CA: Jossey-Bass; 2002.

10. Eoyang G, Holladay R. *Adaptive Action: Leveraging Uncertainty in Your Organization.* Palo Alto, CA: Stanford University Press; 2013.

How to Avoid a Board Complaint and What to Do if You Receive One

Jesse A. Fallon, DVM, PhD

Session #3315

Affiliation: From West Virginia State Board of Veterinary Medicine, Cross Lanes, WV 25313, USA, Cheat Lake Animal Hospital, Morgantown, WV 26508, USA, Avian Conservation Center of Appalachia, Morgantown, WV 26508, USA.

Abstract: Veterinarians work hard to provide the best medical care for their patients and good service for their clients. Unfortunately, there are occasions when a client may be unsatisfied with the service they receive. Most of the time, we can recover from these negative client interactions and regain their trust, or clients may simply choose to take their pet elsewhere; however, some very unhappy clients may choose to submit a complaint to the state Board of Veterinary Medicine, either because of a real or perceived medical error, or simply because of inadequate communication. How do you best respond to a board complaint? What steps can you take to ensure that a board complaint is resolved with no probable cause? What can you do in your day-to-day practice to help prevent a board complaint in the first place? This presentation will walk you through the steps of how to best respond to a client-driven board complaint and what you can do to help prevent a complaint from ever being filed.

Introduction

One of the greatest fears of a licensed veterinarian or veterinary technician is the dreaded board complaint. Whether it is the outcome of a medical error or simply an outlet for a disgruntled client, a complaint against you to your state's Board of Veterinary Medicine causes an enormous amount of worry, feelings of disappointment or shame, and concern for your license. Perhaps there was an undesirable outcome from surgery, a misdiagnosis, a transcription error on a medication label, a patient died under anesthesia, a rabbit fractured its vertebrae during restraint, or simply a misunderstanding in communication. Whatever the sequence of events that lead up to a complaint being filed, it is important to understand how to respond to a complaint, what you can do to ensure a finding of "no probable cause" from the board, and how you can avoid a complaint altogether. To accomplish these goals, you must first understand how your state board functions, identify the key components of a medical record that can help keep protect you, and recognize steps you can take to prevent a complaint from being filed in the first place.

What is the Board of Veterinary Medicine?

Each state has its own Board of Veterinary Medicine, and the composition, precise duties, and appointment process for board members varies only slightly from state to state. Generally, the mission of your state board is to protect the health, safety, and welfare of animals and the general public by enforcing the licensing and regulation of veterinarians, veterinary technicians, veterinary premises, certified euthanasia technicians, and animal crematories. The board has two major areas of regulatory authority. First, it determines which veterinarians, technicians, premises, and animal crematories are qualified to be licensed. Second, when a consumer (pet owner) complaint has been received and an investigation has been initiated, the board serves to determine whether a licensee's actions have fallen below the standard of care. In summary, the board provides qualified individuals the guidelines for acquiring and maintaining licensure in their state. To fulfill this duty, the board is made up of a predetermined number and composition of members.

The board's members are charged with ensuring that the licensees have completed and maintain the legal right to practice in their state. To ensure objective evaluation of a complaint, your board members follow an individual state practice act that defines the roles and requirements of licensees. The practice act, as well as the meetings and actions of the board, are accessible by the public. The practice act is written to withstand legal scrutiny and provide maximum public and animal protection. You should be familiar with your state practice act, as it is the playbook by which the board makes decisions about licensees.

The makeup of the board differs from state to state, but most include a majority of licensed veterinarians from different regions in the state, in some cases include a registered veterinary technician, and finally a minority of members who are not licensed to practice veterinary medicine. This diversity helps maintain a balanced perspective and attempts to adequately represent the profession and the public. Additionally, your board will have paid staff including at least an executive director, an attorney, and in most cases, one or more staff members. The full-time staff members do not have voting authority, but rather oversee the day-to-day administrative duties of the board (e.g., license renewals). The individual board members have specific responsibilities including a board chair and committee assignments.

How is a Complaint Filed and How does the Board Respond?

Any pet owner or representative (e.g., an attorney) can file a complaint to the state board. After a complaint is filed, the board sends you (i.e., the licensee) a certified letter stating that a complaint has been filed against you, requests the medical records, includes the letter from the complainant, and requests that your respond to the complaint. You can author the response yourself or you can have an attorney that is retained by your professional liability insurance (e.g., AVMA PLIT) author the letter. If you elect to use an attorney to help author your response, you provide the liability insurance company with a description of the incident presented within the complaint. Once you have provided your response and other required documents, the board members receive all of these redacted documents (no names, facility identification etc.) in advance of the upcoming board meeting. Prior to the board meeting, the complaint committee reviews the complaint letter, your response, and the medical records. At the prescheduled board meeting, the committee summarizes and reviews the case with the board, and typically makes a recommendation as to whether or not there is probable cause of deviation of standard of care, and the board then votes on the case. They may elect to: 1. Find no probable cause of malpractice; 2. Find probable cause of malpractice; or 3. Request that the complaint committee acquire more information (e.g., incomplete medical records, missing details that are important to the case, etc.)

What Should You Do if a Complaint is Filed Against You?

What do you do when or if you receive a letter from the board stating that they have received a complaint? First, don't panic. Review the medical record and the case details, and then immediately contact your PLIT representative. The liability company will ask you to provide them a description of the case that triggered the complaint. Be honest, open, and describe all details surrounding the case. Do not withhold any information about the case, client, or animal that is in question. Their job is to help defend you from the complaint.

After you provide details on the case, you and your liability company work together to author your response letter. This response is extremely important, and should not be accusatory, inflammatory, contain typos, or emotional overtones. It should include a description of your role in the case using clear and concise language, the diagnostic progression of the case, your recommended and accepted treatments, the client communication, and the outcome of the case. Consider the following when composing your response:

- If a medical error was made and the complaint was clearly justified and represents malpractice, often direct admission is the best approach.
- If the complaint was inaccurate or does not represent an error in judgement or decision making, say so. Describe the motivation and thought process you used in the clinical decision making and include supportive evidence, referring both to the medical record, peer reviewed publication, or expert opinion.
- Explain why the complainant is incorrect in their assessment, and state clearly that you did not make a mistake and that, while you have empathy for the client, they are incorrect in their assessment. It is okay and may be helpful to cite or include statements from your state practice act. You may also elect to include supplementary material in your response. This may include phone or text records, pictures or videos, or notarized statements from technicians, receptionists, or other doctors.

Receiving a complaint is stressful, and you will not be thinking objectively in this situation. You will be angry, frustrated, hurt, scared, and anxious. It is okay to feel this way. You may be inclined to take immediate action to rectify or address the complaint, but you should avoid any of the following:

- Falsify or alter the medical record
- Reach out to the client personally
- Post anything on social media

How Can You Ensure Appropriate Medical Records?

One of the most common problems that we see at the level of the state board is incomplete or inadequate medical records. Common gaps include incomplete or inadequate documentation of physical exam, inadequate or lack of anesthetic monitoring, and not documenting communications. Often there are major discrepancies in how the client perceived versus how the veterinarian remembers the sequence of events surrounding the incident that triggered the complaint. From the board's perspective, the most reliable source of information is the documentation in the medical record. Remember, if it is not documented in the medical record, there is no way that the board can determine whose side of the story is true. Your record does not need to contain copious amounts of writing. In fact, a succinct (but complete) medical record is often easier for the board to come to a decision. Be sure that your medical records contain the following:

1. Complete physical exam (including vital parameters)
2. Adequate anesthetic monitoring
3. Clearly stated treatment plan

4. Thorough communication details including discussion of expectations, risks, and complications

5. Thoroughly documented phone or email communication

6. Signed documents as appropriate (e.g., anesthetic consent, euthanasia consent)

Your thorough, complete, yet succinct medical record makes the job of the board's complaint committee easy. If standard of care is met and the record is complete, the outcome of your complaint will be no probable cause.

What Happens if the Board Finds Probable Cause?

If probable cause is found, the board will respond with a letter stating their findings and offer you a consent agreement. This describes the nature of the malpractice and the adverse action (i.e., punishment) which may include specific CE requirements, financial reimbursement for the board's time and expenses, probation, or revocation (very uncommon) of the license. They also provide a timeline for you to complete the required actions. At this point, you can choose to sign the consent agreement and accept the board's adverse action, or if you feel that the finding is unjustified, you can elect to decline to sign. This will most likely result in a court hearing, where you and your attorney will have to defend your position in front of a judge. If you are found at fault, you will be responsible for attorney and court fees. For more details on this, you should rely on your professional liability attorney.

How Can You Avoid a Complaint in the First Place?

There is no foolproof way to prevent a client from filing a complaint; however, one of the consistent themes in complaint cases is that the client often feels that the veterinarian or veterinary staff did not adequately communicate some aspect of their pets' care. This may include feelings that there was inadequate discussion of risks or expectations of a particular surgery, too slow or inadequate response to phone calls with concerns following the discharge of a patient, or a misunderstanding about a diagnosis or treatment course. A large percentage of the complaint cases that I have worked on could have been prevented with a mutual understanding of the pet's current condition, the diagnostic options, and the treatment plan. Ideally, you should have these conversations in person, ask the client if they understand the plan, and then document this reported understanding. If the owners decline to follow your treatment or diagnostic advice, document this in the medical record.

Conclusions

Keep in mind that your State Board of Veterinary Medicine is made up of people that are sympathetic to your situation. They are not your enemy; however, they have been charged by the governor or state legislature to protect the public, and they work hard to be objective in their assessment of complaint cases. Therefore, be sure to maintain a thorough medical record, have clear and consistent communication directly with your clients, and be diligent to uphold current standard of care for whatever species you are working on. Finally, if you do receive a complaint, don't panic. Many excellent veterinarians have had the misfortune of a complaint against them. You can overcome this challenge and continue to serve your patients and clients.

Section 14

Hybrid Potpourri Session

Renee Schott, DVM, CWR
Session Chair

Pradofloxacin Pharmacokinetics in Two Exotic Pet Species: Rabbit and Eastern Long-Necked Turtle

Shangzhe Xie, BSc/BVM, MVS, PhD, DABVP (Avian Practice), DACCM

Session #3103

Affiliation: From Mandai Wildlife Group, 80 Mandai Lake Road, Singapore 729826.

Abstract: The pharmacokinetics of some fluoroquinolones in rabbits and chelonians are well described but not for pradofloxacin, a new broad-spectrum veterinary fluoroquinolone available as an oral suspension for cats and dogs. In two separate studies, the single dose pharmacokinetic profile of pradofloxacin oral suspension at 7.5 mg/kg was investigated in four rabbits (*Oryctolagus cuniculus*) and 10 Eastern long-necked turtles (*Chelodina longicollis*). Plasma concentrations were measured using high-performance liquid chromatography on numerous samples obtained over time. Rabbit pharmacokinetic results indicated that oral absorption of pradofloxacin was slower and elimination faster compared to other fluoroquinolones in healthy rabbits, as well as relative to cats and dogs. Following treatment with pradofloxacin, fecal microbiota profiling showed some compositional differences between treated and control rabbits. This was the result of a significant decrease in the abundance of Proteobacteria, in particular bacteria belonging to the *Pseudomonas*, *Atopostipes* and *Parabacteroides* genera. Despite these microbiota changes, no adverse effects were observed in the treated rabbits in this study, but further studies utilizing multiple dose protocols (i.e., 7 days) are needed to confirm safety. Similarly, no adverse effects were observed in Eastern long-necked turtles administered 7.5 mg/kg q24hr for 7 days; however, using mammalian MIC data, the AUC0-24/MIC ratios for a range of bacterial isolates suggest that this dose of pradofloxacin is unlikely to be effective against many bacterial pathogens in Eastern long-necked turtles. Further studies are required for both species to establish clinically relevant dosages, but these two studies provide a platform for the understanding of this fluoroquinolone.

Introduction

Pradofloxacin is a third-generation fluoroquinolone developed specifically for veterinary use. This fluoroquinolone has good efficacy against gram-negative, gram-positive and anaerobic bacteria in dogs and cats.[1,2] Its main inhibitory effect on bacterial pathogens is through dual targeting of bacterial DNA gyrase and topoisomerase IV enzymes, and its bactericidal activity is maintained during stationary or slow growing phases of microbial replication.[3] Pradofloxacin has been shown to be effective in treating infections when the pathogen showed reduced susceptibility to other fluoroquinolones.[2] It also has lower potential for inducing antimicrobial resistance than other fluoroquinolones due to its low mutant prevention concentration.[4] Pharmacokinetic investigations in dogs and cats have shown that the maximum plasma concentrations are achieved within 3 hours of dosing and that the half-life is 5 to 10 hours.[5,6] Prior to these studies, there were no published reports of pradofloxacin use in rabbits and reptiles. There were also no previous pharmacokinetic studies for any drug in Eastern long-necked turtles (*Chelodina longicollis*) or any other Australian freshwater turtle (Chelidae, Carettochelydae).

Methods and Materials

In the first study, four female mature New Zealand white rabbits (body weight median 3.1 kg; range: 2.9-3.4 kg), were housed indoors in separate stainless steel cages with free access to feed (i.e., commercial pelleted diet) and water via automated dispensers, including the days of pradofloxacin administration and sample collection.[7] Rabbits were clinically examined daily and weight, food consumption, feces production and presence of any abnormal signs (i.e., ruffled coat, poor posture, abdominal pain, lung sounds, gut sounds, dehydration, diarrhea or lethargy) noted on the clinical record sheet. Each rabbit was administered pradofloxacin (25 mg/ml) oral suspension (Veraflox Oral Solution for Cats, Bayer Australia, Pymble, NSW), at a dose of 7.5 mg/kg via oral syringe.[7] To quantify plasma pradofloxacin concentrations, blood samples (0.5 ml per time point) were collected using a 25-gauge needle via the ear or lateral saphenous vein into lithium heparin tubes at t = 0, and hourly there after until t = 10 hours.[7]

In the second study, 10 captive, adult, Eastern long-necked turtles (*Chelodina longicollis*, body weight 450-1500 g, sex undetermined) were housed in two outdoor partially shaded enclosures.[8] Five turtles were housed in each enclosure, assigned to optimize the similarity of turtle weights between enclosures. Each enclosure comprised a 700-L pool, water temperature was maintained between 26-30 °C, with a small dry haul-out area with leaf-litter substrate and a basking light. A gelatin based pudding made in house with whitebait, beef mince, and Wombaroo Reptile Supplement (Wombaroo Food Products, Glen Osmond, South Australia) was fed once daily. Eight turtles were administered 7.5 mg/kg pradofloxacin oral suspension (Veraflox, 25 mg/ml, Bayer Australia, Pymble, NSW) delivered by syringe into the esophagus under manual restraint prior to feeding. Turtles were held for 1 minute to monitor for regurgitation of drug and then returned to the enclosure. Food was offered 1-2 hours after drug administration. One additional turtle in each tank received no pradofloxacin to evaluate the possibility of environmental drug exposure. Following iodine disinfection over the jugular vein, blood samples (1 milliliter) were collected using a 25-gauge needle and placed into lithium heparin tubes at t = 0, 1, 3, 6, 9, 12, 24, 48, 72, 96, 120, 144 and 168 hours relative to pradofloxacin administration.[8] Total blood volume over the 7-day sampling period from each turtle was limited to 1% body weight. All turtles were sampled at t = 0. Only one turtle was of sufficient size for all 13 blood samples to be collected; the remaining seven treated turtles had their respective allowable number of 1-milliliter samples distributed such that there were seven samples at 1 hour reducing to four samples at 72 hours. Aliquots of pool water from each enclosure were also collected at each time point.[8]

Plasma and water samples from both studies were stored at -80 °C prior to determining pradofloxacin concentration by high pressure liquid chromatography.

Results and Conclusions

In rabbits, the pradofloxacin median (range) T_{max} was 4.50 (2.00-5.00) hours, C_{max} 600.66 (395.85-886.72) ng/ml and $T_{½}$ was 1.27 (0.12-1.39) hours.[7] These results indicate that oral absorption of pradofloxacin was slower, and elimination faster compared with other fluoroquinolones in healthy rabbits, as well as relative to cats and dogs.[7] Following treatment with pradofloxacin, fecal microbiota profiling showed some compositional differences between treated and control animals. This was the result of a significant decrease in the abundance of Proteobacteria, in particular bacteria belonging to the *Pseudomonas*, *Atopostipes* and *Parabacteroides* genera.[7]

In Eastern long-necked turtles, pradofloxacin was quantifiable in plasma and tank water samples for up to 48 hours after drug administration.[8] The T_{max} (9.0 hours) and $T_{½}$ to 48 hours (13.16 hours) were longer, and the C_{max} (0.2 µg/ml) and AUC0-24 (2.2 hr*µg/ml) lower, than previously reported in cats and dogs.8 No adverse effects were observed in six turtles administered 7.5 mg/kg q24h for 7 days.[8] Using mammalian MIC data, the AUC_{0-24}/MIC ratios for a range of bacterial isolates suggest that this dose of pradofloxacin in turtles is unlikely to be effective against many bacterial pathogens.[8]

References

1. Silley P, Stephan B, Greife HA, Pridmore A. Comparative activity of pradofloxacin against anaerobic bacteria isolated from dogs and cats. *J Antimicrob Chemother.* 2007;60(5):999-1003.

2. Lees P. Pharmacokinetics, pharmacodynamics and therapeutics of pradofloxacin in the dog and cat. *J Vet Pharmacol Ther.* 2013;36:209-221.

3. Silley P, Stephan B, Greife HA, Pridmore A. Bactericidal properties of pradofloxacin against veterinary pathogens. *Vet Microbiol.* 2012;157:106-111.

4. Wetzstein H-G. Comparative mutant prevention concentrations of pradofloxacin and other veterinary fluoroquinolones indicate differing potentials in preventing selection of resistance. *Antimicrob Agents Chemother.* 2005;49(10):4166-4173.

5. Sykes JE, Blondeau JM. Pradofloxacin: a novel veterinary fluoroquinolone for treatment of bacterial infections in cats. *Vet J.* 2014;201(2):207-214.

6. Hauschild G, Rohn K, Engelhardt E, et al. Pharmacokinetic study on pradofloxacin in the dog-comparison of serum analysis, ultrafiltration and tissue sampling after oral administration. *BMC Vet Res.* 2013;9(1):1-8.

7. Xie S, Trott DJ, Saputra S, et al. Pharmacokinetic profile and effect on the faecal microbiome of a single dose of pradofloxacin oral suspension in the rabbit (*Oryctolagus cuniculus*). *J Vet Pharmacol Ther.* 2022;45(2):203-212.

8. Taylor E, Trott DJ, Kimble B, et al. Pharmacokinetic profile of a single dose of an oral pradofloxacin suspension administered to Eastern long-necked turtles (*Chelodina longicollis*). *J Vet Pharmacol Ther.* 2021;44:503-509.

VetGyver: How to Apply a Normal Thing to Exotic Pet Equipment

Chaowaphan Yinharnmingmongkol, DVM, Dipl,TBVM

Session #3243

Affiliation: From AnimalSpace Exotic Pet Hospital, 99/30 Puttamonthon Sai2, Taweewuttana, Salatummasop, Bangkok, 10170, Thailand.

Abstract: Exotic pets have unique anatomy and behaviors, which means that when working with multiple species, veterinarians must adapt to each animal's needs. Unfortunately, most veterinary equipment is designed for use with humans, dogs, and cats, making it challenging to find suitable equipment for exotic pets. By using standard equipment or everyday items creatively, veterinarians can often find cost-effective solutions. For example, intubating a rabbit typically requires an endoscope, but a USB ear scope can be used instead, saving thousands of dollars. A 1.5-millimeter silicone endotracheal tube, designed for small avians or mammals, can also be used as a feeding tube for tiny animals. Clear plastic surgical drapes are useful for visualizing small mammals during surgery, but an oven bag can serve the same purpose. Transparent hooks can be used in many scenarios, including hanging an intravenous line, and can be covered with an electronic cord cover to protect against animal biting. A saline bottle can be modified to function as an oxygen or gas supply mask and can even be adjusted to fit an Elizabethan collar for a prairie dog. In some cases, LEGO® toys can be repurposed in certain situations to prevent feces and urine from contaminating the surgical site following a plastronotomy procedure on a tortoise. By thinking outside the box and utilizing a range of everyday items, exotic pet veterinarians can save money while still providing quality care for their patients.

Ecotourism in the Caribbean: The Good, the Bad and the Ugly

Odette Doest, DVM

Session #3493

Affiliation: From Dierenartsenpraktijk Doest, St. Michielsweg 41, Julianadorp, Curacao.

Note: Abstract format selected by author(s). See Front Matter for description of conference proceedings manuscripts submitted as scientific abstracts.

Abstract: The main economic income for most Caribbean destinations is tourism. Exploring nature, hiking, camping, and ecotourism is booming. This phenomenon increases the possibility of conflict between nature conservation and nature exploitation. Ecotourism can be done in various ways and animal encounters are increasingly popular. We compare the different practices in Trinidad and Tobago, Suriname, the Dominican Republic, Mexico and the Dutch Caribbean and the way it affects wildlife.

Definition of Ecotourism

According to the website of the international ecotourism society, ecotourism is now defined as "responsible travel to natural areas that conserves the environment, sustains the well-being of the local people, and involves interpretation and education (The International Ecotourism Society, 2015). Education is meant to be inclusive of both staff and guests."[1]

Impact of Ecotourism

Wildlife observation excursions often are part the Caribbean ecotourism experience and involve tours to flamingo breeding and foraging sites, sea turtle nesting areas by night, and adventurous hikes into bat caves and ecosystems. Feeding wild animals often is a way to entice people to choose for certain tours that "guarantee" wildlife encounters. The feeding of wildlife can alter behaviors and has long-term impacts on animal diets and animal health.[2] Wildlife species are impacted by the presence of human visitors in the area be it on foot, boat or by car.[3] This presentation demonstrates the impact of ecotourists' actions and the importance of educating ourselves and our clientele about how to interact with wildlife species. Veterinarians often participate in wildlife excursions during vacations and should be aware of how to avoid potential harm. Veterinarians in an ecotourism area should know how to guide clients in responsible behavior.

References

1. The International Ecotourism Society. What is Ecotourism? The Definition. Available at: http://www.ecotourism.org. Accessed April 1, 2023.

2. Smulders FOH, O'Shea OR, Christianen MJA. Animal-borne video reveals atypical behaviours in provisioned green turtles: a global perspective of a widespread tourist activity. *Global Ecol Conserv.* 2012;25:e01417.

3. Ouboter DA, Kadosoe VS, Ouboter PE. Impact of ecotourism on abundance, diversity and activity patterns of medium-large terrestrial mammals at Brownsberg Nature Park, Suriname. *PLoS One.* 2021;16(6):e0250390. doi: 10.1371/journal.pone.0250390. PMID: 34077471; PMCID: PMC8171955.

Use of Alfaxalone in Exotic Animal Practice

Elizabeth A. Martinez, DVM, Dipl ACVAA

Session #3224

Affiliation: From Jurox Inc, now part of Zoetis, 1321 Burlington St. Suite D, N Kansas City MO 64116, USA.

Abstract: Alfaxalone is a sedative and anesthetic with a formulation that is indexed by the U.S. Food and Drug Administration for use in multiple minor species commonly encountered in exotic animal practice, including several species of small mammals, reptiles, amphibians, and birds. Following a brief summary of the pharmacology of alfaxalone and the indexing process, discussion of clinical case examples provides information to help veterinary professionals achieve optimal results and patient outcomes.

Introduction

It is often necessary to sedate or anesthetize exotic patients for physical examinations, laboratory sample collection, and diagnostic or surgical procedures. The ideal drug would be safe and versatile, have multiple routes of administration options, and provide a smooth sedation, anesthesia, and recovery process. Alfaxalone has been studied in numerous exotic species and has been shown to be safe and effective with a wide margin of safety.[1,2]

Clinical Pharmacology

Alfaxalone is a synthetic neuroactive steroid with the properties of a general anesthetic. The primary mechanism of action is via positive allosteric modulation of the gamma-aminobutyric (GABA) subtype A receptor ($GABA_A$) in the central nervous system (CNS). The $GABA_A$ receptor is a transmembrane ion channel. When alfaxalone binds to the $GABA_A$ receptor, the effect of endogenous GABA is potentiated, with movement of chloride ions intracellularly and hyperpolarization of the neuron and inhibition of action potential propagation.[3] It is important to remember that alfaxalone has no analgesic properties and that an appropriate analgesic plan is instituted. Other drugs that have a similar mechanism of action include benzodiazepines, propofol, etomidate, and barbiturates.

Indexing Process

Indexing is the process to legally market unapproved new animal drugs for non-food producing minor species. To be placed on the Index, the drug must have its safety and effectiveness affirmed through an alternative U.S. Food and Drug Administration process through the Minor Use Minor Species (MUMS) group. A drug that is Indexed is not registered. To have a product on the Index is especially helpful to individuals treating animals or classes of animals representing markets too small to support the costs of the traditional drug approval process.

A group of Expert Panelists reviewed over 100 peer and non-peer reviewed publications, abstracts and personal communications regarding alfaxalone's use in various species. All the information was compiled into a database and reviewed by the MUMS group for determination of eligibility of alfaxalone to be listed on the Index.

Alfaxalone (Alfaxan® Multidose IDX, Jurox Inc, N Kansas City, MO, USA) was indexed in February 2020 and joins a handful of other medications that can be found on the FDA's website, along with links to the Freedom of Information Summary (FOI) and Product Labeling.[4] To date there are only 15 other products on the FDA's Index.

Clinical Use of Alfaxalone

Exotic companion mammals

Rodents, ferrets (*Mustela furo*), guinea pigs (*Cavia porcellus*), and rabbits (*Oryctolagus*) are some of the most common exotic companion mammals. The use of alfaxalone in these species has been reported. Most commonly, alfaxalone is included in a multimodal sedation or anesthetic plan and administered intravenously (IV) or intramuscularly (IM).

Rodents

In mice (*Mus musculus*), alfaxalone has been administered IV, IM, and subcutaneously (SC).[5-7] Intravenous dosing without premedication was between 15-20 mg/kg producing successful anesthesia and was maintained with a CRI at 0.25-0.75 mg/kg/min.[5] Many may find IV access in a mouse challenging to achieve and the intraperitoneal (IP) route of administration of 80 mg/kg with 10 mg/kg of xylazine produced anesthesia with longer sleep times observed in female versus male mice.[6] In rats, alfaxalone has been administered IV, across a large dosing range, for both induction and maintenance of anesthesia, as well as IP.[2,5,8,9]

Ferrets

Because physical restraint may cause undue stress and discomfort on the animal, ferrets may require sedation to facilitate routine noninvasive medical procedures. Alfaxalone has been administered IM for sedation with butorphanol,[10] or butorphanol and midazolam.[11] Alfaxalone has also been administered IV for both induction and maintenance of anesthesia.[12,13]

Guinea pigs

The intramuscular administration of 5 mg/kg in guinea pigs produced successful sedation when administered alone.[14] Onset of sedation was rapid (2-3 minutes) and physiological variables were maintained within normal ranges. Duration of sedation was approximately 30 minutes. Alternatively, 20 mg/kg of alfaxalone alone has been used for non-painful procedures for less than 60 minutes.[15] The authors also concluded that 15 mg/kg alfaxalone with dexmedetomidine at 0.25 mg/kg and buprenorphine at 0.05 mg/kg subcutaneously would be adequate for non-painful procedures requiring consistent immobility for a minimum of 90 minutes.[15]

Rabbits

Administration of alfaxalone in rabbits is considered off-label since this species is considered a food-producing minor species. Numerous publications are available demonstrating that alfaxalone is a safe and effective drug for sedation, induction, and maintenance of anesthesia.[16-19]

Reptiles

Snakes

Alfaxalone may be administered IV, to effect, for induction of anesthesia in snakes. If IV access is not possible, alfaxalone may also be administered IM, SC, or intracelomic (ICe). In the ball python (*Python regius*), alfaxalone has been shown to provide a rapid smooth induction when administered IM, when injected caudally at 30 mg/kg and cranially at 20 mg/kg.[20] Cranial injection resulted in more successful intubation and significantly greater depth of anesthesia, likely due to hepatic shunting and first-pass metabolism via the renal portal system. In corn snakes (*Pantherophis guttatus*), SC administration of alfaxalone, 15 mg/kg, resulted in consistent anesthesia and is recommended for intubation and short procedures.[21] In the common garter snake (*Thamnophis sirtalis*), ICe administration of alfaxalone, 30 mg/kg, resulted in reliable loss of righting reflex (LRR). The anesthetic depth was greatest between 10 and 30 minutes after LRR.[22]

Lizards

Alfaxalone has been evaluated in numerous lizard species and with various routes of administration. A wide dose range has been published, that is likely due to whether the desired effect was sedation or anesthesia, the route of administration, the anatomical site of injection, and the effects of concurrent drugs before or simultaneous with alfaxalone. An IM injection of alfaxalone, 10 mg/kg, in either the triceps or quadriceps muscles provided reliable chemical restraint in bearded dragons (*Pogona vitticeps*) for short, nonpainful, diagnostic procedures.[23] A combination of alfaxalone, 15 mg/kg, and midazolam, 1 mg/kg, administered SC craniodorsal to the thoracic limb provided appropriate sedation to allow for physical examination and venipuncture in leopard geckos (*Eublepharis macularius*).[24] In veiled chameleons (*Chameleo calyptratus*), alfaxalone, 5 mg/kg IV following butorphanol premedication, provided light anesthesia and allowed for intubation; alfaxalone, 15 mg/kg IV, provided light anesthesia suitable for electroejaculation.[25]

Chelonians

Induction of anesthesia in chelonians with alfaxalone IV or IM has been described. In terrapins and tortoise species, 5 mg/kg IV was suitable for intubation, with full recovery in 30 minutes.[26] In loggerhead sea turtles (*Carreta caretta*), 3, 5, and 10 mg/kg alfaxalone IV was suitable for intubation, with dose-dependent duration of anesthesia and respiratory depression.[27] Alfaxalone, 10 or 20 mg/kg, IM in red-eared sliders (*Trachemys scripta* elegans) resulted in a smooth, rapid induction and uneventful recovery.[28] Depth and duration of sedation was body temperature dependent.

Amphibians

Studies of alfaxalone have investigated its administration IM and via immersion. In American bullfrogs (*Lithobates catesbeiana*), doses of 10-17.5 mg/kg IM resulted in immobilization with similar duration of effect, with higher doses having a shorter time to recumbency.[29] However, no dose prevented responsiveness to noxious stimuli. Immersion in alfaxalone at 2 g/L for 30 min did not produce immobilization or anesthesia.[29] In Oriental fire-bellied toads (*Bombina orientalis*) immersion in alfaxalone at 200 mg/L for 14 minutes resulted in sedation but remained responsive to noxious stimuli and is only suitable for non-invasive short procedures.[30] Immersion with a combination of alfaxalone, 30 mg/L and morphine, 50 mg/L or butorphanol, 25 mg/L improved antinociception.[31] Several case reports have been published describing anesthesia via immersion in axolotls (*Ambystoma mexicanum*) for surgical procedures and radiotherapy.[32-34]

Birds

Psittaciformes

Dose-dependent sedation was seen following IM alfaxalone in quaker parrots (*Myiopsitta monachus*);[35] however, muscle tremors and hyperexcitation was observed. Premedication with midazolam IM 10 minutes prior to alfaxalone produced a smoother induction but prolonged the recovery time compared to alfaxalone alone. The combination of intranasal midazolam and butorphanol and IM alfaxalone was found to be a safe and effective protocol for sedating quaker parrots.[36] Budgerigars (*Melopsittacus undulatus*) administered 5 mg/kg or 10 mg/kg IM achieved sedation adequate for handling nonpainful procedures. Some birds did show excitement during induction and recovery.[37]

Passeriformes

In the Bengalese finch (*Lonchura domestica*), dose-dependent effects on induction time, duration, and depth of anesthesia were observed with SC alfaxalone administration.[38] The addition of midazolam, 0.7 mg/kg, or butorphanol, 1 mg/kg, to 30 mg/kg alfaxalone improved muscle relaxation and increased depth of anesthesia. In black-cheeked lovebirds (*Agapornis nigrigenis*), alfaxalone, 12.5 mg/kg SC, provided nearly 1 hour of excellent sedation appropriate for noninvasive procedures. Rough induction and recoveries, along with minor movements and twitching were observed.

Waterfowl

Rose flamingos (*Phoenicopterus roseus*) undergoing orthopedic surgery, alfaxalone, 2 mg/kg, IV produced a smooth and rapid induction of anesthesia.[40] Alfaxalone, 10 mg/kg IV, was found to be a suitable induction drug for use in mute swans (*Cygnus olor*), producing a smooth a rapid induction adequate for intubation.[41] Agitation and ataxia was observed during recovery, requiring manual restraint in some birds.

Disclosure Statement: The author, Dr. Elizabeth A Martinez, is an employee of Jurox Inc, now part of Zoetis.

References

1. Full Prescribing Information Alfaxan® Multidose IDX. Available at: https://jurox.com/us/product/alfaxan-multidose-idx. Accessed March 28, 2023.

2. Lau C, Ranasinghe MG, Shiels I, et al. Plasma pharmacokinetics of alfaxalone after a single intraperitoneal or intravenous injection of Alfaxan(®) in rats. *J Vet Pharmacol Ther.* 2013;36(5):516-520.

3. Warne LN, Beths T, Whittem T, et al. A review of the pharmacology and clinical application of alfaxalone in cats. *Vet J.* 2015;203(2):141-148.

4. The Index of Legally Marketed Unapproved New Animal Drugs for Minor Species. Available at: https://www.fda.gov/animal-veterinary/minor-useminor-species/index-legally-marketed-unapproved-new-animal-drugs-minor-species. Accessed March 28, 2023.

5. Tremoleda JL, Kerton A, Gsell W. Anaesthesia and physiological monitoring during in vivo imaging of laboratory rodents: considerations on experimental outcomes and animal welfare. *EJNMMI Res.* 2012;2(44):123.

6. Siriarchavatana P, Ayers JD, Kendall LV. Anesthetic activity of alfaxalone compared with ketamine in mice. *J Am Assoc Lab Anim Sci.* 2016;55(4):426430.

7. Higuchi S, Yamada R, Hashimoto A, et al. Evaluation of a combination of alfaxalone with medetomidine and butorphanol for inducing surgical anesthesia in laboratory mice. *Jpn J Vet Res.* 2016;64(2):131139.

8. White KL, Paine S, Harris J. A clinical evaluation of the pharmacokinetics and pharmacodynamics of intravenous alfaxalone in cyclodextrin in male and female rats following a loading dose and constant rate infusion. *Vet Anaesth Analg.* 2017;44(4):865875.

9. White K, Aldurdunji M, Harris J, et al. Alfaxalone population pharmacokinetics in the rat: model application for pharmacokinetic and pharmacodynamic design in inbred and outbred strains and sexes. *Pharmacol Res Perspect.* 2022;10(6):e01031.

10. Milloway MC, Posner LP, Balko JA. Sedative and cardiorespiratory effects of intramuscular alfaxalone and butorphanol at two dosages in ferrets (*Musela putorius furo*). *J Zoo Wildl Med.* 2021;51(4):841-847.

11. Bercier M, Langlois I, Dunn M, et al. Cytological analysis of bronchoalveolar lavage fluid acquired by bronchoscopy in healthy ferrets: A pilot study. *Can J Vet Res.* 2016;80(1):7480.

12. Thas I. Acquired salivary mucoceles in two domestic ferrets (*Mustela putorius furo*). *Vet Rec Case Rep.* 2014;2:18.

13. Giral M, GarcíaOlmo DC, GómezJuárez M, Gómez de Segura IA. Anaesthetic effects in the ferret of alfaxalone alone and in combination with medetomidine or tramadol: a pilot study. *Lab Anim.* 2014;48(4):313320.

14. d'Ovidio D, Marino F, Noviello E, et al. Sedative effects of intramuscular alfaxalone in pet guinea pigs (*Cavia porcellus*). *Vet Anaesth Analg.* 2018;45(2):183189.

15. Doerning CM, Bradley MP, Lester PA, Nowland MH. Effects of subcutaneous alfaxalone alone and in combination with dexmedetomidine and buprenorphine in guinea pigs (*Cavia porcellus*). *Vet Anaesth Analg.* 2018;45(5):658666.

16. Costa RS, Ciotti-McClallen M, Tilley R, et al. Intramuscular alfaxalone with or without buprenorphine or hydromorphone provides sedation with minimal adverse effects in healthy rabbits (*Oryctolagus cuniculus*) in a randomized blinded controlled trial. *J Am Vet Med Assoc.* 2022;261(2):223-228.

17. Knutson KA, Petritz OA, Thomson AE, Balko JA. Intramuscular alfaxalone-butorphanol-midazolam compared with ketamine-butorphanol-midazolam in New Zealand white rabbits. *J Am Assoc Lab Anim Sci.* 2022;61(5):475-481.

18. Bradley MP, Doerning CM, Nowland MH, et al. Evaluation of alfaxalone total intravenous anesthesia in rabbits (*Oryctolagus cuniculus*) premedicated with dexmedetomidine or dexmedetomidine and buprenorphine. *Vet Anaesth Analg.* 2022;49(3):308-312.

19. Reabel SN, Queiroz-Williams P, Cremer J, Hampton CE, et al. Assessment of intramuscular administration of three doses of alfaxalone combined with hydromorphone and dexmedetomidine for endoscopic-guided orotracheal intubation in domestic rabbits (*Oryctolagus cuniculus*). *J Am Vet Med Assoc.* 2021;259(10): 1148-1153.

20. Yaw TJ, Mans C, Johnson SM, et al. Effect of injection site on alfaxalone-induced sedation in ball pythons (*Python regius*). *J Small Anim Pract.* 2018;59(12):747-751.

21. Rockwell K, Boykin K, Padlo J, et al. Evaluating the efficacy of alfaxalone in corn snakes (*Pantherophis guttatus*). *Vet Anaesth Analg.* 2021;48(3):364-371.

22. Strahl-Heldreth DE, Clark-Price SC, Keating SCJ, et al. Effect of intracoelomic administration of alfaxalone on the righting reflex and tactile stimulus response of common garter snakes (*Thamnophis sirtalis*). *Am J Vet Res.* 2019;80(2):144-151.

23. Shippy S, Allgood H, Messenger K, et al. Pharmacokinetics and pharmacodynamics of intramuscular alfaxalone in central bearded dragons (*Pogona vitticeps*): effect of injection site. *Vet Anaesth Analg.* 2023:S1467-2987(23)00044-2.

24. Doss GA, Fink DM, Sladky KK, Mans C. Comparison of subcutaneous dexmedetomidine-midazolam versus alfaxalone-midazolam sedation in leopard geckos (*Eublepharis macularius*). *Vet Anaesth Analg.* 2017;44(5):1175-1183.

25. Ferreira TH, Mans C. Sedation and anesthesia of lizards. *Vet Clin North Am Exot Anim Pract.* 2022;25(1): 73-95.

26. Knotek Z. Alfaxalone as an induction agent for anaesthesia in terrapins and tortoises. *Vet Rec.* 2014;175(13):327.

27. Phillips BE, Posner LP, Lewbart GA, et al. Effects of alfaxalone administered intravenously to healthy yearling loggerhead sea turtles (*Caretta caretta*) at three different doses. *J Am Vet Med Assoc.* 2017;250(8): 909-917.

28. Kischinovsky M, Duse A, Wang T, Bertelsen MF. Intramuscular administration of alfaxalone in red-eared sliders (*Trachemys scripta elegans*)-effects of dose and body temperature. *Vet Anaesth Analg.* 2013;40(1): 13-20.

29. Posner LP, Bailey KM, Richardson EY, et al. Alfaxalone anesthesia in bullfrogs (*Lithobates catesbeiana*) by injection or immersion. *J Zoo Wildl Med.* 2013;44(4):965-971.

30. Adami C, Spadavecchia C, Angeli G, d'Ovidio D. Alfaxalone anesthesia by immersion in oriental fire-bellied toads (*Bombina orientalis*). *Vet Anaesth Analg.* 2015;42(5):547-551.

31. Adami C, d'Ovidio D, Casoni D. Alfaxalone-butorphanol versus alfaxalone-morphine combination for immersion anaesthesia in oriental fire-bellied toads (*Bombina orientalis*). *Lab Anim.* 2016;50(3):204-211.

32. McMillan MW, Leece EA. Immersion and branchial/transcutaneous irrigation anaesthesia with alfaxalone in a Mexican axolotl. *Vet Anaesth Analg.* 2011;38(6):619-623.

33. Modesto F, Nicolier A, Hurtrel C, Benoît J. Excisional biopsy and radiotherapy for management of an olfactory neuroblastoma in an axolotl (*Ambystoma mexicanum*). *J Am Vet Med Assoc.* 2021;260(4):436-441.

34. Vuong KS, Sheley W, Liu M, et al. Chromatophoromas on the tail of a pet axolotl (*Ambystoma mexicanum*) with presumptive metastasis. *J Exot Pet Med.* 2023;45:8-13.

35. Whitehead MC, Hoppes SM, Musser JMB, et al. The use of alfaxalone in quaker parrots (*Myiopsitta monachus*). *J Avian Med Surg.* 2019;33(4):340-348.

36. Conner CM, Hoppes SM, Stevens BJ, Simon BT. Intranasal butorphanol and midazolam administered prior to intramuscular alfaxalone provides safe and effective sedation in quaker parrots (*Myiopsitta monachus*). *Am J Vet Res.* 2022;83(12):ajvr.22.08.0140.

37. Balko JA, Lindemann DA, Allender MC, Chinnadurai SK. Evaluation of the anesthetic and cardiorespiratory effects of intramuscular alfaxalone administration and isoflurane in budgerigars (*Melopsittacus undulatus*) and comparison with manual restraint. *J Am Vet Med Assoc.* 2019;254(12):1427-1435.

38. Perrin KL, Nielsen JB, Thomsen AF, Bertelsen MF. Alfaxalone anesthesia in the Bengalese finch (*Lonchura domestica*). *J Zoo Wildl Med.* 2017;48(4):1146-1153.

39. Greunz EM, Limón D, Bertelsen MF. Alfaxalone sedation in black-cheeked lovebirds (*Agapornis nigrigenis*) for non-invasive procedures. *J Avian Med Surg.* 2021;35(2):161-166.

40. Villaverde-Morcillo S, Benito J, García-Sánchez R, et al. Comparison of isoflurane and alfaxalone (Alfaxan) for the induction of anesthesia in flamingos (*Phoenicopterus roseus*) undergoing orthopedic surgery. *J Zoo Wildl Med.* 2014;45(2):361-366.

41. Baldrey V, Stanford M, Bacon H. A comparison of the pharmacodynamic effects of intravenous ketamine-xylazine with alfaxalone in mute swans (*Cygnus olor*) presenting at a wildlife veterinary hospital. *Vet Anaesth Analg.* 2021;48(5):716-724.

Ooo That Smell! Maggot-Infested Wounds in Wild Mammals

Renée Schott, DVM, CWR

Session #3385

Affiliation: From Wildlife Rehabilitation Center of Minnesota, 2530 Dale St. N, Roseville, MN 55113, USA.

Abstract: Young (and some adult) mammals often present during the summer months with maggot-infested wounds. This gross, but fun, talk covers the basics on what the crud that bug is doing in the wound and what that means for the patient. Although wild mammals are used as case studies, the information in this masterclass can be applied to domestic species.

Introduction to Maggots

Maggots are the larval form of true flies, which are in the order Diptera.[1-5] These true flies have two main wings and two greatly reduced wings and include over 150,000 species. This presentation focuses on maggots in wounds and *Cuterebra*.

Generalized Life Cycle of Flies

Because there are so many species of true flies, and the timing of the life cycle varies greatly depending on ambient temperature, tissue temperature, and a variety of other factors, a very generalized fly life cycle is discussed here.

Egg (Day 0)

Most females lay approximately 2-millimeter-long eggs in clumps up to 300. Some fly species lay eggs on decaying matter, some need to be laid on flesh wounds, and some on intact flesh. Some flies need a living host and others don't. In around a day the egg will hatch into the larva or maggot.

Larva, maggot (Days 1-15)

The larva, called the first instar, initially feeds on fluid exuded from the body and then migrates into the body or wound. The maggot will double in size to approximately 5 millimeters and molt into the second instar. The second instar will double in size again (to 10 millimeters) and molt. This process is repeated via the third instar, which can be up to 20 millimeters long. The third instar molts into a pre-pupa and migrates away from the body, usually burrowing into the soil. There it remains in the soil for approximately 4 days, not feeding, until it develops into the pupa.

Instars (maggots) have hooks to shred and rake in decaying flesh. They have a muscular, segmented body, a simple intestine and a pair of very large salivary glands. These salivary glands secrete digestive enzymes which help to create their soup environment they eat. They have anterior and posterior spiracles to breathe, so they can feed and breathe 24 hours a day.

Pupa

The pupa does not feed and remains in the soil until it emerges as the fly (approximately 10 days).

Fly

Upon emerging from the pupa, the fly mates. Flies feed on protein from body fluids and lay eggs in approximately 2 days.

Cuterebra Life Cycle

Although *Cuterebra* life cycle follows the above generalized fly life cycle, a few differences are worth mentioning. Eggs are laid on grass stems, wood chips, or bark that is near the opening to a rodent or rabbit burrow. The eggs hatch and the larva then move onto the fur of passing rodents and rabbits. Larva enters the body through any body opening (not through the skin). The larva then migrates to the subcutaneous tissue, creating a subcutaneous cyst, it creates a breathing pore and grows. In 3-6 weeks, it drops to the ground once it has developed into a pupa. The pupa burrows into the ground and the adult fly emerges several months to years later.

Types of Myiasis Based on How it Infects

Myiasis can be generally grouped into five categories. Each category is described below.

Cutaneous

Fly eggs are laid on an animal or other object, hatching into the first instar. As an animal brushes against the object, the instar hitches a ride onto the animal and burrows into skin or enters through a body orifice. These larvae produce lesions that look like boils, pimples or furnucular lesions.

Wounds

Maggots are in an open wound that they either helped create, or infested a wound that was already present.

Creeping or larva migrans

Larva migrate around the body. *Baylisascaris* in non-hosts is an example. This group is not addressed here.

Myiasis of body cavities

Myiasis of body cavities is maggot infestation in the eye, nasal passes, ear canal or mouth.

Accidental myiasis

Accidental myiasis happens when eggs or maggots are accidentally eaten by an animal, such as an animal eating carrion with a maggot infestation. Usually this causes only gastrointestinal symptoms. The eggs or larvae may or may not be defecated whole. This grouping is not covered here.

Types of Myiasis Based on What It Eats

Facultative

These eat only dead or decaying organic matter or living matter.

Obligatory

These require a living host for development.

Medicinal Maggots

Medicinal maggots are generally the species *Lucilia sericata*, which is a species that only eats dead tissue. In a laboratory, they are propagated in a sterile environment and fed a strict diet. Modern companies now put these maggots into a porous, sterile bag that goes directly into the wounds. The enzymes the maggots secrete exit the bag to digest the dead tissue and they can suck up the soup through the bag as well. The enzymes this species secretes have antibacterial, antifungal and sometimes antiviral properties. Each bag is left in the wound for approximately 4 days and each would generally need 1-12 sessions.

Cases

Cases from the Wildlife Rehabilitation Center of Minnesota (WRCMN) are used to demonstrate some treatment options when dealing with maggot infested wounds. Prognosis is discussed using release proportion data from WRCMN.

References

1. Gerhardt RR, Hribar LJ. Flies (Diptera). In: Mullen GR, Durden LA, eds. *Medical and Veterinary Entomology.* 3rd Ed. London, UK: Academic Press; 2019. doi:10.1016/B978-0-12-814043-7.00011-X.

2. Ahmadnejad M, Rezazadeh F. Maggot therapy for snakebite necrotic wound in a horse. *Iran J Vet Surg.* 2021;16:156-160.

3. Harvey ML, Dadour IR, Gasz NE. Maggot therapy in chronic wounds: new approaches to historical practices. *Ann Entomol Soc Am.* 2021;114:415-424.

4. Stadler F. The maggot therapy supply chain: a review of the literature and practice. *Med Vet Entomol.* 2020;34:1-9.

5. Cousquer G. Veterinary care of rabbits with myiasis. *In Pract.* 2006;28:342-349.

Spatial Analysis of Wildlife Rehabilitation Data for Optimization of Outreach Efforts

Katherine Y. Choo, SVM,
Krystal M.T. Woo, DVM, MVSC,
Peter Flowers, RVT,
Carlos L. de la Rosa, PhD

Session #3427

Affiliation: From University of California, Davis, School of Veterinary Medicine, 944 Garrod Drive, Davis, CA 94566, USA (Choo), Lindsay Wildlife Experience, 1931 First Avenue, Walnut Creek, CA 94597, USA (Woo, Flowers), Center for Plant Conservation, 15600 San Pasqual Valley Rd., Escondido, CA 92027, USA (de la Rosa).

Note: Abstract format selected by author(s). See Front Matter for description of conference proceedings manuscripts submitted as scientific abstracts.

Abstract: Over the past decade, studies at wildlife hospitals have worked to highlight leading causes of wildlife injuries, providing information to improve local conservation strategies.[1-4] Identifying spatial patterns of the incidence of these injuries for application in outreach and rehabilitation has previously been difficult due to the constraints of wildlife tracking and monitoring. This study visualized and analyzed data from a wildlife hospital using geographic information systems software to evaluate the types of injuries that occur in cities in the California San Francisco Bay area. Of approximately 62,000 cases evaluated, the primary reasons for intake in 60% of cases were identified using specific search features, word processing programs, and manual assessment. Among the categorized cases, leading causes included domestic attack by dogs and cats (25.3%), abandonment/abduction (21.4%), fall trauma (15.0%), window collision (6.6%), and vehicular collision (6.2%). Upon spatial analysis of these causes, the top ten cities contributing to wildlife patients at Lindsay Wildlife Rehabilitation Hospital (LW) were found to have statistically significant differences in their proportions of common reasons for presentation when compared to those of the overall population of cities serviced by LW and to each other. Domestic attacks by dogs and cats ranged from 18.1% to 30.9% of categorized cases from these cities, and window collisions from 3.2% to 13.0%. The results indicate that cities with similar numbers of wildlife patients can differ in the primary types of injuries presented. While linear regression did not reveal significant correlations between the proportions of individual injury categories and city characteristics as defined by distance from LW, city population, and city area, this study demonstrates potential for future studies to determine other combinations of predictors to identify risk probability of these injuries in a city of interest. By identifying these patterns, we aim to focus and optimize prevention efforts and future outreach approaches to better inform local communities.

References

1. Long RB, Krumlauf K, Young AM. Characterizing trends in human-wildlife conflicts in the American Midwest using wildlife rehabilitation records. *PLoS ONE* 2020;15(9):e0238805.

2. Romero F, Espinoza A, Sallaberry-Pincheira N, et al. A five-year retrospective study on patterns of casuistry and insights on the current status of wildlife rescue and rehabilitation centers in Chile. *Revista Chilena de Historia Natural.* 2019;92:6.

3. Schenk AN, Souza MJ. Major anthropogenic causes for and outcomes of wild animal presentation to a wildlife clinic in east Tennessee, USA, 2000-2011. *PLoS ONE* 2014;9(3):e93517.

4. Baker TM. Exploring public opinion of urban wildlife and effective use of urban wildlife education in Tucson, Arizona. Master's Dissertation: Prescott College in Environmental Studies, 2017.

Secret Tips for Intubation in Tiny Exotic Pets (Hamsters, Sugar Gliders, Chinchillas, Hedgehogs, Squirrels)

Chaowaphan Yinharnmingmongkol, DVM, Dipl, TBVM, and
Rapon Kanokpornwasin, DVM

Session #3244

Affiliation: From AnimalSpace Exotic Pet Hospital, 99/30 Puttamonthon Sai2, Taweewuttana, Salatummasop, Bangkok, 10170, Thailand.

Abstract: One of the most critical aspects of small mammal anesthesia is successful intubation. The risk of anesthesia-related complications is higher in small exotic pets, such as guinea pigs, chinchillas, hamsters, hedgehogs, and sugar gliders, and intubation may help facilitate a safer anesthesia.[1,2] During anesthesia, hypoxemia and apnea are common problems that can be alleviated through proper intubation. Endotracheal intubation in small exotic patients can be challenging. It is recommended to secure the airway during anesthesia. While using a reinforcement silicone tube works well for the blind intubation technique in rabbits, small exotic pets may prefer a blind method using a feeding tube. To intubate small exotic pets, the animal should be positioned in sternal recumbency with hyperextension of the neck. The mouth can be opened using a small thread to hook the incisor, and the feeding tube is inserted into the mouth, same as the line technique in rabbits. To ensure success, it is important to cut the feeding tube at a 45-degree angle and reduce the sharpness using a flame. The feeding tube method can be used to intubate hamsters, hedgehogs, sugar gliders, chinchillas, and other small exotic animals. The feeding tube is a cheaper and more flexible alternative to a silicone endotracheal tube, and its head can be connected to an endotracheal adapter. Overall, intubation is a critical aspect of anesthesia in small exotic pets, and proper technique can reduce the risk of complications. By using a feeding tube and blind technique, veterinarians can successfully intubate a range of small exotic pets and provide quality care while minimizing costs. These techniques are demonstrated during the presentation.

References

1. Bennett K, Lewis K. Sedation and anesthesia in rodents. *Vet Clin North Am Exot Anim Pract.* 2022;25(1): 211-255.

2. Gardhouse S, Sanchez A. Rabbit sedation and anesthesia. *Vet Clin North Am Exot Anim Pract.* 2022;25(1): 181-210.

Roles of Veterinarians in Rewilding Projects in Victoria, Australia

Patricia Macwhirter, BVSc (Hons), MA, PhD, FANZCVS, and
Madeleine Rowe, DVM, MANZCVS

Session #3163

Affiliation: From Melbourne Bird Veterinary Hospital, 1 George St, Scoresby, Victoria, Australia, 3179 and Healesville to Phillip Island Naturelink Inc.

Abstract: Human-caused species extinctions and global collapse of biodiversity are critical issues affecting our planet. Rewilding is a comprehensive conservation effort focused on restoring sustainable biodiversity and ecosystem health by protecting core wild areas, providing connectivity between such areas, and protecting or re-introducing keystone and other locally indigenous species. A key element of this work is re-establishing local ecological vegetation communities and their associated ecological webs including indigenous flora and fauna species, both vertebrate and invertebrate. Veterinarians have attributes, alongside their veterinary skills, that make them valuable team members for groups working to halt biodiversity collapse. These include organizational abilities, entrepreneurship, influence within their community, local knowledge and broad networks. Roles complementing traditional services where veterinarians can assist rewilding projects include advocating for landscape-wide rewilding, communicating global and local biodiversity goals, research into conservation assets and threatening processes, engaging in rewilding on private or public land and helping to navigate biodiversity incentives. In this talk the development of biolinks in the face of urban expansion near Western Port, a bay south-east of Melbourne, Victoria, Australia, is used as a case study.

Introduction

Species extinctions and global collapse of biodiversity are critical issues affecting our planet, the magnitude of which has not been seen since the extinction of the dinosaurs 65 million years ago.

- The United Nations (UN) most recent Climate Change 2023: Synthesis Report attributed, with high or very high confidence, changes in ecosystem structure, species ranges and seasonal timing to climate change caused by human activity.[1]

- World Wildlife Fund's Living Planet Report (2022) estimated an average decline of 69% in monitored wildlife populations since 1970.[2]

- The UNs' IPBES Global Assessment Report on Biodiversity and Ecosystems, 2019, noted the rate of global change in nature is unprecedented and direct drivers with the largest global impact have been changes in land and sea use; direct exploitation of organisms; climate change; pollution; and invasion of alien species.[3]

- Following European colonization, Australia has had one of the worst extinction records in the world, including the worst record for mammal extinctions, with greater than 10% of endemic terrestrial mammal species now lost.[4] Alien species, human activity and changed fire regimes, partly induced by climate change, have been key factors in local species declines.

Like other people, veterinarians have contributed to these existential problems. While we are not ecologists, understanding disrupted ecologies is centered in biological sciences and based on the same principles as epidemiology. These are areas where veterinarians are well trained and experienced. Given the magnitude of the problem, there is a compelling 'why' veterinarians should be part of teams finding and applying solutions to halt global species declines.

Rewilding is a comprehensive conservation effort focused on restoring sustainable biodiversity and ecosystem health by protecting core wild areas, providing connectivity between such areas, and protecting or re-introducing keystone and other locally indigenous species.[5] It is a paradigm shift in the way people interact with nature and seeks to address the direct drivers of biodiversity and ecosystem decline identified in the multiple reports noted above. For success, it needs scientific and local knowledge, community support and a focus on restoration of dynamic ecological webs.

Roles for Veterinarians

Veterinarians in private practices and wildlife facilities see injuries and disease in native species caused by human activities, as well as the close inter-species connections that humans form with pets and wildlife. We appreciate the benefits, and challenges, these connections have for people, wildlife, and the broader community. In the Australian context, cats, and dogs, whose ownership our profession encourages, have contributed to biodiversity collapse through predation and as feral populations. So, too, has farming and farm animals that have become feral (including pigs, goats, horses, and camel), and alien species that have acclimatized such as European rabbits, hare, deer, rats, and mice.

Many veterinarians help sick and injured wildlife by providing traditional veterinary care. While this is useful and rewarding on a micro-scale, it generally has negligible impact on global biodiversity collapse. However, veterinarians have additional attributes, alongside veterinary skills, that make them valuable team members for groups working to halt species extinctions. These include organizational abilities; entrepreneurship; influence within their community; local knowledge and broad networks. Roles complementing traditional services where veterinarians can assist include advocating for landscape-wide rewilding projects, communicating global and local biodiversity goals; researching conservation assets and threatening processes; engaging in hands-on habitat repair and helping to navigate biodiversity incentives. Conservation begins where you stand and, in this presentation, biolinks (wildlife corridors that connect larger habitat cores) being developed in the face of urban expansion near Western Port,[6] a bay south-east of Melbourne, Victoria, Australia, close to where the authors work and live, is used as a case study of roles veterinarians can play in rewilding projects.

Advocating For Landscape-Wide Rewilding

Western Port has a remarkable diversity of flora and fauna, many of which are iconic, threatened, or endangered species. It is home to breeding colonies of the little penguin (*Eudyptula minor*), koala (*Phascolarctos cinereus*), platypus (*Ornithorhynchus anatinus*), echidna (*Tachyglossus aculeatus*), southern brown bandicoot (*Isoodon obesulus*), Australian fur seal (*Arctocephalus pusillus*), hooded plover (*Thinornis rubricollis*), short-tailed shearwater (*Puffinus tenuirostris*), growling grass frogs (*Litoria raniformis*), migratory shorebirds and many more. In the 1970s a strong environmental movement emerged in Western Port in response to threats of major industrialization, including plans for a nuclear power plant. Strong activism, with veterinarians being minor players, led to a reversal of government policy. Western Port became recognized as a Ramsar Site of international significance[7] for migratory waders and, in 2002, it was designated the Mornington Peninsula and Western Port UNESCO Biosphere Reserve,[8] part of a global network of 738 reserves in 134 countries world-wide and one of only five Biosphere Reserves in Australia.

Compared with other jurisdictions, citizen advocacy has kept government support for conservation at the forefront in Victoria and, as a result, any direct exploitation of native wildlife is tightly controlled. However, populations are far from secure. Habitat loss from human expansion, development and extractive industries, climate change, pollution, and invasion of alien species (especially European red foxes and cats) are ongoing challenges. These themes are echoed in many places around the world.

Communicating Global and Local Biodiversity Goals

With species decline continuing unabated world-wide, acting both locally and globally is essential if the current extinction trends are to be halted. Online resources, including reports referenced earlier, are useful in identifying why, where, and how conservation efforts might best be prioritized, as well as giving global context for local conservation projects or species recovery plans.

The United Nations (UN) Sustainable Development Goals[9] are a collection of 17 interlinked objectives designed to serve as a shared blueprint for peace and prosperity for people and the planet now and into the future. They emphasize the interconnected environmental, social, and economic aspects of sustainability. There is a plethora of worthy opportunities across the UN Goals where veterinarians, with agile minds and interests in birds, animals, and science, can play positive roles.

Rewilding as a means of safeguarding global biodiversity falls primarily within the Life on Land and Climate Change UN Sustainability Goals. One of the key findings in the 2022 U.N. Environment Program report,[10] reviewing progress with these goals, was that environmental emergencies and human well-being need to be addressed together to achieve sustainability. The report also advocates that everyone has a role to play in ensuring that human knowledge, ingenuity, technology, and cooperation are redeployed from transforming nature to transforming humankind's relationship with nature. For some veterinarians the 'why' for local rewilding will not be as good a match as other worthy causes, but there may be synergies elsewhere on the planet where they might assist. For instance, a U.S. veterinarian helped crowd fund financial aid following the 2020 Victorian bushfires (Driggers T, Facebook and email communications, 2020) while Australian veterinarians have helped Indonesian colleagues in the Friends of the National Parks Foundation in Bali with their work with wildlife conservation.[11]

Successful rewilding projects have proven useful in addressing species declines, but there are significant challenges. These include starvation and native vegetation loss when herbivore populations are not effectively controlled or community backlash if predators dangerous to humans expand from the rewilding site. Based on difficulties with the Oostvaardersplassen project in the Netherlands,[12] which started in 1983 and aimed to recreate natural Pleistocene-like conditions, Kopina et al 2019 listed golden rules of rewilding as Cores, Carnivores, Corridors and Compassion, with an emphasis on putting the welfare of the animals used in rewilding experiments at the forefront.[13] Multiple sites in Australia where predator-proof fences or island introductions have excluded exotic predators such as cats and foxes (*Vulpes vulpes*) but not re-introduced indigenous predators such as dingos (*Canis lupus dingo*) or quolls (*Dasyurus* spp.) have seen population explosions and mass starvation, for instance with koalas on Western Port islands causing eucalyptus die back, or macropod overgrazing causing grassland ecosystem collapse.[14] The International Union for the Conservation of Nature discusses these and other challenges in its publication, Guiding Principles for Rewilding.[15]

Indigenous peoples' knowledge and perspectives, which often co-evolved with local flora and fauna, can offer valuable insights for rewilding projects, for instance First Nations Australians traditionally used fire-stick farming in shaping the Australian landscape. The cessation of regular burning after European colonization has contributed to the increased ferocity of Australian bush fires. Fuel-reduction burns in line with indigenous cultural practice are now commonplace. Managing the risk of fire is a complex challenge in rewilding projects, especially where connectivity is lacking, as fires are both dangerous to human lives and assets and can destroy critical wildlife populations.

Researching Conservation Assets and Threatening Processes

Understanding the unique flora, fauna and other conservation assets of a region, as well as threatening processes that are impacting upon them, is important both for advocacy and in undertaking rewilding projects. Town planning provides for roads and other built infrastructure, but it has not traditionally included provisions for conservation hubs and biolink connectivity for wildlife. This is an area where veterinarians can join with others to press for critical research and change.

In the context of Western Port, urban growth corridors eliminated known wildlife corridors and splintered habitats into isolated reserves. As a result, wildlife was over-exposed to dogs, cats, traffic, and foxes, and some species, such as bandicoots and koalas, became locally extinct and unable to repopulate. This was a catalyst that led a coalition of environmental groups initiating projects to identify possible biolinks (which often corresponded to drains and waterways) across the landscape. Counter-balancing developers, a citizen advocacy group called Healesville to Phillip Island Nature Link Inc, emerged to promote naturelinks across both private and public land so current and future generations will be able to live sustainably with Victoria's unique flora and fauna, mitigate climate change and enjoy positive benefits of connecting with nature.[16]

A landscape-wide, intergenerational, well-connected, citizen advocacy approach has been helpful in encouraging local government to adopt their own biolink plans. The Victorian state government has also supported this, for instance by developing and promoting NatureKit, a publicly accessible online mapping tool for biodiversity data integration and decision support.[17] Using NatureKit and the Victorian Biodiversity Atlas[18] it is possible to generate lists of local flora, fauna and ecological vegetation classes that land owners and managers can use in rewilding projects. Indigenous plant nurseries can supply plants on these lists to recreate historic ecological vegetation communities from farmland or degraded sites. As a result, landowners, including veterinarians, can embark upon rewilding projects even if their own knowledge of indigenous vegetation or ecological webs is limited.

While these are good starting points, knowledge of locally indigenous ecological vegetation classes, including their web of flora and invertebrate and vertebrate fauna, will always be limited. There are many simple opportunities to contribute to greater understanding, such as survey work using traditional or innovative methods (e.g., environmental DNA) to establish baseline information and information regarding response following rewilding efforts. Investigation of potential wildlife pathogens, including ones with zoonotic potential such as *Chlamydia psittaci*, is an area where veterinarians are well trained. Combining skills, it is possible to use both veterinary and ecological approaches to solve complex biodiversity challenges, for instance the discovery of chytrid fugus as the cause of global die-offs in frog populations by the team now based at Melbourne University.[19]

Hands-On Habitat Repair

Rewilding can involve protecting and 'restoring' indigenous ecological vegetation classes and their associated fauna communities or landscape remediation (e.g., of farmland or industrial sites) to re-establish such communities. Both provide healthy opportunities to get away from computers and re-connect with the natural world. Rewilding and fox control over the past 20 years on the author's (PM's) property Harewood has seen the expansion of three endangered species (southern brown bandicoot, growling grass frogs and swamp skink *Lissolepis coventryi*) and eleven indigenous ecological vegetation classes, six of them endangered. Environmental DNA studies are currently underway to better understand how ecological webs function in these different, threatened, communities.

Combining Endangered Species Recovery and Rewilding Programs

While endangered species recovery programs focus on individual species, the focus in rewilding programs is on re-establishing broad, functional ecological webs. These are complementary. Captive breeding programs need suitable habitats for species release if successful re-introduction is to occur and rewilding projects need captive breeding projects to support their work.

From a public relations perspective, it is often easier to promote saving an iconic bird or animal than it is saving a complex ecosystem. Joining forces with local artists, educators, and farmers to promote the plight of multiple species of endangered parrots kept as pets has been used by bird owners and veterinarians in developed countries in trying to protect these species in the birds' home ranges, for instance in the Caribbean and South and Central America where the global extinction crisis is particularly acute. Such efforts need to be backed up with in-depth understanding of the species habitat needs and human social context if protection or re-introduction efforts are to be successful.

In Victoria, veterinarians at Healesville Sanctuary, a state government native wildlife and ecotourism/education facility with a state-of-the-art wildlife hospital, have served in traditional roles as well as fostering, alongside other team members, core habitat areas and biolink connectivity for captive-bred endangered species release. Programs include those for helmeted honeyeaters (*Lichenostomus melanops cassidix*) and orange-bellied parrots (*Neophema chrysogaster*). A privately owned facility, Moonlit Sanctuary, with team support from private veterinarians, has also had success in breeding orange-bellied parrots. These parrots migrate from their natural breeding grounds on the west coast of Tasmania to coastal wintering grounds on the mainland, including Western Port. This species feeds on salt marsh, habitat that has been greatly reduced because coastal areas are prime real estate for human development. Cirocovirus has also been a problem in wild populations. Riparian forested areas near Melbourne, Helmeted Honeyeater's former stronghold, have also been impacted by forest clearing and urban expansion with invasion of more aggressive native and exotic species including noisy miners (*Manorina melanocephala*) and common mynahs (*Acridotheres tristis*). Re-establishing large enough areas of habitat for these endangered species, controlling predators and aggressive competitors, and addressing disease challenges are some of the multiple issues with which recovery teams, including veterinarians, are reckoning.

While these recovery efforts have meet with some success, status of other local species (including swift parrots, *Lathamus discolor,* and blue-winged parrots, *Neophema chrysostoma,* both of whom also migrate across the Bass Strait) have been in decline with re-classifications moving from near-threatened towards endangered. This demonstrates why it is critical to invest effort in near-threatened species before their status deteriorates and to review lists of regionally threatened species in a timely manner. When trying to preserve habitats, a stronger case can often be made when survey data is available, and a site is known to support multiple threatened species.

In the Western Port context, the Victorian Wader Study Group, which includes veterinarians amongst its diverse membership, has surveyed shorebirds since the 1970s. The citizen science data they have produced has been key in encouraging government to protect key areas for conservation.

Helping to Navigate Incentives for Biodiversity Projects

In December 2022, 188 countries adopted the Kunming-Montreal Global Biodiversity Framework (GBF) which set new targets aimed at halting extinction of known threatened species. It offers aid for biodiversity initiatives with financial incentives increasing, both in Australia and globally, through a GBF Fund and other avenues.[20] As one of 17 megadiverse countries that occupy less than 10% of the world's area but have 70% of biodiversity, Australia was a key player and donor.[21] Domestically, Australia committed to zero new extinctions and establishment of a voluntary Nature Repair Market to support landholders to restore and protect nature, which will operate alongside existing sources such as offsets funding provided by developers.

GBF initiatives will only succeed if communities, including veterinarians, take ownership and work to protect their indigenous flora and fauna through projects such as rewilding. This presentation has suggested possible roles to help individuals tweak their own imaginations on how they might work to reverse current extinction trends. It will need a great deal of creative and collective effort by many people around the world if future generations are to experience the beautiful natural areas and extraordinary biodiversity our generation has been so privileged to enjoy.

Acknowledgments: Thank you to Graeme Phipps from the R.A.R.E.S. Foundation for his review of this paper.

References

1. Intergovernmental Panel on Climate Change. Climate Change 2023: Synthesis Report. p. 7. Available at: https://www.ipcc.ch/report/ar6/syr/. Accessed June 1, 2023.

2. World Wildlife Fund. The Living Planet Report. 2022. Available at: https://livingplanet.panda.org/. Accessed June 1, 2023.

3. Brondizio ES, Settele J, Díaz S, Ngo HT, eds. *Global Assessment Report on Biodiversity and Ecosystem Services of the Intergovernmental Science-Policy Platform on Biodiversity and Ecosystem Services.* Bonn, DE: Intergovernmental Science-Policy Platform on Biodiversity and Ecosystem Services; 2019. doi.org/10.5281/zenodo.6417333.

4. Woinarski J, Burbridge A, Harrison P. Ongoing unravelling of a continental fauna: decline and extinction of Australian mammals since European settlement. *Proc Natl Acad Sci.* 2015;112(15):4531-4540. doi.org/10.1073/pnas.1417301112.

5. International Union for Conservation of Nature. Issues Brief: The Benefits and Risks of Rewilding. Available at: https://www.iucn.org/resources/issues-brief/benefits-and-risks-rewilding. Accessed June 1, 2023.

6. Wikipedia. Western Port. Available at: https://en.wikipedia.org/wiki/Western_Port . Accessed June 1, 2023.

7. The Ramsar Convention Secretariat. Wetlands of International Important. Available at: https://www.ramsar.org/about/wetlands-of-international-importance-ramsar-sites. Accessed June 1, 2023.

8. Western Port Biosphere. Mornington Peninsula and Western Port UNESCO Biosphere Reserve. Available at: https://www.biosphere.org.au/. Accessed June 1, 2023.

9. United Nations. Sustainable Development Goals. Available at: https://www.undp.org/sustainable-development-goals. Accessed June 1, 2023.

10. U.N. Environment Programme. Annual Report 2022. Available at: https://www.unep.org/resources/annual-report-2022. Accessed June 1, 2023.

11. Friends of the National Parks Foundation, Bali. Available at: https://www.fnpf.org. Accessed June 1, 2023.

12. Schwartz D. European Experiments in Rewilding: Oostvaardersplassen. Available at: https://rewilding.org/european-experiments-in-rewilding-oostvaardersplassen/. Accessed June 1, 2023.

13. British Association of Nature Conservationists. The golden rules of rewilding - examining the case of Oostvaardersplassen. *ECOS.* 2019;40(6).

14. Whisson D, Ashman K. When an iconic native animal is overabundant: the koala in southern Australia. *Conserv Sci Pract.* 2020;2:5e188. doi:full/10.1111/csp2.188.

15. International Union for the Conservation of Nature. Rewilding Thematic Group, Global Definition and Guiding Principles for Rewilding. Available at: https://www.iucn.org/our-union/commissions/group/iucn-cem-rewilding-thematic-group. Accessed June 1, 2023.

16. Healesville to Phillip Island Nature Link Inc. Available at: https://www.hpnl.org.au/. Accessed June 1, 2023.

17. State Government of Victoria. Energy, Environment and Climate Action. NatureKit. Available at: https://www.environment.vic.gov.au/biodiversity/naturekit. Accessed June 1, 2023.

18. State Government of Victoria. Energy, Environment and Climate Action. Victorian Biodiversity Atlas. Available at: https://www.environment.vic.gov.au/biodiversity/victorian-biodiversity-atlas. Accessed June 1, 2023.

19. Winthrope S. The invasive fungus threatening earth's biodiversity. Pursuit. 2019. Available at: https://pursuit.unimelb.edu.au/articles/the-invasive-fungus-threatening-earth-s-biodiversity. Accessed June 1, 2023.

20. Convention on Biodiversity. COP15: Final Text of Kunming-Montreal Global Biodiversity Framework. Available at: https://www.cbd.int/article/cop15-final-text-kunming-montreal-gbf-221222. Accessed June 1, 2023.

21. The High Ambition Coalition for Nature and People. Available at: https://www.hacfornatureandpeople.org/home. Accessed June 1, 2023.

How Non-Traditional Companion Animals Impact Climate Change

Gianluca Deli, MRCVS, DVM, CertAVP (ZM), GPCert (ExAP), PgCert (EAS), GpAdvCert (SMP), PgCert (HE), FHEA, VetGD Advisor, and Emanuele Lubian, DVM, GPCert (ExAP)

Session #3292

Affiliation: From Medivet Faygate, Tower road, Faygate, Horsham, RH12 4SD, UK (Deli), Clinica MyPet, Viale D. Ranzoni, 10 20149 Milano, IT (Lubian).

Note: Abstract format selected by author(s). See Front Matter for description of conference proceedings manuscripts submitted as scientific abstracts.

Abstract: This presentation will go through a wide-ranging bibliographic review and a reflection on how patients, non-traditional companion animals (NTCAs), can directly and indirectly affect climate change. The intent is to show how the maintenance and care of these animals can significantly affect the balance of our planet, much more than one might expect. The problem also is seen from the human point-of-view, where various factors come into play, such as natural disasters, epidemic, war, which further aggravates the problem. This is to underline how responsible management can also bring benefits and positive changes in the long term: preferential exploitation of local resources, reduction in the movement of goods (production of food in developing countries), breeding of mutations and their potential liberation (species allochthonous), reduction of the impoverishment of flora and fauna. Veterinary surgeons can play an important role both in daily clinical activity and in guiding and educating clients towards a more sustainable management of NTCAs.

References

1. United Nations. UN News: 5 things you should know about the greenhouse gases warming the planet. Available at: https://news.un.org/en/story/2022/01/1109322?gclid=Cj0KCQjwk7ugBhDIARIsAGuvgPY86dtb4iALK66YWyBiuQIke-C3kXQxpEewJQOLdMpMdniY0DUKBv8aAp7bEALw_wcB. Accessed June 1, 2023.

2. Protopopova A, Ly LH, Eagan BH, Brown KM. Climate change and companion animals: identifying links and opportunities for mitigation and adaptation strategies. *Integr Comp Biol.* 2021;61(1):166-181.

3. Benessaiah K, Sengupta R. How is shrimp aquaculture transforming coastal livelihoods and lagoons in Estero Real, Nicaragua? The need to integrate social-ecological research and ecosystem-based approaches. *Environ Manage.* 2014;54(2):162-179. doi:10.1007/s00267-014-0295-x. PMID: 24912580.

4. Nemecek T, Kägi T. Life Cycle Inventories of Agricultural Production Systems. 2007. Available at: https://db.ecoinvent.org/reports/15_Agriculture.pdf. Accessed June 1, 2023.

5. World Association of Zoos and Aquariums. A Guide. Reducing, Measuring, and Offsetting Carbon at Your Zoo or Aquarium. Available at: https://www.waza.org/wp-content/uploads/2022/09/WAZA-Carbon-Guide_WEB.pdf. Accessed June 1, 2023.

Section 15

Hybrid Special Session

Use of 3D Printed Surgical Models to Aid Exotic Animal Surgeries

Angela M. Lennox, DVM, DABVP (Avian Practice), DABVP (Exotic Companion Mammal), DipECZM (Small Mammal), and
Sean Bellefeuille

Session #3191

Affiliation: From Avian and Exotic Animal Clinic, 9330 Waldemar Road Indianapolis, IN 46268, USA (Lennox), Med Dimensions, LLC, Ontario, NY 14519, USA (Bellefeuille).

Abstract: The use of 3D printed anatomic and surgical models are useful for specific skill learning and can be used for complicated surgeries where a clear understanding of anatomy is critical for surgical success. A systematic review and meta-analysis of medical students learning with models showed increased satisfaction and equal or improved test scores than students learning from cadavers.[1] 3D print models are used successfully and being investigated in human surgery, for example for planning patient specific liver resection where disease and anatomic variation make surgical decisions challenging.[2] In veterinary medicine, a study showed similar learning results for veterinary students using horse limb bones of an actual specimen versus a 3D printed model.[3] Use of 3D model for teaching purposes and for surgical management of several cases are presented.

Models

Models were printed from computed tomography (CT) scans by Med Dimensions (med-dimensions.com). Med Dimensions uses patient medical images (CT, magnetic resonance imaging [MRI]) to create models, tools or implants to improve surgical accuracy and clinician confidence. Volumetric data from CT or MRI scans are used to extract anatomy of interest and create 3D reconstructions that can be analyzed and manipulated digitally. The 3D model is used to create either a 3D print for surgical preparation/rehearsal or a model in a computer-aided design software to create patient-specific tools. These tools can be cutting/drilling guides or can be custom implants created to fit the patient perfectly. The tools/implants are then 3D printed as well, typically in a different material than the models.

Models are typically 3D printed in plastic using an additive manufacturing method called fused deposition modeling, commonly known as extrusion based printing or filament printing. The materials can be multicolored to point out regions of interest or specific lesions like a tumor. Intraoperative tools like guides or implants are typically printed using more advanced methods such as stereolithography and they are typically made out of resins, metals or polymers. Most of these materials can be autoclaved which is very important to bring them into the operating room safely.

Applications

The applications for this technology are very broad. Orthopedics, neurosurgery or oncology come to mind for surgical preparation and execution. Educational models have been increasingly popular and in demand. With 3D printing and other associated technologies it has become easier to create models that are anatomically accurate and cost-effective. Recently, Med Dimensions has been exploring the combination of silicone molding with 3D printed parts to create multi material (multiple tissue type) models that reflect the anatomy and feel more realistic. An example of this application would be intubation models that have a realistic silicone airway.

References

1. Zhen Y, Aisher D, Hanming J, et al. The role of 3D printed models in the teaching of human anatomy: a systematic review and meta-analysis. *BMC Med Educ.* 2020;20(1):335.

2. Huber T, Hanke LI, Boedecker C, et al. Patient-individualized resection planning in liver surgery using 3D print and virtual reality (i-LiVR)-a study protocol for a prospective randomized controlled trial. *Trials.* 2022;23(1):403.

3. De Alcantara Leite dos Reis D, Gouveia BLR, et al. Comparative assessment of anatomical details of thoracic limb bones of a horse to that of models produced via scanning and 3D printing. *3D Print Med.* 2019;5(1):10-13.

Section 16

Hybrid
Marla Lichtenberger
Emergency and Critical Care Session

Salina Locke, DVM, DABVP (Avian Practice)
Session Chair

Stayin' Alive 3. You Can't Always Get What You Want: Brass-Tacks versus Gold-Standard Diagnostic and Treatment Options

Meera Music, BVM&S, MRCVS,
Alicia McLaughlin, DVM, CertAqVet,
Anneliese Strunk, DVM, DABVP (Avian Practice)

Session #3268

Affiliation: From Center for Bird and Exotic Animal Medicine, 11401 NE 195th St. Bothell, WA 98011, USA.

Abstract: Exotic animals commonly present to veterinary clinics on an emergency basis. Both acute and chronic diseases may manifest with an acute onset of symptoms due to the tendency of exotic species to hide signs of illness until it is advanced. Accurate diagnosis of the underlying cause for illness is essential to provide appropriate treatment, but not all owners can do all the recommended testing. In this presentation, authors discuss how to optimize the clinical approach to the critically ill exotic animal regardless of budget provided, and review common emergency presentations and recommended approaches for exotic mammals, reptiles, and birds.

Introduction

This presentation is intended to highlight the broad spectrum that exists within exotic emergency medicine between the brass-tacks approach needed to establish prognosis and guide treatment plans, and the gold-standard approach that often allows for a deeper understanding of a patient's disease process and improved outcomes for clients and patients. There is no one-size-fits-all approach that will work in every case, and it is possible to practice evidence-based medicine on a budget.[1-11] Flexibility and adaptability are important.

Pet owners who are dealing with financial hardship are often very connected to their pet(s). While they may not have monetary resources, they may be able to commit a significant amount of time and home management to their pets' medical care. When faced with a pet's illness, some of these pet owners may elect euthanasia preemptively to try to prevent pain and suffering, or from a lack of understanding of what long-term care may entail, even if there are inexpensive management options available. Clinicians who have less experience dealing with complicated health issues/diagnostic approaches on a budget may also jump to recommending euthanasia prematurely. While quality of life concerns must always be assessed carefully, and explicitly, it is the veterinarian's duty to advocate for the animal when appropriate. This includes discussing multiple management options and not making assumptions about what treatments owners are willing to consider. Clinicians have access to a wealth of information from the detailed history/husbandry assessment and physical examination alone. Thoughtfully interpreting these findings based on current scientific literature will allow the clinician to prioritize the case approach based on evidence.

Euthanasia

There are some situations where it is reasonable to consider euthanasia without diagnostics or attempting treatment. These can include the following situations.

- Severe disease or injury resulting in unrelenting pain and suffering, regardless of an owner's financial capability.
- When pet owners are unable or unwilling to commit the time and resources for even the bare minimum of diagnostics or ongoing care.
- Animals with multiple concurrent diseases identified on physical exam that are showing signs of declining quality of life.

Many diseases can be managed adequately with less than gold-standard care as long as a considerate approach is taken. While euthanasia may be a consideration in a budget case, it is possible for many exotic pets to survive even catastrophic disease processes with inexpensive case management, if the owner is dedicated to providing appropriate home care and follow ups as recommended. For example, an animal with traumatic exophthalmia or ocular rupture may be able to heal with medical management and develop a phthisical or comfortable eye without extensive surgical care.

Managing Expectations

A long-term plan for monitoring and treatment must be discussed from the outset. For example, dental disease in rabbits and rodents typically requires lifelong management with ongoing anesthetized occlusal adjustments and if owners are not properly prepared for this, it can result in decreased quality of life for the pet, significant disease progression, and client dissatisfaction. From the moment of diagnosis, owners should be provided with details about the need for and costs of the long-term management required for the rest of the patient's life.

Embarking on treatment for diseases that require significant home management (e.g., wounds and metabolic diseases like chronic renal failure and insulinoma) should include a full discussion about what home management and ongoing bare minimum monitoring entails. In some cases, rechecking blood work may be a vital part of disease management (e.g., monitoring glucose levels in a ferret with insulinoma). A stable cardiac patient with disease that has been historically well managed may be able to skip monitoring blood work every 6 months, even if this is the usual standard of care. This is dependent on the species involved, medical history of the individual patient, and the disease being treated. It is important to clarify to the client that management plans may need to be modified depending on disease progression.

Minimizing Stress

Minimizing stress should continue to be a part of baseline case management, including the use of minimal stress handling and medicating techniques and sedation when appropriate. Fear-free methods can be utilized by staff in hospital and be a part of owner home care. An owner with the ability to commit time towards training can make a huge difference in the quality of life for a pet requiring ongoing home care. For example, there are some bird owners who will not medicate their birds directly and use inadequate measures such as putting the medication on food items, which will likely be incompletely consumed, or lead to consumption of excessive treat foods. Prescribing training for owners to learn how to teach their birds to accept medication willingly allows for consistent and low-stress management of diseases.

Antibiotic Usage

Exotic animal medicine on a budget should not entail indiscriminate antibiotic usage. Every veterinary clinic should develop established guidelines for antibiotic use.[1,2] Although empirical medications may be necessary in some cases, boundaries and expectations must be set for the owner at the initiation of empirically used medications, to promote responsible use. If an infection is suspected, and culture or DNA sequencing of potential pathogens are out of budgetary range, cytology should be performed of lesions as a low-cost, high-yield diagnostic test to help guide antimicrobial usage, if any.

Management Considerations

When a client expresses concern about finances impacting medical care of an animal, the following steps can be very helpful.

- Get a budget from the client to assess whether it is reasonable to attempt management of the disease. Find out if there is any flexibility for financial input later or long term for the pet.

- Be sure to obtain clear, detailed histories including going through husbandry with a fine-toothed comb. Little things like variations in husbandry or timing of medication administration, among others, can make a big impact on treatment outcomes.

- Review the scientific literature, talk to colleagues, and search VIN and other online exotic forums for advice on managing the medical condition inexpensively.

- When making decisions about prescribing medications, think about the timing around starting or stopping administration of medications so you can easily evaluate if the treatment is helping the pet (e.g., therapeutic trials).

- Thoughtfully time follow-up visits based on the ability to have maximum impact for the visit; plan the recheck around an expected time to recheck diagnostic testing, and/or before the course of medications has been completed so that you can decide whether treatment should continue. Always provide written communication to clients. Consider video demonstrations of medication administration or other home treatments so that clients can refer to these and minimize the risk of mistakes in implementation.

- Provide the client with clear, objective measurement methods (e.g., weight, respiratory rate, quality of life assessments) to help keep track of how the pet is doing at home.

- Don't stop recommending and discussing high-quality medicine, even if a client expresses financial constraints. The goal is not to pressure the client, but to make sure that they are educated about possible negative outcomes associated with missing a concurrent or primary diagnosis due to the current case limitations. This will minimize surprises for them along the way, improve satisfaction in the communication process, and give the client the option to change their mind over time about the level of medicine they desire to pursue. Document all recommendations thoroughly in the medical record.

Different diagnostics can range significantly in significance of information gathered to treatment decisions, cost to the client, and possible risk to the patient (e.g., sedation or anesthesia needed, phlebotomy). Being able to convey clearly and unambiguously the pros and cons of different diagnostics is a vital skill for any clinician. This will help an owner to prioritize their finances in the face of a significant illness, as well as helping the clinician prioritize order of testing. Table 1 describes considerations for diagnostic tests that can be useful in management of budget cases.

Table 1. Diagnostic pros and cons.

Test	Cost[a]	Risk to patient	Time to take diagnostics, scheduling	Need for sedation, anesthesia	Potential uses	Other considerations
Single lateral radiograph (including entire caudal aspect of the patient)	$	Minimal	Immediate	Likely not needed.	Mammals straining to urinate, defecate or with hematuria.	May be stressful if patient is dyspneic (orthopneic posture exhibited). Lack of ability to triangulate (i.e., in the case of a ureteral stone that overlies the bladder).
Single ventrodorsal/ dorsoventral radiograph / Single "bird in a box" or "plop" radiograph	$	Minimal	Immediate	Likely not needed.	Useful to identify presence of mineralized, metallic structures or get assessment of respiratory tract, cardiac size. Helps minimize stress on dyspneic animals because it does not require handling.	Without orthogonal view, definitive interpretation of superimposed tissues/bone is not possible (e.g., stone in caudal urinary tract may be difficult to identify).
Orthogonal radiographs (2-3 view)	$$	Moderate, depending on species and stability of patient.	Immediate	May be needed.	Important to prevent misdiagnosis of a variety of disorders.	May be stressful if patient is dyspneic.
In-house ultrasound (ranging from AFAST/TFAST/FLASH or single organ scans to full abdominal or thoracic scans depending on clinician skill, comfort	$	Minimal	Immediate	Likely not needed.	Rabbits with anemia/elevated ALT; animals with urinary signs; coelomic or abdominal distention; increased respiratory effort.	Useful to identify source of mass effects, presence or absence of blood flow in an organ or mass, bladder or renal abnormalities, cystic or follicular structures, or free fluid within the thorax or abdomen.

Table 1 Continued. Diagnostic pros and cons.

Test	Cost[a]	Risk to patient	Time to take diagnostics, scheduling	Need for sedation, anesthesia	Potential uses	Other considerations
Echocardiogram, cardiologist consult	$$$	Moderate, depending on species and stability of patient.	Can take a week or more to coordinate with specialist.	May be needed.	Help confirm diagnosis and guide treatment of a variety of cardiac disorders.	May be difficult to arrange, depending on specialist availability. Specialists may require clinic to facilitate sedation/anesthesia for any procedures performed.
Full abdominal or thoracic ultrasound (by a radiologist or with consultation with a radiologist)	$$$	Moderate, depending on species and stability of patient.	Can take a week or more to coordinate with specialist.	May be needed.	Useful when searching for subtle lesions.	May be difficult to arrange, depending on specialist availability. Specialists may require clinic to facilitate sedation/anesthesia for any procedures performed.
Computed tomography (CT)	$$$	Moderate, depending on species and stability of patient.	Immediate if machine in house. Can take a week or more to coordinate with another facility.	Likely needed (aside from most non-contrast chelonian studies).	Useful for diagnosing a wide range of subtle disorders in patients of all sizes.	External facilities may require clinic to facilitate sedation/anesthesia for any procedures performed.
Cytology (in-house)	$	Minimal	Immediate	May be needed to obtain sample.	Can help guide antimicrobial therapy and inform prognosis.	Skilled staff needed, or doctor clinician may need to read themselves.
Fecal occult blood	$	None	Immediate	None needed.	Can confirm presence of GI bleeding in non-carnivores.	There may be interference from carnivore diets.
Urinalysis	$	Minimal-moderate, depending on if sedation or anesthesia is needed to obtain sample.	Immediate	May be needed.	Useful for assessing presence of pyuria, crystalluria, renal function, ketoacidosis.	None

Table 1 Continued. Diagnostic pros and cons.

Test	Cost[a]	Risk to patient	Time to take diagnostics, scheduling	Need for sedation, anesthesia	Potential uses	Other considerations
Cultures, next-generation DNA sequencing (MiDog)	$$$	Minimal-moderate, depending on if sedation or anesthesia is needed to obtain sample.	Prolonged time until results (7-10 business days in most cases).	May be needed to obtain sample for submission.	May be necessary to achieve good outcomes in cases with resistant infections.	MiDog provides resistance patterns based on genetic sequencing rather than sensitivity testing. Both culture and sensitivity and MiDog should be performed if a resistant or atypical infection is suspected.
In-house blood smear evaluation	$	Minimal-moderate, depending on if sedation or anesthesia is needed to obtain sample.	Immediate	Does not typically require sedation, depending on phlebotomy site.	Useful as a patient-side diagnostic test to determine if antibiotic therapy should be considered.	Skilled staff needed, or doctor clinician may need to read themselves. Very small sample needed in most cases (< 100 µl).
CBC (send out)	$	Minimal-moderate, depending on if sedation or anesthesia is needed to obtain sample.	Immediate with in-house point-of-care machine. 1-3 days for mammals, 3-5 business days for reptiles and avian patients.	Does not typically require sedation, depending on phlebotomy site. Very small sample needed in most cases (< 100 µl).	May be more accurate than in-house analysis depending on skill level of staff reading in-house labs. Useful as a second opinion.	Does not typically require large volumes of blood.
Point-of-care analyzer (iStat 1, Abbott Point of Care Inc. Princeton, NJ, USA)	$$	Minimal-moderate, depending on if sedation or anesthesia is needed to obtain sample.	Immediate	May require sedation, depending on phlebotomy site.	Useful for assessing specialized analytes immediately (glucose, ionized calcium, blood gasses).	Very small sample needed (< 100 µl) but must be unclotted.

Table 1 Continued. Diagnostic pros and cons.

Test	Cost[a]	Risk to patient	Time to take diagnostics, scheduling	Need for sedation, anesthesia	Potential uses	Other considerations
Glucometer	$	Minimal	Immediate	May require sedation, depending on phlebotomy site.	Rapid assessment possible.	Significant variations in precision and accuracy depending on species and machine.
Chemistry (in-house)	$$	Minimal-moderate, depending on if sedation or anesthesia is needed to obtain sample.	Immediate with in-house point-of-care machine.	May require sedation, depending on phlebotomy site.	Excellent way to assess prognosis quickly for some patients on emergency.	Will not include some analytes, depending on which rotors/test slides/cartridges are stocked/methodolog y use. May be challenging to get enough blood to run in house depending on patient size and machine in use.
Chemistry (send out)	$$	Minimal-moderate, depending on if sedation or anesthesia is needed to obtain sample.	1-3 days in most cases.	May require sedation, depending on phlebotomy site.	Same uses as in-house chemistry, but delayed results.	May be challenging to get enough blood to run in house depending on patient size. Volume of blood obtained will determine number of analytes able to be tested.
Surgery (mass removal, exploratory surgery with biopsies	$$$	High	Soon after need is identified, depending on availability of on-site surgical services.	Yes	May be the only way to diagnose and treat the underlying disease.	Underutilized tool. Consider surgical intervention as primary diagnostic in all-or-nothing cases. Risk may outweigh benefits in some patients.

[a]$ = < $100; $$ = $100-350; $$$ = > $350.

References

1. McLaughlin A, Music M, Strunk A. Stayin' alive: current topics in exotic emergency medicine. *Proc ExoticsCon.* 2019;20-33.

2. Music M, McLaughlin A, Strunk A. Stayin' alive 2: triage, medicine, and surgery. *Proc ExoticsCon.* 2022;489-499.

3. Music M, Strunk A. Reptile critical care and common emergencies. *Vet Clin North Am Exot Anim Pract.* 2016.

4. McLaughlin A, Strunk A. Common emergencies in small rodents, hedgehogs, and sugar gliders. *Vet Clin North Am Exot Anim Pract.* 2016;19(2):465-499.

5. Graham JE, Doss GA, Beaufrere H. *Exotic Animal Emergency and Critical Care Medicine.* 1st Ed. Ames, IA: Wiley-Blackwell; 2021.

6. Block G. Doing the most for our patients with the least: evidence-based veterinary care on the cheap. Available at: https://hsvma.memberclicks.net/assets/pdfs/veterinarycareonthecheap_handout.pdf. Accessed June 1, 2023.

7. Enzler MJ, Berbari E, Osmon DR. Antimicrobial prophylaxis in adults. Mayo Clin Proc 2011;86(7): 686-701.

8. Buisson C. Getting comfortable with hospice and palliative medicine. *Proc Southwest Vet Symp.* 2021.

9. Lichtenberger M, Lennox A. Critical care of the exotic companion mammal (with a focus on herbivorous species): the first twenty-four hours. *J Exotic Pet Med.* 2012;21:284-282.

10. Costello M. Principles of cardiopulmonary cerebral resuscitation in special species. *Semin Avian Exot Pet Med.* 2004;13(2):132-141.

11. Lichtenberger M. Shock and cardiopulmonary-cerebral resuscitation in small mammals and birds. *Vet Clin North Am Exot Anim Pract.* 2007;10:275-291.

12. American Veterinary Association. Antimicrobial use in veterinary practice. Available at: https://www.avma.org/resources-tools/one-health/antimicrobial-use-and-antimicrobial-resistance/antimicrobial-use-veterinary-practice/. Accessed March 24, 2023.

13. American Veterinary Association. AAFP/AAHA antimicrobial stewardship guidelines. Available at: https://www.avma.org/resources-tools/avma-policies/aafpaaha-antimicrobial-stewardship-guidelines/. Accessed March 24, 2023.

Containing Contagion and Understanding Zoonotic Disease Potential of Wildlife Species

Julia K. Whittington, DVM

Session #3404

Affiliation: From University of Illinois College of Veterinary Medicine, 1008 W. Hazelwood Dr., Urbana, IL 61802, USA.

Abstract: Avian influenza, salmonellosis, plague, mange, tularemia...and the list goes on. Exposure to infectious diseases, some of which are zoonotic, is likely in wildlife rehabilitation and care as many wild animals are affected by or carry these pathogens. Understanding disease transmission and preparing for the introduction of a potential pathogen into a health care facility is of utmost importance for biosafety and biosecurity. This presentation introduces important concepts in preventing disease transmission during wildlife care.

Wildlife, Infectious Diseases, and Zoonoses

Over 60% of known human pathogens and 75% of emerging diseases that affect humans are zoonotic. More than 50 zoonotic diseases are found in the United States alone, and this list is published as an appendix to the Compendium of Veterinary Standard Precautions for Zoonotic Disease Prevention in Veterinary Personnel.[1] Wildlife species are more likely to be associated with zoonotic diseases than domestic animals and over 80% of the zoonotic diseases found in North America are also found in wildlife.[1,2] Exposure to zoonotic diseases is an occupational hazard of veterinary professionals, and those working with live or dead wildlife species are at greater risk of coming into contact with potentially zoonotic pathogens.[3] Zoonotic diseases commonly associated with North American wildlife include hantaviruses, plague, rabies, and tularemia.[3]

Transmission of infectious pathogens requires a source of infection, a susceptible host, an effective route of transmission and a point of entry (Table 1).

Table 1. Route, point of infection, and high-risk activitie3s associated with transmission of zoonotic diseases of wildlife.

Transmission route	Point of infection	High-risk activities	Examples of disease exposure
Direct contact	Ingestion, contamination of mucous membranes, contact through non-intact skin	Handling animal without proper restraint or use of personal protective equipment (PPE)	Brucellosis, plague, rabies, *Salmonella* spp., tularemia, scabies, avian influenza
Indirect contact	Ingestion, contamination of mucous membranes, contact through non-intact skin	Handling contaminated equipment or material without proper PPE	Brucellosis, plague, rabies, *Salmonella* spp., tularemia, scabies, avian influenza
Aerosol	Inhaled particles or droplets deposited on mucous membranes	Power washing contaminated surface without PPE, handling animals with infectious respiratory disease without proper PPE	Hantaviruses, highly pathogenic influenza, plague
Vector-borne	Bites from an infected arthropod	Working with animals or in an environment without appropriate vector control	Lyme disease, plague, tularemia, Rocky Mountain spotted fever, West Nile virus

Source: Taylor T, Buttke D. Safe work practice for working with wildlife. In: Richgels KLD, Gibbs SEJ, Wild MA, eds. *Field Manual of Wildlife Diseases.* Reston, VA: U.S. Geological Survey; 2020:1-26.

Wildlife is susceptible to a myriad of infectious pathogens. In some cases, large-scale die-offs may be experienced during an outbreak event, and in other cases, individual animals may experience clinical signs while most of the population remains unaffected. Asymptomatic carriers of infectious disease are common in wildlife species. Therefore, it is prudent to maintain the mindset that a wildlife patient has the potential to be at risk for transmitting a zoonotic agent. This index of suspicion is the driving factor during outbreak investigations where the initial steps involve determining if the problem is an individual incident or a population problem. Early identification and quick implementation of a planned intervention can reduce the risk and severity of the incident or outbreak. During this process of case definition and classification, the veterinarian must define the cause of illness in the animal, confirming the diagnosis, while also looking for trends that pertain to the animal type, timing of the presentation, and location, that may indicate that the problem is at the population scale.[4] It is important to be familiar with the diseases of wildlife found in the region and any signs or symptoms seen in animals and humans. It is also prudent to stay abreast of local wildlife trends, including unusual mortality events. Maintaining open communication with wildlife care professionals in the area, as well as with local natural resources managers, will allow for networking and more rapid detection of an emerging issue.[3]

An incomplete summary of zoonotic agents of concern in wild animals is found in Table 2.

Table 2. Incomplete list of zoonotic diseases associated with wildlife of North America. Causative agent, host, route of transmission and associated symptoms are listed.

Disease or agent	Host	Transmission, incubation	Signs and symptoms of disease in humans
Campylobacteriosis *C. jejuni*	Multiple	Fecal-oral, 1-10 days	Diarrhea, ± mucous and blood, nausea, fever
Cryptosporidiosis *Cryptosporidium* spp.	Multiple	Fecal-oral, 3-7 days	Abdominal pain and cramping, profuse water diarrhea, anorexia
Dermatomycosis *M. canis, Trichophyton* spp.	Multiple	Direct, indirect contact, 4-14 days	Focal, flat, spreading circular skin lesions with peripheral crusted, scaly redness
Erysipiloidiosis *E. rhusiopathiae*	Rodents	Direct contact, 1-7 days	Local raised, reddened skin lesions, progressive, painful and itchy
Giardiasis *Giardia* spp.	Multiple	Fecal-oral, 1-4 weeks	Abdominal cramps, bloating, anorexia, nausea, intermittent diarrhea
Hantavirus Bunyaviridae	Rodents	Aerosolized, direct contact, 12-16 days	Severe pulmonary syndrome, fever, weakness, cough
Histoplasmosis *H. capsulatum*	Bats, birds	Aerosolized, 3-17 days	Asymptomatic, acute respiratory disease with chronic pulmonary pathology
Leptospirosis *L. interrogans*	Multiple	Direct contact, 7-12 days	Fever, headache, weakness, red and watery eyes
Lyme disease *B. burgdorferi*	Multiple	Vector-borne, 1-2 weeks	Bull's-eye rash, fever, muscle and joint pain
Lymphocytic choriomeningitis virus (LCMV) Arenaviridae	Rodents	Direct contact, Aerosolized, vector-borne, 1-3 weeks	Fever, weakness, headache, lymphadenopathy
Orf Poxvirus	Wild hoof stock	Direct, indirect contact, 3-6 days	Solitary skin lesion, discolored or pustular, progressing to weeping proliferative
Pasteurellosis *Pasteurella* spp.	Multiple	Direct, indirect contact, aerosolized, variable incubation	Local redness, swelling, pain, fever, can lead to septicemia
Plague *Y. pestis*	Rodents	Direct contact, vector-borne, 2-6 days	Fever, lymphadenopathy, septicemia
Psittacosis *C. psittaci*	Birds	Direct contact, aerosolized, 4-15 days	Fever, headache, weakness, respiratory disease, ± hepatitis
Q fever *C. burnetii*	Wild hoof stock	Direct contact, aerosolized, 2-4 weeks	Fever, headache, muscle aches, sweating
Rabies *Lyssavirus*	Mammals	Direct contact, 3-8 weeks	Headache, fever, malaise, sensory changes, excitability, behavior changes, hydrophobia
Salmonellosis *Salmonella* spp.	Multiple	Fecal-oral, 12-36 hours	Abdominal pain, diarrhea, headache, vomiting
Tularemia *F. tularensis*	Multiple, lagomorphs, rodents	Direct contact, aerosolized, 3-5 days	Dependent on point of entry, fever, headache, weakness, vomiting, diarrhea, pneumonia
Tuberculosis *Mycobacterium* spp.	Multiple	Aerosolized, direct contact, 4-12 weeks	Cough, weight loss, fatigue, fever

Biosecurity

Biosecurity refers to both management practices and physical measures taken to reduce the risk of introduction, establishment and spread of disease-causing biological agents, either from natural or intentional exposures, to humans, animals, plants, and the environment.[5] Biosecurity is based on the principles of preparation, prevention, response, and recovery. Basic to any biosecurity protocol are two main tenets: containment and prevention. These are achieved through identification of potential infectious agents, prevention of their introduction into populations or facilities, and control of their transmission through isolation, treatment, quarantine, or removal of affected subjects. Wildlife care facilities are presented with a unique challenge. Caregivers must protect hospitalized patients and staff from potential infectious disease exposure while providing treatment for injured, ill, or orphaned wild animals that may be the source of exposure.

Facility preparedness for dealing with the risk of infectious disease or zoonotic disease exposure will depend on its biosecurity, the monitoring of disease trends in wildlife populations, and open communication with staff, diagnostic laboratories, wildlife rehabilitators, wildlife biologists, and government agencies.

Standard Precautions

Standard precautions are a set of practices put in place to prevent the transmission of infectious diseases through blood or bodily fluids. These standards are universal in that they typically refer to basic measures needed to prevent human-to-human transmission of infection; however, the standards are considered best practice in the clinical veterinary setting as well. These include the following actions.

- Use protective barriers such as gloves and protective outerwear.
- No eating, drinking or smoking when working with animals or in animal care areas.
- Practice good hand hygiene.
- Avoid activities that increase risk of bites, scratches, needle sticks or other injuries.
- Provide good ventilation of the work environment.
- Disinfect soiled and contaminated surfaces and equipment.
- Handle tissues and biological samples safely.
- Seek medical attention when ill and advise care provider of potential infectious disease exposure.

Minimum standards for biosecurity are available; however, individual facilities must undergo their own risk assessment to personalize procedures. It is important to note that no biosecurity measure is 100% effective and each measure, if adhered to, adds layers of protection to the total plan (Fig 1). Modification of the physical design of an environment, although difficult and costly, is the most effective means of biosecurity enforcement as it does not rely on human behavior or compliance. Control measures used to mitigate the potential for zoonotic disease transmission vary in effectiveness and are cumulative. Those that remove the risk or replace the risk with a less risky option are preferred to those that simply manage the risk. Prevention strategies that utilize control measures to interrupt potential pathogen transmission at multiple points along the continuum of source, susceptible host, effective route, and point of entry are preferred.[1]

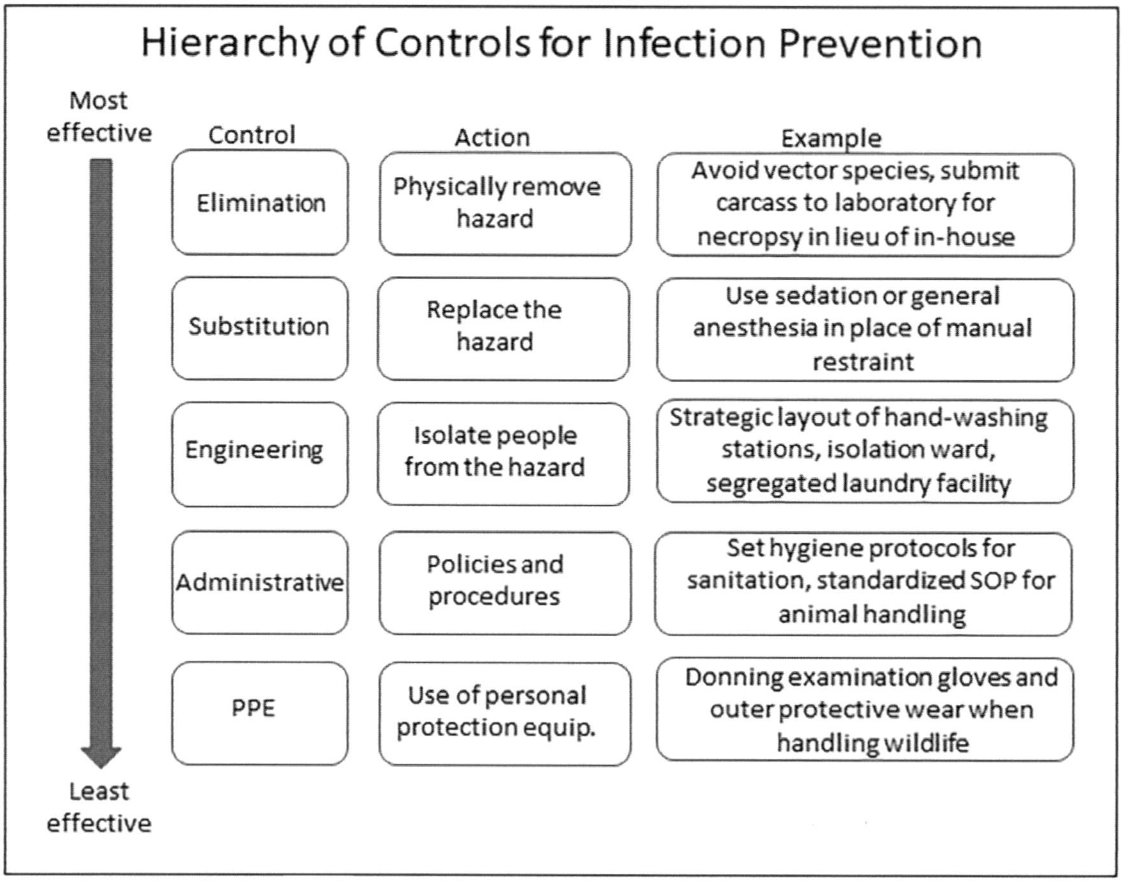

Figure 1. The hierarchy of controls for infection prevention indicating most effective to least effective control measure. Control measures are additive in effectiveness and should be implemented as needed to mitigate risks of exposure to infectious pathogens.

Hand hygiene

Hand hygiene is the single most important veterinary standard precaution in the prevention of disease transmission. Hand hygiene includes hand washing with soap and water or the use of alcohol-based hand rubs. Hand hygiene is important when handling potentially infectious animals or being in contact with contaminated materials or surfaces. It is important to practice good hand hygiene between handling individual animals or groups (e.g., litter or clutch).

Hand washing

Most zoonotic pathogens are transient flora and reside in the upper layers of the stratum corneum of the skin. The mechanical abrasion experienced during hand washing in conjunction with the detergent properties of soap effectively remove or deactivate most potential pathogens present on the hands. Commercial soap is appropriate for hand-washing stations, but bacterial overgrowth is possible with refillable containers that are not emptied and cleaned well prior to refilling. The temperature of the water used during hand washing is only important in promoting good compliance with the technique but has no effect on actual microbe removal. Moisturizing soaps are a reasonable option that may help hands from becoming chaffed. Nails should be kept short and clean. Artificial nails and hand jewelry that can harbor potential pathogens should be avoided when working with wildlife. The correct hand washing procedure involves wetting the hands with running water, placing soap in the palms, and rubbing the hands together to make lather. All surfaces should be engaged for a total of 20-30 seconds. The hands are then rinsed with running water and dried using a disposable towel. The towel can be used to turn off the faucet if there is concern that the faucet handle is contaminated.

Alcohol rubs

When soap and a water source is unavailable, alcohol-based hand rubs may be used for hand hygiene. Alcohol rubs should be between 60-95% ethyl or isopropyl alcohol to allow for fast-acting denaturing of the microbial protein. Alcohol rubs have broad microbial spectrum and are well tolerated due to the presence of emollients in the products. Organic debris and visible soiling on the hands should be removed prior to applying the alcohol rub. The correct procedure for using an alcohol hand rub is to remove all gross debris, then place the alcohol-based hand rub in the palms. Apply the rub to all surfaces of the hands and rub the hands together until dry.

Personal protective equipment

Even though most veterinarians have knowledge and concern about risks associated with zoonotic pathogens, a large number do not engage in the use of appropriate personal protective equipment (PPE), even when working with animals displaying signs of infectious disease.[6] No PPE is 100% effective. Protective measures must be augmented with good behavioral practices including appropriate hand hygiene. The effectiveness of using PPE depends on its consistent use and accessibility.[1] If the risk of zoonotic pathogen exposure exists, then employers are required to provide equipment to mitigate that risk and training on the appropriate use of that equipment. Training should include what PPE to use and when, how to use it including putting it on and taking it off, how to maintain and dispose of the PPE, and what limitations are inherent in the use of PPE. Non-disposable PPE must be disinfected before its storage or reuse.

Gloves

Hand protection should be appropriate for the activities. Disposable gloves made of lightweight nitrile or vinyl offer protection against infectious disease while heavy gloves are more appropriate for protection from injury. Ideally, exam gloves should be worn inside any outer glove being used for restraint. Gloves are barrier protection between the potential pathogen and the hands of the handler. They should be used when handling biological specimens, when the handler has lost skin integrity, and when performing procedures that have a higher risk for contacting potential infectious organisms. These include dental procedures, obstetric procedures, resuscitation, and necropsy. Gloves should also be worn when cleaning cages, litter boxes, contaminated equipment or surfaces, and handling soiled laundry. Unlike in human medicine, veterinary practitioners very often do not wear gloves when examining healthy animal patients unless the index of suspicion for the presence of a zoonotic disease is high. This should not be the case for wildlife; gloves should be worn whenever handling a wild animal. Additionally, gloves should be changed between examination of animals, between performing 'dirty' and 'clean' procedures, and when torn. Care must be taken to avoid contact between contaminated gloves and mucous membranes or non-contaminated surfaces. Glove removal is performed by pulling the glove off each hand inside-out, avoiding contact with the exterior of the glove. Hand washing must still be performed after removing the gloves.

Outerwear

If possible, protective clothing such as lab coats or smocks should be worn over street clothes to limit the transfer of potential pathogens via direct or indirect contact. Avoid wearing protective outerwear outside of the clinical environment and change or clean the garment when it becomes soiled. Disposable gowns should be used any time infectious disease is suspected and the garment should be used solely for that individual animal to avoid indirect transmission to another. Foot protection is important in a comprehensive biosecurity plan and should be suitable for the specific environment. Disposable foot covers add a layer of protection over street shoes and should be removed when leaving the contaminated environment.

Facial and respiratory protection

Eye protection in the form of safety glasses or goggles protects eyes and conjunctiva from trauma as well as exposure to potential pathogens. It should be used when splashes or sprays are likely. Respirators and face shields are recommended for all persons potentially exposed to air-borne pathogens. The National Institute of Occupational Safety and Health has approved disposable particulate respirators (N95, N99, or N100) as a minimum standard of respiratory protection against viral pathogens such as avian influenza. These require individual health assessment, training, and fit testing to ensure effective protection. Surgical face masks do not offer adequate protection against particulate air-borne pathogens.

Animal management

A fundamental concern for wildlife care facilities is cross-contamination or co-mingling of species. The potential for transmission of zoonotic diseases between wild and domestic or zoo species increases with their proximity and interaction. Human traffic and common treatment areas raise the risk for pathogen dissemination. Patient intake and movement are key points of control regarding biosecurity. If the safe and effective handling of a wild animal in the clinical care setting is impractical, while also maintaining biosecurity for other patients and humans, do not attempt to provide the care and refer the patient to a more appropriate facility.

Patient intake

Ideally, wildlife and domestic animals should be segregated at all times. Wildlife being admitted should bypass the waiting room and be moved directly to a quarantine area within the clinical space dedicated to wildlife. Classic signs that the animal is suffering from a potential infectious disease include neurologic signs, diarrhea, respiratory tract signs, fever, wounds or chronic infections, and irritated or inflamed skin. Assume there is a potential for an infectious etiology even if clinical signs are not immediately evident. If a common clinical space is used for triage or exam, do not use the space for a different patient until it has been thoroughly disinfected. Limit contact of animals with suspected zoonotic disease to essential personnel.

Patient screening

Establishment of a minimum health screening protocol for admitted patients is essential to identify infected animals and disease trends. Diagnostic procedures should include physical examination, baseline bloodwork, fecal evaluation, and survey radiographs. Further testing should be based on individual animal presentation and species. Consideration of diseases endemic to the facilities' region must be considered when developing specific screening protocols. Screening does not preclude the need for quarantine as incubation periods for infectious diseases vary. Medical records for patients should include information such as the age, sex, and species of the animal as well as the location the animal originated to facilitate disease monitoring.[7]

Most reported injuries among veterinary personnel occur during animal handling or treatment.[8] When performing animal handling, be prepared. If manual restraint is risky, consider sedation or anesthetic restraint of the patient. If working with large wildlife species, always have an escape plan ready. Preventing animal-related injuries requires pre-planning, appropriate equipment, and good communication among the clinical care team. Have a post-exposure/injury management plan established for use in an emergency.

Necropsy and handling biological samples

Useful information can be learned from a necropsy, but necropsies can be a high-risk procedure due to the possibility of exposure to infectious agents in body tissue, fluids, and aerosols, as well as the possibility of injury.

In-house necropsy

Only essential personnel should participate in a necropsy to limit exposure. Minimally, protective gloves, facial protection, and impermeable protective outerwear should be worn while performing the necropsy. When there is a concern for aerosolization of particles, eye wear and respiratory PPE should be used. The area in which the necropsy is to be performed should be well ventilated and have good lighting.

Submission to outside laboratory

If the necropsy is performed at an outside laboratory, or if biological samples are being submitted to an outside lab for processing, a thorough history with the submission is of critical importance. Be sure to clearly label all samples. Assume that infectious disease is a differential diagnosis for wildlife and wear gloves when handling samples. If there is information that raises the index of suspicion for an infectious and/or zoonotic etiology, clearly note that in the submission information. Underreporting of zoonotic disease potential significantly increases the risk of exposure for laboratory personnel.[9]

The U.S. Department of Health and Human Services, in partnership with the Public Health Service, Centers for Disease Control and Prevention (CDC), and National Institutes of Health publishes the Biosafety in Microbiological and Biomedical Laboratories (BMBL) as an advisory document recommending best practices for the safe conduct of work in biomedical and clinical laboratories from a biosafety perspective. The document defines four primary biosafety levels for laboratories working with animal pathogens and details facility design features and safety equipment, facility practices and procedures, and personal protective equipment to mitigate risks.[10] Since these standards are already in place for diagnostic laboratories, it is prudent to submit the carcass of an animal suspected of having a zoonotic disease to a commercial laboratory for post-mortem evaluation. In addition, the CDC, in conjunction with the U.S. Department of Agriculture, administer the Federal Select Agent Program that determines which agents and toxins have the potential to pose a severe threat to both human and animal health.[11] Listed among these are several pathogens known to cause zoonotic disease in wildlife including *Francisella tularensis* and *Yersinia pestis*, both Tier 1 agents of gravest concern, as well as *Coxiella burnetii*, Eastern equine encephalitis virus, avian influenza virus, Newcastle disease virus, *Brucella abortis*, and *B. melitensis*. If any of these infectious agents are suspected or identified, maintain biosecurity and contact your state veterinary and/or public health officials.

Facility management

Isolation, quarantine, and segregation

Isolation is a critical aspect of biosecurity that can contain the spread of infectious diseases in wildlife care facilities. To protect hospitalized patients, animals suspected of infectious disease should be placed in an isolation room with appropriate precautions and signage. Equipment and PPE should be exclusive to this area, and protocols should be clearly posted. Ideally, isolation rooms have negative air pressure in relation to the adjacent rooms and should exhaust air directly outside or have the air recirculated using a HEPA filter. If a true isolation area is not available, a room dedicated to treating infectious animals can be used. The purpose of quarantine is to segregate an animal that may be incubating a disease from the main population while monitoring it for signs of clinical disease. Clinical signs of infectious disease are often vague and non-specific. All animals should be considered potentially infectious and placed in quarantine before grouped with other patients. Wildlife should be kept segregated from domestic/captive patients. Whenever possible, dedicated handling equipment should be used for wildlife and those suspected of infectious diseases. Equipment that can't be dedicated for use (e.g., ultrasound unit) should be disinfected between patients.

Ventilation

Controls to prevent the spread of airborne contaminants focus on proper ventilation of clinical space including the dilution and removal of contaminants, the pressurization of wildlife clinical space in relation to other spaces, and the pressurization of the building to the outdoors.[11] Despite the importance of proper ventilation, most veterinarians are unfamiliar with the ventilation in their facility. Clinic design should ensure that air contaminants are being filtered or exhausted and not introduced to surrounding clinic space. To maximize the dilution and removal of contaminants, air changes per hour should equal or exceed twelve.[11] Similar to isolation areas described previously, air pressure should be negative with relation to the rest of the clinic.

Sanitation

Surface contamination with some pathogens may persist for months. Disinfectants effective against a broad range of pathogens are ideal for veterinary facilities that care for wildlife. Common disinfectants with good spectrum of activity for veterinary use include quaternary ammonium compounds, hypochlorite (e.g., household bleach), and hydrogen peroxide-based oxidizing agents.[1,13] Each of these should be stored, diluted, and used according to recommended standards. The CDC maintains a list of all disinfectants labeled to be efficacious against potential pathogens and should be consulted when preparing a biosecurity plan.[14] Frequent routine cleaning and removal of gross contamination from surfaces and equipment, regardless of the disinfectant used, is essential for optimal sanitation. Be sure to include high-touch non-animal surfaces such as doorknobs, cage door handles, and faucet handles. It is prudent to create a facility sanitation plan and schedule. If a contaminated surface is porous or the material is organic (e.g., soil or bark for burrowing, branches for perching, etc.), it is best to discard the item and replace it.

Foot baths

Disinfectant footbaths and footpads, using peroxygen-based disinfectant, have been shown to reduce bacterial numbers on footwear. Both mechanical removal of organic debris and the antimicrobial activity of the disinfectant help achieve this effect. Adhesive mats offer an alternative and remove contaminants through their adhesive and antimicrobial properties. These measures do not sterilize footwear and should not be relied upon solely to prevent pathogen transmission. Bioaccumulation will require frequent changes of footwear disinfectants. In wildlife care facilities where prevention and containment are equally important, the use of disposable shoe coverings may be useful. Personnel training to ensure proper removal and disposal is essential to avoid hand contact with contaminated shoe covers.

Medical waste

The disposal of medical waste is regulated at the state level by multiple agencies including the Environmental Protection Agency, which categorizes medical waste as infectious, hazardous, radioactive, and other. Veterinary hospitals, including wildlife care facilities, most often generate infectious or hazardous waste including sharps, tissues, contaminated materials, and dead animals. Dedicated biohazard receptacles, including sharps containers, should be made available and used routinely in wildlife care facilities. Standard operating procedures to prevent needle sticks should be followed and include proper restraint of the animal, keeping needles capped until in use, never placing needles or syringes in the mouth, using the scoop method to recap needles, and removing uncapped needles from a syringe using an instrument.

Laundry

Soiled laundry may also pose a hazard for exposure to potential pathogens and proper handling is required to minimize the risk of disease transmission. Separate clean and dirty laundry bins should be used, and gloves should be worn when handling soiled laundry. Machine washing with standard laundry detergent followed by drying is effective in removing contaminants. Wash personal laundry separately from laundry soiled during animal care.

Vector control

Rodents and insects are important vectors for transmission of many diseases. Patient triage often includes implementation of routine or as needed ectoparasite control. Additionally, insect vectors such as flies and gnats must be controlled. Effective prevention and control of rodents is dependent on limiting access to food and nesting material. Food stores should be stored in air-tight containers and access to paper, clutter, or hay should be minimized. Food for animals should be stored and prepared separately from food intended for human consumption.

Personnel management

It is essential that worker education is a priority. Occupational health and safety plans, based on identification of facility hazards, should be developed. All personnel must be trained regarding zoonosis awareness, infectious disease prevention and control practices, safe animal handling and behavioral cues of animals they will be handling, the proper use and disposal of PPE, safe handling of drugs and equipment, proper handling and storage of biological tissues, basic first aid, and emergency protocols. In addition, employees must understand the potential risk in working with wildlife and the risk posed to other animals that they interact with (e.g., domestic patients, personal pets, production animals). Providing baseline training for new employees, as well as regular refresher training for current personnel, is critical. Training completion should be documented, and the employee provided a written copy of all relevant material.[1]

Vaccination against rabies, tetanus, and seasonal influenza are often recommended or required for persons working with wildlife. Additionally, veterinary practitioners who are at high risk for unrecognized exposure to rabies should have their serum anti-rabies virus antibody titer checked every two years. The Advisory Committee on Immunization Policies recognizes a titer of ≥ 0.5 IU/ml as correlating with protection for rabies. Those with potential exposure that is recognized should pursue post-exposure prophylaxis.[15] Vaccination against hepatitis B is recommended for those working with nonhuman primates, and individuals working with marine mammals and nonhuman primates should speak with their healthcare provider about annual tuberculosis testing. Immunocompromised and pregnant personnel should be encouraged to speak with their healthcare provider to understand the risks and necessary measures to mitigate those risks when working with wildlife.

Creating and Implementing an Infection Control Plan

Providing care for wildlife patients is rewarding on both professional and personal levels. However, there is risk to the individuals working with these animals, and to animals that may come into direct or indirect contact with pathogens of wildlife patients. Wildlife care facilities are in a unique position to identify emerging disease or bioterrorism through care of wild animal sentinels.[8] Preparedness planning for infectious disease outbreaks and developing biosecurity protocols will help ensure the safety of personnel while continuing to care for hospitalized wildlife.

The Occupational Safety and Health Administration (OSHA) sets specific standards for employer responsibility to provide PPE and respiratory protection for workers who are potentially exposed to infectious agents.[16] Some states also set standards (e.g., California, General Industry Safety orders exclusively for veterinarians). Infection control plans that address these requirements should be developed prior to being needed and must be reviewed regularly to ensure that the plan is meeting evolving needs. Each plan should be tailored to the individual practice and should provide clear, thorough guidance for how risks associated with infectious disease exposure will be mitigated. Templates for such plans are available (Williams) and often follow standardized veterinary precautions as discussed above. The plan must indicate which personnel are responsible for each area discussed and the activities assigned. It will be necessary to identify essential personnel responsible for oversight of facility operation in the event of emergencies and these people will need to be fitted with appropriate PPE. Contact and resource information for staff to use if questions arise should be provided. Every staff member should certify that they have received and understand the training. Once a plan is established, there should be a non-punitive mechanism for reporting untoward events that resulted from the plan not being followed or failing. This allows for each team member to have ownership in the plan's success and for trends to be identified that indicate the plan needs to be modified. Integrating risk management processes into daily clinic operation promotes a positive culture of safety.

References

1. Williams CJ, Scheftel JM, Elchos BL, et al. Compendium of veterinary standard precautions for zoonotic disease prevention in veterinary personnel: National Association of State Public Health Veterinarians: Veterinary Infection Control Committee 2015. *J Am Vet Med Assoc.* 2015;247(11):1252-1277.

2. Jones KE, Patel NG, Levy MA, et al. Global trends in emerging infectious diseases. *Nature.* 2008;451: 990-993.

3. Taylor T, Buttke D. Safe work practice for working with wildlife. In: Richgels KLD, Gibbs SEJ, Wild MA, eds. *Field Manual of Wildlife Diseases.* Reston, VA: U.S. Geological Survey; 2020:1-26.

4. Lim ZJK, Xie S. Application of epidemiology and principles of herd/flock health or the exotic animal veterinarian. *Vet Clin North Am Exot Anim Pract.* 2021;24:495-450.

5. Terrestrial Animal Health Code. World Organisation for Animal Health (OIE). 2022. Available at: https://www.woah.org/en/what-we-do/standards/codes-and-manuals/terrestrial-code-online-access/. Accessed April 2, 2023.

6. Wright JG, Jung S, Holman RC, et al. Infection control practices and zoonotic disease risks among veterinarians in the United States. *J Am Vet Med Assoc.* 2008;232(12):1863-1872.

7. Miller E, ed. *Minimum Standards for Wildlife Rehabilitation.* 3rd Ed. St. Cloud, MN: National Wildlife Rehabilitators Association; 2000.

8. Nordgren LD, Gerberich SG, Alexander H, et al. Evaluation of factors associated with work-related injuries to veterinary technicians certified in Minnesota. *J Am Vet Med Assoc.* 2014;245:425-433.

9. Morse S, Henkel R. *Francisella tularensis*: understanding reported occupational exposure and laboratory methods used for the identification of *Francisella tularensis*. *Appl Biosafety.* 2018;23(1):11-18.

10. Centers for Disease Control and Prevention. Biosafety in microbiological and biomedical laboratories (BMBL), 6th Ed. CDC Laboratories. Strengthening Lab Safety. Available at: https://www.cdc.gov/labs/BMBL.html. Accessed April 3, 2023.

11. Centers for Disease Control and Prevention. Select Agents and Toxin List. Available at: https://www.selectagents.gov/sat/list.htm. Accessed April 2, 2023.

12. Centers for Disease Control and Prevnetion. Background C. Air. Environmental Infection Control Guidelines. Infection Control. Available at: https://www.cdc.gov/infectioncontrol/guidelines/environmental/background/air.html. Accessed April 3, 2023.

13. The Center for Food Security& Public Health. Iowa State University. The Antimicrobial Spectrum of Disinfectants. 2018. Available at: https://www.cfsph.iastate.edu/pdf/antimicrobial-spectrum-of-disinfectants. Accessed April 2, 2023.

14. Centers for Disease Control and Prevention. Chemical Disinfectants. Infection Control. Available at: https://www.cdc.gov/infectioncontrol/guidelines/disinfection/disinfection-methods/chemical.html. Accessed April 2, 2023.

15. Rao AK, Briggs D, Moore SM, et al. Use of a modified preexposure prophylaxis vaccination schedule to prevent human rabies: recommendations of the Advisory Committee on Immunization Practices – United States, 2022. *Morb Mortality Wkly Rep.* 2022;71(18):619-627.

16. U.S. Department of Labor Occupational Safety and Health Administration. Safety and Health Topics. Healthcare. Standards. Available at: https://www.osha.gov/healthcare/standards. Accessed April 3, 2023.

Wildlife: Coping with Common to Crazy Emergencies

Heather Wilson Barron, DVM, DABVP (Avian Practice), CertAqV

Session #3221

Affiliation: From Loggerhead Marinelife Center, 14200 US Hwy 1, Juno Beach, FL 33408, USA.

Abstract: Wildlife rehabilitation centers and veterinary hospitals that commonly admit wildlife patients often encounter a wide variety of wildlife from a large geographic area. As a result, these entities may be well-suited to participate in disease surveillance and One Health objectives. Obstacles to participation may include permit restrictions, financial disincentives, staff safety, absence of mandates to collect relevant data, and lack of contact between wildlife and public health agencies. Practitioners may be greatly rewarded for investment of time and other resources by gaining more knowledge about wildlife diseases that impact humans or companion and food animals.

Introduction

Wildlife may present to veterinary hospitals for a number of emergent conditions. Although trauma is the primary reason that wildlife species are admitted, a variety of pathogens and toxins should also be considered. Additionally, the health of wildlife species has received growing attention both from a conservation perspective and from the global 'One Health' initiative, which integrates human, animal, and ecosystem health. Aside from their intrinsic value, wildlife can be important as sentinel species for both the health of the environment and people. Of the 1,461 infectious diseases now recognized in humans, approximately 60% are due to multi-host pathogens characterized by their movement across species lines. Over the last three decades, approximately 75% of new emerging human infectious diseases have been zoonotic; however, routine surveillance for pathogens in free-ranging wildlife presents financial, logistical, welfare, and other constraints. Veterinary hospitals and wildlife rehabilitative centers may present an alternative source to facilitate the monitoring of wildlife populations for a wide variety of pathogens. Data collected this way may be biased, but if the ecology of the tested species and the specific temporal, spatial and age group distribution of admissions are considered, then the information can be a useful window into the health of regional wildlife.

Wildlife Triage

Triage is the process of prioritizing patients based on presenting condition and the resources available to deal with that condition. Therefore, triage of wildlife may necessarily vary to some degree based on individual philosophy, case load, local, state, and federal regulations, and the presenting situation (i.e., oil spill affecting hundreds of animals versus the normal daily load of injury and illness). In any case, the primary objective to keep in mind during the triaging process is eventual return of the patient to the wild. If triaging wildlife in a private practice setting, there are some considerations that may help set triage policy. Guidelines about what the practice is willing to accept should be made clear to staff and to the public. Recognize when the animal should be returned to the wild, as is often the case with abducted neonates. Also be aware of reasons why a wild animal may not be releasable. The goal of wildlife medicine is always eventual release and the possibility of euthanasia needs to be kept in mind during triage if chances of release are remote. Realistic resource allocation is especially important in a for-profit clinic that is underwriting wildlife treatment.

Pathogens Under Surveillance

Surveillance is critical for identifying and monitoring pathogen activity in wildlife populations, and although there are some limitations, wildlife and veterinary hospitals have been shown to be useful in monitoring for a variety of pathogens, including West Nile virus, Newcastle disease virus (NDV), avian influenza, bacteria such as *Escherichia coli* and *Salmonella* spp., and the following pathogens listed here.[1-9]

Ranavirus

Ranavirus typically exists at low prevalence (less than 5 percent of individuals infected in a population), and then rapidly emerges, with mortality often exceeding 90 percent in multiple species. With one in three amphibian species and over 40 percent of turtles at risk of extinction, *Ranavirus* represents a significant threat to herpetofaunal biodiversity. Ranaviral disease has been likened to Ebola virus for amphibians because their bodies swell and bleed profusely. The reasons for *Ranavirus* emergence at a site vary, but often are related to stressors that can be natural or anthropogenic (human-related) in origin. For example, outbreaks might signal that local ponds and wetlands are becoming more susceptible to disease under the stressors of climate change, pollution and development. A *Ranavirus* epizootic event was documented involving several Florida box turtles (*Terrapene carolina bauri*) from Sanibel and Captiva islands. Clinical signs included facial swelling, stomatitis, open-mouth breathing, mucopurulent nasal discharge, bilateral chemosis, dehydration, emaciation, and skin sloughing over the head and neck. The animals were either dead or euthanized upon presentation to the clinic. Postmortem findings in all carcasses examined included buphthalmos, oronasal discharge, fibrinonecrotizing stomatitis, glossitis, and red-tinged fluid and blood within the coelomic cavity. Histologic findings included fibrinonecrotizing and heterophilic stomatitis and glossitis, as well as marked skeletal muscle degeneration. *Ranavirus*-specific polymerase chain reaction (PCR) was done on blood samples and oral swabs when available, from all turtles (n = 8) examined. Sequencing and phylogenetic analysis confirmed FV3 in 4/7 oral swabs and 3/8 blood samples. Immunohistochemical staining for *Ranavirus* sp. confirmed the presence of antigen in the oral mucosa and peri-oral skin (three turtles) and in the kidney (one turtle). Diagnosing *Ranavirus* infection in free-ranging box turtles can be challenging in chronic stages of disease. Box turtles may live for a hundred years or more, but their populations are increasingly threatened by disease, habitat loss, vehicle strikes, and illegal collection. Reptiles can be good indicators of ecological health because of their reliance on the environment for appropriate physiologic functions, such as immunity, digestion, and thermoregulation.

Newcastle disease virus

Paramyxoviruses can be widespread in avian populations, but do not always cause illness. NDV was previously synonymous with avian paramyxovirus type 1 (APMV-1); however, due to changes in taxonomy is now referred to as avian avulavirus. The virus strains can be differentiated as lentogenic, mesogenic, or velogenic based on a mean death time in chicken embryos. Currently, velogenic strains of NDV are not endemic in U.S. domestic poultry; however, these strains are present in other countries and are occasionally detected in wild birds in the United States, which could be potentially devastating for the poultry industry, making wildlife reservoirs a concern. This disease is listed by the World Health Organisation for Animals (OIE) as vitally important for avian species and products. Virus detection in a specific geographical location often leads to trade restrictions and embargos and thus virulent NDV is listed as a Tier 1 U.S. Department of Agriculture Select Agent as it is a pathogen of national concern and a significant threat to animal agriculture in the United States.

Double-crested cormorants (*Phalacrocorax auritus*) (DCCO) are the wild avian species most associated with NDV illness; however white pelicans and ring-billed gulls have also been affected during outbreaks. Depending upon the strain, signs of infection can vary from no signs to dehydration, emaciation, weakness, diarrhea, and neurologic signs such as depression, wing or leg paralysis, clenching of toes, torticollis, ataxia, and head tremors. NDV can spread from wild avian populations to domestic poultry populations; however, thanks to good biosecurity, this has been documented very rarely and only in association with DCCO virus outbreaks. NDV is not considered a risk to human health, with signs of conjunctivitis and mild flu-like symptoms being the most severe clinical signs demonstrated.

North American river otter parasites of clinical concern

North American river otters (*Lontra canadensis*) (LOCA) may have a variety of parasites that can have an impact on health, especially in a rehabilitation setting or in relocation efforts. *Dirofilaria lutrae* and *Dirofilaria immitis* have been reported in LOCA with the former species considered otter-specific. In-house ELISA tests for heartworm disease may be positive with both infections because female antigens are conserved across species. Confirmation of a positive result, either by an echocardiogram or examination of blood for microfilaria or a Knott's test, is needed. Adult *Di. lutrae* worms are found subcutaneously and the microfilaria in the blood are much larger than those of *D. immitis*. The subcutaneous abscesses that can develop due to *Di. lutrae*, may be lanced and the worm gently removed. The wound should be flushed, and the animal treated with doxycycline and ivermectin. Otters in an outdoor rehabilitative setting should ideally be placed on monthly heartworm preventative. Another differential for worms found subcutaneously in NA otters is *Dracunculus lutrae* (or Guinea worm). When these worms or nodules are lanced, larvae with long taping tails will be detected vs. the microfilaria that would be present with *Di. lutrae* infections. Also, no *Dr. lutrae* larvae enter circulation. Local hair loss and skin damage due to self-inflicted trauma (scratching) may occur with *Dr. lutrae*. There is no treatment other than physical removal and because *Dracunculus* have no evidence of *Wolbachia* endosymbionts, treatment with doxycycline would probably not be indicated. Animals with evidence of pneumonia should be checked for an eosinophilia on CBC. If present, this may indicate a high probability for *Crenosoma goblei* (lungworm). Although *C. goblei* larvae may be found on a fecal exam, bronchoscopy or bronchoalveolar lavage may be more sensitive. In the author's hands, treatment with fenbendazole (50 mg/kg PO q24h for 3 days) appeared to be more effective than ivermectin (0.3 mg/kg PO) in treating several individuals.

Anticoagulant Rodenticides

In some populations, more than 80% of individuals test positive for anticoagulant rodenticides (AR), pointing to the potential for ecosystem-wide effects. This secondary exposure can be directly lethal to some mammalian and avian predators, including endangered species. These poisons are mostly derived from coumarin and function primarily by interrupting the vitamin K cycle leading to failure to synthesize clotting factors I, II, VII, IX, and X in the liver. Today, most commercially available rodenticides are second-generation anticoagulant rodenticides (SGARs). Relay toxicosis is a problem in many different species of wildlife and may manifest clinically in a variety of ways.

For example, SAGRs have been shown to contribute to immunosuppression and mange in bobcats and mountain lions in California.[9,10] Bobcats, like domestic cats, seem tolerant to the vitamin K antagonistic effects of ARs, as indicated by normal clotting times in AR-exposed bobcats; however, this tolerance may lead to chronic sublethal exposure that influences other physiological parameters, as recently indicated in one study looking at genome-wide expression of multiple systemic effects.[11] Other documented non-target affected wildlife commonly include deer, raccoons, foxes, squirrels, weasels, voles, and skunks.[12]

Symptoms in raptors may include weakness, depression, hemorrhage, pale mucous membranes, peripheral vascular collapse, and shock. Normal thrombocyte counts in the face of prolonged bleeding or clotting times and petechial hemorrhages are highly suggestive of anticoagulant rodenticide intoxication. Assessment of coagulation times in avian patients is an area that has only been established in select species. Commercial coagulation tests are unavailable for avian patients, and those currently used have been modified from available mammalian products. In birds, the intrinsic pathway is thought to play only a minor role in coagulation; therefore, the prothrombin time would be the most clinically useful parameter in diagnosing a coagulopathy but has limited clinical availability. Activated coagulation (aka clotting) time (ACT) is a test that has long been used in mammalian patients as a method to evaluate the intrinsic and common pathways of the coagulation cascade. ACT is a more economical and easily performed method in any hospital, and in cases of wildlife patients can be used to assess for coagulopathies affecting the intrinsic and common pathways. Anticoagulant rodenticide screening tests are available and practical for use in some species, but the risk of encouraging further hemorrhage from venipuncture must be considered.

Treatment consists of intravenous fluids, iron dextran (10 mg/kg IM), vitamin B12 (0.5 mg/kg IM), with oxygen and heat support. The mainstay of treatment used by the author is administration of vitamin K1 1.0 mg/kg IM or SC q8-12h until stable (usually 1-2 days), then q24h PO for 7-30 days. This author has not found it necessary to go longer than that for most cases; birds make their own vitamin K and usually appear to go back to clotting normally quickly. The injectable vitamin K1 can be given orally in patients too small for the oral capsules/tablets; there is also a pediatric oral solution that is 120 **µg**/0.8 ml). Blood transfusions are given in more severe cases (this author uses a PCV < 15% as a guideline, but this should always be assessed in light of the clinical picture; patients too BAR to tolerate an intravenous catheter are probably not going to get a blood transfusion).

Lead

Lead is a persistent inorganic environmental pollutant that affects humans and animals worldwide. Virginia opossums (*Didelphis virginiana*) were found to have elevated lead levels relative to other mesomammals in the southeastern study area.[13] Median blood lead level was 7.3 **µg**/dL (n=104) in free-ranging raccoons in New York City.[14] Recent studies suggest that lead poisoning is estimated to kill a million wildfowl a year and cause sub-lethal poisoning in another ≥3 million in Europe alone. Modelling and correlative studies have supported the potential for population-level effects and lead clearly inhibits reproductive capabilities in adult birds as well as altering growth and development of hatchlings.[15] Interestingly, even sub-lethal exposure has been shown to significantly affect behavior leading to increased mortality risk in eagles.[16] Chronic low levels of exposure can cause immunosuppression as well as build up in the tissues and eventually manifest similarly to that of an acute exposure, especially in birds in the breeding season as hens mobilize calcium from their bones for their eggs and free stored lead along with it.

Lead toxicosis in wildlife may be diagnosed by sending whole blood to a diagnostic lab or in house with something like the LeadCare II system, which has been shown to be highly accurate in domestic animals and is used frequently in wildlife centers.[17] Heparinized whole blood concentrations greater than 20 µg/dl are suggestive. Levels greater than 50 µg/dl are diagnostic. A recent study in bald eagles has shown that histologic lesions compatible with lead toxicity occurred within the heart (76/93 birds), brain (59/93 birds), and eyes (24/87 birds) caused by ischemic necrosis of the small to medium arteries in these organs and that lead levels of 4 ppm or greater are more likely to lead to lesions in the heart.[18]

Treatment involves using chelation therapy. The author prefers starting with calcium EDTA (Calcium disodium versenate, 3M Pharmaceuticals) as the initial chelator for lead. It is important to maintain hydration as the bound molecules are large and may cause renal insult if the patient is dehydrated. This drug should not be given orally as it may increase lead absorption from GI tract. [19] The usual dose used by the author is 30-35 mg/kg IM or

SC q12h for 3-5 days, off 3-5 days, then repeat as needed. In extremely depressed patients where there may be a concern about efficacy of subcutaneous or intramuscular absorption, it can be used intravenously short term (48 hours) at 20-35 mg/kg diluted from 200 mg/ml to 60 mg/ml in saline as per the VIN formulary when given to mammals. It is not known if this is necessary in birds or reptiles but the author has used it safely for the past 30 years in many different species. Dimercaptosuccinic acid (DMSA) is often used as a after initial treatment with CaEDTA but can also be used in conjunction with it at a dose of 25-35 mg/kg PO q24h for as many days as needed to resolve elevated blood lead levels. Removal of any lead particles from the gastrointestinal tract can be an important part of treatment and can be managed in a variety of creative ways.

Euthanasia and Assessing Life

Veterinarians, veterinary staff, and rehabilitators care for wildlife for a number of professional and personal reasons. Many wild animals are fully releasable once recovered from their illness or injury. Individual animals may be ecologically important in areas of declining wildlife populations. Members of the general public expect veterinarians and rehabilitators to save the wild animals they have found. Medical and surgical skills developed from working on wildlife can be adapted and used to help treat companion animals; however, for those animals with severe injuries and illnesses that negate ultimate release and where permanent placement is not available, practical, or humane, euthanasia must be considered. Ultimately, veterinarians and rehabilitators are advocates for wild animal welfare and well-being. Euthanasia should not be considered a failure.

Euthanasia should be conducted without undue patient stress or restraint. Premedication and a multi-modal drug delivery approach to euthanasia should strongly be considered. For example, mammals can be premedicated with drugs such as dexmedetomidine and midazolam prior to induction with a gas anesthetic before administration of sodium pentobarbital. If sodium pentobarbital or other barbiturates are used, proper carcass disposal must be practiced, as fatal secondary barbiturate poisonings have recently been documented in species including golden eagles (*Aquila chrysaetos*). Veterinarians have been prosecuted for such poisonings under the federal Eagle Protection Act.

Conclusions

Wildlife rehabilitation centers and veterinary hospitals that commonly admit wildlife patients often encounter a wide variety of wildlife from a large geographic area and so may be ideally suited to participate in disease surveillance and One Health objectives. Obstacles to participation may include permit restrictions, financial disincentives, staff safety, no mandate to collect relevant data, and lack of contact between wildlife and public health agencies; however, practitioners may be greatly rewarded for investment of time and other resources by being more knowledgeable about wildlife diseases that may impact humans or companion and food animals.

References

1. Randall NJ, Blitvich BJ, Blanchong JA. Efficacy of wildlife rehabilitation centers in surveillance and monitoring of pathogen activity: a case study with West Nile virus. *J Wildl Dis.* 2012;48(3):646-653.

2. Camacho M, Hernández JM, Lima-Barbero JF, Höfle U. Use of wildlife rehabilitation centers in pathogen surveillance: A case study in white storks (*Ciconia ciconia*). *Prev Vet Med.* 2016;130:106-111.

3. Ana A, Perez Andrés M, Julia P, et al. Syndromic surveillance for West Nile virus using raptors in rehabilitation. *BMC Vet Res.* 2017;13(1):368.

4. Willette MM, Ponder JB, McRuer DL, Clark EE Jr. Wildlife health monitoring systems in North America: from sentinel species to public policy. In: Aguirre AA, ed. *New Directions in Conservation Medicine: Applied Cases of Ecological Health.* New York, NY: Oxford Press; 2012:552-562.

5. Sleeman JM, Clark EE, Jr. Clinical wildlife medicine: a new paradigm for a new century. *J Avian Med Surg.* 2003;17:33-37.

6. Farley JM, Romero CH, Spalding MG, et al. Newcastle disease virus in double-crested cormorants in Alabama, Florida, and Mississippi. *J Wildl Dis.* 2001;37(4):808-812.

7. Sleeman JM. Use of wildlife rehabilitation centers as monitors of ecosystem health. In: Fowler ME, Miller RE, eds. *Zoo and Wild Animal Medicine: Current Therapy.* Vol 6. St. Louis, MO: Elsevier Saunders; 2008.

8. Brown JD, Sleeman JM. Morbidity and mortality of reptiles admitted to the Wildlife Center of Virginia, 1991-2000. *J Wildl Dis.* 2002;38:699-705.

9. Serieys LEK, Lea AJ, Epeldegui M, et al. Urbanization and anticoagulant poisons promote immune dysfunction in bobcats. *Proc Biol Sci.* 2018;31;285.

10. Riley SPD, Bromley C, Poppenga R, et al. Anticoagulant exposure and notoedric mange in bobcats and mountain lions in urban southern California. *J Wildl Manag.* 2007;71:1874-1884.

11. Fraser D, Mouton A, Serieys LEK, et al. Genome-wide expression reveals multiple systemic effects associated with detection of anticoagulant poisons in bobcats (*Lynx rufus*). *Mol Ecol.* 2018;27(5):1170-1187.

12. Stone WB, Okoniewski JC, Stedelin JR. Poisoning of wildlife with anticoagulant rodenticides in New York. *J Wildl Dis.* 1999;35(2):187-193.

13. Lockhart JM, Siddiqui S, Loughry WJ, Bielmyer-Fraser GK. Metal accumulation in wild-caught opossum. *Environ Monit Assess.* 2016;188(6):317.

14. Rainwater KL, Marchese K, Slavinski S, et al. Health survey of free-ranging raccoons (*Procyon lotor*) in Central Park, New York, New York, USA: implications for human and domestic animal health. *J Wildl Dis.* 2017;53(2):272-284.

15. Pain DJ, Mateo R, Green RE. Effects of lead from ammunition on birds and other wildlife: A review and update. *Ambio.* 2019;48(9):935-953.

16. Ecke F, Singh NJ, Arnemo JM, et al. Sublethal Lead Exposure Alters Movement Behavior in Free-Ranging Golden Eagles. *Environ Sci Technol.* 2017;51(10):5729-5736.

17. Bischoff K, Gaskill C, Erb HN, et al. Comparison of two methods for blood lead analysis in cattle: graphite-furnace atomic absorption spectrometry and LeadCare(R) II system. *J Vet Diagn Invest.* 2010;22(5):729-733.

18. Manning LK, Wünschmann A, Armién AG, et al. Lead intoxication in free-ranging bald eagles (*Haliaeetus leucocephalus*). *Vet Pathol.* 2019;56(2):289-299.

19. Bradberry S, Vale A. A comparison of sodium calcium edetate (edetate calcium disodium) and succimer (DMSA) in the treatment of inorganic lead poisoning. *Clin Toxicol (Phila).* 2009;47(9):841-58. doi: 10.3109/15563650903321064. PMID: 19852620.

Addressing Dogma in Exotic Animal Practice

Nicolle Lofgren, VMD, DACEPM

Session #3394

Affiliation: From Mountain West Veterinary Specialists, Layton, UT 84041, USA.

Abstract: Exotic animal care suffers from reliance on dogma likely due to a paucity of exotic-exclusive clinics and veterinarians. This requires veterinarians to rely on handed down information and treatment protocols, which may be out of date or no longer clinically relevant. Treatment of exotic animal patients is just as complex as any species, with no single treatment fitting all presenting patients in a single species. This masterclass addresses some of the commonly held dogmas in exotic animal practice to help provide more effective diagnostics and treatments, as well as more accurate information for clients.

Introduction

Dogma is common in exotic animal practice, with many treatment and diagnostic recommendations being based on long-held beliefs with little or no scientific backing, potentially setting a dangerous precedent. These dogmas range from the assumption that feather plucking is caused by boredom to treating all rabbit infections as if they are *Pasteurella*. Dogmas addressed in this lecture include peas as a causative agent for bladder stones in ferrets, treatment of feather plucking as a behavioral disease, calcium restriction as bladder stone prevention in guinea pigs, gastrointestinal (GI) stasis as a disease rather than a symptom, and others. Recent literature on each topic is summarized to help break each dogma. This lecture serves as an update to exotic pet medicine to help practitioners make better-informed treatment and diagnostic decisions and better guide their clients.

Breaking Down Common Dogma

Bladder stones in guinea pigs are caused by high-calcium foods, and should be prevented with calcium restricted diets

Bladder stones in guinea pigs are classically thought of as a problem of genetics and calcium metabolism. Calcium carbonate stones are the most common stone identified in guinea pigs[1] and owners are frequently advised to restrict calcium and oxalates in their guinea pig's diet to prevent calciuria. This calcium restriction can increase calciuria by triggering secondary calcium mobilization by the parathyroids in response to dietary restriction. More recent studies in humans have shown nitrate ingestion and excess sodium are more likely the culprits for calciuria and secondary stones.[2] Excessive ascorbic acid supplementation may also contribute as the breakdown of NO2- into NO is catalyzed by ascorbic acid, and NO can then inhibit sodium transport in the proximal convoluted tubule, reducing calcium reabsorption. Reducing foods high in nitrates such as spinach[3,4] and advising against calcium restriction is likely key to stone prevention in guinea pigs.

Ferret cystine bladder stones are caused by peas

Ferret urinary obstruction due to cystine urolithiasis has become a common presenting emergency in the last decade.[5] Ferret owners are often guided to avoid grain free diets due to grain replacers such as peas, however recent studies have not shown any correlation between peas or specific legumes and cystine stone formation.[6,7]

As noted in one review, not all ferrets fed a grain free diet form cystine stones, and the disease is likely more complex than a simple ingredient. The ideal diet for a cystine stone former is an alkalinizing, low protein, high water content and low sodium diet. Unfortunately, ferrets require a high protein level and as carnivores have acidic urine normally. Although we are not yet sure what component of certain diets is causing this stone formation in ferrets, the assumption that peas and legumes are the culprit is unfounded and misleading. Because we cannot screen for risk, grain free foods should be discouraged, but owners should be made aware that research has not yet elucidated the specific components of grain free foods and low-quality dog/cat kibble causing stone formation.[8] Until further information is available, the safest preventative option is to maintain ferrets on a non-grain free diet formulated specifically for ferrets, and non-grain free treats labelled for ferrets. Additionally it is recommended not to mix multiple brands of food to limit risk.

GI stasis is a disease

Rabbits frequently present on emergency for having GI stasis. Many GI stasis protocols commonly aim to encourage gut motility, without taking into account that often gut motility is affected by elevated cortisol levels in response to stressors[9,10] and treating GI stasis as a disease rather than a symptom can reduce the efficacy of treatment for the animal. Framing GI stasis as a disease which can be treated with motility agents may also encourage owners to try to treat at home rather than ruling out potential causes for anorexia in their pet. Finding the underlying disease process is imperative to treating the animal's discomfort. Veterinarians are encouraged to reframe these symptoms to owners as a sign of distress, and to narrow down a differential list via diagnostic testing prior to administering a medication protocol.

Differentials vary based on an animal's age, husbandry, recent introduction to new pets, vaccine status, spay/neuter status and other medical history, but generally will start with a CBC and chemistry and radiographs to rule out renal failure, liver lobe torsion, anemia, sepsis, foreign body ingestion and gas distension of the GI indicating enteritis. Further work up can include *Encephalitozoon cuniculi* testing, advanced imaging for thorough evaluation of dental structures, joints and internal anatomy, and clostridium toxin testing, etc. Commonly, GI stasis protocols will include metoclopramide or other gut motility agents and meloxicam, without evaluation for underlying cause or normal renal function. Practitioners treating rabbits are encouraged to consider their medications similarly to an older cat when evaluated for anorexia, judging medications based on safety and chances of improving comfort.

Respiratory rabbits have snuffles, which can be assumed to be *Pasteurella* infection

Rabbits presenting with signs of respiratory disease are frequently diagnosed with snuffles, a layman's term assumed to be *Pasteurella* infection, without culture. As with any potential bacterial infection, empirical antibiotic use without culture and susceptibility testing goes against antibiotic stewardship and should be avoided when able. Additionally, assuming the identity of an infection may lead to frustration if not responding to empirical antibiotic choice. Terming any respiratory signs snuffles additionally simplifies cases which are potentially more complex than primary infection. Any diagnosis in a respiratory rabbit should take into account signalment and potential source of infection- what is the patient's dental disease risk based on age, breed, and diet, and is a dental abscess possibly the source of infection? Is this pneumonia or localized to the upper respiratory tract? Are the patient's naso-lacrimal ducts patent and what are the implications for long-term symptoms and treatment? These considerations may be sidelined by the assumptions made when diagnosing snuffles.

Retrospective reports of cultures in cases of oral and respiratory disease in rabbits show that infections are often mixed, composed of aerobic and anaerobic bacteria. Most studies find infections to mainly be composed of *Fusobacterium*, *Bacteroides*, *Pseudomonas*, *Prevotella*, *Pasteurella*, *Moraxella*, *Mycoplasma*, *Actinomyces*, *Staphylococcus* and *Streptococcus*.[11-13] Culture and susceptibility often indicate the necessity of multi-modal antibiotic therapy. Veterinarians are encouraged to work up any infection based on presenting symptoms and to the letter of antibiotic stewardship, including imaging, culture, and blood work to fully evaluate origin and extent of infection when able.

In this author's search, few studies in the literature include anaerobic culture,[14,15] potentially, causing over-reporting of aerobic and facultatively anaerobic species such as *Pasteurella*. Because rabbits form caseous pus and encapsulated abscesses, and due to the normally anaerobic inclusive flora of the rabbit oral cavity,[13] anaerobic infection should be considered in all respiratory rabbits and either aerobic/anaerobic culture, or next-gen DNA sequencing culture are recommended.

Feather plucking is caused by boredom

Feather destructive behavior (FDB) is classically considered a sign of boredom in avian species; however, this is a diagnosis of exclusion. Reducing stressors ultimately reduces FDB in many cases, however underlying causes are extensive and should be evaluated as FDB can indicate underlying pathology. Considerations such as age, recent introduction of new birds, boarding facility stay, access to household toxins and diet should be evaluated and a differential list should be formed. Primary differentials include infectious disease pathology (e.g., avian ganglioneuritis caused by bornavirus, chlamydiosis, circovirus, polyomavirus, parasitism, and *Aspergillus*), toxins such as heavy metals or chronic exposure to aerosolized toxins such as smoke, heart disease, hormonal and reproductive causes and causes for pain such as arthritis or gout. In animals presenting for or with FDB, a workup is recommended with priority based on signalment and history. In reproductive cases with no other indicators of disease or negative testing presumed to be behavioral this author will sometimes employ a Lupron trial. A single dose of Leuprolide acetate (Lupron Depot, AbbVie, Chicago, IL, USA) is administered and the owner is instructed that if signs improve after the initial week a deslorelin implant (Suprelorin, Virbac Animal Health, Fort Worth, TX, USA) may be considered. In cases responsive to hormonal control, testing for ganglioneuritis caused by bornavirus is still recommended as reproductive/ hormonal activity can be a trigger for bornavirus symptoms.[16]

Furosemide is not effective in reptiles due to the lack of a loop of Henle

As reptiles become more common pets and our husbandry improves, reptiles are living longer lifespans and we are beginning to see diseases of age. Heart failure in reptiles is occasionally reported in the literature, and medical management is rarely pursued. Although reptiles lack a loop of Henle, furosemide has been used successfully to cause experimental dehydration in bearded dragons, as well as having successful therapeutic use in one reported case.[17,18] Despite lacking a loop of Henle, furosemide is thought to work in the Na-K-Cl transporters in the distal tubules of the reptile nephron, or possibly during post-renal processing in the colon.[18,19] Regardless of mechanism, medical management of heart failure in reptiles can be offered in a clinical setting, with similar monitoring recommendations to mammals.

References

1. Hawkins MG, Ruby AL, Drazenovich TL, Westropp JL. Composition and characteristics of urinary calculi from guinea pigs. *J Am Vet Med Assoc.* 2009;234(2):214-220.

2. Damasio PC, Amaro CR, Cunha NB, et al. The role of salt abuse on risk for hypercalciuria. *Nutr J.* 2011;10:1-4.

3. Nuñez de González MT, Osburn WN, Hardin MD, et al. A survey of nitrate and nitrite concentrations in conventional and organic-labeled raw vegetables at retail. *J Food Sci.* 2015;80(5):C942-C949.

4. Brkić D, Bošnir J, Bevardi M, et al. Nitrate in leafy green vegetables and estimated intake. *Afr J Tradit Complement Altern Med.* 2017;14(3):31-41.

5. Hanak EH, Di Girolamo N, DeSilva U. Composition of ferret uroliths in North America and Europe: 1055 Cases (2010-2018). *Proc ExoticsCon.* 2019;403-404.

6. Pacheco RE. Cystine urolithiasis in ferrets. *Vet Clin North Am Exot Anim Pract.* 2020;23(2):309-319.

7. Lamglait B, Brieger A, Rainville MP, et al. Retrospective case control study of pet ferrets with cystine urolithiasis in Quebec, Canada: epidemiological and clinical features. *J Vet Med Surg.* 2021;5(1):31.

8. Stockman J, Malka S, Lofgren N, et al. Determining the urine cystine and amino acid concentrations in domestic ferrets (*Mustela putorius furo*). *J Exot Pet Med.* 2023.

9. Jang SJ, Kang SS, Son SJ, et al. Cortisol levels and gastrointestinal disorders after stressful surgery in rabbits. *In Vivo.* 2017;31(4):637-640.

10. Schoemaker N. Recurrent gut stasis: advanced therapeutic options and how to diagnose and treat the underlying causes. *Proc BSAVA Congr.* 2016;148-149.

11. Gardhouse S, Sanchez-Migallon Guzman D, Paul-Murphy J, et al. Bacterial isolates and antimicrobial susceptibilities from odontogenic abscesses in rabbits: 48 cases. *Vet Rec.* 2017;181(20):538.

12. Taylor WM, Beaufrère H, Mans C, Smith DA. Long-term outcome of treatment of dental abscesses with a wound-packing technique in pet rabbits: 13 cases (1998–2007). *J Am Vet Med Assoc.* 2010;237(12): 1444-1449.

13. Tyrrell KL, Citron DM, Jenkins JR, Goldstein EJ. Periodontal bacteria in rabbit mandibular and maxillary abscesses. *J Clin Microbiol.* 2002;40(3):1044-1047.

14. Abd-Algawad R, Mowafy R, Salem SM, et al. Molecular characterization of snuffle-causing bacterial pathogens in rabbits in terms of prevalence and pathogenicity. *Egypt J Agr Res.* 2021;99(4):475-485.

15. Pellett S, Harcourt-Brown F, Pinborough M. Bacterial and fungal culture and sensitivity from rabbits presenting with nasal discharge. *Proc BSAVA Congr.* 2016;553-553.

16. Rossi G, Dahlhausen R, Orosz S. Avian ganglioneuritis in clinical practice. *Vet Clin North Am Exot Anim Pract.* 2018;21:33-67.

17. Parkinson, Lily A, Mans C. Effects of furosemide administration to water-deprived inland bearded dragons (*Pogona vitticeps*). *Am J Vet Res.* 2018;79(11):1204-1208.

18. Lofgren N, Malka S, Vecere G. Therapeutic use of furosemide in a bearded dragon with pericardial effusion. *Proc ExoticsCon.* 2021;475.

19. Goncharevskaya OA, Monin Y, Natochin Y. The influence of furosemide and Co 2+ on electrolyte and water transport in newt distal tubule and frog skin. *Pflügers Arch.* 1986;406(6):557-562.

Section 17

Hybrid Session 2

Dan Johnson, DVM, DABVP (Exotic Companion Mammal Practice)
Session Chair

No Patient is Too Small to Help: Microsurgery for Exotic Pets

Zoltan Szabo, Dr med vet, Dipl ACZM, DABVP (Exotic Companion Mammal Practice), DABVP (Avian Practice), GpCert(ExAP), MRCVS

Session #3175

Affiliation: From Concordia Pet Care, No. 5-7, Blue Pool Road, Happy Valley, Hong Kong.

Abstract: Similar to the increasing popularity of small mammals as pets, owners' demands for state-of-the-art surgical treatments are also increasing. Guinea pigs, chinchillas, hedgehogs, and hamsters are the smallest mammals that most veterinarians see; however, the small size of these patients has some unique surgical challenges. Microsurgical techniques require the surgeon to work under magnification with delicate tissues in a limited space. The masterclass discusses the recommended surgical instruments, magnification methods, and surgical techniques necessary to perform successful soft tissue and orthopedic surgeries on the smallest of exotic pets.

Introduction

The anesthetic risk of rodents is higher than that of dogs and cats. These small mammals are especially prone to complications such as hypovolemia from blood loss, hypothermia, and respiratory compromise. The owners' lack of knowledge and experience with the species can also lead to unrealistic expectations and complications. Some clinical signs, like anorexia, are mild in dogs but can indicate a potentially life-threatening problem in rodents. Common diseases, like abscesses or intestinal obstruction, can be straightforward in dogs and cats, but the prognosis is usually worse in rodents. Postoperative complications are also more frequent in these species.[1]

Preoperative Period

Ideally, every patient should undergo preoperative diagnostic testing, including hematology and blood biochemistry, radiography, or contrast-enhanced computer tomography. Subclinical renal disease is common in rabbits, and respiratory disease in guinea pigs. Perioperative antibiotic therapy for routine, sterile surgeries (e.g., neutering and skin mass removals) is unnecessary.[2]

Preoperative fasting

Rabbits and rodents cannot vomit, so fasting before surgery is unnecessary; however, the syringe feeding of anorectic patients should be ceased for 1-2 hours before anesthesia to ensure the oral cavity is empty. If the mouth is full of food, that makes endotracheal intubation in rabbits difficult, and it could lead to aspiration pneumonia in guinea pigs or chinchillas. To prevent this complication the oral cavity should be cleaned at them time of the premedication.[3]

Fluid therapy

Parenteral fluid administration is recommended for every surgical procedure and should be initiated before surgery. Intravenous catheter placement is surprisingly easy in rabbits and hedgehogs because their skin is relatively soft. Guinea pigs are the opposite; the thick skin on their limbs makes catheter placement challenging. The standard recommendation for intravenous fluid support during anesthesia and surgery is 10 ml/kg per hour; however, IV catheter placement could be impossible in smaller patients, and intraosseous catheters are rarely used in clinical practice because most clinicians don't find them practical. Therefore, subcutaneous fluid administration is the norm. If maintaining vascular access is possible after surgery, continue administering intravenous fluids until the patient is discharged from the hospital.[4]

Thermal support

Smaller patients have a relatively larger body surface than larger ones. Therefore, hypothermia during anesthesia and surgery is very common in small rodents. Therefore, every anesthetized patient's core temperature should be monitored continuously, and warming devices must be used during every surgery.[4]

Hemodynamic support

The total circulating blood volume of these tiny patients is minimal; therefore, seemingly small amounts of blood loss can have disastrous effects. In addition, blood from a conspecific for transfusion is usually not available. Therefore, strict attention to hemostasis and prevention of blood loss is critical. During surgery, bipolar electrosurgical forceps should be used to seal any bleeding blood vessels immediately. In addition, metal hemostatic clips can be used for bigger blood vessels and in areas that are difficult to reach with conventional ligatures.[4]

Analgesia and Anesthesia

Because of the fragile nature of rodents, sedation prior to any stressful intervention is essential. In the perioperative period, multimodal anesthesia, using a combination of systemic and local anesthetic drugs that target different steps of the body's pain transmission pathway, is essential. Gas anesthesia is usually delivered via tight-fitting facemasks, which is the preferred method to maintain anesthesia in small rodents. While endotracheal intubation is technically possible, it is rarely done in clinical practice. The physiologic parameters (heart rate, respiratory rate, oxygen saturation, body temperature) should be monitored and maintained during the surgery.[5]

Surgical Preparation

The skin is aseptically prepared prior to surgery. The skin of most exotic mammals is thin, and the fur is dense; therefore, clipping should be performed carefully to prevent skin damage. Gentle and atraumatic skin handling reduces postsurgical pain and self-inflicted wound trauma. Water-based disinfectants should be avoided to reduce heat loss, and alcohol-based ones should be used instead. The organs should be handled gently and examined in situ during abdominal exploration without unnecessary manipulation. Regular lavage with warm, sterile saline, as well as the use of moistened gauze squares, helps prevent drying of the abdominal organ surface of the organs. Wound infection is usually due to contamination or self-inflicted wound trauma. It can be prevented with gentle surgical site preparation, atraumatic tissue handling, wound closure by apposing edges correctly and with no tension, minimally reactive suture materials, and perioperative local anesthesia.

Surgical Instruments

Most instruments used for eye surgeries are unsuitable for small rodents' microsurgery. The ophthalmic instruments are short, making them challenging to handle and manipulate tissues within a body cavity. Full-length, dedicated microsurgical instruments are designed so that only their tip is miniaturized. The instruments' tips extend beyond the surgeon's fingers far enough to use within the patient's body cavity. Rounded handles are important for the needle holders, where a curved needle must be moved through the tissue by rolling the needle holder with the fingers.

A microsurgical kit minimally should consist of a microsurgical needle holder, micro scissor, and two thumb forceps. Ring-tipped microsurgical forceps are of a standard length, counterbalanced, and delicate, with a tip that has a ring to distribute the pressure over a wider surface, making them less traumatic but allowing the surgeon to hold the delicate strand of suture for knot tying.

Sterile cotton-tipped applicators are helpful for gentle tissue dissection and manipulation, and they can also be used to absorb blood or fluids. Moisten the applicators for tissue manipulation and dissection and use them dry to absorb fluids. An absorbable gelatin sponge and other absorbable hemostatic materials help to control bleeding.

Magnification

Magnification is recommended because the surgeon's inability to see small structures limits dexterity and manipulation. In addition, with magnification, small amounts of hemorrhage appear more important, and individual vessels are more easily identified. Binocular loupes are available in different styles, with cheap hobby loupes being the simplest but having several disadvantages, including a set focal distance, no focal light source, and the inability to look above the lenses when magnification is not needed. In general, the higher the magnification, the closer the lenses must be to the object to be in focus. Anything outside of the specific focal distance of the lenses will not be in focus, resulting in a blurry image.

Surgical loupes eliminate most of the problems. They are designed to be used with a good neck position, the lenses do not cover the whole visual field so the surgeon can look above the lenses, they have a focal LED light source, and the lenses have a wider focal range rather than a set focal distance, which generally allows the entire patient to be in focus during surgery.[6]

Sutures

Sutures and needles used in the average dog and cat practice are too large for use in small mammals. Suture sizes 4-0 to 7-0 are commonly used in small rodent species. Skin closure is best accomplished with an intradermal and subcuticular technique to reduce the chance of wound chewing.

References

1. Lee HW, Machin H, Adami C. Peri-anaesthetic mortality and nonfatal gastrointestinal complications in pet rabbits: a retrospective study on 210 cases. *Vet Anaesth Analg.* 2018;45(4):520-528.

2. Szabo Z. Soft tissue surgery: rodents *In:* Quesenberry KE, Orcutt CJ, Mans C, Carpenter JW, eds. *Ferrets, Rabbits, and Rodents: Clinical Medicine and Surgery.* 4th Ed. St. Louis, MO: Elsevier; 2021:467-482.

3. Szabo Z, Bradley K, Cahalane AK. Rabbit soft tissue surgery. *Vet Clin North Am Exot Anim Pract.* 2016;19(1):159-188.

4. Bennett RA. Rabbit Soft Tissue Surgery. In: Bennett RA, Pye GW, eds. *Surgery of Exotic Animals.* Ames, IA: Wiley-Blackwell; 2021:240-276.

5. Bennett K, Lewis K. Sedation and anesthesia in rodents. *Vet Clin North Am Exot Anim Pract.* 2022;25(1): 211-255.

6. Phillips H. Magnification surgery. In: Bennett RA, Pye GW, eds. *Surgery of Exotic Animals.* Ames, IA: Wiley-Blackwell; 2021:23-34.

How to Plan and Write a Case Series: Tips from the Editor of the Journal of Exotic Pet Medicine

Nicola Di Girolamo, DMV, PhD, DipECZM (Herpetology), Dipl ACZM, Dipl ACVPM

Session #3501

Affiliation: From Exotic Pet Service, Department of Clinical Sciences, College of Veterinary Medicine, Cornell University, 930 Campus Road, Ithaca, NY 14853, USA.

Abstract: Proper development and reporting of original research are fundamental to avoid research waste and ensure maximum effectiveness of resources invested in research. Case series can be defined as original research articles in which information from individual patients that share something in common is collected. This lecture covers how to plan, write, and interpret a case series.

Introduction

Proper development and reporting of original research are fundamental to avoid research waste and ensure maximum effectiveness of resources invested in research. Case series can be defined as original research articles in which information from individual patients that share something in common is collected.[1] Although case series are considered studies lower on the overall evidence pyramid,[2] they may provide important information on new or poorly described disorders.[1] Considering the paucity of research on clinical diseases in exotic patients, case series are still among the most common study designs used for clinical articles in exotic pet medicine. Typical topics suited for a cases series, include a novel or poorly described condition, or diagnostic imaging findings with a specific diagnostic (perhaps new) technique in a specific clinical condition, or outcomes of a specific intervention for a specific clinical condition.[1] Case series cannot establish effectively the effectiveness of a treatment, or the diagnostic accuracy of a technique, and as such, investigators should be careful about the conclusions drawn when interpreting the results.[1] Case series can be either prospective (i.e., when the collection of the information from the patients is planned at a certain time and is then performed over a period of time) or retrospective (i.e., when collection of information from the patients is planned at a certain time when the information is already available in a database or similar). Retrospective case series are by far the most common design, and this presentation focuses primarily on this specific study design.

How to Set Up a Retrospective Case Series

The first step in setting up a case series is typically to identify a knowledge gap. A knowledge gap is a specific piece of clinical information that is currently unknown. A knowledge gap could be something as simple as the rectal temperature of rabbits presented to a hospital, or as complex as the change in a specific analyte after a specific surgery. In exotic pet medicine, many simple knowledge gaps are currently present, and research should be tailored to answer them.

The second step consists of drafting a protocol (i.e., a document that contains background information) that will become the introduction of the final manuscript and the methods. While drafting the protocol and working on the background information, a comprehensive literature search should be performed. At this stage, it is not uncommon to find out that your knowledge gap already has been answered in another study. Previous studies should be carefully evaluated and the need for research should be reconsidered or aims may need to be slightly edited to provide useful information. For example, if interested in describing the rectal temperature of rabbits presented to a hospital, and a previous research article already reported similar data, one might either decide to avoid performing the study and employ the information found from the other study in clinical activity, or look to a certain detail (e.g., the rectal temperature during a certain season or during certain clinical conditions). It is also important to describe and record the literature search(es) performed. A common template that is advised for use is the following:

"The following databases (XX, XX and XX) have been searched with the following keywords XX XX on [date; e.g., 01/01/2020]; the following textbooks have been consulted (References). No other reports of XX have been found doing these searches."

The second part of the protocol typically includes in detail the methods that are going to be used to carry on the case series. Important methodologic aspects that need to be considered when performing a case series can be summarized in those that pertain to inclusion of patients, such as population of interest, inclusion criteria, medical record search, and those that pertain to extraction of information from the records, such as what data were extracted and how data were extracted. It is critical at this stage to define exactly how and when the medical record database will be searched, to evaluate the need for multiple assessors of the medical records, in case extraction of certain information may be debatable, and to create definition for data that will be extracted. For example, something as simple as the presence of fever in a patient may be extracted in many ways from a medical records and appropriate definition is important to avoid imprecisions.

Reporting a Case Series

Introduction

The introduction explains the context of the case series and should clarify (1) what is the current scientific knowledge on the topic and (2) why the topic deserves further investigation. Keep the introduction brief and focus only on previous original research articles. Typically two to three paragraphs summarizing relevant research are sufficient. The final paragraph of the introduction presents the specific objectives. The terminology "To our knowledge…" is not acceptable anymore; instead to replace it with a thorough literature search.

Materials and methods

The following headings (or a variation thereof) are expected to be included when considering a case series for publication.

- Study design
- Population (inclusion and exclusion criteria, recruitment)
- Medical record search
- Data extracted
- Variables of interest and outcomes
- Data reporting ± statistical analysis

Results

The following headings (or a variation thereof) are expected to be included when considering a case series for publication.

- Population
- Signalment and history
- Presenting complaints
- Physical examination
- Diagnostics
- Treatments
- Outcome

These headings should be adapted to the topic of the case series. For a case series focusing on diagnostic imaging findings in a specific condition, it would be natural to expect slightly different headings like the following:

- Population
- Signalment, complaints & PE
- Diagnostics
- Radiographs
- Ultrasound
- Computed tomography
- Treatments and outcome

Discussion

In the discussion the results of the case series should be reported in the context of existing evidence. Cautiously interpret the results (and methods) compared to existing literature. In terms of paragraphs, a typical discussion could be composed as follows:

(1) three to four paragraphs that summarize the main three or four results considering the study objectives;

(2) limitations of the study, including sources of potential bias and imprecision; and

(3) generalizability (external validity) of the results.

One of the most common mistakes made when writing the discussion and conclusions section of a case series is to reach conclusions based on the results of the study without considering the lack of a control group. For example, if a case series on appendicitis in rabbits found that 90% of the appendicitis occurred during summer and fall, this does not automatically mean that appendicitis as a disease is more common during those seasons. Since the higher percentages of cases reported during those seasons could be a mere consequence of overall increased caseload in that season, or even something more trivial, such as data extraction that for some reason excluded cases from other seasons. As such, authors of case series should understand that only limited conclusions can be drawn from this uncontrolled type of research.

References

1. Albrecht J, Meves A, Bigby M. Case reports and case series from Lancet had significant impact on medical literature. *J Clin Epidemiol.* 2005;58(12):1227-1232.

2. Murad MH, Asi N, Alsawas M, Alahdab F. New evidence pyramid. *Evid Based Med.* 2016;21(4):125-127.

Section 18

Hybrid Session 3

Byron de la Navarre, DVM
Session Chair

Electrochemotherapy in Exotic Animals: A Review of the Literature

Brynn McCleery, DVM, DABVP (Avian Practice)

Session #3135

Affiliation: From Red Bank Veterinary Hospital, 197 Hance Avenue, Tinton Falls, NJ 07722, USA.

Abstract: Electrochemotherapy (ECT) is an ablative therapy for tumors that combines electroporation with the use of hydrophilic chemotherapeutics, commonly bleomycin or cisplatin, either locally or systemically. ECT allows the tumor cells to become more permeable to the chemotherapeutic agents resulting in high intracellular concentrations and, subsequently, increased antitumor activity. Use of ECT in human medicine is well established, while this therapy is still novel in veterinary medicine. Preclinical and clinical veterinary studies have shown ECT to be an effective and safe therapeutic option for multiple cutaneous and subcutaneous tumor types in dogs, cats, and horses. This therapy has also been utilized in multiple zoological species. This master class serves as a review of ECT and its use in exotic animal species.

Introduction

To reach their biological targets, chemotherapy drugs must overcome a major obstacle: the cell membrane of tumor cells.[1,2] Electroporation (i.e., electropermeabilization) is a method utilizing high-voltage electric field pulses to induce an electrically mediated reorganization of the plasma membrane of cells. This leads to formation of aqueous pathways (i.e., electropores or transient permeant defects), and, therefore, leads to targeted permeabilization permitting the cross flow of material.[1,3,4] Electrochemotherapy (ECT) combines electroporation of tumors with local or systemic administration of chemotherapeutic drugs that have poor membrane permeability on their own and greatly increases uptake of drugs by the tumor cells, leading to more effective cellular death.[1,3,4]

ECT has been added to the armamentarium available for local and systemic cancer management, along with surgery, radiation therapy, and chemotherapy. In veterinary oncology, late referrals and/or rapid tumor growth often prevent achievement of local control with surgery alone. A multimodality approach involving adjuvant therapies is often required.[1] ECT, along with chemotherapy and radiotherapy, can help achieve better control in these cases. Chemotherapy can be used to reduce tumor volume permitting wide excision, reduce the probability of local recurrence after surgical excision of some tumor types, and reduce the chances of distant dissemination (its major application).[1] The clinician must balance attaining sufficiently high levels of chemotherapeutic drug in the tumor (through high dose intensity and dosing frequency) to achieve cytotoxicity of tumor cells without causing systemic toxic effects, such as bone marrow suppression or gastrointestinal symptoms.[2,4] This problem can be avoided with ECT, which utilizes lower doses of chemotherapy drugs and is associated with greatly reduced side effects. Radiation therapy is utilized for the treatment of solid tumors as an adjuvant therapy or for treatment of inoperable neoplasms. Compared to radiation therapy, ECT provides more targeted tumoricidal therapy and spares healthy tissue with less side effects to collateral tissues.[2]

Overall, ECT can be used as a curative, palliative, adjuvant, neoadjuvant, or concomitant therapy for local control of solid neoplasms.[1,5]

Mechanism of Action

The principal mechanism of action of ECT is electropermeabilization of cells in tumors allowing chemotherapeutic drugs to reach their intracellular targets (i.e., increased drug permeability).[3] Cells have two possible fates after electroporation: (1) reversal of the process and return to the previous steady state or (2) impossibility to reverse the ion fluxes, activation of the caspase apoptotic or necrotic pathways, and, ultimately, cell death.[1] Reversible electroporation is performed most commonly to introduce drugs or genetic material into the cells without affecting cell viability.[6] Membrane sealing in vitro typically occurs within minutes, depending on the electrical parameters used. In a mouse model, 63% of membrane resealing occurred within 9 minutes of the applied electrical pulses.[2]

Other mechanisms exist in addition to electropermeabilization that increase the effectiveness of ECT. These include the following mechanisms.

Vascular lock phenomenon

A transient but reversible reduced blood flow occurs immediately following electrical pulse application leading to drug entrapment in the tissue for several hours.[3,4] The reduced blood flow occurs due to reflex vasoconstriction of the afferent arterioles and is medicated by the sympathetic nervous system.[2] The hypoperfusion is transient in normal tissue but lasts up to 12 hours with electroporation only and 5 days with electrochemotherapy.[3] This mechanism traps the chemotherapy drug in the treated area, providing more time for the drug to act, and also prevents bleeding from the tissue, which is an important feature when treating hemorrhagic tumors.[2,3]

Vascular disrupting action

Cytotoxic damage extends beyond just the tumor cells and affects stromal cells, including endothelial cells of tumor blood vessels, resulting in cell death. This leads to disruption of tumor blood flow and death of tumor cells surrounding the vessel.[3,4]

Massive tumor antigen shedding

After ECT, tumor antigen shedding leads to induction and upregulation of systemic immunity to neoplastic cells via biological response modifiers including interleukins (ILs) 2 and 12, granulocyte-macrophage colony stimulating factor, and tumor necrosis factor alpha.[3,4] The immune response appears to play a crucial role in the efficacy of ECT. Immunocompromised laboratory mice had a lower response to treatment compared to those with normal immune systems.[5]

Development

ECT has received considerable attention over the past 30 years. The term was first coined in 1991 in a study evaluation the modality in nude mice.[7] Further in vivo, preclinical studies in various laboratory animal models (including mice, rats, hamsters, rabbits) showed antitumor effectiveness of ECT on subcutaneous tumors grown in muscle, brain, or liver and of different histologic type (including sarcomas, carcinomas, gliomas, and melanomas). These studies showed minimal toxicity and good safety of ECT.[3] Based on the preclinical studies, clinical studies were started in human patients. Clinical studies were initiated in veterinary medicine soon after the first human clinical trials were underway, and the first veterinary clinical trials involved treatment of soft tissue sarcomas in feline patients.[1,3] The modality became more routine in 2006 after the European Standard Operating Procedures for Electrochemotherapy (ESOPE) published a standard operating procedure for ECT in human patients and a medical grade electroporator was approved (Cliniporator).[5,7]

Effectiveness

Clinical studies in humans have demonstrated a pronounced antitumor effectiveness of ECT. In a prospective, nonrandomized, multi-institutional study, an objective response rate of 85% (74% complete and 11% partial responses) of cutaneous and subcutaneous tumors was reported, regardless of tumor type (histologic origin), drug utilized, route of drug administration, or type of electrodes employed for tumors 3cm in diameter or smaller.[2,3,8] Additionally, a good response was noted in tumors that were previously resistant to chemotherapy or radiotherapy.[2]

Clinical studies in veterinary patients (i.e., dogs, cats, horses) have also shown high efficacy of ECT in the treatment of cutaneous and subcutaneous tumors of different histologic types with an approximately 80% long-lasting objective response rate.[3] Types of tumors that have been successfully treated include mesenchymal tumors (soft tissue sarcomas, fibromatosis), epithelial tumors (perianal and anal sac tumors, carcinomas, ganglioneuroblastoma, trichoblastoma), round cell tumors (melanoma, mast cell tumors, lymphoma, Sticker's sarcoma), and equine sarcoids.[1]

Indications

ECT is indicated for the following cases[5]:

- Primary or metastatic cutaneous or subcutaneous tumors that cannot be sufficiently treated with their respective first-line treatments
- Tumors affecting quality of life due to bleeding, ulceration, and/or pain
- Oral or nasal tumors as a solo treatment or in combination with surgery
- Incompletely resected tumors or as a method to extend the safety margins during surgery
- As a primary therapy if first-line therapies are possible but rejected by the owner
- To reduce cancer burden in primary or metastatic tumors that are accessible to the electrodes (superficial or via surgery)
    - Patients showing a poor response to systemic therapy
    - Large tumors prior to surgery to improve relapse-free survival
    - Palliative intent if no other treatment options exist
- Curative intent for tumors up to 3cm[3] to avoid post-operative complications or if surgery would be difficult based on location

Chemotherapeutic Drugs Utilized

Drugs used with ECT cannot cross the cell membrane on their own due to chemical properties or lack of a transport mechanism (nonpermeant or poorly permeant).[2,3] Increased permeability of the tumor cells through electroporation allows the use of drugs that would otherwise require high systemic doses to be effective, but the doses utilized in ECT are so low as to have no systemic toxicity.[3,4] Drug doses with minimal or no antitumor effectiveness produced nearly an 80% complete response when delivered by ECT.[3]

Bleomycin

Bleomycin in normally transported across the cell plasma membrane by protein receptors and cannot diffuse across the membrane due to its larger size.[1,2,4] Carrier proteins internalize the drug via an endocytotic pathway.[4] Transport is therefore limited by a low number of carrier proteins exposed at the cell surface and withdrawal of the proteins from the membrane during endocytosis.[2,4] Due to this required transport system, uptake in tumor cells is slow and restricted under normal conditions. Furthermore, loss of membrane receptors is a main mechanism of chemoresistance again bleomycin. These obstacles are bypassed by electropermeabilization, which allows almost free diffusion of bleomycin into cells for as long as they remain permeable (up to 1 hour).[1,2] Once inside the cell, bleomycin's mechanism of action is through fragmentation of DNA resulting in G2-M phase arrest of the neoplastic cell cycle.[1,5] The number of DNA lesions dramatically increases leading to apoptosis when bleomycin is used in combination with ECT.[1]

Bleomycin can be administered intravenous (IV) or intratumorally. ECT with bleomycin has a comparable effect despite the route of drug administration used.[3] Typical doses for bleomycin are: 15-30 mg/m^2 (15,000-30,000IU/m^2).[1,3] Electropulsation of cells increases the cytotoxicity of bleomycin 100- to 5000-fold in vitro, depending on the study.[2,3,9] The author could not find any pharmacokinetic studies for bleomycin in avian or reptile species; a few PK studies exist in laboratory rodents.

Cisplatin

Cisplatin has poor cellular permeability with only approximately 50% of drug transported through the plasma membrane by passive diffusion.[2,4] The platinum complexes in cisplatin bind DNA bases leading to cross-linking of DNA and subsequent cell death.[1,2] Electroporation enhances the transmembrane passage of cisplatin by a factor of 2-8 times, and, therefore, increases the number of cross-links noted in tumor cells.[1,4] Although typically not tolerated well by cats systemically, cisplatin used intratumorally with ECT is well tolerated.[1] Electropulsation increases the cytotoxicity of cisplatin 8- to 80-fold in vitro.[2-4] Cisplatin is only administered intratumorally when used for ECT in veterinary patients.[3] It is considered an ideal drug for intratumoral injection as it does not cause necrosis.[10] However, necrosis was seen in a case of ECT used in a cockatiel with a facial fibroma and has also been reported in canine patients. In the cockatiel, necrosis occurred approximately 2-3 days after the first treatment and sloughing followed.[11] The pharmacokinetics of cisplatin have been evaluated in an avian species (sulphur-crested cockatoos), rabbits, and rodents.[12-16]

Combination

Bleomycin and cisplatin can also be combined during a single ECT session, typically with bleomycin administered by bolus IV and cisplatin administered intratumorally.[1,9]

Other Drugs Reported

Doxorubicin

Doxorubicin belongs to the anthracycline class. Multiple mechanisms of action are reported including interactions with tumor DNA (base intercalation, DNA strand breakage, and topoisomerase II inhibition), DNA polymerase activity inhibition, altered regulation of gene expression, and release of free radicals. Preclinical studies suggest that ECT may improve the efficacy of doxorubicin, especially in the treatment of mammary cancer in veterinary patients.[1]

Mitoxantrone

Mitoxantrone is a synthetic anthracycline analogue with a similar mechanism of action as doxorubicin but less cardiotoxicity than other members of this drug family. Reports of improved efficacy of mitoxantrone when used with ECT exist for both human and veterinary patients. Dosing of 5 mg/mm^2 IV has been reported.[1]

Benefits

ECT is a highly effective treatment of various primary tumors and metastases of different histologic types across species. It can be used with curative intent for solitary or multiple cutaneous or subcutaneous tumors or as an adjunct treatment to surgery or radiotherapy to prevent localized recurrence by removing occult micrometastases or as a neoadjuvant therapy (facilitating other modalities by reducing the size of the mass prior to surgical resection or radiotherapy).[1,3,4] Results of treatment are comparable to other standard therapeutic approaches with a lower cost and relative ease of application.[1,3] Electroporation is conducted in seconds to minutes. Along with preparation and post-treatment observation, ECT is a quick therapy that can be performed on an out-patient basis in most cases.[2]

ECT has a high specificity for targeting cancer cells (sparing to healthy tissue) and high degree of localization of treatment effect. It also preserves innate immune response and structure of the extracellular matrix.[5,8] Compared to surgery, ECT is often less invasive.[7] Furthermore, ECT utilizes lower drug doses and is a nonthermal ablation modality.[1-6] These characteristics lead to lower morbidity and side effects. Since tumor cells are not dying secondary to application of physical energy (as with irreversible electroporation, radiotherapy, and many other ablative therapies), the cells die off in a more controllable way from apoptosis or other programmed types of cell death. This results in a slower shrinkage without the development of severe necrosis and reduces the risk of infection.[7] ECT can also be repeated without the development of chemoresistance.[7]

Adverse Effects and Contraindications

Overall, side effects of ECT are limited.[1] The most common local adverse side effect is contraction of muscles underlying the treatment area. The contractions are only present during application of the electrical pulses and resolve at the completion of treatment.[3,4] However, this modality causes discomfort, and heavy sedation or general anesthesia is required during treatment of veterinary patients.[1,3,5] In humans, ECT is typically performed under local anesthetic.[1]

Other reported side effects are mostly local and transient including erythema and edema of the skin.[6] Alopecia and pigment change have also been reported.[5] For post-operative patients, delayed healing of surgical beds and wound dehiscence are reported side effects.[17] Formation of a fistula, extensive necrosis and anatomic defects may occur if the tumor compromises the whole thickness of the tissue.[5] Severe radiation recall has also been reported in a cat.[17] If repeated or used in conjunction with standard chemotherapy, cumulative effects of bleomycin are possible and may lead to lung fibrosis. Maximum doses should be calculated and monitored (200,000 IU/m^2 in dogs, unknown in cats and exotics).[5]

ECT is contraindicated in pregnant and lactating animals.[5]

Treatment Protocol

Electropermeabilization protocols vary in terms of electrical shapes, voltage, amperage, frequency, and interpulse duration.[1] Only very brief electrical pulses are required for electroporation.[4] Protocols usually use trains of 8 square or biphasic wave electrical pulses with amplitudes (voltage to distance ratios) of approximately 1000-1300 V/cm and a pulse duration of up 100 μs (50+50 μs for biphasic pulses, interpulse interval of 300 μs to 1 ms) with a frequency of 1-5000 Hz.[1-3,5] Square wave pulses offer the advantage of controlling the pulse amplitude and length.[2,5]

Electrodes

Multiple types of electrodes exist for electroporation. These include the following electrodes.

Parallel plate

Effective for surface tumors of different sizes with limited depth.[3,5] Plate electrodes provide a more uniform electrical field distribution, but plates typically have a maximum treatment depth of < 3 mm.[3,5]

Needles

Allow for deeper application of pulses in tissues but a more heterogenous electrical field distribution. Needle probes are more invasive and are difficult to use on tough skin (e.g., horses).[3] One group of authors recommends use of needle electrodes for most tumors as needles can help surpass the electrical impedance from thicker skin and allow a more adequate distribution of the electric field to the tumor.[5]

Wire

Wire electrodes are easy to bring into contact with shaved skin and can be moved with ease in different orientations. However, a limited amount of tissue is affected by field discharge, and successive treatments are usually required for eradication of tumors.[3]

Time interval after injection

The time interval between drug injection and application of electrical pulses to the tumor is very important. Sufficient drug must be in the tumor at the time of electroporation.[3,4] During intravenous administration of cisplatin (5 mg/kg) or bleomycin (0.5 mg/kg) to laboratory rabbits and rodents, an interval of only a few minutes is required to reach maximum drug concentration in the tumor. After intratumoral administration of cisplatin (2 mg/cm^3) or bleomycin (3 mg/cm^3), the interval is even shorter, and electrical pulses should follow administration of the drug immediately (within 1 minute).[3] In larger animals, it is recommended to deliver electrical pulses to the tumor site during the pharmacokinetic peak of the medication if delivered IV; this is typically 8-28 minutes later in humans.[4] After intratumoral application, electroporation is performed 1-10 minutes later in humans, dogs, cats, and horses.[4]

Protocol details

Pre-treatment

Blood work, evaluation of anesthetic risk, cytology or histopathology confirming tumor type, and oncological staging is recommended for patients prior to ECT.[5]

Anesthesia or heavy sedation

Similar protocols should be used as for surgery in the same area.

Chemotherapy injection

IV bolus of bleomycin (over 30-45 seconds) is recommended in dogs and cats as this route almost always achieves adequate distribution and concentration of the drug in the tumor and its margins when timing of electroporation is appropriate, is safer due to less leakage after injection, avoids areas of tumor with insufficient drug concentrations secondary to errors in intratumoral injection technique, and is an immune system activator that contributes to a local immune response.[5] However, the intratumoral route is preferred in exotics due to lack of information on dosing and effects of systemic bleomycin.[5]

Time interval

Electrical pulses are applied 5-8 minutes after IV bleomycin or immediately to 5 minutes later after intratumoral injection of bleomycin or cisplatin in most species.[5] As noted above, timing in laboratory rabbits and rodents was shortened with peak drug concentrations in the tumor occurring within minutes after IV dosing and < 1 minutes for intratumoral application.[3]

Electroporation pulses

Electroconductive gel improves contact of the external electrodes to lesions.[1] ECT only affects cells directly around the electrode, so electroporation should be repeated several times until the clinician is satisfied that the full surface of the tumor and tumor margins have been treated, taking care to avoid treatment overlap.[4]

Post-treatment

Monitor patients after therapy for a couple of hours.[2] Post-treatment NSAIDs have been recommended by some clinicians.[5]

Repeat

ECT is often performed at 2-week intervals for two sessions for post-operative adjunctive therapy or until complete response/tumor regression if for direct ablation of gross disease.[1] A minimum treatment interval of 4 weeks had been recommended for ECT with IV bleomycin.[5] Maximum therapeutic effect may occur 6-8 weeks later or longer as tumors can enter a quiescent state and shrink again after 2-3 months once cells try to divide again (as bleomycin's mechanism of action affects dividing cells). Based on this delay, some clinicians recommend monitoring and only treating again if the tumor is not sufficiently treated in the first session, starts to grow, or new lesions develop to prevent overtreating and development of necrosis.[5] Stagnant lesions can be biopsied to evaluate for persistent tumor versus scar tissue. Frequent monitoring is recommended after treatment, often at 2 weeks, monthly for 3 months, then every 3 months (or a similar variation).[5]

Evolutions

Four evolutions have been reported for tumors after treatment with ECT. Understanding the evolutions can help determine if repeated treatment is needed for a particular neoplasm.[5]

1. *Common evolution:* The tumor shrinks after initial swelling and continues to shrink to the final response.
2. *Two times evolution:* The tumor shrinks, becomes quiescent, then shrinks again.
3. *No change evolution:* No shrinking is seen due to a high population of quiescent cells.
4. *Tumoral escape evolution:* After an initial regression, the tumor starts growing again. These masses should be retreated.

Future Application

Newer research on ECT involves treatment of new tumor types (e.g., feline clear cell thymoma, nasal cell tumors, rectal neoplasms), use of ECT as an adjunct to novel therapies (e.g., cold plasma in a mouse melanoma model), evaluation of new or nonvalidated chemotherapy drugs (e.g., doxorubicin, 5-fluorouracil), and development of new electrodes (e.g. grid, single needle, endoscopic). There are also newer studies evaluating the use of electroporation with gene-based products (electrogene therapy), immunomodulatory agents, and other candidate compounds, such as calcium, oligonucleotides, plasmids, and small proteins.[1,2,4,7]

New electrodes have been developed to allow ECT therapy to be used on visceral and intracavitary tumors using endoscopic guidance or an intraoperative approach.[1,2,4] The endoscopic electrodes attach to standard endoscopes and allow treatment of surgically inoperable neoplasms and cancers recalcitrant to radiation therapy.[2] One of the endoscopic electrodes in development has a vacuum feature which sucks tissue into contact with the electrodes during treatment.[2]

Cases in the Literature

A summary of published cases of ECT in exotic animals is provided in Table 1.

Table 1. Published reports of electrochemotherapy in exotic animals.

Species (Age, sex)	Tumor	Treatment (Intratumoral)	Outcome
Central bearded dragon[a] *Pogona vitticeps* (7.8-year-old male)	Fibrosarcoma (oral mucocutaneous) 10 × 20 mm	Cisplatin (10 mg/ml) 0.1 mg/cm^3 0.5- to 1-second intervals Six treatments q2wk	Reduced size 90% after sixth treatment, resolved at 18 months No recurrence at 2 years
Yellow bellied slider[b] *Trachemys scripta scripta* 5-year-old (female)	Squamous cell carcinoma (cutaneous on neck, well differentiated incompletely excised) 2 × 1.5 cm	Bleomycin (1 mg/ml) 1 mg, 1 ml 5 minutes after injection, 1200V/cm, 1Hz, 100µsec (50 + 50 µsec) Two treatments q2wk	Treated surgical scar and 1-cm margin No recurrence at 1 year
Green sea turtles[c] *Chelonia mydas* n = 2	Fibropapillomas (herpesvirus) a. 1.43 × 1.4 cm carapace b. 1.52 × 1.12 cm left anterior flipper	Bleomycin 1U/cm^3 4 minutes after injection, needle electrode, 1000V Two treatments, q33d	Response starting at 4 weeks, complete regression at 10 months No recurrence at 1 year Note: local anesthetic (lidocaine) only
Cockatiel[d] *Nymphicus hollandicus* 12-year-old female)	Squamous cell carcinoma (uropygial, ulcerated) 9 × 7 × 4 mm	Cisplatin (2 mg/ml) 1 mg/kg (first) Bleomycin 0.5 mg/kg (second, third[d]) q1mo Bleomycin 1 mg/kg (fourth) 3 months later 1-2 minutes after injection, plate (first) and needle electrodes, trains of eight pulses, 1300V/cm, 5 Hz, 100 µsec	No change in tumor size with cisplatin Mass necrotic and regressed with two treatments of bleomycin (90% reduced at 4 weeks, 100% after third treatment) Recurrence 3 months later- fourth treatment No recurrence at 18 months
Cockatiel[e] *Nymphicus hollandicus* (11-year-old male)	Fibroma (cutaneous, left face) 10 mm diameter	Cisplatin 1 mg/kg, 0.05 ml 1-2 minutes after injection, 1300 V/cm, 5 Hz, 100 seconds Two treatments, q3mo	Regressed to 7 mm diameter after first treatment. Regressed to 4 mm after second Unchanged after 3 years
Budgerigar[f] *Melopsittacus undulatus* (10-year-old male)	Epithelioid hemangioendothelioma (subcutaneous, low-grade, left wing, incompletely excised) 2 × 1 cm	Bleomycin (2.5 mg/ml) 0.25 mg, 0.1 ml 5 minutes after injection, 1200 V/cm, 1 Hz, 50+50 µsec Two treatments q2wk	Complete regression. No recurrence at 12 months
African pygmy hedgehog[g] *Atelerix albiventris* (5 years old)	Squamous cell carcinoma (oral, ulcerated) 1 × 1.5 cm	Bleomycin (1.5 mg/ml) 1.5 mg total, 5 minutes after injection, 1200 V/cm, 1Hz, 50+50 µsec Five treatments, q1mo	Partial remission (25% regression) 5 months until tumor progression and euthanasia

Table 1 Continued. Published reports of electrochemotherapy in exotic animals.

Species (Age, sex)	Tumor	Treatment (Intratumoral)	Outcome
African pygmy hedgehog[h] *Atelerix albiventris* (5.5-year-old male)	Hemangiosarcoma (cutaneous, left scapula/forelimb, incompletely excised) 3 × 3.3 cm	Bleomycin 1U (first) and 0.7U (second) 5 minutes after injection, 1200 V/cm Two treatments, q2wk	Euthanized 2 months later for oral/cervical masses (poorly differentiated sarcomas, thyroid carcinoma), no recurrence of HSA on necropsy
Ferret[i] *Mustela putorius furo* (5-year-old castrated male)	Mast cell tumor x2 (cutaneous, tail base 0.8 × 0.8 × 0.5 cm; lateral right hock 1.2 × 1 × 0.7 cm)	Bleomycin (3 mg/ml) 0.5 mg, 0.1 ml 1-2 minutes after injection, 1300 V/cm, 5 kHz, 100 µsec One treatment	Complete regression (day 30 first, day 67 second), no recurrence at 15 months for first tumor and 11 months for second
Ferret[i] *Mustela putorius furo* (7-year-old spayed female)	Sebaceous adenoma (cutaneous, lateral left hock) 0.7 × 0.7 × 0.5 cm	Bleomycin (3 mg/ml) 0.5 mg, 0.1 ml 1-2 minutes after injection, 1300 V/cm, 5 kHz, 100 µsec One treatment	Complete regression (day 70), no recurrence at 8 months
Ferret[i] *Mustela putorius furo* (7-year-old spayed female)	Squamous papilloma (cutaneous, dorsolateral left thorax) 0.4 × 0.4 × 0.2 cm	Bleomycin (3 mg/ml) 0.5 mg, 0.1 ml 1-2 minutes after injection, 1300 V/cm, freq 5 kHz, 100 µsec One treatment	Complete regression, no recurrence at 8 months
Rats[j] *Rattus norvegicus* 18 months old, 24 months old n = 2	Mammary carcinoma (axillary, incompletely excised) 3 cm, 3.5 cm (max diam)	Cisplatin (0.5 mg/ml), 1 mg, 2 ml 5 min after injection, 1200 V/cm, freq 1 Hz, 50 + 50 µsec Two treatments q2wk	Treated surgical scar and 1-cm margin Disease free at 10 months (older rat, no tumor at necropsy) and 14 months (younger rat)

Source: [a]Doneley RJT, Sprohnle-Barrera CH, Jones EE, et al. The use of electrochemotherapy to treat an oral mucocutaneous fibrosarcoma in a central bearded dragon (*Pogona vitticeps*). *J Exot Pet Med.* 2022;41:38-41.
[b]Lanza A, Baldi A, Spugnini EP. Surgery and electrochemotherapy for the treatment of cutaneous squamous cell carcinoma in a yellow-bellied slider (*Trachemys scripta scripta*). *J Am Vet Med Assoc.* 2015;246(4):455-457.
[c]Brunner CHM, Dutra G, Silva CB, et al. Electrochemotherapy for the treatment of fibropapillomas in *Chelonia mydas*. *J Zoo Wildl Med.* 2014;45(2):213-218.
[d]Racnik J, Svara T, Zadravec M, et al. Electrochemotherapy with bleomycin in the treatment of squamous cell carcinoma of the uropygial gland in a cockatiel (*Nymphicus hollandicus*). *J Exot Pet Med.* 2019;29:217-221.
[e]Racnik J, Svara T, Zadravec M, et al. Electrochemotherapy with cisplatin for the treatment of a non-operable cutaneous fibroma in a cockatiel (*Nymphicus hollandicus*). *N Z Vet J.* 2019;67(3):155-158.
[f]Lanza A, Baldi A, Rossi G, Spugnini EP. Electrochemotherapy for the treatment of an incompletely excised subcutaneous low-grade epithelioid hemangioendothelioma in a budgerigar parakeet (*Melopsittacus undulatus*). *Open Vet J.* 2019;9(3):269-272.
[g]Spugnini EP, Lanza A, Sebasti S, et al. Electrochemotherapy palliation of an oral squamous cell carcinoma in an African pygmy hedgehog (*Atelerix albiventris*). *Vet Res Forum.* 2018;9(4):379-381.
[h]McCleery B, Morges M. Electrochemotherapy for the treatment of an incompletely excised hemangiosarcoma in an African pygmy hedgehog. *Proc ExoticsCon.* 2021;2788.
[i]Racnik J, Svara T, Zadravec M, et al. Electrochemotherapy with bleomycin of different types of cutaneous tumours in a ferret (*Mustela putorius furo*). *Radiol Oncol.* 2018;52(1):98-104.
[j]Lanza A, Pettorali M, Baldi A, Spugnini EP. Surgery and electrochemotherapy treatment of incompletely excised mammary carcinoma in two male pet rats (*Rattus norvegicus*). *J Vet Med Sci.* 2017;79(3):623-625.

Reptiles

ECT has been reported in multiple reptile species. Use of ECT may be due to difficulties associated with vascular access for chemotherapy and relative radiation resistance in these species, making traditional therapies more difficult or less effective. Treatment has been reported in a bearded dragon with an ulcerated, low-grade (based on a canine grading scheme) fibrosarcoma of the rostral maxillary mucocutaneous region of several months duration; yellow-bellied slider with an incompletely excised, cutaneous, well-differentiated squamous cell carcinoma; and two green sea turtles with herpesvirus associated fibropapillomas on the carapace and flipper.[4,8,18] The only side effects reported in these cases was muscle contraction during electric pulse application. The green sea turtles received ECT under local anesthetic only due to concerns for systemic debilitation.[18]

Birds

Various tumors have been treated with ECT in avian patients including an ulcerated, rapidly growing squamous cell carcinoma of the uropygial gland of a cockatiel; non-operable cutaneous fibroma in a cockatiel; and an incompletely excised, subcutaneous, low-grade epithelioid hemangioendothelioma on the left wing of a budgerigar.[11,19,20] Not surprisingly, all of the reports of ECT in avian patients are in small birds where complete surgical excision may not be possible, ease of administration of other therapies may be more difficult, and risks associated with traditional therapies may not be acceptable to owners.

Exotic companion mammals

ECT has been reported in two hedgehogs. The first was diagnosed with oral squamous cell carcinoma of 1-month duration and was treated with palliative ECT using bleomycin.[21] The second hedgehog received ECT with bleomycin for an incompletely excised hemangiosarcoma on the left scapula and mid-brachium.[22] Other reports of ECT in exotic companion mammals include two ferrets each treated for two cutaneous tumors (two mast cell tumors, a squamous papilloma, and a sebaceous adenoma) and two rats treated post-operatively for incompletely excised mammary carcinomas in the axillary region.[17,23] All four skin tumors in the two ferrets treated with ECT became necrotic after treatment (between days 6-15 after therapy) but regressed completely (within 30-70 days from treatment).[23]

Additional reports in zoological species:

- Asian elephant: Treated for a malignant mammary mass at the El Paso Zoo.
- Asian small-clawed otter (*Aonyx cinereus*):[24] A 7-year-old female was diagnosed with oral squamous cell carcinoma. The mass was nonresectable on CT scan and involved the tongue, ventral pharynx, and hyoid bones. The patient received three ECT treatments every 2-4 weeks using bleomycin (20 mg/m^2) IV 10 minutes prior to intralesional cisplatin (0.2 mg/kg). Reversible and irreversible electroporation was performed with biphasic electrical pulses. The rostral portion of the tumor regressed with poor response caudally. The procedure was reported to provide palliation with good quality of life (follow up or 7 months after diagnosis).
- Binturong (*Arctictis binturong*):[25] A 3-year-old male binturong at the Naples Zoo was diagnosed with squamous cell carcinoma of the left jaw (5 cm diameter) that was invasive on CT scan. ECT was performed with bleomycin (10 mg/kg IV). Repeat CT showed metastasis to the lungs and spleen, and the patient was euthanized.

- Vietnamese pot-bellied pig (*Sus scrofa*):[26] A 9-year-old FS pot-bellied pig was treated for multiple cutaneous squamous cell carcinoma lesions and a history of pruritus. The patient had raised, hemorrhagic crusts and ulcerated lesions on the flanks (bilaterally), dorsum, ventral abdomen, pelvic limbs, and caudal to the ears. ECT was performed eight times over the course of 1 year (recommended q2wk but up to 4-month delays due to travel constraints of the owners) using bleomycin IV (15 IU/m^2) 5 minutes prior to electroporation (1200-1300V/cm). Partial remission and marked improvement in pruritus was sustained for 18 months post-diagnosis. A new lesion in an untreated location was present at 2 months. Side effects of the ECT included mild erythema 1 week after treatment, mucopurulent discharge from the lesions 3-5 days after treatment, and transient weight loss.

One review[5] of ECT in veterinary medicine mentioned use of the modality in fishes and snakes; however, no references were provided and no published reports could be found on literature review.

Conclusions

Electrochemotherapy increases the internalization of nonpermeant or poorly permeant chemotherapy drugs into the cytosols of target cells.[2] ECT has a good cost-benefit ratio and can be used alone or as a neoadjuvant, adjuvant, or concomitant treatment with standard therapies for the curative or palliative treatment of primary or metastatic cutaneous and subcutaneous tumors of various histologic types in many species with low cost and side effects.[5,7] In exotic animals, ECT is an alternative option to other ablative locoregional therapies for tumors in sensitive or difficult areas to treat, when other therapies are considered too invasive or risky by the owner, where mutilation may occur, or when cost of standard therapies is too high.[7]

References

1. Spugnini EP, Baldi A. Electrochemotherapy in veterinary oncology: state-of-the-art and perspectives. *Vet Clin North Am Small Anim Pract.* 2019;49(5):967-979.

2. Impellizeri J, Aurisicchio L, Forde P, et al. Electroporation in veterinary oncology. *Vet J.* 2016;217:18-25.

3. Cemazar M, Tamzali Y, Sersa G, et al. Electrochemotherapy in veterinary oncology. *J Vet Intern Med.* 2008;22:826-831.

4. Doneley RJT, Sprohnle-Barrera CH, Jones EE, et al. The use of electrochemotherapy to treat an oral mucocutaneous fibrosarcoma in a central bearded dragon (*Pogona vitticeps*). *J Exot Pet Med.* 2022;41:38-41.

5. Tallado M, Mir LM, Maglietti F. Veterinary guidelines for electrochemotherapy for superficial tumors. Front Vet Sci. 2022;9:868989.

6. Fiorentzis M, Viestenz A, Siebolts U, et al. The potential use of electrochemotherapy in the treatment of uveal melanoma: in vitro results in 3D tumor cultures and in vivo results in a chick embryo model. *Cancers.* 2019;11:1344.

7. Cemazar M, Sersa G. Recent advances in electrochemotherapy. *Bioelectricity.* 2019;1(4):204-213.

8. Lanza A, Baldi A, Spugnini EP. Surgery and electrochemotherapy for the treatment of cutaneous squamous cell carcinoma in a yellow-bellied slider (*Trachemys scripta scripta*). *J Am Vet Med Assoc.* 2015;246(4): 455-457.

9. Torrigiani F, Pierini A, Lowe R, et al. Soft tissue sarcoma in dogs: a treatment review and a novel approach using electrochemotherapy in a case series. *Vet Comp Oncol.* 2019;17:234-241.

10. Graham JE, Kent MS, Theon A. Current therapies in exotic animal oncology. *Vet Clin North Am Exot Anim Pract.* 2004;7:757–81.

11. Racnik J, Svara T, Zadravec M, et al. Electrochemotherapy with cisplatin for the treatment of a non-operable cutaneous fibroma in a cockatiel (*Nymphicus hollandicus*). *N Z Vet J.* 2019;67(3):155-158.

12. Najjar TAO, Saad SY. Cisplatin pharmacokinetics and its nephrotoxicity in diabetic rabbits. *Chemotherapy.* 2001;47l;128-135.

13. Siddik ZH, Newell DR, Boxall FE, Harrap KR. The comparative pharmacokinetics of carboplatin and cisplatin in mice and rats. *Biochem Pharmacol.* 1987;36(12):1925-1932

14. Filippich LJ, Charles BG, Sutton RH, Bucher AM. Carboplatin pharmacokinetics following a single-dose infusion in sulphur-crested cockatoos (*Cacatua galerita*). *Aust Vet J.* 2004;82(6):366-369.

15. Filippich LJ, Bucher AM, Charles BG, Sutton RH. Intravenous cisplatin administration in sulphur-crested cockatoos (*Cacatua galerita*): clinical and pathologic observations. *J Avian Med Surg.* 2001;15(1):23-390.

16. Zehnder A, Graham J, Reavill DR, McLaughlin A. Neoplastic diseases in avian species. In: Speer BL, ed. *Current Therapy in Avian Medicine and Surgery.* St. Louis, MO: Elsevier; 2016;107-141.

17. Lanza A, Pettorali M, Baldi A, Spugnini EP. Surgery and electrochemotherapy treatment of incompletely excised mammary carcinoma in two male pet rats (*Rattus norvegicus*). *J Vet Med Sci.* 2017;79(3):623-625.

18. Brunner CHM, Dutra G, Silva CB, et al. Electrochemotherapy for the treatment of fibropapillomas in *Chelonia mydas*. *J Zoo Wildl Med.* 2014;45(2):213-218.

19. Racnik J, Svara T, Zadravec M, et al. Electrochemotherapy with bleomycin in the treatment of squamous cell carcinoma of the uropygial gland in a cockatiel (*Nymphicus hollandicus*). *J Exot Pet Med.* 2019;29:217-221.

20. Lanza A, Baldi A, Rossi G, Spugnini EP. Electrochemotherapy for the treatment of an incompletely excised subcutaneous low-grade epithelioid hemangioendothelioma in a budgerigar parakeet (*Melopsittacus undulatus*). *Open Vet J.* 2019;9(3):269-272.

21. Spugnini EP, Lanza A, Sebasti S, et al. Electrochemotherapy palliation of an oral squamous cell carcinoma in an African pygmy hedgehog (*Atelerix albiventris*). *Vet Res Forum.* 2018;9(4):379-381.

22. McCleery B, Morges M. Electrochemotherapy for the treatment of an incompletely excised hemangiosarcoma in an African pygmy hedgehog. *Proc ExoticsCon.* 2021;2788.

23. Racnik J, Svara T, Zadravec M, et al. Electrochemotherapy with bleomycin of different types of cutaneous tumours in a ferret (*Mustela putorius furo*). *Radiol Oncol.* 2018;52(1):98-104.

24. Anderson KB, Siegal-Willott J, Merrick CH, et al. Electrochemotherapy for the treatment of oral squamous cell carcinoma in an Asian small-clawed ottern (*Aonyx cinereus*). *Proc Annu Conf Am Assoc Zoo Vet.* 2019.

25. Silvestre P, Prisco F, Capasso M, et al. Oral squamous cell carcinoma in a captive-born bearcat (*Arctictis binturong*). *Proc Zoo Wildl Health.* 2019;153.

26. Weissman M, Donnelly LL, Branson K, Glahn A, Nagy D, Havis B, Johnson G, Flesner B. Electrochemotherapy for a cutaneous squamous cell carcinoma in a Vietnamese pot-bellied pig (*Sus scrofa*). *J Exot Pet Med.* 2020;30:37-43.

Introduction to Beekeeping

Elizabeth B. Mackey, DVM

Session #3368

Affiliation: From Mackey Exotic Animal Clinic, 1760 Antioch Church Road, Watkinsville, GA 30677, USA.

Abstract: This masterclass is designed to introduce veterinarians to the science of beekeeping and their role in working with the beekeepers. The unique aspects of their eusocial behavior are discussed as well as common diseases and parasites. Equipment used in the colony is also presented. The goal is to have the attendees comfortable with developing a veterinary-client-patient relationship with keepers.

Introduction

There are over 20,000 species of bees worldwide,[1] but it is *Apis millifera*, the western honey bee, that is the predominant species in beekeeping. Native to Africa, the Middle East and northern Europe, these bees are often managed so we can take advantage of their products, such as beeswax, honey, pollen and propolis. We also use their behavior of foraging for nectar, which results in pollination, as a way of boosting the yield of our commercial crops, resulting in an estimated increased yield of $15 billion.[2]

As with other species, there are bacterial, viral and fungal diseases of the honey bee. In 2017 FDA required Veterinary Feed Directives for food producing animals, and in June of 2013 over-the-counter antibiotics will no longer be available.[3] While involvement of veterinarians and their understanding of appropriate use of antibiotics is a positive change, many of us do not understand the basics of hive management and the terminology used by beekeepers. As with any other animal we treat, we need to understand the husbandry of the patient, as well as the anatomy, common diseases and common parasites.

Become familiar with your state's regulations for honey bees and how they define a veterinary-client-patient relationship. The focus with food animals tends to be having a relationship with the owner and familiarity with their apiary, whether a few hives or hundreds or more. Diseases that are reportable, as well as options to treat, may have state variations. It is also important to understand the beekeeper's experience level. Individuals may have an interest and purchase impulsively but have no understanding of the basic health and husbandry requirements. They may have poor frame handling technique and be unable to evaluate brood patterns. They may also be uncomfortable with alcohol washes that sacrifice bees for diagnostic purposes. Commercial beekeepers, on the other hand, have vast knowledge and experience. Do not be offended if they will not allow you to remove frames and insist on handling the bees themselves. The health of their queen is everything, and if they don't want to risk you rolling the queen, they can remove the frames and present them to you for assessment.

Your examination with the hive begins as with any other species you see. Questions about feeding and treatments for parasites are important, as well as details on how the colony originated. Take time to observe the bees and assess their behaviors. Are they showing normal levels of activity? Are there normal behaviors being observed, such as orientation flights, trophallaxis and waggle dances? Look closely at the frames. Are there adequate stores of nectar, pollen and honey? Is the brood pattern sufficient for raising new worker bees? Is there evidence of a failing queen, as may be seen with poor brood patterns or the presence of supercedure cells?

A key difference that must be remembered with honey bees is that the colony is the patient, not the individual bee. Honey bees, along with termites and ants, are eusocial, with generations that overlap and care for the young of others. Often referred to as a superorganism,[1] they work collectively for survival and reproduction of the colony. Understanding their natural behaviors is essential for beekeepers, who can then develop a management plan to work with, or against these behaviors.

Although the colony comprises many bees working together, it is easier to understand them if you think of them as a single organism. The colony begins small at approximately 3-5 pounds of bees. It grows during times of plenty, with the goal of storing enough food to survive the winter, as well as growing large enough to be able to reproduce. An adult colony may weigh as much as 20 pounds and be comprised of 50-60,000 individual bees. Reproduction involves the colony dividing, with the queen and approximately two-thirds of the bees leaving with her to a new location of their choice. Although this swarming process is natural for the bees, we as beekeepers do not want to lose our bees. Beekeeping requires managing.

The Bees

As an insect, the honey bee has a life cycle of complete metamorphosis. It begins as an egg, hatches into a larva, pupates, and then emerges as an adult bee. The details of that are variable, though, as within the colony not all bees are created equally! The cell size, diet, and days to reach maturity are different for queens, workers and drones.

The worker bee begins as an egg that was laid by the queen. The diploid egg (16 pair of chromosomes) is laid at the bottom of a cell of comb. The egg hatches at day three and the larva eats royal jelly for the next 3 days. This high protein substance is produced by the mixing of pollen and the secretions from the mandibular and hypopharyngeal gland of the nurse bee.[4] She is then transitioned to brood food, which is a combination of pollen and nectar that is slightly fermented to allow for longer storage. The cell is then capped and she continues to develop until ready to emerge as a full-sized adult around day 21.

Worker bees have numerous jobs and as they age they rotate through these various positions.[4] They start as being nurse bees, cleaning up the cell from which they just emerged, then taking care of feeding other developing bees. These bees are all developing in the same region of the hive, and this is collectively referred to as the brood. It is the worker bees who care for the queen and guard the entrance of the hive to protect from pests and robbers. It is the workers who communicate with the foragers (older workers) to let them know what supplies are needed and then relieve them of those loads on their return. These supplies must then be properly stored. It is the workers who determine when the colony is ready to swarm, agitating the queen until she is ready to leave. Workers remove the dead from the hive and remove the males from the colony when they are no longer needed. They protect the hive, stinging an invader if necessary. The stinging will only occur once, though, as the barbed stinger will pull the venom sac from the abdomen of the bee, resulting in her death. The last job a worker will have is that of a forager. This is a high-risk job as they must leave the protection of the hive and may travel several miles in search of nectar, exposing themselves to predators and the elements. The average lifespan for these workers is ~ 4-6 weeks. When the temperatures drop in the fall, brood production will begin to decrease, and then longer-lived workers bees will remain through the winter months. Now their focus is to keep the colony warm and survive until they can begin the cycle again in the spring.

Drones are the male bees that develop from unfertilized eggs (haploid) and they have no stinger, but do have a reproductive endophallus.[4] They are much larger than the worker bees and are readily distinguishable by their size and larger eyes. They have only one job- mate with a queen. In the spring when drone brood emerges (24 days from egg), the males from multiple hives will join together in drone congregations. Here they spend their time waiting for a queen. If one arrives they will attempt to mate with her. This is their last act as the process will remove their endophallus and cause death. If they do not find a queen they will return to the hive for the night and then go back to the congregation area the next day. When the days turn cool and the focus of the colony is winter survival, the drones are no longer necessary and are cast from the hive.

The queen has one responsibility: lay eggs. Within a few days of emerging from her cell (16 days from egg) she will go on her maiden flight in search of a drone congregation. She will mate with 15-20 drones and then return home. If she doesn't find a congregation or not many drones are present she will seek them out the next day. This mating process provides all of the sperm she will ever have, which she will store in her spermotheca.[4] The queen is recognized by her elongated abdomen. She is usually found in the vicinity of the brood as she is focused on laying egg, producing as many as 1500 in a day. She is capable of stinging multiple times, as the barbs are minimal and it is not a fatal process for her. She does not engage in hive protection, so her stinging is directed at rival queens, even if they have not emerged yet.

The Hive

The hive is the man-made physical structure we put colonies in. There are numerous types but the most common is the Langstroth hive. This hive was designed by Lorenzo Langstroth and provided removeable frames and, extremely important, the preservation of bee space.[5] This is the critical distance of one-fourth to three-eighths of an inch between structures in a hive through which a bee navigates. If the space is too small, it is not navigable, and the bees will fill the gap with propolis. If the space is wider, then there is likelihood of the bees using the space for comb. The traditional Langstroth hive is placed on a stand or bricks to elevate it above the ground. This helps to protect against insects and mice as well as being more comfortable for the beekeeper. The base is called a bottom board and can be screened (additional ventilation) or solid. On this is placed the hive body, or deep super. The structure of the bottom board results in a gap between it and the hive body that is placed on it, and this then becomes the entrance to the hive. Additional hive bodies/supers are added on top and ultimately there is a lid to protect from the elements.

Within the hive bodies are frames. These are placed in the hive body like index cards in a file box. Hive bodies can be different depths (deep, medium, shallow) and can hold different numbers of frames (e.g., 10 frames, 8 frames, 5 frames). Correct bee space is maintained regardless of the size of the box. A frame holds the wax honeycomb that is the foundation on which a colony lives. It is here that they store nectar and pollen, create and store honey, and raise the brood. Most cells are the same size, but workers can build other sizes and types as needed. If the queen detects a larger cell, she will lay an unfertilized egg which will develop into the larger drone bee. Swarm cells are designed to raise new queens. When the original queen leaves with many of the workers to start a colony elsewhere, the new queen from the swarm cell will emerge, go on her maiden flight, and return to re-build the colony. Supercedure cells, another type of queen cell, are built when the workers perceive a new queen is needed. It may be that the original one has died, or run out of sperm, or is ill. For whatever reason, they have determined that the old queen needs to be replaced.

Feeding of the colony is often provided in the form of syrup (sugar and water in a ratio of 1:1 to 2:1), which can be in containers placed inside the hive or outside the hive, or even as a feeding station well away from the hive. This may be done with a new hive that doesn't have much comb. Building the wax comb is very energy intensive, so keepers can provide support with syrup when the nectar flow isn't strong enough. Feeding might also be necessary at other times if the colony doesn't have enough reserves. Starvation is a leading cause of death in colonies, and part of beekeeping is monitoring to assure they have adequate reserves or sources of their protein and carbohydrate needs.

Protective gear is recommended when working with bees. This should always include a veil to protect the face. It is natural for bees to go for the face of an intruder. Other protective gear may include full body suits or just jackets, gloves (heavy or lightweight disposable) and a smoker. The smoke is used to block the pheromone communication and is used just enough to provide a distraction but not so much the bees fear there is a fire and abscond.

References

1. Seeley TD. *Honeybee Democracy.* Princeton, NJ: Princeton University Press. 2010;20.

2. U.S. Food and Drug Administration. Helping Agriculture's Helpful Honey Bees. 2018. Available at: https://www.fda.gov/animal-veterinary/animal-health-literacy/helping-agricultures-helpful-honey-bees. Accessed June 1, 2023.

3. U.S. Food and Drug Administration. Using Medically Important Antimicrobial in Bees. 2023. Available at: https://www.fda.gov/animal-veterinary/development-approval-process/using-medically-important-antimicrobials-bees-questions-and-answers. Accessed June 1, 2023.

4. Winston ML. *The Biology of the Honey Bee.* Cambridge, MA: Harvard University Press; 1987.

5. Delaplane KS. *First Lessons in Beekeeping.* Hamilton, IL: Dadant and Sons; 2007:29-39.

Section 19

Technician Session 1

Christina Miller, RVT, BSc
Session Chair

Gimme Shelter (and UVB and Tube Feeding): Caring for Hospitalized Reptiles

Jovanna Berls, CVA, LVT, FFCP (Veterinary), FFCP (Avian)

Session #3152

Affiliation: From ABC Animal & Bird Clinic, 11930 Highway 6 South, Sugar Land, TX 77498, USA.

Abstract: Hospitalized reptiles in some ways are similar to other species commonly seen in veterinary clinics; like all patients, they need appropriate food, caging, and medication. But they do have some specialized needs, such as ultraviolet lighting and particular attention to thermal support. This presentation reviews methods for enhancing comfort and welfare for reptile patients during a hospital stay, with attention to the different requirements for commonly seen species of lizards, snakes, and chelonians. Techniques for providing heat, light, nutrition, and hydration will be covered, along with tips for the care and feeding of live food items.

Introduction

Like any other hospitalized patient, reptiles need housing, food, water, and medication; however, there are nearly 12,000 known species of reptiles,[1] originating from across the world, with a wide range of natural histories, feeding strategies, and environmental requirements. It is simply not possible for even the most expert of veterinary technicians to know from memory what is needed for every species, but a basic grounding in commonly seen pet species will go a long way toward preparing technicians to deal with the myriad of possible reptile patients. This presentation focuses on frequently seen species of lizards (bearded dragons [*Pogona vitticeps*], leopard geckos [*Eublepharis macularius*], and veiled/Yemen chameleons [*Chamaeleo calyptratus*]), snakes (ball pythons [*Python regius*], corn snakes [*Pantherophis guttatus*]), boa constrictors [*Boa constrictor*]), and chelonians (red-eared sliders [*Trachemys scripta elegans*], box turtles [*Terrapene carolina*], and Russian tortoises [*Testudo horsfieldii*]).

These patients have specialized requirements that set them apart from other species commonly seen in veterinary hospitals. As ectotherms, they often need thermal support and specific humidity levels; many need provision of ultraviolet light; they may also require live foods. These things are necessary to maintain homeostasis, a complex process that requires a balance of appropriate nutrition, hydration, temperatures, and activity. Homeostasis is essential for normal animals to remain healthy, and ill or injured animals are less likely to recover well without them.

Effects of Stress on Healing and Recovery

For many exotic species that are essentially wild animals, captivity alone can be a source of stress, making appropriate husbandry techniques especially critical for their health, reproductive success, and even survival. If an exotic pet has been stressed in its usual home, it is likely to be more so in a hospital environment in which it will be handled by unfamiliar people and subjected to sometimes unpleasant procedures, such as venipuncture or administration of medications.[2]

Stress can be defined as a response to anything that disturbs homeostasis and the body's response as it tries to adapt and restore homeostasis.[3] It is a physiologic response related to release of catecholamine and, in reptiles, corticosterone from the adrenal glands. This chemical release has many effects, some helpful and some not, such as elevated heart and respiratory rates, reduced digestive function, and redirection of blood flow from visceral organs to skeletal muscle.[4] In addition, elevated levels of corticosterone can alter metabolism and suppress the immune system.[5] Altered metabolism, in turn, may affect the absorption, distribution, and elimination of pharmaceutical agents used to treat the patient.[3,5]

Detrimental effects of stress may present acutely, as in a rabbit that collapses suddenly as a result of stress-induced cardiomyopathy,[4] or they may appear very slowly, as in a reptile that has suffered low-level, chronic thermal stress before developing an infection.[6] It is important to remember that corticosterone release can occur very rapidly. Recent research shows that cockatiels can experience elevated levels of plasma corticosterone in as little as 3 minutes after a stressful event.[7]

Thus, maintaining an environment that is as comfortable and stress-free as possible is extremely important for any hospitalized patient. When dealing with an unfamiliar species, always check the patient's natural history to determine the least stressful conditions in which to house it.

Also keep in mind that one of the most stressful events to which patients will be exposed is capture and restraint, whether for administration of medication or simply moving the animal for cleaning of its enclosure. Research shows that, for many animals, handling can be far more stressful than an actual medical procedure.[8] Thus, minimizing restraint and using low-stress techniques can significantly reduce the negative effects of stress; this is especially important for shy species, such as chameleons and ball pythons.

General Principles of Reptile Hospital Care

Housing and substrate

Reptiles should be kept in a secure enclosure to prevent escape. Small reptiles can be wily and squeeze through surprisingly tiny openings. Tearing a ward apart in an effort to find them is not a pleasant way to spend a day! Spartan caging, defined as caging without decorative furnishings or a particulate substrate, is generally preferred, as it facilitates monitoring of the patient. A particulate substrate, such as coconut fiber, may be needed if encouraging egg laying or other natural behaviors, such as burrowing behavior in fossorial species. A paper substrate, such as a paper towel, may be preferable for monitoring purposes, but bear in mind that it can interfere with a small reptile's ability to find and ingest prey by allowing food items to hide underneath it. Remove the paper and leave out the substrate if needed.[9]

Aquatic or semi-aquatic patients will require at least part of the enclosure to be filled with water. Some chelonian patients will only feed, urinate, and defecate in water, so it is important to carefully monitor water quality.[9]

Temperature

As mentioned above, all reptiles are ectotherms, dependent on the environment for the temperature necessary to support their physiological functions, such as heart rate, respiratory rate, digestion, and elimination. Every species has what is called a preferred optimal temperature zone (POTZ), and it is necessary for a reptile patient to be maintained within that range for adequate physiological function and healing. These ranges can differ markedly even between similar species. For example, the daytime POTZ for a veiled chameleon is 22.2-37.8 °C (72-100 °F),[10] with a night-time drop of about 12.2 °C (10 °F), while a Jackson's chameleon *(Trioceros jacksonii)* must be kept in a lower daytime POTZ of 20-28.3 °C (68-83 °F), with a drop to 10-15 °C (50-59 °F) at night, and it should not be kept above 28.9 °C (84 °F) for long periods of time.[11]

The ideal situation is to provide a temperature gradient, with a warm area and a cooler area available in the same enclosure, so the patient can choose the temperature that feels most comfortable; however, that may not be practical in smaller enclosures, and a sick reptile may not be willing or able to move around. Providing a steady temperature at the upper end of the POTZ will provide adequate care.[9]

A wide variety of lamps and other heating equipment is available, and it may take some research to determine which of them will work best in an individual hospital situation. Keep in mind that it can be even more dangerous to overheat a patient than to keep him/her too cool, so good control of the temperature is essential.

Hydration and humidity

Because reptiles originate from a wide variety of climates from tropical to desert, it is critical to provide an appropriate amount of humidity for proper hydration and ecdysis. Humidity should be monitored regularly, and depending on the species, a humid hide and/or water bowl large enough for the animal to soak in may be needed. Appropriate humidity for the macrohabitat of many desert species is approximately 30%, while for tropical species, 60% or higher humidity will be preferred.[12] Note that even desert species will need access to water to remain adequately hydrated and often will take advantage of a more humid microhabitat, such as a humid hide (in captivity).

A variety of misting and drip systems are available or can be made from inexpensive materials; humid hides can be fashioned from small plastic boxes with moist paper towels or moss placed inside.[9] Providing a shallow warm-water bath for 20-30 minutes once or twice a day can assist with hydration, as well as stimulating digestion and metabolic activity. Take care to ensure the water is neither too hot nor too cold. What feels barely warm to humans, whose normal body temperature is 37 °C (98.6 °F), may feel too hot for an animal accustomed to a normal temperature of 29.4 °C (85 °F).[12]

Some reptiles, especially chameleons, typically drink water clinging to leaves, rather than lapping water from a bowl. These patients will need leaves, natural or artificial, that can be sprayed with water in order to remain hydrated.[10] Bearded dragons sometimes can be persuaded to drink by dripping water on their heads. Many species, especially crepuscular or nocturnal species such as leopard geckos, will appreciate provision of a humid hide, for maintaining hydration and/or because it is natural for them to stay concealed during the day.

Ultraviolet light (UVA and UVB)

In the wild, reptiles use every part of the spectrum of light provided by the sun, from infrared to ultraviolet, even if they are crepuscular or nocturnal.[13] When kept indoors, however, that spectrum is unavailable, at least in the same form. Many reptiles who are housed indoors must be provided with artificial ultraviolet light in the range of 290-320 nanometers, as well as warmth, to produce the vitamin D necessary for processing calcium. Without calcium supplementation and ultraviolet light (whether natural or artificial), these reptiles are highly susceptible to nutritional secondary hyperparathyroidism, also known as NSHP or metabolic bone disease. Even reptiles that do not spend much time in the sun during the day and carnivorous species that ingest adequate amounts of calcium from their diet, such as ball pythons, can benefit from provision of ultraviolet light. Hospital accommodation for these patients should provide ultraviolet light whenever possible.

Most reptiles see colors, and UVA light is a crucial component of color vision. Light in this spectrum also can reach nonvisual receptors in the reptile retina, pineal body, and parietal eye (if present), which transmit information to a complex neuroendocrine network that controls daily and seasonal behaviors. Thus, red and blue daylight or night light bulbs should not be used at night because they are visible to the reptile and can disrupt the animal's circadian rhythms.[13]

The length of time to which an animal should be exposed to ultraviolet light also varies by species. Animals originating from regions near the equator generally need approximately 12 hours of light and 12 hours of darkness, while those whose natural habitat is in more temperate latitudes may do better with a more variable light cycle, from 9-14 hours, depending on the season.[14] Guidelines on the appropriate amount of exposure are purely experimental, however.[13] Never leave the lights on 24 hours a day, as that can be a significant source of stress.[14] Provide a hide in the enclosure so that the patient can get away from the light if it prefers to do so.

A wide variety of ultraviolet lamps are available, and some research will be necessary to determine the best set up for an individual hospital's situation. Keep in mind that most such lamps only emit ultraviolet light for about six months, regardless of what the label says, and that any screening or other material between the lamp and the animal may reduce the amount of light being received. In addition, the depth to which the ultraviolet light will penetrate varies greatly from one lamp to another, so the size of the enclosure may impact the effectiveness of the lamp.

Feeding

Reptiles in the wild have a variety of feeding strategies; the reptiles that are the focus of this presentation are primarily herbivores (Russian tortoises), insectivores (leopard geckos, veiled chameleons, juvenile bearded dragons), omnivores (adult bearded dragons, adult box turtles, adult red-eared sliders), and carnivores (ball pythons, boa constrictors, corn snakes). Some insectivorous species, such as bearded dragons, eat primarily animal protein (e.g., insects and worms) when they are young, then become more omnivorous as they age, eating various greens and fruit. This is also true of box turtles and red-eared sliders. In addition, some mostly insectivorous species, such as veiled chameleons, occasionally eat some plant material, so if live plants are placed in the enclosure, care should be taken to ensure they are non-toxic. Leaves and branches of ficus (*Ficus benjamina*) or hibiscus (*Hibiscus* spp.) are appropriate. It is generally preferable to give frozen and thawed small mammal prey to species that will eat them, such as snakes, as opposed to live mice or rats that may fight back against attacks and injure the reptile. For insectivorous lizards, such as chameleons, leopard geckos, or omnivorous chelonians, it is acceptable to feed live insects and insect larvae; however, attention must be paid to the nutrients provided, especially calcium, when feeding insects to reptiles that are housed indoors. Nearly all common feeder insects have a negative calcium:phosphorus ratio, and reptiles that consume them will need calcium supplementation to compensate.[15]

For reptiles that are not self-feeding while in the hospital, force feeding may be indicated. Keep in mind that force or tube feeding can be extremely stressful for the patient, so it should not be undertaken unless it is truly needed. Reptiles' feeding often is influenced by physiological factors such as gravidity, season, brumation, or ecdysis. Thus, there often is no rush to ensure the reptile is fed, especially every day, and it is essential to ensure the patient is well-hydrated and warmed to its POTZ before attempting to feed. A useful rule of thumb is to feed if the patient has experienced a weight reduction of 10% or more. [16]

The rate for maintenance fluid replacement in reptiles has been estimated at 10-15 ml/kg/day, given over 72 to 96 hours.[17] Fluids can be given via intracoelomic (ICo), intraosseous (IO), intravenous (IV), or subcutaneous (SC) routes and should be administered in the caudal one-third of the body. Position the patient in dorsal recumbency to reduce the likelihood of introducing fluids into the lungs.[5,17]

In general, daily energy needs are estimated at $SMR = 32 (BW^{0.75})$ for a reptile kept at 30 °C (86 °F), where SMR represents standard metabolic rate in kcal/day and BW represents weight in kg. Reptiles are subject to "refeeding syndrome," in which potentially life-threatening hypophosphatemia and hypokalemia can occur. Start by feeding 40-75% of daily energy requirements for the first two to four days, then gradually increase to the full calculated energy requirements. Note that the stomach capacity of most adult reptiles is approximately 2-5% of body weight.[16]

Either a metal gavage tube or a red rubber catheter can be used to tube feed most snakes and lizards. Make sure the food is warm before it is given. Cold food can result in decomposition rather than digestion of food in the gastrointestinal tract. Measure the distance from the mouth to the stomach, which generally is about mid-body on the left side. Be gentle when opening the mouth, especially in small reptiles or if there is any suspicion of NSHP, as it is possible to fracture the jaw. Lizards' teeth and the keratin surfaces of chelonians' beaks also can be fragile, so exercise caution with metal gavage tubes and/or specula used to hold the mouth open. It can be helpful to hold the patient upright for a minute or two after feeding to minimize the risk of regurgitation, particularly if the tube used was too short to reach the stomach. Oral medications can be administered through the gavage tube along with food if desired.[17]

Feeding chelonians can be considerably more challenging, especially if the patient is large. Even a small, sick turtle or tortoise can be very strong and have defenses such as withdrawing firmly into the shell that are not easily overcome. If the patient must be gavage fed repeatedly, it may be more advantageous to place a pharyngostomy or esophagostomy tube, especially if the patient is aggressive.[19]

As with other species, changing a reptile patient's diet while in hospital is not advisable, even if the diet fed at home is deficient, because doing so runs the risk of triggering anorexia. Patients who have been eating solely live food are unlikely to accept frozen-thawed mice or rats immediately. If the patient is accustomed to exclusively eating mealworms, it may refuse crickets or vegetables, and vice versa. That said, it may be possible to make minor adjustments that would have minimal or no effect on the appearance or taste of the food, such as providing appropriately gut-loaded insects or adding a powdered calcium supplement to greens or vegetables.

Medications

The renal portal system in reptiles directs blood from the caudal region of the body into the kidneys, especially at times of water conservation. Research indicates that a valve in the abdominal vein closes, redirecting blood through the iliac veins and into the kidneys. This can have an impact on the absorption and distribution of injected medications in the caudal half of a reptile patient. The extent of that impact is a subject of some controversy, but it is generally agreed to be safer and more effective to administer parenteral medications in the cranial half of the body.[19]

Reptile morphology differs significantly from one species to another, which can affect where and how medications are administered. Intravenous access can be difficult, especially in very small reptiles, and giving medications per os can be challenging in some species (chelonians). Research indicates oral administration may be of little value because the reptile gastrointestinal tract is quiescent between meals, resulting in little or no absorption of the drug.[20] The intramuscular (IM) route is generally preferred in most species, although there are some limitations on the volume that can be administered in a single site. Preferred sites for IM injections in chelonians are the upper forelimb (deltoid or triceps muscles); in lizards, the forelimbs (avoid the cranial surface of the limb to minimize the risk of damage to the radial nerve); for snakes, epaxial muscles in the cranial half of the body.[21]

Subcutaneous space is available for injections, but it is generally smaller and less well vascularized than in mammals. It is not necessary to tent the skin to give SC injections. Preferred sites are the axillary and inguinal spaces in chelonians; for lizards, the lateral body wall and the inguinal space; for snakes, the lateral body wall.[20]

Intracoelomic injections can be used but limit the amount of fluids being introduced to the coelomic cavity, as excessive amounts can cause lung compression. This route also requires taking care not to penetrate visceral structures with the needle, which has prompted some experts to recommend avoiding it unless absolutely necessary.[19]

When IV access is available, preferred sites in lizards are the tail vein and the ventral abdominal vein; for chelonians, the right jugular vein; in snakes, the tail vein is most commonly used.[20]

Monitoring

Hospitalized reptiles require at least daily monitoring, and that includes monitoring of the environmental conditions in the enclosure to ensure they remain appropriate. Changes in a reptile's condition can be subtle and easily missed without careful and regular monitoring. Minimum monitoring parameters include weight, mentation, appetite, and eliminations. More critical patients may require more frequent monitoring.[9]

Body weight is an important indicator, as continued weight loss is a poor clinical sign; however, continuing weight gain is also a concern if it does not correlate well with the amount of food and water being consumed or improvement in hydration. Increasing weight gain may indicate retention of fluid or constipation/obstipation or both, which can ultimately result in compression of other visceral organs, including the lungs. Production of feces may not occur every day, and that is not necessarily abnormal.[12]

Species-Specific Data

Bearded dragons[10]

- *Temperature in hospital incubator:* 31.1 °C (88 °F)[5]
- *Feeding strategy:* Insectivore (juvenile); omnivore (adult). For juveniles, offer live insects or worms with a body length no more than the space between the lizard's eyes.[10] Salad greens can be offered even to young dragons so they become accustomed to eating vegetable matter early. Adult dragons can eat larger insects/larvae and larger amounts of vegetables. Feed in the morning.
- *Supplementation:* Gut load insects/larvae; powdered calcium and multivitamin supplements can be provided on salad.
- *Humidity:*30-50%. Provide a water feature large enough for the dragon to soak itself in if possible.
- *UVB and UVA light:* Yes.
- *Diurnal*

Veiled or Yemen chameleons[10]

- *Temperature in hospital incubator:* 29.4 °C (85 °F)[5]
- *Feeding strategy:* Insectivore but will eat some plant material. Feed appropriately gut-loaded live insects in the morning. Additional supplementation may be needed, depending on the nutrients provided by the gut-loading formula used.
- *Humidity:* 40-60%. Spray leaves in the enclosure at least q12h. Enclosure should have no loose substrate.
- *UVB and UVA light:* Yes.
- *Diurnal*

Leopard geckos[10]

- *Temperature in hospital incubator:* 29.4 °C (85.0 °F)
- *Feeding strategy:* Insectivore. Feed appropriately gut-loaded live insects no greater in length than the length of the lizards' head.[10] Additional supplementation may be needed, depending on the nutrients provided by the gut-loading formula used. Feed at night.
- *Humidity:* 40-60%, with a humid hide and a water feature provided.
- *UVB and UVA light:* Optional but may do better with it, depending on the patient, such as NSHP patients.
- *Crepuscular/nocturnal*

Ball pythons[10]

- *Temperature in hospital incubator:* 30.5 °C (87 °F)
- *Feeding strategy:* Carnivore. Feed frozen-thawed rodents no wider than the widest part of the snake's body.[10] Feed at night. Acceptance may be improved by ensuring the feeder rodent is warm.
- *Humidity:* 60-80%; provide a water feature large enough for the patient to soak in it if possible.
- *UVB and UVA light:* Optional but may do better with it, depending on the patient.
- *Nocturnal*

Corn snakes[10]

- *Temperature in hospital incubator:* 29.4 °C (85 °F)
- *Feeding strategy:* Carnivore. Feed frozen-thawed rodents no wider than the widest part of the snake's body.[10] Feed at night. Acceptance may be improved by ensuring the feeder rodent is warm.
- *Humidity:*40%-60%. Provide a water feature large enough for the patient to soak in it if possible.
- *UVB and UVA light:* Optional but may do better with it, depending on the patient.
- *Diurnal*

Boa constrictors[10]

- *Temperature in hospital incubator:* 30.5 °C (87 °F)
- *Feeding strategy:* Carnivore. Feed frozen-thawed rodents no wider than the widest part of the snake's body.[10] Feed at night. Acceptance may be improved by ensuring the feeder rodent is warm.
- *Humidity:* 60-80%; provide a water feature large enough for the snake to coil its entire body in it if possible.
- *UVB and UVA light:* Optional but may do better with it, depending on the patient.
- *Nocturnal*

Red-eared sliders[10]

- *Temperature in hospital incubator (dry-docked):* 28.8 °C (84 °F)
- *Water temperature:* 23.8-30 °C (75-86 °F)
- *Feeding strategy:* Insectivore and or carnivore (juvenile, up to 40% protein), omnivore (adult, 25-50% may be commercially available turtle pellets). Appropriate sources of protein include earthworms, mealworms, and feeder fish. Appropriate plant matter includes finely chopped fresh dark leafy greens that are high in calcium or mixed vegetables.
- *Humidity:* Semi-aquatic, needs water feature unless dry-docked because of condition.
- *UVB and UVA light:* Yes.

Box turtles[10]

- *Temperature in hospital incubator:* 30.0 °C (86 °F)
- *Feeding strategy:* Omnivore. Appetite is stimulated by color and odor/ripeness of fruits and vegetables, and movement of prey items. Recommended foods include dandelion leaves, sweet potatoes, flowers such as hibiscus and geraniums (*Pelargonium* spp.), earthworms, slugs, and snails. Ensure food is obtained from a pesticide-free source. Feed when most active, early morning and late evening.
- *Humidity:* 50-70%; provide a water feature large enough for the patient to soak in it.
- *UVB and UVA light:* Yes.

Russian tortoise[10]

- *Temperature in hospital incubator:* 27-30 °C (80-85 °F)
- *Feeding strategy:* Herbivore. Dry leafy grass such as timothy hay should be available at all times. Fresh greens and vegetables also may be offered, but do not feed fruit. Dust greens with calcium and multivitamin supplement.
- *Humidity:* 30-50%. Provide fresh, clean water at all times.
- *UVB and UVA light:* Yes.
- Diurnal

Tips on Husbandry for Live Feeder Worms and Insects

Feeding insectivores can pose particular challenges because most insects and larvae contain little calcium and are low in fat-soluble vitamins, both of which are essential for maintaining health in these species.[22,23] The recommended solution for this is to gut load the insects. In other words, feed the insects with a nutrient-rich diet that then is transferred to the reptile when the insect or larva is eaten.

Preferred gut loading diets are those that provide a positive calcium:phosphorus ratio (calcium higher than phosphorus). For most vertebrate animals, a ratio of 2:1 or 3:1 is generally considered adequate, but it is unknown whether that is sufficient for health in rapidly growing juvenile reptiles.[22]

The most common insects used are house crickets (*Acheta domesticus*), mealworms (*Tenebrio molitor*), black soldier fly larvae (*Hermetia illucens*), dubia roaches, sometimes sold as tropical spotted roaches (*Blaptica dubia*), and tomato hornworms (*Manduca sexta*).

Not every veterinary hospital will keep its own colonies of feeder insects, but it is usually not necessary to gut load feeder larvae/insects for more than about 24 hours, so it is possible to keep small numbers of them or ask clients to bring food for their reptiles and gut load as needed. It is recommended to fast the insects for 24 hours before gut loading to encourage them to eat the offered diet. Several different gut-loading formulas are available, but it is important to select one that provides at least 8% calcium to improve the calcium:phosphorus ratio to a beneficial level.[22] Gut-loading diets that provide fat-soluble vitamins are also available. Note that some insectivorous species, such as chameleons and leopard geckos, are unable to convert beta-carotene to vitamin A, so will need insects fed a diet that includes preformed vitamin A.[15]

Crickets

Research has shown that crickets are a good source of moisture, protein, and fat but are deficient in calcium and fat-soluble vitamins. They should be gut loaded with a high-calcium diet (8% or more) and fed to reptiles within 6-12 hours (no more than 24 hours) of gut loading. Beyond that time, the crickets may stop eating the gut-loading formula because of its high mineral content, at which point the calcium:phosphorus ratio will revert to an undesirable level.[22]

Mealworms

These feeder insects tend to be significantly higher in fat than crickets; if fed excessively, they can lead to obesity in reptiles. They should be gut loaded for about 24 hours before being fed to patients.[22]

Black soldier fly larvae

Sometimes appearing in pet stores as Calci Worms or Phoenix Worms®, these feeder larvae have a naturally high calcium:phosphorus ratio; however, they have poor digestibility and may pass through the reptile's digestive tract without the calcium being absorbed. They can be fed the same gut-loading formula as mealworms, which will increase provision of fat-soluble vitamins to the reptile, but they are unlikely to eat the formula unless provided with a water source.[22]

Dubia roaches

Similar to crickets, these insects are available in some locations in various sizes and provide adequate amounts of protein, fat, and moisture. They can be fed a gut-loading formula used for crickets and/or mealworms for a period of about 48 hours to achieve a beneficial calcium:phosphorus ratio.[22]

Tomato hornworms

Little has been reported about the nutritional content of these worms, and they also have an inverse calcium:phosphorus ratio. Research has shown that gut loading with the only commercially available species-specific diet for them does not provide an adequate level of calcium; however, their moisture content is approximately 85%, so they may be beneficial if there is concern about the reptile's hydration.[22]

Conclusions

Watching a reptile patient recover and leave the hospital after some time in your care is a wonderful experience. Ensuring those patients' special needs are met during their hospital stay is the key to creating those positive outcomes.

References

1. Uetz P. Species Numbers (as of December 2022). The Reptile Database. Available at: http://www.reptile-database.org/db-info/SpeciesStat.html. Accessed March 2, 2023.

2. Chitty J. Hospitalization of birds and reptiles. *J Exot Pet Med.* 2011;20:102-103.

3. Hunt C. Stress and welfare. In: Divers S, Stahl S, eds. *Mader's Reptile and Amphibian Medicine and Surgery.* 3rd Ed. Philadelphia, PA: Elsevier; 2019:380.

4. Minh H, Boyeaux A, Pignon C. Assessment and care of the clinically ill rabbit. *Vet Clin North Am Exot Anim Pract.* 2016: 19(2): 105.

5. Mitchell M. Managing the reptile patient in the veterinary hospital: establishing a standards of care model for nontraditional species. *J Exot Pet Med.* 2010: 19(1);56-72.

6. Rossi J. General husbandry and management. In: Divers S, Stahl S, eds. *Mader's Reptile and Amphibian Medicine and Surgery.* 3rd Ed. Philadelphia, PA: Elsevier; 2019:114-115.

7. Turpen K, Welle K, Trail J, et al. Establishing stress behaviors in response to manual restraint in cockatiels (*Nymphicus hollandicus*). *J Avian Med Surg.* 2019;33:41.

8. Heidenreich B. Reducing stress for avian capture and restraint. *Proc Annu Conf Assoc Avian Vet.* 2014;367.

9. Cutler D, Divers S. Hospitalization. In: Divers S, Stahl S, eds. *Mader's Reptile and Amphibian Medicine and Surgery.* 3rd Ed. Philadelphia, PA: Elsevier; 2019:432-436.

10. Association of Amphibian and Reptile Veterinarians. Companion Reptile Care Series. Available at: https://arav.site-ym.com/page/CompanionCareGuides. Accessed March 2, 2023.

11. Healey M. Lighting and Temperature Requirements for Jackson's Chameleons. Available at: https://reptifiles.com/jacksons-chameleon-care/jacksons-chameleon-lighting-temperatures/. Accessed March 2, 2023.

12. Pees M, Girling S. Emergency care. In: Girling S, Raiti P, eds. *BSAVA Manual of Reptiles.,*3rd Ed. Quedgeley, UK: British Small Animal Veterinary Association; 2019:110.

13. Baines F, Cusack L. Environmental lighting. In: Divers S, Stahl S, eds. *Reptile and Amphibian Medicine and Surgery.* 3rd Ed. Philadelphia, PA: Elsevier; 2019:131-138.

14. Varga M. Captive maintenance. In: Girling S, Raiti P. *BSAVA Manual of Reptiles.* 3rd Ed. Quedgeley, UK: British Small Animal Veterinary Association; 2019:43-45.

15. Boyer T, Scott P. Nutrition. In: Divers S, Stahl S, eds. *Mader's Reptile and Amphibian Medicine and Surgery.* 3rd Ed. Philadelphia, PA: Elsevier; 2019:210-215.

16. Pollock C. Feeding the Hospitalized Lizard. LafeberVet. Available at: https://lafeber.com/vet/feeding-the-hospitalized-lizard/. Accessed March 2, 2023.

17. Petritz O, Son T. Emergency and critical care. In: Divers S, Stahl S, eds. *Reptile and Amphibian Medicine and Surgery.* 3rd Ed. Philadelphia, PA: Elsevier; 2019:971.

18. Pollock C. Nutritional Support in Reptiles. LafeberVet. Available at: https://lafeber.com/vet/how-to-guide-nutritional-support-in-reptiles/. Accessed March 2, 2023.

19. Pollock C. Feeding the Hospitalized Turtle or Tortoise. LafeberVet. Available at: https://lafeber.com/vet/feeding-the-hospitalized-turtle-or-tortoise/. Accessed March 2, 2023.

20. Pollock C. Administration of Medications in Reptiles. Lafebervet. Available at: https://lafeber.com/vet/administration-of-medication-in-reptiles-2/. Accessed March 2, 2023.

21. Perry S, Mitchell M. Routes of administration. In: Divers S, Stahl S, eds. *Mader's Reptile and Amphibian Medicine and Surgery.* 3rd Ed. Philadelphia, PA: Elsevier; 2019:1130-1138.

22. Boykin K, Mitchell M. Serving up a healthy plate of evidence-based nutrition ensures good digestion for reptiles. *Proc ExoticsCon.* 2022;364-370.

23. Boyer T, Scott P. Nutritional diseases. In: Divers S, Stahl S, eds. *Mader's Reptile and Amphibian Medicine and Surgery.* 3rd Ed. Philadelphia, PA: Elsevier; 2019:939.

Rabbit Anesthesia 101 for the Veterinary Technician

Stefanie Kotschwar, RVT, BScH (Zoology, Nutrition), MSc (Zoo Conservation), AAS (Vet Tech), Diploma (Web Design and Equine Science)

Session #3234

Affiliation: From Lynwood Animal Hospital, 107-30 Stafford Rd, Ottawa, ON, K2H 1W8, Canada.

Abstract: Many veterinary technicians find rabbit anesthesia overwhelming and scary. The purpose of this talk is to go over all the steps from intake to discharge of rabbit patients presenting to the clinic for surgery. Topics covered include history, pre-surgical exam, anesthetic drugs, anesthetic equipment, monitoring equipment, monitoring patient and assessing anesthetic depth, monitoring during recovery, post-surgical in hospital care, post-surgical home care client education There is emphasis on areas of risk and how reduce them along the way. The focus is on what veterinary technicians can do at each step of the way for a successful outcome.

Introduction

Rabbits have become wildly popular as pets in North America and more are brought to the veterinary clinic for regular check-ups.[1] The veterinary team might encounter these patients on a regular basis for procedures that require sedation or general anesthesia. Many pet owners and even veterinary professionals may be under the impression that rabbits have a high anesthetic risk and procedures should be avoided. While it is true that they have a higher anesthetic risk than cats or dogs, they can do well under general anesthesia as well. One published article went over studies that have shown the anesthetic risk in rabbits can range anywhere from 1.39-4.8%.[1] These studies have also shown that deaths not only occur during the surgical procedure but up to 48-hours post anesthesia.[1] Successful anesthesia rates require understanding of rabbit anatomy and husbandry, all components of an anesthetic procedure, our limitations, and staff training. While there are no general anesthetic guidelines published for exotic companion animals, the veterinary team can refer to the 2020 American Animal Hospital Association (AAHA) Anesthesia Monitoring Guidelines for Dogs and Cats as a basic outline and apply it to not-so-typical patients. Any strategies used to keep dogs and cats safe under general anesthesia can be applied to rabbits with some modifications.[2] Individual components for each procedure are described below.

Patient History

The rabbit owner should have ideally been counseled on appropriate care and nutrition of their pet at an exam prior to the day of anesthesia. Nevertheless, the admitting support staff personnel should obtain a brief history on the patient's behavior, appetite, and stool production, as this can help identify any changes in health (or husbandry) that may influence anesthetic morbidity.[2,3] The focus should be on questions regarding alertness, eating and defecating habits. Any changes should be investigated further. Provide stabilization first, if required.

Pre-Anesthetic Examination

Prior to handling the rabbit for a pre-anesthetic examination, observe it in the carrier for any signs of distress. Rabbits are prey species that try to mask clinical signs of illness when not feeling well. An examination prior to sedation is important to ensure a proper anesthetic plan can be made and adjusted based on the rabbit's health status. Not only should vital signs be taken including heart rate, respiratory rate, and temperature, but also check the color of mucus membranes and capillary refill time.[2,3] Perform an abdominal auscultation to detect borborygmi, as the absence of these sounds suggests that the patient may be stressed or painful. Table 1 highlights normal vital signs for rabbit patients.[1-3]

Table 1. Normal vital signs for rabbits.

Vital sign (unit)	Normal range
Heart rate (beats per minute)	200-300
Respiration rate (breaths per minute)	32-60 (higher rates are not uncommon in stressed patients)
Temperature (°C)	38.5-39.5
Weight (kilograms)	1-10

Hospital Cage Set-Up

Rabbits are a prey species so keeping them in an area specifically for small exotic animals away from any predatory species can help reduce patient stress.[2,3] If no separate ward is available, provide hiding shelters as well as litter boxes to make the cage more comfortable and place them as far away from predator species as possible.

Anxiolytics such as gabapentin are commonly used in cats and dogs to be given at home prior to the veterinary visit or surgical procedures to ensure their fear, anxiety and stress are reduced.[1] One study has investigated the efficiency of a one dose use of gabapentin in rabbits and has had some positive outcomes. It seems that even one dose of gabapentin can have relaxing effects on rabbits, but further research in this area is needed.[4]

Pre-Anesthetic Blood Work

Ideally a complete blood cell count, plasma biochemistry along with a blood smear evaluation should be performed to rule out any underlying issues prior to any anesthetic events, especially on senior patients. If full blood work is not possible, performing a packed cell volume and total protein at minimum should be considered.[1]

Pre-Anesthetic Fasting

Fasting is only recommended 30 minutes to 1 hour prior to sedation to reduce the incidence of ingesta remaining in the oropharyngeal space that could be accidentally introduced into the respiratory tract, particularly if a blind intubation technique is used. This fasting time can also help reduce the volume of gastrointestinal contents which could put pressure on the diaphragm which may result in the stomach applying pressure to the diaphragm, affecting the patient's ability to ventilate under anesthesia. Longer fasting times are not recommended as withholding food from rabbits can result in an increased risk of gastrointestinal stasis, or ileus. Rabbits cannot vomit or eructate due to a muscular cardiac sphincter.[1,3]

Procedure and Duration

Consider what type of surgical procedure is being performed, anticipated length, type of heat support, fluid support and how painful it could be.[2]

American Society of Anesthesiologists Risk Status

The American Society of Anesthesiologists created a physical classification status that has been widely used not only in human medicine but also in veterinary medicine. This system is useful in assessing the patient and any comorbidities that could contribute to anesthetic risks.[5] Table 2 reviews the risk assessment grades.

Table 2. American Society of Anesthesiologists (ASA) physical status classifications.

ASA classification	Definition
ASA I	A normal healthy patient
ASA II	A patient with mild systemic disease
ASA III	A patient with severe systemic disease
ASA IV	A patient with severe systemic disease that is a constant threat to life
ASA V	A moribund patient who is not expected to survive without the procedure

Source: American Society of Anesthesiologists. ASA Physical Classification Status. Available at: https://www.asahq.org/standards-and-guidelines/asa-physical-status-classification-system. Accessed June 1, 2023. Cripwell D, Dugdale A, Michou J. ASA Physical Classification Status: A Guide for Veterinarians. Available at: https://www.thinkanesthesia.education/sites/default/files/2021-03/US%20-%20ASA%20physical%20Status%20Classification%20Chart%20-%20%20V2.pdf.

Anesthetic Drugs

A multimodal anesthesia approach is always preferred as a combination of drugs is used to maximize their positive effect while minimizing any negative effects.

Pre-anesthetic drugs

The goal is to sedate patients and to provide preemptive analgesia. The use of premedication facilitates catheter placement, decreases the amount of induction agent needed and decreases minimum alveolar concentration (MAC) during the procedure. The protocol use should be tailored to each individual patient based on the type of procedure, their age, breed, and any existing comorbidities.[1] Having a preset plan that can be used as a starting base and modified as needed is appropriate.[2]

Midazolam is a commonly used sedative and anxiolytic belonging to the benzodiazepine class. It is a short acting drug that is water soluble and can be given intramuscularly (IM), subcutaneously (SC), intravenously (IV), or intraosseously (IO). This drug can be used alone for minor, noninvasive and non-painful procedures, and in combination with ketamine, opioid, alpha-2 agonists for extensive procedures.[1]

Alpha-2 agonists include drugs such as xylazine, detomidine, medetomidine and dexmedetomidine and they provide excellent sedation and muscle relaxation. They can have significant cardiovascular effects that include vasoconstriction, decreased cardiac output, second degree heart block and bradyarrhythmia.[1,2]

Ketamine is an NMDA antagonist and provides analgesia and dissociative anesthesia. It is considered a good adjunct in premedication but can cause muscle rigidity and should be used with muscle relaxants. It can also be used as induction agent alongside midazolam.[1]

Anticholinergic drugs may be used to minimize salivary and bronchial secretion and reduction of vagally-induced bradyarrhythmia. They can increase the viscosity of airway secretions. Their use has been questioned in human medicine and there are no controlled studies in rabbits. About 61%[1] of rabbits have significant atropine esterase activity which necessitates higher doses of atropine compared to other mammalian species. Glycopyrrolate has a duration of effect of 60 minutes, but due to lack or proven efficacy and effect on gastrointestinal motility routine use is not recommended.[1,3]

Opioids

Opioids can provide potent analgesia. Their receptor affinities can be used to classify this group of drugs.

Pure μ agonists

Pure μ agonists are preferred for procedures where moderate to severe pain is expected (e.g., limb amputation, laparotomy). These drugs include hydromorphone, fentanyl, morphine, and methadone.[1]

Partial opioid agonists, agonists and antagonists

This subgroup of opioids can provide sedation (and potentiate the effect of other sedatives) but is not highly effective for analgesia. Butorphanol, for example, is not appropriate for painful procedures, but can be used in conjunction with other drugs (example, midazolam) to improve sedation for minor procedures such as skull radiographs that require adequate patient immobilization. Butorphanol in particular has a notable ceiling effect to its analgesic and sedative properties at higher doses, but minimal MAC sparing effects are observed. This kind of drug is only appropriate for mild pain and as a sedative for minor diagnostic procedures such as skull radiographs. It can cause significant respiratory depression when combined with other sedatives. It can be given IM or IV.[2]

Buprenorphine is a partial μ agonist and effective analgesic to treat mild to moderate pain. When used alone it only provides mild sedation. It does not have any cardiovascular effects but causes mild respiratory depression. If combined with midazolam the sedative effects can last up to 90-120 minutes.[1]

Local anesthetics

Local anesthetics such as lidocaine and bupivacaine are valuable tools in the anesthetic protocol as they reversibly block transmission of nociceptive signals from reaching the central nervous system. Local anesthetics are applied via topical, infiltration, intraarticularly or epidural route.[1,7]

Epidural provides preemptive analgesia by injecting the drug into the epidural space that surrounds the spine. It allows for anesthesia without systemic effect, reduced recovery time, as well as decreased use of gas anesthesia. The lumbosacral junction site is the most used site with a technique like in dogs and cats. This type of anesthesia is beneficial in abdominal surgeries and orthopedic procedures. A study using morphine and lidocaine proved effective analgesia for limb orthopedic surgery. Potential side effects include transient paresis or paralysis.[1,7]

Induction

Anesthetic induction can be achieved either via injectable or inhalant drugs. Inhalant or volatile anesthetics are noxious and irritating to the airway and may cause laryngospasm. If used for induction, adequate pre-anesthetic medication is necessary to provide moderate to deep sedation to avoid producing an excitement phase.

Induction agents are ideally administered with an IV catheter in place. Propofol produces rapid induction without an excitement phase, however apnea is common so preoxygenate the patient and ensure that a visual intubation technique is possible. Side effects are dose dependent, short lived and might decrease cardiac contractility. Side effects can be exacerbated by rapid administration.[1]

Alfaxalone is effective as an IV induction agent. It has a rapid onset and short duration of action. The cardiovascular effects of this drug has not been well studied in rabbits. Alfaxalone can also be used via IM or SC route for sedation. Like with propofol, apnea can occur when administered IV, so it is important to pre oxygenate and visual intubation is needed.

Etomidate is an ultrashort-acting non-barbiturate hypnotic anesthetic agent. It does not provide analgesia. It is reported to cause hemolysis, thrombophlebitis, and pain on injection.[1]

Ketamine and midazolam can be given together to effect for a smooth induction. Apnea can occur but it is not common.

Equipment

All equipment needed should be prepared prior to sedating a patient to ensure readiness.[2]

Vascular access supplies

Intravenous access should be obtained in all anesthetic patients and should be considered for patients undergoing sedation. Applying a topical local anesthetic cream at least 20 minutes prior can reduce discomfort when placing an IV catheter. Sites that can be used for IV catheter placement include cephalic vein, marginal ear vein and the lateral saphenous. The IV catheter sizes are selected based on patient size and can range from 22-26 gauge. Necrosis of the ear veins can occur with irritating drug administration. Some drugs will require to be diluted prior to be given IV as they can cause irritation.[1]

The use of an IO catheter is an option, especially for sick patients but may not be needed for stable surgical patients.

Airway access supplies

Ideally any rabbits undergoing any surgical procedures, especially lengthy ones, should have airway access available to support them as needed. There are several methods for securing airway access. It is important to remember that rabbits are obligate nasal breathers.[1,3] Once the rabbit has been sedated, oxygen via a facemask should be provided to improve oxygen saturation.[3]

Oral endotracheal intubation

This method provides multiple challenges due to the rabbit's narrow oral cavity, long tongue, narrow gape and potential for laryngospasm. These anatomic features make placement difficult and could lead to laryngeal trauma and tube displacement. There are different methods that can be used to place an endotracheal tube in a rabbit: Blind, direct visualization and with the use of an endoscope. Being proficient in more than way is advantageous. Use the largest possible endotracheal tube size to reduce airway resistance. Appropriate tube placement can be confirmed as in other mammals: Fogging of the tube observed on inspiration, positive capnography readings, and auscultating the lungs for breath sounds with intermittent positive-pressure ventilation. Complications that can arise post extubating include tracheal injury, ulceration, post-intubation tracheal stricture.[1]

Supraglottic airway devices

Another available option is supraglottic airway devices which are fast to place (14-38 seconds[1]), cause minimal trauma and allow for intermittent positive pressure if needed. Whenever this airway device is used a capnograph to measure $ETCO_2$ is a must. They are also not recommended for extended procedures due to potential lingual edema and cyanosis, gastric tympany and potential dislodgement.[1]

Incubators

With smaller patients, hypothermia is a major concern due to the large surface-area-to-volume ratio. Thermoregulation can be further depressed due to the different anesthetic drug use, so consider placing patient in an incubator once sedated to reduced heat loss.[1] The patient should also be placed in the incubator during recovery.

Anesthetic machine

Most rabbit patients weigh under 10 kilograms, so the use of the non-rebreathing system would be most appropriate. Select a reservoir bag that is appropriate for the size of rabbit undergoing anesthesia. Check the system for any leaks and ensure the pop off valve is open. Ensure there is enough inhalant anesthetic in the machine.

Fluid Therapy

Gut hydration is needed to ensure proper functioning and motility, otherwise ileus can develop, which can be fatal in some cases. Daily water intake ranges from 50-150 ml/kg of BW. This high requirement makes fluid therapy especially important.[1]

Monitoring

A quick summary of the available monitoring tools that should be used will be reviewed below. The best monitoring tool is a highly trained and observant technician.

Positioning

Always monitor the patient's head position and ensure it is straight and there are no kinks in the endotracheal tube. The chest should also be slightly elevated to reduce the pressure from the gastrointestinal tract on the diaphragm.[1,3]

Anesthetic depth

Anesthetic depth is an indicator to determine if the patient is at an appropriate anesthetic plane and should be monitored by a trained veterinary technician at least every 5 minutes. Anesthetic depth can be monitored by assessing reflexes including righting, palpebral, corneal, toe withdrawal and pinnal reflex.[1,7] The author finds the jaw tone to be a reliable indicator of light anesthetic depth.

Thoracic auscultation

Evaluate the heart rate and rhythm with a stethoscope of even with the use of a Doppler flow detector.[3]

Pulse strength

Peripheral pulses can be palpated in rabbits at the central auricular and metatarsal arteries. The femoral pulse can also be palpated, but in many surgical procedures this region may not be accessible to the anesthetist.[3]

Mucous membrane color and capillary refill time

The color of the mucous membranes should always be pink and any change could indicate an airway obstruction or other issues like decreased perfusion or internal bleeding.[3]

Blood pressure

While direct blood pressure measuring is the most accurate method, it is not always feasible in a general practice environment. Direct blood pressure is an invasive method that requires special equipment, arterial catheter placement and maintenance that can also be time consuming. Indirect blood pressure measurements with a Doppler is preferred over non-invasive oscillometry inaccurate with rapid heart rates. The use of the forelimb preferred.[1,8]

Electrocardiogram

The electrocardiogram (ECG) allows for continuous monitoring of the heart rate and rhythm of the patient. Atraumatic clips are preferred for rabbits to avoid iatrogenic injury.[1,3,10]

Pulse oximeter

This is a very useful tool but can be unreliable in some animals due to their thick fur. The probe can be attached to ears, tongues, feet and tails.[3,11]

Capnograph

This is an extremely helpful monitoring tool to ensure the endotracheal tube is properly placed as well as an early indication that there is a potential obstruction. Normocapnia in small mammals is associated with $ETCO_2$ values of 35-45 mmHg.[1,9]

Temperature

Rectal or esophageal thermometers are appropriate to use in surgery. Use of various heating sources, IV fluid warmer, socks and bubble wrap to maintain body temperature is recommended.[3,12]

Post-Surgical In-Hospital Care

Many anesthetic mortalities occur in the post-anesthesia period, so close monitoring is required. The patient should be maintained on oxygen until it is ready to be extubated and movement occurs. Once extubated, keep the patient in a stress free and warm recovery cage or even incubator. Ongoing monitoring of temperature, heart rate and respiration rate is recommended. The IV catheter should be maintained in case of an emergency until close to discharge time. Provide food and water as soon as possible to avoid hypoglycemia and to ensure proper gastrointestinal motility. Monitor fecal output closely. Provide syringe feeding for any patient that is not eating hay after a surgical procedure.[3,11,13]

Pain Assessment

Rabbits tend to be alert and active in their normal environment. As they are prey species, they tend to hide signs of disease and discomfort for as long as possible, so it is imperative to monitor them closely for any signs of pain. Changes in their normal behavior such as lack of activity, decreased food consumption, hunched posture, hiding, signs of aggression, bruxism, and lack of grooming can be indicators of pain.[13] A validated resource that can help all staff members identify whether a rabbit is experiencing pain is the grimace scale from N3CRs.[14] The grimace scale is based off easily identifiable facial expressions which provide a score that can indicate pain level.[15]

Post-Surgical Home Care Client Education

As studies have shown, post-anesthetic complications can occur days after the procedure, so it is important to thoroughly go over the home care with owners. As with dog and cat patients, explaining how the procedure went and highlighting the importance of reduced exercise and daily incision checks is key. Next ensure the owner knows how to measure out medication and knows how to give it to their rabbit. It can be beneficial to break down the home care into different sections. Focus on the first 24-48 hours where the priority is pain control, return of appetite and fecal production. Once the owner is comfortable with the care for the first 2 days, go over the details for the next 1-2 weeks of home care. As needed show the owner and provide tips and tricks on how to feed and medicate the rabbit at home. If possible, show the owner what the incision looks like on discharge and even have them take a picture, so they have a reference.

Conclusions

While rabbits are known to have higher anesthetic-related deaths compared to cats and dogs, there are many tools that can be implemented to ensure a positive outcome. It is important as veterinary technicians to be skilled in using and implementing as many tools as possible to ensure rabbit patients do well under anesthesia and to keep our knowledge up to date.

References

1. Gardhouse S, Sanchez A. Rabbit sedation and anesthesia. *Vet Clin North Am Exot Anim Pract.* 2021;25: 181-210.

2. Grubb T, Sager J, Gaynor JS, et al. AAHA anesthesia and monitoring guidelines for dogs and cats. *J Am Anim Hosp Assoc.* 2020;56(2):1-24.

3. Longley L. Rabbit anesthesia. In: Longly L, ed. *Anaesthesia of Exotic Pets.* Philadelphia, PA: Elsevier Saunders; 2008:27-35.

4. Burton M, Conway R, Mishkin N. et al. Pharmacokinetics of gabapentin after single, oral administration in domestic rabbits (*Oryctolagus cuniculus*). *J Exot Pet Med.* 2023;45:1-5.

5. American Society of Anesthesiologists. ASA Physical Classification Status. Available at: https://www.asahq.org/standards-and-guidelines/asa-physical-status-classification-system. Accessed June 1, 2023.

6. Cripwell D, Dugdale A, Michou J. ASA Physical Classification Status: A Guide for Veterinarians. Available at: https://www.thinkanesthesia.education/sites/default/files/2021-03/US%20-%20ASA%20physical%20Status%20Classification%20Chart%20-%20%20V2.pdf.

7. Pollock C, Lichtenberger M, Echols MS. Epidural anesthesia in small mammals. LafeberVet Website. 2013. Available at https://lafeber.com/vet/epidural-anesthesia-in-small-mammals/. Accessed June 1, 2023.

8. Lafferty K, Pollock CG. Blood pressure monitoring in exotic animal species. LafeberVet Website. 2018. Available at https://lafeber.com/vet/blood-pressure/. Accessed June 1, 2023.

9. Lafferty K, Pollock CG. Capnometry in exotic animal species. LafeberVet Website. 2018. Available at https://lafeber.com/vet/capnometry/. Accessed June 1, 2023.

10. Lafferty K, Pollock CG. Electrocardiography in exotic animal species. LafeberVet Website. 2018. Available at https://lafeber.com/vet/electrocardiography/. Accessed June 1, 2023.

11. Lafferty K, Pollock CG. Pulse oximetry in exotic animal species. LafeberVet Website. 2018. Available at https://lafeber.com/vet/pulse-oximetry/. Accessed June 1, 2023.

12. Lafferty K, Pollock CG. Monitoring vital signs in exotic animal species. LafeberVet Website. 2018. Available at https://lafeber.com/vet/monitoring-vital-signs-in-exotic-animal-species/. Accessed June 1, 2023.

13. van Zeeland YRA, Schoemaker NJ. Analgesia, anesthesia, and monitoring. In: Graham JE, Doss GA, Beaufrère H, eds. *Exotic Animal Emergency and Critical Care Medicine.* Hoboken, NJ: Wiley-Blackwell; 2021:95-108.

14. National Centre for the Replacement Refinement and Reduction of Animals in Research. The Rabbit Grimace Scale. Available at: https://www.nc3rs.org.uk/3rs-resources/grimace-scales/grimace-scale-rabbit. Accessed June 1, 2023.

15. Keating SCJ, Thomas AA, Flecknell PA, et al. Evaluation of EMLA cream for preventing pain during tattooing of rabbits: changes in physiological, behavioural and facial expression responses. *PLoS ONE.* 2012;7(9):1-11.

Section 20

Technician Session 2

Brittany Skinner, RVT
Session Chair

Introduction to Zoology for Veterinary Technicians

Christina Miller, RVT, BSc

Session #3510

Affiliation: From Clayton Park Veterinary Hospital, 278 Lacewood Dr, Halifax, NS, B3M 3N8, Canada.

Abstract: Veterinary hospitals working with exotic species encounter a wonderful diversity of animals compared to the standard canine and feline patients in the average small animal practice. Consider how differently dog and cat patients are treated: drug doses, behavioral principles, vital signs, test values, and common diseases differ from these two ubiquitous companion animals but how can technicians prepare to care for an even greater range of unique and distinctive species? Understanding some key concepts in zoology will increase understanding of the patient's nature and needs.

Introduction

Zoology is the scientific study of the behavior, structure, physiology, classification, and distribution of animals.[1] While medicine is a key focus in the day-to-day operations of technicians and nurses, zoology undeniably has a place in our work. Understanding the natural history, phylogenetic relationships, anatomy, and physiology of patients makes us better able to provide appropriate care for them.

A few other terms that are important to define are as follows:

- Animal: Organisms that belong to kingdom Animalia (also called Metazoa), that are multicellular and heterotrophic organisms.[2]
- Heterotroph: An organism deriving its nutritional requirements from complex organic substances[1] (e.g., from other organisms). They lack the ability to chemically produce their own food from inorganic molecules.
- Taxonomy: The branch of science concerned with classification, especially of organisms.[1]
- Species: A group of living organisms consisting of similar individuals capable of exchanging genes or interbreeding. The species is the principal natural taxonomic unit, ranking below a genus and denoted by a Latin binomial.[1]
- Phylogenetics: The study of organisms' shared evolutionary history.[3]
- Evolution: Descent with inherited modification, including changes in gene frequency in a given population over time.[4]
- Evolutionary fitness: The effectiveness of a particular genotype in producing successful offspring in its environment.[4]

Classifying Animals

Organisms are currently classified based on *shared ancestry* and not *shared characteristics*. Classically, Linnaean taxonomy was used to organize animal life based on the shared characteristics of animals, meaning "like" was grouped with "like." This was based on physician, botanist, and zoologist Carl Linnaeus' (as well as a few other early biologists) work, *Systema Naturæ*, first published in 1735. With the advent of molecular biology and greater accessibility to means of analyzing the DNA of animals and other life forms, organisms are now classified based on shared ancestry to better illustrate their evolutionary relationships.

Animal organisms are named following the International Code of Zoological Nomenclature, which is published by the International Commission on Zoological Nomenclature (ICZN). The code regulates a uniform system that maintains both universality as well as continuity in naming animal organisms so that scientific communication regarding these organisms can remain clear across all languages. The rules include how names are established with appropriate binomial nomenclature, which name will be officially used if there are conflicts in the literature, and how scientific literature must cite animal names. Note that the naming of plants is dictated by a different authority, and overlap exists between plants and animals (for example, there are both plant and animal genera *Abronia* and *Dracaena*).

Perhaps one of the most fluid concepts in animal classification is the designation of species. While the textbook definition of species that is listed above often applies, there are many examples of gray areas that are more challenging to definitively decide upon. For example, the presence of animal hybrids, such as fertile offspring of the tiger (*Panthera tigris*) and lion (*Panthera leo*), breaks this rule as it suggests that these two very distinct species would "technically" be the same species. It has instead been suggested that "species" should refer only to naturally occurring interbreeding animals.

Examples exist of naturally occurring hybridization, such as Algonquin or Eastern wolf (*Canis lupus lycaon*) naturally hybridizing with their close relative, the coyote (*Canis latrans*). Where the Eastern wolf is a protected species, but coyotes are not (and are often considered a pest animal in many regions), where is the line drawn with legal protection of wolf-coyote hybrids?[5] It has even been determined that many of these "coywolf" hybrids not only have both wolf and coyote genes, but a 2016 study found domestic dog (*Canis familiaris,* or *Canis lupus familiaris*) genes in these populations. It has been suggested that this hybridization be treated as a speciation event, and be recognized as a new species (*Canis oriens*).[6]

Zoogeography and Phylogenetics

Zoogeography is the science of the geographic distribution of animals,[1] and is best interpreted when paired with phylogeny. Whereas zoogeography aims to answer "what lives where, and why does it live there," phylogenetics adds a temporal scale to these spatial questions to provide a better resolution of the answers.

An example of questions that can be answered by these sciences, is "Why are there so many marsupials in Australia?" Nearly 70% of extant marsupial species live in Australia; the remaining 30% live in the Americas, and they are not distributed on any other continent.[7] The inquisitive naturalist may ask, why is this the case? Are the marsupials in Australia closely related to the marsupials in the Americas? Have they lived anywhere else, and if so, why has that distribution changed?

Much of our understanding of modern phylogenetic relationships comes from the theory that Earth during the Paleozoic and Mesozoic eras was made up of one mostly fused supercontinent, Pangea. There is significant geological evidence supporting this theory. Evolutionary relationships between taxa are supported by this theory, as well[8]: Zoogeographic regions with spatial patterns of animal distributions have been recognized, and the patterns demonstrated by these regions vastly match the geological evidence for the supercontinent theory.[9]

Why Relationships Matter

Understanding that shared characteristics do not necessarily equate to shared ancestry is a key concept in modern zoology. While many species look related (i.e., have shared characteristics), this can be the result of convergent evolution. This can be the result of similar environmental pressures and adaptive mechanisms that allow for better survival (or better evolutionary fitness), or simply random chance.

A complex example is the evolution of the camera eye in both vertebrates and cephalopods, which are invertebrates. These eyes have both evolved an iris around a circular lens, a light-sensitive retina with photoreceptors, as well as anterior and posterior chambers containing humours.[10] The cephalopod eye differs in that it develops embryonically as an invagination of the body surface instead of from brain tissue, the cornea is congruent with the epidermis instead of compromising the surface of the globe, and the cephalopod eye also lacks a retinal blind spot.[11]

Understanding relationships between species can be immensely helpful in veterinary practice when data are deficient on a particular patient that the clinician and technician or nurse are treating. While cautious judgment must always be exercised, it can be far more relevant to extrapolate data from a closely related species, compared to a very distant relative. Normal physiologic parameters, clinical chemistry and hematologic values, and even drug doses can be relevant in many scenarios.

Conversely, some species are closely related, but inhabit differing ecological niches and have demonstrable differences in their environmental needs. In the absence of published care guidelines, examining other species within similar taxa that have comparable natural behaviors and habitats can yield life-saving information in the hospital.

Where Do Patients Fit on The Tree of Life?

A cladogram can be used to illustrate the phylogenetic relationships between organisms. This helps to determine the significance of shared characters, which may be anatomic, behavioral, or genetic.

Synapomorphies are shared derived characters, thought to be evolved in an ancestral species to a group and then shared with the descendants. Symplesiomorphies are shared characters between distant, related groups. Homoplasies are characters that are a result of convergent evolution, or a regression to an ancestral or primitive version of a character.[1]

Conclusions

The exotic animal veterinary technician or nurse will be exposed to a plethora of different species. Understanding relationships between species helps to deepen the appreciation for this wonderful diversity of animals and provide better care for unusual patients.

References

1. Oxford University Press. Oxford English Dictionary. Available at: https://www.oed.com. Accessed February 2, 2023.

2. Animal Diversity Web. Animalia (animals). Available at: https://animaldiversity.org/accounts/Animalia/. Accessed February 2, 2023.

3. Stanford Encyclopedia of Philosophy. Phylogenetic Inference. Available at: https://plato.stanford.edu/entries/phylogenetic-inference/. Accessed February 2, 2023.

4. University of California, Berkeley. Understanding Evolution. Available at: https://evolution.berkeley.edu/evolution-101/mechanisms-the-processes-of-evolution/evolutionary-fitness/. Accessed February 2, 2023.

5. Zimmer C. What is a species? *Scientific American*. 2008. Available at: https://www.scientificamerican.com/article/what-is-a-species/ Accessed February 2, 2023.

6. Way JG, WS Lynn. Northeastern coyote/coywolf taxonomy and admixture: a meta-analysis. *Canid Biol Cons*. 2016;19(1):1-7.

7. Wikipedia. Marsupial. Available at: https://en.wikipedia.org/wiki/Marsupial. Accessed February 4, 2023.

8. Cox CB, Moore PD, Ladle RJ. *Biogeography: An Ecological and Evolutionary Approach*. Hoboken, NJ: Wiley-Blackwell; 2016.

9. Holt BG, Lessard J-P, Borregaard MK, et al. An update of Wallace's zoogeographic regions of the world. *Science*. 2013;339(6115):74-78. doi:10.1126/science.1228282.

10. Serb JM. Toward developing models to study the disease, ecology, and evolution of the eye in Mollusca. *Am Malac Bull*. 2008;26:3-18.

11. Hanke FD, Kelber A. The eye of the common octopus (*Octopus vulgaris*). *Front Physiol*. 2019;10:1637.

Getting a Bird's Eye View on Avian Anesthesia and Analgesia

Liz Vetrano, BS, CVT, VTS (Exotic Companion Animals)

Session #3263

Affiliation: From Mount Laurel Animal Hospital, Mount Laurel, NJ 08054, USA.

Abstract: Anesthesia has always had a reputation for being an intimidating and difficult aspect of avian medicine. With appropriate knowledge and training regarding anatomy, multimodal analgesia, and anesthetic monitoring, these cases can be both challenging and successful. This presentation discusses the process of avian anesthesia from start to finish. This discussion includes several types of common avian procedures and how they may make a difference in surgical protocols including multimodal pain management. Preoperative management is also discussed to create the best outcome for the patient. It is essential to evaluate the case in collaboration with a clinician to determine stability and appropriate timing of any anesthetic event. This evaluation also relates to determining the most accurate American Society of Anesthesiologists' status and preparing for potential complications both during and post operation. During the presentation a procedure is followed while reviewing correct set up, sedation, pain management (including constant rate infusions), preparation (operating room set up and pre-operating table), anesthetic monitoring, and management, including use of a ventilator, and recovery. All these details can simultaneously contribute to setting up the patient for a successful anesthetic event and overall recovery.

Introduction

Avian anesthesia is often viewed as a challenging aspect of veterinary medicine instilling hesitation in most support staff, and occasionally clinicians. With the appropriate knowledge and skills set, any avian anesthetic event can have a successful outcome with a favorable prognosis and good quality of life. Evaluating and performing appropriate diagnostics prior to sedation or anesthesia is the proper way to fully evaluate the patient's current condition and any other possible underlying conditions that may affect the overall anesthetic protocol and treatment plan. Once the patient is determined to be a stable anesthetic candidate, an appropriate protocol should be deliberated between the clinician and technician, including suitable analgesia. Analgesia can vary depending on species, type of procedure, patient status, or availability of drugs. Once the protocol has been solidified, it is the technician's responsibility to be knowledgeable about surgery preparation including catheter placement, preparing the surgical site, and equipment/instruments required. During anesthetic monitoring, it is important to be familiar with normal vital ranges, injectable sedation, gas anesthesia, monitoring equipment, and how to manage the patient if there are anesthetic complications. Recovery can be considered one of the most important steps in this whole process as it has been thought that most patients decline during this time frame during the surgical case. Observing the patient and providing appropriate care can determine the outcome. Overall, it is important for the entire team to collaborate and communicate to determine the most appropriate plan, for both analgesia and anesthesia, for the individual case that will provide a positive prognosis.

Pre-Surgery

Surgical preparation for avian cases can be very diverse. A young, stable chicken could be presenting for a simple laceration repair, or a critical, egg-bound, 30-year-old Moluccan cockatoo could be presenting for an emergency salpingohysterectomy (and don't forget the patient's history of atherosclerosis). Each of these cases should be evaluated based on their species, status, and history. An important part of having a successful anesthetic event is evaluating the patient's current condition. Typically, radiographs and blood work are baseline diagnostics required for clinicians to make a reasonable list of differentials and determine if any other treatments need to be completed prior to the procedure. In addition to these, sometimes the clinician will determine that the patient would benefit from an ultrasound or computed tomography scan. This can help then solidify a more detailed surgical approach depending on the procedure (e.g., mass removal, exploratory coeliotomy, endoscopy). Depending on the patient's status, starting treatments prior to the procedure can be an imperative part of ensuring a successful anesthetic event. For example, an egg-bound patient typically benefits from fluid therapy, antibiotics, and very importantly, pain management, prior to surgery. Fluid therapy is a valuable treatment used to maintain blood pressure and increase hydration. Administering a warmed crystalloid fluid, such as Normosol R® or Lactated Ringers®, subcutaneously can correct fluid deficits and improve perfusion. Antibiotics starting pre-operatively or intra-operatively can also be valuable if an infection is suspected due to their current condition or reported white blood cell count. Getting ahead of the pain, otherwise known as avoiding the wind up, is also a key feature in preparing for any procedure. Starting pain management as quickly as possible before a procedure can not only improve the patient's level of comfort, but also the anesthesia protocol itself, and the recovery process. Specific analgesia and protocols will be discussed separately.

On occasion with the critical patient, checking a packed cell volume/total solids (PCV/TS) in addition to other baseline blood work can determine if the patient requires a blood transfusion. For most species, a normal PCV range can be between 40-55%[1]. If a patient presents with a PCV of 15% or lower, or if the patient is clinically weak from anemia, a whole blood transfusion may be required.[1] Birds can tolerate lower PCVs if the condition is chronic. Since birds do not possess specific blood types, whole blood from other species may be used if that is all that is available; however, homologous whole blood has a much longer half-life. The author has used chicken blood for psittacine patients requiring emergency surgery and experiencing hemorrhage. Ten percent of the donor's body weight may be taken and administered to receiver over 1-4 hours.[1]

Analgesia and Sedation Plan

Providing appropriate analgesia for any exotic species typically has had many challenges, including lack of pain recognition and appropriate pain medications capable of providing appropriate therapeutic levels. With an increase in the number of studies performed supporting the suspicion that all of these exotic companion animals experience pain as much as our canine and feline patients it is imperative to provide the appropriate analgesic protocols. Several factors play a role in determining the correct protocol, including species, procedure, comorbidities, age, and medications. One of the first steps in determining a protocol is evaluating each individual case and referring to a pain scale. There have been several scales developed for exotics species which show specific signs indicating levels of pain. For birds, these signs may include but are not limited to orbital tightening, clenched feet, increased respiratory effort, fluffed plumage, not perching, among other signs. Some of the more common types of analgesics and how they can be applied within the anesthetic protocol before, during, and after the procedure are described below.

Opioids are a common type of analgesic used regularly in the medical field. They can be used for cases experiencing mild to moderate amounts of pain. It is important to refer to pain scales and determine which specific opioid is most suitable for the signs of pain the patient is experiencing, as well as the species. Buprenorphine is a partial mu-agonist which has proven to be a suitable choice for some avian species including American kestrels (*Falco sparverius*) and Columbiformes.[2] This is acceptable for mild pain in these species as these animals initially present to the hospital for a painful condition or as a surgical pre-medication. Butorphanol is typically the preferred opioid of choice for most avian species due to its affinity for kappa receptors. Most species have been demonstrated to have predominantly kappa receptors, so butorphanol provides excellent pain management. This drug can be administered either intramuscularly into the pectoral muscle, or intranasally (which is less invasive and avoids bruising). The only disadvantage of butorphanol in some cases is the frequency of dosing, which can be anywhere from every 2-4 hours to maintain appropriate therapeutic levels.[2] This means more handling, stress, and risk of iatrogenic injury which can be detrimental for a critical patient. One of the major advantages is that butorphanol can be used as a constant rate infusion during hospitalization and/or procedures. Not only will it provide adequate pain relief, but also help reduce the amount of gas anesthesia required to keep the patient at an appropriate plane of anesthesia. A loading dose is typically delivered in conjunction with a sedative, in order to improve the amount of sedation, and the CRI is started shortly afterwards. Tramadol is another opioid with an affinity for mu receptors. Although it does not target kappa receptors, this provides mild pain management for birds, simply because of its ability as a serotonin reuptake inhibitor.[3] This can be administered both by mouth or by intramuscular injection for mild pain.

Midazolam is a water-soluble benzodiazepine typically used in conjunction with butorphanol, or by itself for avian sedation. It does not provide any analgesia when used by itself. It is important to note that just like any other medication, its effects can vary between species and individuals. It can be delivered intramuscularly or intranasally and is an essential part of the avian pre-medication protocol. Doses can be anywhere from 0.1-2 mg/kg IM or IV and 2 mg/kg IN (species may vary).[2] Flumazenil is the antagonist for midazolam which is typically administered after the procedure at 0.02-0.1 mg/kg IM/IV for most species.[2] For very non-invasive or brief procedures (i.e., prolapse repair, diagnostics), a dose of 1-2 mg/kg can be paired with 2 mg/kg of butorphanol and can be a great alternative for gas anesthesia.[2]

Alfaxalone is a neuroactive steroid used for sedation in many species; however, there has been very little literature on its use in avian patients. Due to the lack of peer-based reviews and evidence-based recommendations. The author has not utilized this medication in avian patients.

Ketamine is a dissociative anesthetic that when used in collaboration with a sedative, such as midazolam, it not only can provide significant sedation, but some pain management as well, as it has some affinity for the NMDA receptor. Although there are some clinicians that utilize ketamine in avian patients, the author has no experience using this medication in them.

Gabapentin is a GABA analogue used to treat human neuropathic pain.[2] This can be a useful analgesic both pre and post any surgical event. Studies have been completed involving Senegal parrots

Meloxicam is a popular non-steroidal anti-inflammatory drug (NSAID), typically administered post-operatively in a multi-modal analgesic plan. It preferentially inhibits cyclooxygenase-2 and is administered after gas anesthesia in case of renal dysfunction or hypovolemia. The NSAIDs can play a major role in limiting the amount of inflammation that can occur from any traumatic procedure. Celecoxib is another NSAID that can be useful post-operatively; however, it can only be administered by mouth. Like meloxicam, it is a selective COX-2 inhibitor and helps reduce inflammation.

Surgery Preparation and Operating Room Setup

It is imperative to not only prepare for what is to be expected, but also for the unexpected during a surgical procedure. Prior to the procedure, it is necessary to collaborate with the clinician to determine what specialized equipment or instruments they need to streamline that specific procedure. A procedure such as a cloacal prolapse repair would require different instruments than a salpingohysterectomy, so it is important to familiarize yourself with common tools used in the more frequent surgeries being performed. Smaller instruments are typically more desirable for any avian procedure. This includes hemostats, towel clamps, needle holders, forceps, and scissors. It is important to note that certain instruments are preferred for avian surgeries due to their extremely thin and friable tissue. This includes Brown-Adson forceps (not rat-toothed) and suture with a tapered needle. Cautery is also a very useful piece of equipment for cutting and cauterizing the friable tissue throughout all stages of the procedure. Providing the clinician with bipolar cautery and very small, thin cautery forceps are essential to streamline cutting and provide coagulation as to limit hemorrhaging. Hemostatic clips (e.g., Weck® Hemo-clip® Metal Ligation System, Teleflex, Morrisville, NC) are another essential tool for hemostasis during an avian procedure. There are small, medium, and large options depending on the size of the patient. Right-angled hemostatic clips are also available and sometimes more useful for the clinician on smaller patients. Statinsky Debakey vascular clamps are also useful in these patients, as they are designed for occluding large, deep vessels. Hemostatic sponges (Vetspon®, Elanco Animal Health, Greenfield, IN) are another useful item to have available in case of excess bleeding as they are easily placed onto the site to give the clinician some time to evaluate the source of bleeding.

In addition to discussing specific instruments and equipment needed, it is also important to discuss positioning. Positioning of the patient is an essential part of the surgery itself and setting the surgeon up for success. For example, salpingohysterectomies should always be placed in right lateral recumbency since the reproductive tract is located on the left side of the patient. Different surgical procedures require different positions which can vary greatly. Knowing how the surgeon would like the patient to be placed on the table ahead of time can streamline the preparation process and limit the patient's anesthetic time.

The surgical prep area also plays a valuable role in streamlining the surgical process, prior to moving the patient into the operating room. Sedation, induction, intubation, intravenous catheter placement, and plucking (if necessary) typically take place during this phase. All equipment and items should be set up prior to the patient being sedated. This includes heat support. Heat support is one of the most essential pieces of equipment in any bird procedure. Birds have very naturally high body temperatures ranging from 38-45 °C (100.4-113.0 °F). Failing to provide heat support or even administering room-temperature fluids can result in a compromising decrease in body temperature and potentially shock. Providing adequate heat support, such as a forced air warmer (3M™ Bair Hugger™ System, Saint Paul, Minnesota) during all aspects of the procedure is a requirement.

Streamlining surgical preparation once the patient has been sedated is clearly one of the major players in ensuring a shorter, and more efficient anesthetic time. In addition to heat support, having equipment prepared for intubation and intravenous or intraosseous (IO) catheterization is also important. This includes various IVC size options, hypodermic needle options (if IO access is a possibility), local anesthetic if needed, aseptic scrub, suture, tape, bite block, and different Cole endotracheal (ET) tube sizes. Once the patient has been sedated with pre-medications, received pre-oxygenation, and placed under gas anesthesia, securing an airway is the number one priority. Birds possess a trachea with fully encircled cartilaginous rings. This puts them at a much higher risk for trauma and strictures due to intubation. Only uncuffed tubes should be used. Cole tubes are the most commonly used endotracheal tubes in birds. They allow a good sealing of the trachea, and they limit the dispersion of air with no risk to the tracheal epithelium.[4] To ensure appropriate placement, a capnograph should be utilized to detect an appropriate end-tidal CO_2 ($ETCO_2$) reading. A bite block is also necessary to prevent the patient from biting through the ET tube. The author prefers to make these from tongue depressors or cotton-tipped applicators wrapped in cohesive bandage based on the size of the patient. Bite blocks also provide additional support for securing the ET tube to the patient. Once an airway has been secured, vascular access should also be secured.

Like in any other species, vascular access is essential for any procedure to allow for fluid therapy and/or emergency medications. The size and status of the patient are both factors that could help determine whether an IV or IO catheter is preferred. For medium to larger-sized patients, an IV catheter is preferred as they are less invasive, painful, and easier to place. Potential IV sites include the basilic (ulnar or wing) vein, jugular vein (preferred in more critical patients as this site can put the patient at risk for hemorrhage), medial metatarsal vein, or pedal vein. Any vascular access in the caudal half of the patient's body may pass through the renal portal system before entering systemic circulation. For smaller patients, an IO catheter may be the only option. The preferred site for this is the ulnar bone as this is easily accessible, is not a pneumatic bone, and can be wrapped to keep comfortably in place post-surgery. Once vascular access has been achieved, fluid therapy should be started as soon as possible. Again, it is essential that these are warmed and never at room temperature since it could cause them to go into shock.

For site preparation, it is necessary to pluck the feathers since feather follicles are still intrusive to the surgical site. It is not always necessary to pluck for certain procedures, such a cloacal prolapse repair, but it is for any invasive procedure involving entering the coelomic cavity (e.g., salpingohysterectomies, mass removals, biopsies via endoscopy). Plucking can either be achieved simply by hand or with hemostats. The author recommends using hemostats for feathers with larger shafts. It is also important to hold the skin taught with the other hand while plucking since their skin is very thin and at risk for tearing. Confirm with the surgeon that the area is appropriately prepped prior to moving the patient in the OR.

Surgery and Monitoring

After moving the patient into the operating room, ensure that the patient is in the optimal position prior to taping and securing them to the table. Be mindful and gentle where possible when taping the patient down, especially when positioning the wings. Another technician should simultaneously be attaching the patient to the anesthetic machine. The author prefers sevoflurane as an inhalant anesthetic over isoflurane for avian patients due its high potency, lack of odor, and fast-acting effects. Depending on the procedure and the size of the patient, a ventilator is a great option for avian procedures. Due to their advanced respiratory systems and their inability to perform gas exchange adequately under anesthesia, it is essential to provide intermittent positive pressure ventilation every 6-12 seconds.[5] It is important to also remember the differences in species such as Anseriformes (waterfowl) that possess a dive reflex allowing them in instinctively hold their breath. Intermittent positive pressure ventilation bypasses their ability to shunt. The ventilator is extremely convenient and allows technicians to perform other tasks such as monitor or circulate the operating room. Ventilators also allow for tight control of the delivered tidal volume with long and deep breaths allowing for beneficial gas exchange and better planes of anesthesia. A ventilator is typically not recommended for procedures that require the clinician to enter the air sac. Once the air sac is punctured, the pressure that the ventilator requires to operate is no longer possible, and the ventilator will fail to function. If a ventilator is not available (or if an air sac needs to be punctured), a non-rebreathing circuit is also acceptable. This requires a single dedicated staff member who will provide IPPV during the entire procedure.

Monitoring smaller exotic patients has always been a challenge due to the lack of usable equipment available and their small size. Thankfully, more equipment has been developed to allow for more precise monitoring options other than just a Doppler probe. For respiration, attaching a side-stream capnograph, such as an Airmate®, (Sentier, Waukesha, WI, USA) is necessary to not only monitor respiratory rate but $ETCO_2$, as well. This will also ensure appropriate gas exchange, confirm tube placement throughout the procedure, and that the breaths being delivered are adequate. A study on Congo African grey parrots (*Psittacus erithacus*) demonstrated that a normal $ETCO_2$ range is 30-40 mmHg.[6] Although this is a good reference point, it is important to remember that this can vary amongst species. If a clinician is required to enter an air sac, the $ETCO_2$ readings are invalid until the air sac is closed.

Evaluating an accurate electrocardiogram (ECG) can also be difficult in these species due to their small size and high heart rates. Despite these challenges, Sentier has developed a very small, light, and mobile machine, called a Vetcorder® (Sentier, Waukesha, WI, USA) that can accommodate these patients and potentially produce an ECG. Not only does it come with small metal clips for mammals, but they also provide electrode needles suitable for avian patients. These are placed gently through the skin. An ECG reading can be obtained by placing the right atrial lead on the right side of the cranial sternum and the left atrial lead on the left side of the caudal sternum. Both atrial leads may also be placed on the caudal sternum in cases where it can disrupt the surgical site. Once the needles are in place, small amounts of isopropyl alcohol 70% can be used to help with the reading. Be mindful to only use small amounts as too much can lower the body temperature since alcohol is very volatile and this may contribute to significant evaporative heat loss. The normal ECG in the avian patient can have the appearance of ventricular tachycardia, primarily because of the large negative S wave. This QRS morphology is caused by the high density of sympathetic and parasympathetic nerve fibers in the avian ventricles and atria when compared to mammals.

A Vetcorder® (Sentier, Waukesha, WI, USA) is also capable of providing SPO_2 values. The pulse oximeter is used to measure the oxygen saturation of hemoglobin in the bloodstream. Like any other species, values should be anywhere between 96-100% for normal perfusion. Pulse oximetry also provides a heart rate reading. Normal heart rates can have a wide range, depending on the patient's body size (the smaller the patient, the higher the heart rate. See Table 1).[7] Just like with any other vital sign, it is always important to gather them prior to the procedure to have a general understanding of what is normal for that individual or to ensure their values are normal prior to an anesthetic event. The pulse oximeter may be placed on various sites. The most common is the foot or toes depending on the patient size. Other options for the pulse oximeter probe are the basilic (wing vein) or the medial metatarsal vessel. Pulse oximetry uses infrared lighting to shine through and determine the amount of oxygen traveling in the bloodstream. Because of this, it can have difficulty providing an accurate reading on patients that have darker skin pigmentation. In these cases, a Doppler probe is always great to have as a backup to evaluate heart rate.

Table 1. Normal avian heart rates based on weight.

Weight (grams)	Heart rate, resting (bpm)
25	380
100	329
200	245
300	226
400	213
500	203
1000	175
2000	152

Source: Speer B. *Current Therapy in Avian Medicine and Surgery.* St. Louis, MO: Elsevier; 2015.

In addition to ECG and SPO_2, a digital temperature probe is a necessity. This is a great way to evaluate the patient's internal temperature and ensure appropriate heat support is being provided throughout the procedure. Note that a patient's temperature is more likely to drop when there is a large, open incision versus no incision required. Regardless, heat support is always a necessity when the patient is sedated. Hypothermia can be one of the most common causes of death for avian patients under anesthesia. The author prefers forced air warmers for avian patients as they provide a large amount of consistent heat support without risk of burning. Heat lamps should be available to provide additional heat support if needed. Providing warmed IV fluids is also necessary to maintain normal body temperature. There are small fluid warmers available specifically for IV fluid administration and should be placed as close to the patient as possible.

Blood pressure is another value that is typically measured during anesthetic procedures. This is probably the most challenging value to obtain due to the limited small cuff sizes available that could gather an accurate reading. The pressure cuff can be placed on the distal humerus, femur, or tibiotarsus. Normal pressures range between 90-180 mmHg in conscious psittacine birds.[8] It is important to recognize that normal values can vary based on species. Unfortunately, certain studies have shown that, in birds less than 2 kilograms, direct blood pressure values do not tend to correlate with indirect values.[8] As a result, indirect blood pressure is typically an unreliable value, especially on smaller species. Since it may be possible to still obtain values, it is still valuable to monitor trends.[8]

In addition to all the monitoring equipment, it can be argued that a technician with a stethoscope is the most valuable implement in the operating room. Sometimes the equipment can be faulty or unable to read. It is imperative to learn how to troubleshoot all the equipment, including the anesthesia machine, while simultaneously evaluating the patient. In the event of a true emergency such as respiratory or cardiac arrest, it is important to be prepared if the patient has been labeled as a cardiopulmonary resuscitation. Having emergency drugs calculated and already drawn up can streamline any code.

Recovery

Recovery can be the most vulnerable time for an avian patient after an anesthetic event. This is the time that hypothermia can occur, or they can seemingly suddenly go into cardiac or respiratory arrest. During closing, it is important to communicate with the surgeon and if appropriate, begin to taper the gas anesthesia. Once completed and the gas turned off, disconnect the patient, flush the machine (to rid the remaining anesthetic gas from the line), and re-hook the patient to the machine so that they can continue to receive oxygenation while waking up. Continuing fluid therapy (with warmed fluid) and heat support is necessary. Reversal agents should be given at this time (in regard to sedation, not analgesics). Birds have the reputation of waking up very quickly. It is important to be prepared with a towel for restraint (and to prevent injury) and to extubate efficiently when appropriate. The bite block should provide protection if the bird wakes up faster than the technician can extubate. The tube should be inspected for any blood. Sometimes a small amount of trauma, simply from movement, can be a possibility. Once the patient is ambulatory, they should be placed in an enclosure that can provide both heat support and oxygen therapy. Depending on the invasiveness of the procedure, it is sometimes recommended that they remain hospitalized and under observation for an additional 24 hours. Fluid therapy, gavage feedings, heat support, and other treatments may be necessary to ensure a positive prognosis.

Conclusions

With the appropriate skills and knowledge, avian anesthesia does not have to bear the stigma of resulting in death. As a team, it is imperative to collaborate from start to finish. This includes determining an appropriate ASA status, surgical protocol, preparation plan, and recovery plan. Like any other treatment plan, communication between all parties is essential throughout each phase. Ensure that everyone is playing a role. With teamwork, practice, continuing education, and communication, bird procedures can have a positive outcome with a good prognosis. It is important to remember that all animals, including birds, deserve gold-standard medicine to the best of our ability. In veterinary medicine, it is our job to challenge ourselves enough to be able to provide these services.

References

1. Graham JE, Doss GE, Hugues B, eds. *Exotic Animal Emergency and Critical Care Medicine.* Hoboken, NJ: Wiley-Blackwell; 2021.

2. Plumb DC. *Plumb's Veterinary Drug Handbook.* 6th Ed. Ames, IA: Blackwell Publishing; 2008.

3. Carpenter JW, ed. *Exotic Animal Formulary.* 5th Ed. St. Louis, MO: Elsevier; 2018.

4. Crosta L. LafeberVet. Avian anesthesia webinar. Available at: https://lafeber.com/vet/avian-anesthesia-webinar/. Accessed June 1, 2023.

5. Longley LA. *Anaesthesia of Exotic Pets.* Philadelphia, PA: Elsevier Saunders; 2008.

6. Edling TM, Degernes LA, Flammer K, et al. Capnographic monitoring of anesthetized African grey parrots receiving intermittent positive pressure ventilation. *J Am Med Vet Assoc.* 2001;219(12):1714-1718.

7. Speer B. *Current Therapy in Avian Medicine and Surgery.* St. Louis, MO: Elsevier; 2015.

8. Lafferty KA, Pollock C, Ambros B, Nugent-Deal J. Blood pressure monitoring in exotic animal species. LafeberVet Website. Available at: https://lafeber.com/vet/blood-pressure/. Accessed June 1, 2023.

Section 21

Technician Tutor Session

Lorelei D'Avolio, LVT, VTS (Clinical Practice- Exotics), CVPM
Session Chair

Avian Pain Assessment for Veterinary Technicians

Jovanna Berls, CVA, LVT, FFCP (Veterinary), FFCP (Avian)

Session #3151

Affiliation: From ABC Animal & Bird Clinic, 11930 Highway 6 South, Sugar Land, TX 77498, USA.

Abstract: Pain control is an essential part of nursing care for all species of veterinary patients, but managing pain appropriately depends on being able to determine whether a patient is experiencing discomfort. Validated methods of assessing pain have been developed for some types of animals, including some exotic species, but so far, no such pain scale has been devised for birds. As in other species, evaluation of pain in avian patients relies heavily on behavioral cues such as self-grooming and social interactions. These behaviors vary from one species of bird to another, and birds are very adept at using masking behaviors to avoid being perceived as ill or weak. This presentation reviews behaviors likely to be observed in birds suffering discomfort and discusses some of the ways technicians can act as advocates for their patients' healing and welfare.

Introduction

There is ample evidence that birds have the physiologic and anatomic mechanisms to experience pain, but validated methods of measuring it have not yet been developed. Assessing the severity of pain in avian species is challenging for several reasons. Some assessments that have proved helpful for mammals, such as grimace scales available for rabbits and rodents, rely on facial expressions that are much more limited in avian species. Physiological indicators, such as changes in heart rate, respiratory rate, and blood pressure, have been investigated, but some of these parameters are more reliable than others, and measuring them can itself induce stresses that can mimic a pain response. Hypothermia has been reported as interfering with accurate pain assessment.[1,2]

Thus, as in other species of exotic pets, pain assessment relies heavily on behavioral cues, and there have been some efforts to develop a pain scale based on such signs that would be applicable to birds. These behavioral cues vary from one species of bird to another: A scale that works for chickens may not be accurate for parrots, for example. In addition, birds are adept at using masking behaviors to avoid exhibiting signs of discomfort.

Behavioral Cues

Most research into behavioral indicators of pain in birds has focused on poultry, not the psittacine species that often are the bulk of the exotic veterinary technician's caseload. A recently published integrative literature review of the subject showed that, out of 132 total articles on pain-related behaviors in birds, only 5 articles (3.8%) involved parrots, with 73 articles (55.4%) devoted to chickens and turkeys.[3]

Before attempting to evaluate whether behavioral cues indicate pain, it is important to be familiar with the normal behavior of the species being studied. For example, *Pionus* spp., when excited, frequently make a wheezing or panting sound, which should not be mistaken for a sign of pain or illness. In addition, because of the limited usefulness of facial expressions, it is useful to assess the individual as a whole, noting posture, position of the tail and the head crest (in species that have them), and how the wings are carried. Keep in mind also that stress and the presence of an observer can alter behavior regardless of whether pain is present.[1]

In general, birds tend to respond to noxious stimuli with a fight-or-flight response (e.g., attempts to escape, vocalization, thrashing) or with conservation and withdrawal reactions (e.g., making no effort to escape, immobility, no vocalization). Pet birds who are in pain may allow interactions with humans that they normally would not permit.[4]

Some behaviors that may indicate pain include struggling, vocalizing, aggressive behaviors (especially in a normally tolerant bird); reduction of normal behaviors (such as eating, perching, ambulating, preening); changes in interactions with conspecifics or human handlers; new abnormal behaviors; and conservation or withdrawal responses (e.g., hiding, decreased self-grooming, hyporexia). In studies of musculoskeletal pain, Hispaniolan Amazons (*Amazona ventralis*) and green-cheeked conures (*Pyrrhura molinae*) with experimentally induced arthritis were found to ambulate less on the affected limb, groom the affected limb more than needed for general feather maintenance, and have a ruffled or unkempt appearance to the feathers.[1] African grey parrots (*Psittacus erythacus*) subjected to thermal or electrical stimuli were seen to vigorously flinch their wings when the stimulus was perceived to be noxious.[3]

Behavioral cues that have been recommended for consideration when evaluating pain in birds include:

- Dull mentation
- Eyes closed or squinting in a novel environment
- Rapid or labored breathing
- Change in environmental interaction
- Hiding
- Decreased or abnormal perching
- Increased or decreased maintenance behaviors
- Decreased locomotion
- Focal preening of a specific area
- Fluffed appearance or unkempt feathers
- Abnormal hunched body posture
- Decreased appetite or change in food choice.[1,5]

It is important to observe the avian patient both before and after administration of any analgesic drugs.[2] After administration of analgesics, a marked, easily noticed change toward more normal posture and behaviors should be observed; if it is not, pain management should be re-evaluated.[4] It is interesting to note that chickens with limb pain were observed to self-select food containing an analgesic medication when offered a choice of foods.[1]

Types of Pain: Acute, Chronic, Visceral, Neuropathic

Behavioral signs of acute pain, such as trauma, may differ from those exhibited in cases of chronic pain, such as osteoarthritis. Acute pain may result in restlessness, hunched body posture, change in appetite and physical activity, and/or lameness. Also, in cases of severe, acute pain, changes in the eyes may be noted, including pupil size, squinting, and abnormal tear production.[1]

For chronic pain, the bird may engage in increased guarding of a painful area, avoiding behaviors that the bird may perceive as exacerbating pain, self-isolation, feather destructive behavior and/or self-mutilation, inappetence, and weight loss.[1] Chronic pain also may be manifested by a reduction in social grooming behaviors, such as decreased dust bathing in chickens, and in a reduction in the performance of comfort and self-maintenance behaviors.[2] By contrast, chronic pain also may result in overgrooming or feather destructive behavior or both.[4]

Visceral pain may be more difficult to notice, as it may involve more nonspecific signs such as anorexia, lethargy, and weight loss. Neuropathic pain has been reported less often but may result in dull mentation and overgrooming at the site of the pain.[1]

Severity of Pain

In the absence of numerical pain scores for avian patients, pain can be characterized as mild, moderate, or severe.[1] It can be useful to keep in mind how you might feel if you were experiencing a similar medical condition.[2]

There have been some efforts to develop actual pain scores that may be useful as well, such as one developed in 2012 to evaluate fracture pain in pigeons (*Columba livia*). This scale set numerical values from 0-5, with 0 indicating no pain and 5 indicating extreme pain, with values assigned to four criteria: Pigeon's attitude in the presence of the observer, fracture limb position in the presence of the observer, observer's evaluation of pigeon's degree of pain (i.e., overall assessment), and pigeon's motor activity during 10 minutes of video recording.[6]

Keep in mind that a scale validated for pigeons may not be accurate for psittacine birds or other species.

The lack of scientifically validated methods of pain assessment for avian patients increases the importance of frequent and thoughtful monitoring. Always keep an eye out for subtle signs of abnormal behavior and remain aware that the patient may expend considerable effort to prevent observation of discomfort. If all else fails, the familiar standard applies to birds as much as it does to other veterinary patients: If it would be painful for you, assume it is painful for your patient.

References

1. Mikoni N, Sanchez-Migallon Guzman D, Paul-Murphy J. Pain recognition and assessment in birds. *Vet Clin North Am Exot Anim Pract.* 2023;26:65-81.

2. Speer B. Basic anaesthesia. In: Chitty J, Monks D, eds. *BSAVA Manual of Avian Practice: A Foundation Manual.* Quedgeley, UK: British Small Animal Veterinary Association; 2018:234-235.

3. Mikoni N, Sanchez-Migallon Guzman D, Fausak E. Recognition and assessment of pain-related behaviors in avian species: an integrative review. *J Avian Med Surg.* 2022;36(2):153-172.

4. Hawkins M, Paul-Murphy J, Sanchez-Migallon Guzman D. Recognition, assessment, and management of pain in birds. In: Speer B, ed. *Current Therapy in Avian Medicine and Surgery.* St. Louis, MO: Elsevier; 2016:616.

5. Turner T. Pain recognition in exotic species: a training guide for veterinary technicians. *Proc Annu Conf Assoc Avian Vet.* 2009:103-106.

Caudal Intravenous Catheter Placement in Squamate Reptiles

Christina Miller, RVT, BSc

Session #3511

Affiliation: From Clayton Park Veterinary Hospital, 278 Lacewood Dr, Halifax, NS, B3M 3N8, Canada.

Abstract: Vascular access in reptile patients can be challenging. Small patient size, absence of obvious peripheral veins in many animals, and the presence of thick skin all make identifying suitable vessels for either venipuncture or intravenous (IV) catheterization difficult. The caudal vein (ventral coccygeal vein) is a suitable site for IV catheter in some patients. Appropriate patient candidate selection, placement technique, as well as catheter maintenance will be reviewed.

Introduction

Reptiles are reputed as being challenging patients to obtain vascular access due to a lack of obvious peripheral veins in many species, the presence of thick and often scaled skin, and sometimes diminutive patient size. Recognizing techniques that are effective in a given species and appropriate for a given patient is a critical skill in treating these animals. Caudal vein (ventral coccygeal vein) catheterization has been discussed in the literature since as early as 2004.[1]

Who Are the Squamate Reptiles?

Order Squamata (the scaled reptiles) is the most speciose of class Reptilia; as of December, 2022 the total number of reported squamate reptiles is 11,549.[2] Squamata traditionally contained three sub-orders that were grouped morphologically: Serpentes (snakes), Sauria (lizards), and Amphisbaenia (worm-lizards); these classifications, although convenient, are now considered defunct in light of genetic data that has provided better resolution to the squamate reptile phylogeny.[3-5] In spite of this, squamate reptiles that are considered snakes and lizards are discussed as these animals can be readily identified as such in the veterinary practice.

What Sites Exist for Vascular Access?

Sites for catheter placement are relatively limited in these species compared to more traditional companion animals, particularly mammals:

- Cephalic: Accessible in lizards, this requires a surgical cut-down. Sedation with or without local anesthesia is recommended in patients except perhaps those that are moribund.

- Jugular: Accessible to varying degrees in lizards and snakes, this site also requires a surgical cut-down. General anesthesia is required for placement in snakes.

- Intracardiac: An emergency technique in snakes, a suitably-sized catheter can be introduced directly into the ventricle of the heart.[6,7,8] These can reportedly be left in place for up to 24 hours,[7] although long-term consequences of placement do not appear to have been studied.

- Intraosseous (IO): Several sites exist in most lizards, however there are no suitable IO catheter sites present in snakes or legless lizards species.

The caudal or ventral coccygeal vein is present in both lizards and snakes, and is found along the length of the tail. The vein is located immediately ventral to the caudal vertebrae.

Selecting Suitable Candidates

Vascular access is advantageous when treating critically ill and/or anesthetic patients, not only for rapid parenteral administration for emergency drugs, but also for providing fluid support. According to the American Animal Hospital Association, intravenous (or intraosseous) fluids are advantageous for hospitalized patients that are not taking in food or water voluntarily, patients undergoing anesthesia, and patients that require relatively large and/or rapid fluid bolus administration.[9] Reptile patients that are over 8% dehydrated would ideally have intravenous or intraosseous access for fluid administration, as absorption from other routes may be ineffective or increase intracoelomic pressure (subcutaneous and intracoelomic, respectively).[10]

When is intraosseous NOT an option?

While intraosseous catheters are a viable option for many patients, a few contraindications exist:[11,12]

- Pyoderma at the insertion site

- Pre-existing fracture

- Pre-existing osteomyelitis

- Nutritional secondary hyperparathyroidism

- Recent (within 48h) IO catheterization of the same site

Caudal autotomy

Many lizard species (and a few snakes) can perform caudal autotomy (tail drop), an anti-predator behavior that involves self-amputation of the tail at pre existing fracture planes. The amputated tail will continue to twitch and writhe, which is suspected to be a distraction for the would-be predator while the animal escapes. A ring of muscle constricts around the remaining caudal tissue, limiting bleeding. Some species can regenerate the tail in the form of a rod of cartilage in lieu of new vertebrae; the new tail often has differing colouration and scalation compared to the original appendage.[13] Caudal autotomy occurs in most lizard taxa, excluding true chameleons (Chamaeleonidae), most agamids (Agamidae), beaded lizards (Helodermatidae), monitor lizards (Varanoidea: Varanidae and Lanthanotidae), and the knob-scaled lizards (Xenosauridae). In some lizard species, autotomy and caudal regeneration is only present in juveniles, such as some members of Iguanidae (notably, *Ctenosaura* sp., *Cyclura* sp., *Iguana* sp., and *Sauromalus* sp.) and some members of Scincidae (*Tiliqua* sp.).[13]

Tail manipulations (including catheter placement) in these species should be limited to avoid inducing autotomy. Tail loss is not without consequences, as many species rely on the tail for balance when locomoting, social signaling, and energy storage.[14] It has been demonstrated that tail regrowth is energetically costly,[15] which is an additional stressor for already sick patients.

The renal portal system

Reptiles exhibit a circulatory mechanism absent in mammals, the renal portal system. Blood circulating in the caudal half of the body, including blood from the caudal, iliac, or hypogastric veins, may circulate through the renal parenchyma and be excreted via the renal tubules prior to transiting through general circulation.[16,17] This is hypothesized to be a compensatory mechanism for dehydration, sequestering blood to the kidneys could prevent ischemic injury.[17]

A few studies have demonstrated variable circulation through the renal portal system.[18-21] It is thought that reptiles have valves present that control blood flow into the renal portal veins as has been demonstrated in birds, but this has not been thoroughly evaluated in reptiles.[22,23]

The presence of the renal portal system in reptiles creates a potential complication: Drugs injected into the caudal half of the body could be concentrated in the kidneys prior to entering general venous circulation. It is considered a reasonable precaution to avoid administering potentially nephrotoxic or drugs excreted through the renal tubules (but not the glomeruli) in the caudal half of the body, including via the caudal vein.[16]

Technique and Maintenance

The need for sedation depends on the patient's demeanor. The author finds that in patients where a lateral approach to the tail is used, mild sedation may be needed. For patients where a ventral approach is taken, deep sedation or anesthesia is advised to eliminate the righting reflex. Over-the-needle catheters are suitable for this technique. Patients should be warmed to their preferred optimum temperature zone prior to placement to avoid peripheral vasoconstriction.

Site selection

A distance of 25-50% of the tail length from the cloaca is preferred to avoid accidental trauma of the hemipenes, hemiclitori, or scent glands.[24]

The author prefers a ventral approach in species with a dorsoventrally compressed body shape, and the lateral approach in species with a laterally compressed tail. Species with a round cross-section of the tail are amenable to either approach.

Ventrally-placed caudal catheters can be difficult to maintain in active patients due to the drag that this catheter will experience with patient movement. The site can also easily become contaminated with fecal and urate matter. The author typically reserves this site for anesthetic patients where the catheter is unlikely to require long-term maintenance (e.g., beyond the inpatient procedure), or in inactive patients where the lateral approach is less appealing. This catheter is not manageable in snakes that are active in any capacity.

Site preparation

Appropriate hand hygiene is performed,[25] followed by skin preparation of the insertion site. The site is prepared aseptically; the author routinely uses a combination of chlorhexidine 2% scrub to remove all obvious organic debris, followed by a rinse of isopropyl alcohol 70%. The scrub phase should be quite vigorous to clean underneath the scales, and depending on the ruggedness of the scales, a scrub brush may be needed instead of gauze. Preoperative surgical scrub brushes can be very effective. Hand hygiene is performed again prior to catheter placement, and sterile or clean gloves should be donned.

Catheter insertion, ventral approach

The ventral approach is performed on the midline of the ventral aspect of the tail. Gentle digital pressure may be placed at the base of the tail by the restrainer to temporarily impede venous return. The catheter is inserted at approximately a 30° angle, and slowly advanced until either a flash of blood is observed in the catheter hub, or the presumed anatomical location of the vein is reached. Because of the relatively low blood pressure and slow heart rate of reptiles, catheter filling may be very slow, or effectively nonexistent in volume-depleted patients. The catheter is threaded into the vein, and no resistance should be encountered. Physiologic saline can be used to flush the catheter: There should be no resistance, and no backflow around the catheter hub. It can be difficult or impossible to palpate a small fluid bolus proximal to the placement site, unlike in mammals. Aspirating blood from the catheter should be performed slowly with very gentle negative pressure to avoid causing venous collapse. A PRN port or pediatric infusion port (such as a "T-port") can be attached to the catheter hub. Both of these implements should be primed with sterile physiologic saline to avoid causing air emboli during catheter use.

Catheter insertion, lateral approach

The lateral approach is performed immediately ventral to the midline on the lateral aspect of the tail. Gentle digital pressure may be placed at the base of the tail, as in the ventral approach. The transverse processes of the caudal vertebrae may be palpated as a guide, and catheter insertion should be planned ventral to the level of the processes. Some species, like the green iguana (*Iguana iguana*) have a divet between the dorsal and ventral muscle groups of the tail, and this can be used as a placement guide. The tail may be slightly bent laterally away from the operator to better expose the vein.[8,12] The catheter is inserted as per the technique described above.

Securing the catheter

The catheter can be secured using cloth tape as per typical technique (the author finds that paper tape and transpire tape are both ineffective on reptile skin). In very active patients where a catheter is projected to be maintained for several days, the author has had success placing simple dermal stay sutures (using non-absorbable suture) on a tape butterfly around the catheter hub. For anesthetic patients that will be immobile, tape alone is usually effective in securing the catheter. For catheters maintained in awake patients, a layer of gauze bandage followed by self-adhesive bandage can be applied, with the partial circumference of the cranial margin of the bandage secured to the skin with tape or other adhesive bandage.

Alternatively, transparent film bandage can be used to secure catheters for anesthesia. Padding between the catheter hub and PRN or extension port and the tail can be used if desired, such as a piece of gauze or cotton ball.

Maintaining the catheter

The catheter insertion site should be checked at least daily for signs of infection, phlebitis, or extravasation. When flushed to verify patency, the animal's total fluid intake should be considered, as reptiles have a much lower daily fluid requirement compared to mammals and birds. Other possible complications include hematoma and thrombosis. If any complications are observed, the catheter should be promptly removed.[8]

Conclusions

The caudal vein is an acceptable site for intravenous catheter placement in many lizard and snake patients, particularly where intraosseous catheters are not possible. This site is best avoided in patients that perform caudal autotomy as an anti-predator behavior. Catheter maintenance is similar to other intravenous catheters.

References

1. Wellehan JFX, Lafortune M, Gunkel C, et al. Coccygeal vascular catheterization in lizards and crocodilians. *J Herpetol Med Surg.* 2004;14(2):26-28.

2. Reptile Database. Species Statistics Aug 2019 [sic]. Available at: http://www.reptile-database.org/db-info/SpeciesStat.html. Accessed March 5, 2023.

3. Vidal N, Blair Hedges S. The molecular evolutionary tree of lizards, snakes, and amphisbaenians. *Comptes Rendus Biologies.* 2009;332(2-3):129-139. doi:10.1016/j.crvi.2008.07.010.

4. Pyron RA, Burbrink FT, Wiens JJ. A phylogeny and revised classification of Squamata, including 4161 species of lizards and snakes. *BMC Evol Biol.* 2013;13:93. doi:10.1186/1471-2148-13-93.

5. Wiens JJ, Hutter CR, Mulcahy DG, et al. Resolving the phylogeny of lizards and snakes (Squamata) with extensive sampling of genes and species. *Biol Lett.* 2012;8(6):1043-1046. doi:10.1098/rsbl.2012.0703.

6. Mader DM. DVM 360: Critical care techniques in reptiles. Available at: https://www.dvm360.com/view/critical-care-techniques-reptile-patients-proceedings. Accessed March 5, 2023.

7. Divers SJ. Clinical procedures for reptiles. In: Winter AL, ed. *Merck Veterinary Manual.* Available at: https://www.merckvetmanual.com/exotic-and-laboratory-animals/reptiles/clinical-procedures-for-reptiles. Accessed March 5, 2023.

8. Pardo MA, Divers SJ. Catheter placement. In: Divers SJ, Stahl SJ, eds. *Mader's Reptile and Amphibian Medicine and Surgery.* 3rd Ed. St. Louis, MO: Elsevier; 2019:422-428.

9. American Animal Hospital Association. AAHA/AAFP Fluid Therapy Guidelines for Dogs and Cats: Implementation Toolkit. Available at https://www.aaha.org/globalassets/02-guidelines/fluid-therapy/fluid-therapy_guidlines_toolkit.pdf. Accessed March 5, 2023.

10. Herrin KV. Clinical techniques and supportive care. In: Doneley B, Monks D, Johnson R, Carmel B, eds. *Reptile Medicine and Surgery in Clinical Practice.* Hoboken, NJ: Wiley-Blackwell; 2018:159-173.

11. Lee JA, Guieu LS, Bussières G, Smith CK. Advanced vascular access in small animal emergency and critical care. *Front Vet Sci.* 2021;8:703595. doi:10.3389/fvets.2021.703595.

12. Kanda I, Brandão J. Herptiles: catheterization and venipuncture. In: Graham JE, Doss GA, Beaufrère H, eds. *Exotic Animal Emergency and Critical Care Medicine.* Hoboken, NJ: Wiley-Blackwell; 2021:722-731.

13. Funk RS, Johnson R, Divers SJ. Tail abnormalities. In: Divers SJ, Stahl SJ, eds. *Mader's Reptile and Amphibian Medicine and Surgery.* 3rd Ed. St. Louis, MO: Elsevier; 2019:1347-1348.

14. Durso AM, JC Maerz. Natural behavior. In: Divers SJ, Stahl SJ, eds. *Mader's Reptile and Amphibian Medicine and Surgery.* 3rd Ed. St. Louis, MO: Elsevier; 2019:90-99.

15. Naya DE, Veloso C, Muñoz JL, Bozinovic F. Some vaguely explored (but not trivial) costs of tail autotomy in lizards. *Comp Biochem Physiol A Mol Integr Physiol.* 2007;146(2):189-193. doi:10.1016/j.cbpa.2006.10.014.

16. Girling S. Vascular, hematopoietic, and immune systems. In: Divers SJ, Stahl SJ, eds. *Mader's Reptile and Amphibian Medicine and Surgery.* 3rd Ed. St. Louis, MO: Elsevier; 2019:917-921.

17. Holz PH. The reptilian renal portal system - a review. *Bull Assoc Reptilian Amphib Vet.* 1999;9(1):4-9.

18. Kummrow MS, Tseng F, Hesse L, et al. Pharmacokinetics of buprenorphine after single-dose subcutaneous administration in red-eared sliders (*Trachemys scripta elegans*). *J Zoo Wildl Med.* 2008;39:590-595.

19. Holz P, Barker IK, Crawshaw GJ, Dobson H. The anatomy and perfusion of the renal portal system in the red-eared slider, *Trachemys scripta elegans*. *J Zoo Wildl Med.* 1997;28:378-385.

20. Holz PH, Barker IK, Burger JP, Crawshaw GJ. The effect of the renal portal system on pharmacokinetic parameters in the red-eared slider (*Trachemys scripta elegans*). *J Zoo Wildl Med.* 1998;28(4):386-393.

21. Benson KG, Forrest L. Characterization of the renal portal system of the common green iguana (*Iguana iguana*) by digital subtraction imaging. *J Zoo Wildl Med.* 1999;30:235.

22. O'Malley B. *Clinical Anatomy and Physiology of Exotic Species.* Edinburgh, UK: Elsevier Saunders; 2005.

23. Stephens T, Rosenwax A. Disease of the cardiovascular system. In: Doneley B, Monks D, Johnson R, Carmel B, eds. *Reptile Medicine and Surgery in Clinical Practice.* Hoboken, NJ: Wiley-Blackwell; 2018:287-297.

24. Olsson A, Simpson M. Analgesia and anaesthesia. In: Doneley B, Monks D, Johnson R, Carmel B, eds. *Reptile Medicine and Surgery in Clinical Practice.* Hoboken, NJ: Wiley-Blackwell; 2018:369-381.

25. American Animal Hospital Association. Protocol for hand hygiene using soap and water, or alcohol-based hand sanitizer Available at: https://www.aaha.org/aaha-guidelines/infection-control-configuration/protocols/hand-hygiene2/. Accessed March 5, 2023.

26. Barten S, Simpson S. Lizard taxonomy, anatomy, and physiology. In: Divers SJ, Stahl SJ, eds. *Mader's Reptile and Amphibian Medicine and Surgery.* 3rd Ed. St. Louis, MO: Elsevier; 2019:63-74.

Technique Tutor: Intravenous Catheter Placement in Small Mammals

Stefanie Kotschwar, RVT, BScH (Zoology, Nutrition), MSc (Zoo Conservation), AAS (Vet Tech), Diploma (Web Design and Equine Science)

Session #3335

Affiliation: From Lynwood Animal Hospital, 107-30 Stafford Rd, Ottawa, ON, K2H 1W8, Canada.

Abstract: Small exotic mammals might require intravenous (IV) access while being in the hospital to receive fluids while sick and dehydrated, administration of medications, induce anesthesia or provide support during surgical procedures. In this lecture we review various locations where an IV catheter can be placed, the supplies needed and go over the steps on how to place it with some tips and tricks along the way.

Introduction

Intravenous (IV) catheters are commonly placed in small exotic mammals that require fluid support while sick, in emergency situations or while undergoing a surgical procedure. Any small exotic mammal having a surgical procedure under general anesthesia should have IV access. IV access should even be considered for patients undergoing a procedure under sedation.

Catheterization techniques like those used in cats and dogs are applied to small exotic mammals, the main difference being the patient size. Another factor to keep in mind is that most small exotic mammals are considered prey species and can therefore experience higher levels of stress from handling.[1] If possible, consider using some level of sedation prior to catheterization. Ensure the treatment area used has no predators and keep the noise level as low as possible.

Applying a topical local anesthetic cream at least 20 minutes prior can provide comfort when placing an IV catheter. Sites that can be used for IV catheter placement include cephalic vein, marginal ear vein and the lateral saphenous. The IV catheter sizes are selected based on patient size and can range from 22-26 gauge. Some drugs will require dilution prior to IV administration to avoid vasculitis.[1-5]

Supplies

Supplies needed for intravenous catheter placement are listed below.

- Clippers
- Indwelling IV catheters of varying sizes: 22-26 gauge
- 25-18 gauge hypodermic needle
- Intermittent infusion plug

- White porous tape
- Topical anesthetic cream like EMLA® (lidocaine 2.5% and prilocaine 2.5%; Astra Pharmaceuticals LP, Wayne, PA)
- Tourniquet
- Gauze squares
- Sterile sodium chloride 0.9%
- Chlorhexidine gluconate 0.5% solution
- Self-adherent flexible bandage
- Clean gloves

IV Catheter Placement Sites

Placement sites for IV catheters in multiple small mammal species are described below.

Rabbits

- *Marginal ear vein*
- *Cephalic*
- *Saphenous*

Guinea Pigs

- *Cephalic*
- *Saphenous*

Hedgehogs

- *Cephalic*
- *Saphenous*

Ferrets

- *Cephalic*
- *Saphenous*

Rats

- *Cephalic*
- *Saphenous*
- *Tail*

Other rodents

- *Cephalic*
- *Saphenous*
- *Consider IO*

General IV Catheter Steps

Steps for placing a catheter are outlined below.

- Place the patient on a non-slip surface.
- Maintain the patient in sternal recumbency.
- Shave catheter site – rabbit fur can be difficult to shave!
- Place topical anesthetic cream if available.
- For small mammals, the use of a tourniquet is recommended to help raise the vein. It can be difficult for an assistant to hold off the vein while restraining the patient.
- Prepare the site aseptically.
- Make a small cut down hole with a hypodermic needle: Size will depend on patient size, but a 21-gauge is a good rule of thumb.
- Preload your catheter with sterile 0.9% sodium chloride as this will help you visualize a small blood flush easier and avoid inadvertent air embolism.
- If the IV catheter does not have wings, place some of the white tape across to make tape butterflies as this will make the taping easier. Experiment with this step as not everyone likes IV catheters with wings, but it is the author's preference.
- Hold the IV catheter at an about 20-degree angle to the site. Insert IV catheter slowly until you see a small flash of blood in the catheter. Advance another 1 millimeter, before advancing the IV catheter.
- Place at least one strip of tape to stabilize the catheter, especially if working with a non-sedated patient.

- Prime the intermittent infusion plug with sterile 0.9% sodium chloride and then attach it.
- Continue securing the IV catheter with at least one or two more strips of white tape.
- Flush the IV catheter with sterile 0.9% sodium chloride to confirm placement.
- Wrap the site with Self-adherent flexible bandage. A layer of cling gauze is optional.

Considerations for IV Catheter Placement in Other Sites

Multiple considerations for placing IV catheters in other sites are discussed below.

Saphenous vein

This vein can be a little harder to stabilize, try stretching the skin to stabilize the vein.

Marginal ear vein, rabbits

Ensure you are using the marginal ear vein and not the central auricular artery! The central auricular artery runs along the midline of the ear.

Reports or ear necrosis when using this vein have been reported. Keep that in mind if you are injecting any medication that could be irritating if injected perivascular.

This site does not require a cut down prior to placing the IV catheter.

Roll the pinna around a roll of gauze or rolled up gauze squares to help tape and stabilize the IV catheter.

IV Catheter Maintenance

If the catheter is placed on a hospitalized patient, check it regularly for patency and check the site for erythema, inflammation, and discharge. If there are any concerns, remove the IV catheter and place a new one in a different site.[2-5]

Conclusions

While placing an IV catheter in small exotic mammals can prove to be challenging, there are some tips that can facilitate the process. In addition, continual practice is beneficial.

References

1. Gardhouse S, Sanchez A. Rabbit sedation and anesthesia. *Vet Clin North Am Exot Anim Pract.* 2022;25: 181-210.

2. Nugent-Deal J. Intravenous Catheter Placement in Rabbits. LafeberVet Website. 2021 Available at: https://lafeber.com/vet/intravenous-catheter-placement-in-rabbits/. Accessed March 1, 2023.

3. Echols MS, Lichtenberger M, Pollock C. Intravenous catheter placement in small mammals. LafeberVet website. May 7, 2013. Available at: https://lafeber.com/vet/intravenous-catheter-placement-in-small-mammals/. Accessed March 1, 2023.

4. Nugent-Deal J. Exotic anesthesia and analgesia. In: Ballard B, Cheek R, eds. *Exotic Animal Medicine for the Veterinary Technician.* 3rd Ed. Hoboken, NJ: John Wiley & Sons; 2017:11-30.

5. van Zeeland YRA, Shoemaker NJ. Catheterization and venipuncture. In: Graham JE, Doss JA, Beaufrère H, eds. *Exotic Animal Emergency and Critical Care Medicine.* 1st Ed. Hoboken, NJ: John Wiley & Sons; 2021:51-69.

Technique Tutor: Feeding the Hospitalized Small Mammal

Stefanie Kotschwar, RVT, BScH (Zoology, Nutrition), MSc (Zoo Conservation), AAS (Vet Tech), Diploma (Web Design and Equine Science)

Session #3336

Affiliation: From Lynwood Animal Hospital, 107-30 Stafford Rd, Ottawa, ON, K2H 1W8, Canada

Abstract: This session reviews nutritional support for hospitalized small exotic mammals, with a focus on rabbits. It includes a brief review of digestive anatomy, available feeding formulas, supplies required and tips and tricks on how to feed patients with the least amount of stress.

Introduction

Nutritional support is required for any small mammal presenting to the clinic that is anorexic or hyporexic, due to disease or recovering from a surgical procedure. These animals have a high metabolic rate and a decrease in appetite can very quickly lead to glycogen store depletion and hypoglycemia. In addition, sick patients have an increased energy requirement that can lead to catabolism of fat and muscle if there is decreased caloric intake.[1]

Gastrointestinal Anatomy of Small Exotic Mammals

There is a wide variety of species encompassing small exotic mammals with a variety of feeding strategies. Thus, there are variations in gastrointestinal anatomy among small exotic mammals.

Most small exotic mammals encountered in veterinary practice have herbivorous diets and are monogastric hindgut fermenters. They have a single stomach and a large cecum that provides for breakdown of fiber from their diet, and these species engage in cecotrophy or coprophagy. Species such as rabbits, guinea pigs and chinchillas are hindgut fermenters.[2,3]

Smaller rodents have different adaptations. For example, hamsters have a compartmentalized stomach with fermentation taking place in one of those compartments, the forestomach.[2,3]

Ferrets are obligate carnivores and this is reflected in their gastrointestinal anatomy. They have a simple digestive tract that allows food to pass quickly. They have a rapid gastrointestinal transit time, so they should not be fasted for extended periods of time.[2,4]

Feeding Requirements

While individual nutrient requirements vary among species, daily nutrient requirements can be calculated. Basal metabolic rate ($BMR = 70x\ BW^{0.75} kcal/kg/d$ where BW is the patient's body weight in kilograms) can be used to calculate energy requirements for small mammals. This can then be multiplied by a factor of 1.2-2 depending on their health status.[1]

Feeding Formulas

There are many products available that can be used to provide nutritional support to small mammals. Diet selection depends on the preference of the patient and its requirements. Below are examples of products available from several different companies.

Oxbow Animal Health (Papillion, NE)

- Critical Care® for herbivores comes in apple/banana or anise flavor and is considered a complete maintenance diet. Not appropriate for nasogastric tube feeding as it will clog the tube due to its high fiber content.
- Critical Care® for herbivores fine grind is available for nasogastric feeding tubes.
- Critical Care® for carnivores is a complete maintenance diet.
- Critical Care® for omnivores is a complete maintenance diet.

Lafeber Company (Cornell, IL)

- Emeraid® IC intensive care that is intended for debilitated patients, and animals with digestive disorders that require highly digestible ingredients. It should not be used long term as the fiber content is limited.
- Emeraid® sustain for herbivores is a good maintenance diet.
- A maintenance diet for carnivores and omnivores is also available.
- Supreme™ Pet Foods (Hadleigh, Ipswitch, United Kingdom).
- Recovery for small herbivores.
- Recovery plus has a higher level of fibers and contains pre- and probiotics.

Sherwood® Pet Health (Ephraim, UT)

- Line of recovery foods to support sick animals until appetite has been restored.

Feeding Routes

How a patient is fed depends on its status and compliance. While several options exist, including use of an orogastric tube, nasogastric tube and esophagostomy tube, the focus of this session is syringe feeding.

Supplies

- Bowl
- Feeding formula
- Water, unsweetened applesauce or baby food
- Appropriately sized syringe for the patient

Feeding Tips and Tricks

Once the food is prepared as indicated on the package, there are things to keep in mind and to try when feeding the small exotic mammal.

- Ensure that you are knowledgeable in the husbandry, behavior, and restraint of the species you are working with.
- Always approach slowly and talk in a gentle manner.
- Preparing the food with unsweetened applesauce or baby food adds a little bit of flavor that could entice the patient to eat.
- Always offer food in a bowl first to determine if they will lick it off directly.
- If using a syringe, place the tip of the syringe in the diastema of herbivores. For carnivores, place the tip of the syringe in the corner of their mouth.
- Give little bits of food at a time.
- Try feeding the patient inside the hospital cage first. Most feel safer in that space.
- Use the least amount of restraint necessary. Only use a towel for restraint and do the feeding on a treatment table if all else fails.
- Use a lot of appeasing and stress-reducing techniques like a body massage or even a face rub.
- Be patient and try different techniques.
- Talk to the patient! Bunnies especially like to be told they are a king or queen and that they are the best bunny.
- Giving the patient a face rub and then lifting the lip slightly to get the syringe into the mouth and giving a little bit of food works most of the time.
- Letting the patient move around the cage in between the feedings will give them the opportunity to slowly chew and swallow their food.

- If you have an uncooperative patient that needs to be swaddled in a towel, try keeping them in the cage or feeding them on the floor. Use the treatment table as the last resort as herbivores do not like to be picked up and carried around.

- Be flexible! Some well socialized rodents like rats like to be fed while perching on your shoulder.

- Clean their faces after feeding. No one likes a dirty face and if any food residue is left on their face, it will dry up and make them uncomfortable.

- Sometimes you must let someone else try; do not take it personally.

Conclusions

There are many reasons why small exotic mammals need to be syringe-fed while in the hospital. There is not one way of doing this because each animal is an individual. There are many techniques that can be employed to find the best way to feed each patient to ensure their comfort and low stress. A lot of patience and creativity is required by the veterinary technician to accomplish this process.

References

1. van Zeeland YRA, Schoemaker NJ. Nutrition and fluid therapy. *Exotic Animal Emergency and Critical Care Medicine.* Hoboken, NJ: John Wiley & Sons. 2021;109-122.

2. Bullen LE. Nutrition for pocket pets (ferrets, rabbits, and rodents). *Vet Clin North Am Exot Anim Pract.* 2021;15:583-604.

3. Grant K. Rodent nutrition: digestive comparisons of 4 common rodent species. *Vet Clin North Am Exot Anim Pract.* 2014;17:471-483.

Section 22

Technician Emergencies Session

Debbie Sykes, CWR, VT, FFCP
Session Chair

Exotic Animal Emergencies for Technicians, Part 1: Evaluation of the Emergent Exotic Mammal Patient

Christina Miller, RVT, BSc

Session #3367

Affiliation: From Clayton Park Veterinary Hospital, 278 Lacewood Dr, Halifax, NS, B3M 3N8, Canada.

Abstract: Exotic companion mammals commonly present to the veterinary hospital in a state of marked illness. This masterclass reviews emergency presentations in common exotic companion mammal species. It covers the technician's role in the primary survey, obtaining vitals efficiently and safely, obtaining airway and vascular access, and proceeding with cardiopulmonary and cerebral resuscitation.

Introduction

Exotic pets present some unique challenges in both general practice and the emergency room. Most exotic animals encountered in the veterinary practice exhibit some degree of masking phenomenon, either because they are not truly domesticated, they are prey animals, or both. Masking phenomenon is an instinct to hide signs of illness with the goal of preventing predation, as predators may exhibit a preference towards old, weak, or sick-looking prey since they may be more efficient to capture and subdue (this theory is partially supported by ecological literature; however, predator-prey interactions can be complex and dependent on the species as well as the environment).[1-3]

Regardless, there are some presentations that customer service staff can generally assume are at a high triage level, including:

- Recumbent or non-responsive patient
- Bloated abdomen or coelom, especially if acute
- Dyspnea
- Seizures, muscle tremors
- Toxic substance ingestion or exposure
- Stranguria, hematuria
- Trauma (blunt, crushing, falling, attacked by predator)
- Active bleeding
- Hypothermia, hyperthermia
- Exophthalmos, proptosis, other ocular trauma

- Large open wounds, evisceration
- Ataxia

Unfortunately, acute presentations of chronic conditions are common in exotic pets. This may be due to a combination of masking phenomenon (e.g., the patient is hiding early signs of illness) and poor pet owner knowledge of how health problems present in exotic taxa. Captive exotic species are also very susceptible to husbandry-related disease, which may lend to a slow development of some health problems that reinforces client perception that a pet's behavior or anatomy is normal.

Because of all of these factors, a general rule has been suggested that if a client is calling with a concern about their exotic pet, that the pet should be examined (ideally promptly).[4,5] This is operating on the assumption that if the client has appreciated something wrong or abnormal, that the disease process may be quite advanced.

The Exotic Mammal Primary Survey

The primary survey starts with an initial general impression of the patient, which can be observed in the pet's carrier, or the pet may be allowed to explore a secure exam room (particularly rabbits, ferrets). Try to observe the pet covertly: What is their mentation like? Most exotic mammals should be bright, alert, and responsive to your presence. Rabbits and guinea pigs will adopt a freeze-and-assess response to your presence. This behavior is reflected in wild prey animals, who upon detecting a possible threat, will freeze (to potentially evade predator detection) and assess whether hiding, fight, or flight is appropriate. Sugar Gliders are quite strictly nocturnal and may be asleep or drowsy, but should be rousable. Ferrets may similarly be sleepy or drowsy, as they typically sleep between 14-18 hours per day, but should likewise be rousable. If the patient is dull, dyspneic, or non-responsive to touch, proceed to the ABCDE protocol. The ABCDE protocol is a systematic approach to patient assessment in order of the most acutely life-threatening problems to least life-threatening problems.

The author has personally observed numerous small mammal patients initially appear to be stable during the initial impression observation, then decompensate during history-taking (e.g., patient becomes dyspneic when they initially presented with a normal respiratory pattern). This highlights the amazing ability of these animals to mask their illness.

A: Airway

Airway and breathing assessment are effectively performed simultaneously; however, signs of airway obstruction should be considered immediately, including: Paradoxical thoracic and abdominal movements, increased respiratory effort (especially on inspiration), cyanosis, stertor (low-pitched sound, suggests nasal or pharyngeal obstruction) or stridor (high-pitched sound, suggests lower airway obstruction), and open-mouth breathing in obligate nasal breathers. If there are signs of airway obstruction, oxygen therapy should be implemented immediately.[6]

Anxiolysis and sedation may be considered to help lower the patient's oxygen demand and help to lower their stress. If it is possible to localize and remove the obstruction, it should be done as quickly as possible to hopefully alleviate the dyspnea. Rodents and lagomorphs are obligate nasal breathers, and nasal secretions can cause significant obstruction. Mild suction using a nasal bulb syringe intended for human infants can be very effective; however, if the secretions are dried, the humectant action of humidified oxygen may be needed prior to removal. Oral foreign bodies can be challenging to identify and retrieve due to the long and narrow oral cavity with an infolding of the cheek tissue, the inflexa pellita. Use of an otoscope, bivalve speculum, or endoscope can be used to visualize the oral cavity. Guinea pigs should have ingesta in their mouths, which can make visualization difficult.[6]

If respiratory arrest occurs or the obstruction cannot be identified and relieved, airway access must be secured. If there is jaw tone present, use of sedation and/or induction agents may be necessary. Airway access will be discussed below.

Oxygen therapy

There are several techniques for providing oxygen therapy that can be considered in these animals:

- Flow-by oxygen, oxygen flow rate = 2-3 L/min: This method may be used during initial triage, but it is wasteful to use for longer-term therapy since the flow rate needs to be very high. The oxygen outflow must be held within 2 centimeters of the patient's nares to be effective.[7] The handling required to maintain this technique is inherently stressful to many small mammal patients.

- Tight-fitting face mask, oxygen flow rate => 100 ml/kg/min. FiO_2 is better with a tight-fitting facemask compared to a loose mask or mask lacking a diaphragm, or an oxygen cage. This can be very stressful in some animals, as rodents and lagomorphs have a very sensitive face. Restraint is required to keep the mask on in all but the most compliant of patients, further adding to the patient's stress. If the seal on the mask is very tight, there is less CO_2 escape (less so on a non rebreathing anesthetic circuit), and prolonged use may contribute to rebreathing of waste gas and subsequent hypercapnia, as well as accumulation of heat.[8]

- Loose face mask or oxygen cage, oxygen flow rate => 300 ml/kg/min. The FiO_2 can reach up to 40% using an appropriate cage.[9] Commercial units that allow for temperature and humidity control, and monitor FiO_2 and CO_2 are ideal. House-made oxygen chambers (e.g., using modified steel kennel doors, or smaller containers with an oxygen inlet and passive scavenging ventilation) can still be useful. If using a more rudimentary oxygen chamber, ensure that a bubbler is used to maintain a humid atmosphere (dry oxygen can desiccate tiny airways very quickly) and monitor environmental temperature closely (overheating can occur rapidly with some setups and species like chinchillas and ferrets).[10]

B: Breathing

The quality and rate of the patient's respirations should be assessed, and oxygen therapy initiated immediately if abnormal. Abnormal respiratory depth and rate (e.g., tachypnea, bradypnea, hyperpnea, hypopnea) and patterns (e.g., Cheyne-Stokes, Kussmaul, Biot, ataxic) should be noted. Normal respiratory rates for a selection of species is available in Table 1. Accessory respiratory movements may be appreciated. Marked abdominal effort resulting in paradoxical abdominal and thoracic movements can occur with markedly increased effort, and this is often paired with orthopneic posturing (e.g., head pointed up, elbows out). Obligate nasal breathers may exhibit nasal flaring, which may precede open-mouth breathing, which is a sign of severe dyspnea in these species.[6]

The lung fields may be assessed next via auscultation. Referred upper airway sounds can be normal in obligate nasal breathers. The presence and location of any abnormal sounds (e.g., rales, crackles, wheezing, stridor) should be noted. Thoracic percussion is not typically helpful due to the small size of most exotic mammal patients encountered in practice.[6]

Table 1. Normal vital signs in exotic companion mammals.[6]

Species	Rectal temperature °C	°F	Heart rate (bpm)	Respiratory rate (rpm)
African pygmy hedgehog (*Atelerix albiventris*)	35.4-37.0	95.7-98.6	180-280	25-50
Chinchilla (*Chinchilla lanigera*)	37.0-38.0	98.6-100.4	100-300	40-80
Domestic ferret (*Mustela putorious furo*)	37.8-39.2	100-102.6	180-250	30-60
Domestic rabbit (*Oryctolagus cuniculus*)	37.8-39.2	100-102.6	150-300	30-60
Fancy rat (*Rattus norvegicus domestica*)	37.2-39.2	99-102.6	280-500	60-120
Gerbil (*Meriones unguiculatus*)	37.4-38.2	99.3-100.8	260-600	70-120
Guinea pig (*Cavia porcellus*)	37.2-39.5	99-103.1	220-300	40-120
Hamsters (Cricetidae)	36.1-38.9	97-102	280-420	100-250
Mouse (*Mus musculus*)	37.1	98.8	500-725	80-230
Sugar glider (*Petaurus breviceps*)	32.0 (cloacal)	89.6	200-300	16-40

C: Circulation

Pulse quality, rate, and regularity can be appreciated at the femoral artery, which is palpable in the inguinal region where the limb meets the ventrum, in the dip between the muscle bellies. The author finds that with a gentle touch and using your pinky fingers, this pulse can be palpated in animals as small as guinea pigs and sometimes rats (obesity makes palpation challenging in any species). The central auricular artery can be palpated in most rabbits (exceptions being particularly diminutive dwarf breeds with very small pinnae). Pulse deficits can be very difficult to appreciate in animals with a rapid heart beat (over 200 bpm in the author's experience).[6]

The mucous membranes can be assessed at the conjunctiva, oral mucosa, and vulva or prepuce in most species. The mucous membranes should be medium pink and moist, as in other companion mammals. Capillary refill time can be assessed at the mucous membranes, or on unpigmented footpads, or the inside of the pinnae in rabbits. The extremities should be relatively warm to the touch.[6]

Cardiac auscultation may be performed simultaneously with pulmonary auscultation. Pediatric or infant stethoscopes are essential in exotic mammal practice for more precise localization of heart and lung sounds.

The apex beat can be located at the left hemithorax at about the 5th intercostal space. The exception to this is mustelid species such as the domestic ferret, where the apex beat is located more caudally (6-8th intercostal space). Heart rate, rhythm, and the presence of abnormal sounds should be characterized.[6] Normal heart rates are described in Table 1. Localizing heart murmurs to specific valves is challenging if not impossible in small exotics; however, left-sided versus right-sided can sometimes be identified. Ferasin and associates[11] describe iatrogenic murmurs caused by excessive pressure placed on the thorax of kittens with the stethoscope diaphragm, due to their relatively compliant thorax compared to adult cats. While this phenomenon has not been described in the literature in exotic mammals, the author recommends remaining cautious with the pressure placed on the small mammal thorax.

Both arrhythmias and murmurs can be difficult to appreciate with high heart rates. Electrocardiogram evaluation can be helpful.

If hemorrhage is present, bleeding can be controlled in a similar manner to other species. Non-adherent dressing pads can be used to place pressure. Alternatively, sterile cotton-tipped swabs can be applied to bleeding vessels on very small patients, although removal may dislodge a clot.

D: Disability

Patient mentation and reflexes are examined next. Corneal or palpebral reflexes should be present. A study evaluating neurologic examination in healthy rabbits found that pupillary light reflex (PLR) was slow and weak, and consensual PLR was often absent.[12] Menace reflex is often absent in rabbits and rodents. Dazzle reflex may be absent in normal chinchillas.[13]

Common neurologic deficits in small mammal herbivores (e.g., rabbits, chinchillas, guinea pigs) as well as myomorph rodents (e.g., mice, rats) include head tilt, also referred to as wryneck or torticollis, circling and rolling, ataxia, and nystagmus. The causes for these signs are varied, but an effort should be made to reduce potential trauma by offering a padded environment, ideally in a low light and quiet region of the hospital. Rabbits with these clinical signs may be infected with *Encephalitozoon cuniculi*, which may be zoonotic, so appropriate biosecurity measures are advised until the patient's serologic status for this pathogen is known.[14]

E: Exposure of environment

Body temperature is ideally taken from the rectum, or cloaca where applicable, such as in marsupials. Rectal thermometers intended for companion mammal use are effective. Male guinea pigs have prominent rectal folds, and improper placement in this cloacal region results in a lower temperature reading. Ensure that the thermometer is directed dorsally in these folds to enter the anus (the author finds that gently pulling dorsally on the skin dorsal to the coccyx helps to position the thermometer). Tympanic or auricular thermometers are not considered reliable in most exotic mammal species because the ears are a thermoregulatory surface in rabbits and chinchillas, and the ears of most other exotic mammals are too small for their use. A 2021 study[15] suggests that a tympanic thermometer can be used in the interscapular region in ferrets; however, this has not been validated in patients that are not normothermic.

A low rectal temperature is a strong predictor of a poor prognosis in several species, even if the patient is bright, alert, and responsive. Rabbits that present with a rectal temperature below 36.7 °C[16] and guinea pigs that present with a rectal temperature below 37.9 °C[17] should be treated as if they are in shock.

Survey the body for skin lesions, fractures, bruises, and other injuries.

Airway Access

Airway access should be secured in apneic patients. Intubation of the trachea is ideal, as this gives staff the most control over oxygenation and ventilation, but unfortunately not always possible depending on the species and equipment at your disposal. The author recommends reinforcing these skills with stable patients (i.e. during routine surgeries) so that your staff builds confidence and competence for emergent patients.

With any patient that is not apneic, support staff can attempt preoxygenation to improve/increase the patient's oxygen saturation, in the event apnea does occur, with the goal of preventing hypoxemia. The relatively high or fast metabolic rate of these species results in a greater need for oxygen per unit of body weight compared to larger mammals, so hypoxemic events can rapidly result in ischemic injury.[18] In any dyspneic patient, assume that apnea can occur, and prepare intubation equipment as per the species' needs. The methods of providing oxygen support are described in section A, above.

Ferrets

Ferrets have similar anatomy to canine or feline patients, and orotracheal intubation is comparable to intubating a pediatric cat or kitten. The glottis can be directly visualized with a laryngoscope. Ferrets have incomplete tracheal rings, and a cuffed or uncuffed endotracheal (ET) tube is acceptable. Most ferrets will accommodate an ET tube between 2.0-3.5 millimeters in diameter. The use of a stylet is a personal preference.

The tube length is measured from the rostrum to the thoracic inlet, and lubricated with a water-soluble lubricant. Local anesthetic (e.g., lidocaine) can be applied to the arytenoid cartilages, and then allowing 15-30 seconds for the anesthetic to take effect will help to prevent laryngospasm during intubation attempts (and reduce the risk of traumatic iatrogenic laryngeal edema). The ET tube can be secured using cling gauze tie around the back and arms in a figure-of-8 pattern, as ferrets have a very floppy neck and a simple tie around the back of the skull is ineffective. Take note that ferrets will maintain a jaw tone until a moderate plane of surgical anesthesia is achieved, and may retain the jaw tone for a couple of minutes once they reach this plane. Take care not to get bitten while intubating these patients.

Hedgehogs and sugar gliders

Orotracheal intubation technique in these species is, once again, similar to in canine and feline patients, albeit much smaller in size. A pediatric blade on a laryngoscope (e.g., Miller size 0 blade), an otoscope, or small rigid endoscope may be used to aid in visualizing the glottis. Small uncuffed ET tubes with a diameter of 1.0-2.0 millimeters are appropriately-sized.

The tube length is once again measured from the rostrum to the thoracic inlet, lubricated, and local anesthetic may be applied to the arytenoids (take care with using small volumes to avoid accidental overdose). The tube may again be secured with a cling gauze tie around the back of the head and arms in a figure-of-8 pattern.

Van Zeeland and Schoemaker[6] describe a technique using the V-Gel® supraglottic device (Docsinnovent, Hemel Hempstead, United Kingdom) intended for use in rabbits: Size R1 may fit appropriately in a hedgehog's oral cavity. Capnography must be used to ensure placement.

Rats and mice

Small myomorph rodents follow a similar protocol as above; however, their anatomy is considerably smaller in size. Rats can accommodate an uncuffed ET tube with a diameter of 1.0-1.5 millimeters. Alternatively, an over-the-needle IV catheter with the stylet removed can be modified using a scalpel blade to create a Murphy eye. Size 14-16 gauge can fit comfortably in a rat patient. The catheter hub will fit snugly on a small ET tube adapter. Likewise, the mouse glottis can accommodate a 20-gauge IV catheter.

Rabbits

Rabbits have a long and narrow oropharyngeal cavity, the glottis is hidden by the soft palate in a natural posture, and visualizing the glottis without specialized equipment is challenging if not impossible. Several methods of airway access will be discussed.

Orotracheal intubation

A small uncuffed ET tube (size 2.0-3.5 millimeter will be appropriate for a 2-4 kg rabbit, and a PVC tube is preferred for reasons discussed below) is measured from the rostrum to the thoracic inlet, and lubricated. Local anesthetic should be applied to the arytenoids. The tube may be secured with a cling gauze tie or rubber band around the back of the head. Oxygenation via the nares before and during intubation is ideal.

- Blind method: This is a technically difficult intubation method. There is no visualization of the glottis, and relies on breath sounds, the ET tube fogging (PVC tubes work best), or a capnograph for placement. This technique is not effectively possible during apnea or arrest. The head is positioned in dorsoflexion and the tube is fed along the roof of the mouth, and directed towards the glottis. Local anesthetic can be applied to the glottis via the ET tube once a visual (tube fogging) or audio (breath sounds) cue is obtained. The tube is then introduced into the glottis during a breath. Heavy sedation is required as the animal's swallowing reflex and any jaw movement must be absent for this technique.[19] Unsuccessful intubation attempts leads to traumatic edema of the pharyngeal region, which can contribute to post-extubation dyspnea and respiratory arrest.[20] This technique also does not allow for inspection of the oral cavity for ingesta, which may accidentally be pushed into the glottis.[19]

- Endoscopic: The use of an endoscope allows for visualization of the glottis but necessitates specialized equipment. An oral gag should be used to prevent damage to the scope. A rigid endoscope may be used for a side-by-side technique, where the glottis is visualized, then the ET tube (or a stylet, such as a polypropylene catheter) is introduced into the oropharyngeal space parallel to the scope, then into the glottis. A flexible endoscope may be passed through an ET tube and used effectively as a stylus with direct visualization of the glottis for intubation. The latter technique is limited with very small rabbits.[19]

- Laryngoscope or otoscope: The device is introduced into the side of the patient's mouth behind the incisors, and used to gently move the soft palate dorsally so that the glottis may be visualized. The ET tube or stylus is then directed into the glottis. A neonatal or pediatric straight laryngoscope blade (such as a Miller size 0) is needed for many smaller patients if using the laryngoscopic technique.

Supraglottic airway devices (SAD)

These are species-specific devices that are designed to form a seal around the larynx while occupying space in the oral cavity to prevent accidental esophageal intubation. The V-Gel® (Docsinnovent, Hemel Hempstead, United Kingdom) is designed for rabbit use, and at the time of preparing this presentation, there are six sizes available for rabbits weighing between 0.6 to over 4.5 kg.[21] These devices are best used with a capnograph to confirm placement, as the device may shift if the patient is moved, or if it is not appropriately sized. Docsinnovent has excellent training materials available on their website (https://docsinnovent.com/products/v-gel-rabbit/). The SAD may be used in patients who present in respiratory arrest, as it does not require patient respirations for placement, and may be used to administer positive pressure ventilation.[19] The SAD is relatively easy to use and placement is generally atraumatic.[22] Note that gastric bloat/tympany can occur with IPPV, and the resulting increased abdominal pressure can impede diaphragmatic expansion.

Nasotracheal intubation

This method can be used to secure an airway without obstructing the oral cavity, or if a prolonged recovery is predicted as the tube cannot be chewed on once jaw tone returns. A 1.0- 2.5-millimeter diameter flexible tube (ET tube, or other product suitable as a nasoesophageal tube) can be used in most rabbits. The tube is measured from rostrum to thoracic inlet. Lidocaine (0.1-0.2 milliliters) is instilled into the nostril to be used (maximum

2 mg/kg dose). The head is positioned in dorsoflexion, and the lubricated tube is directed ventromedially. The narrow airway may provide some drag on the tube; however, there should not be any significant resistance. A crunchy sensation suggests that the tube has been directed into the turbinates. Both of these events can result in edema and bleeding of the nasal passage. It is recommended to only attempt nasotracheal intubation of one nostril, since rabbits are obligate nasal breathers. Contraindications include the presence of an upper respiratory tract infection (as this technique may carry pathogens into the lower airway; Devalle[23] demonstrated that this may not be a concern in healthy rabbits) or pre-existing edema or stenosis of the nasal passages (e.g., patients with rhinitis, or maxillary dental disease).[24]

Guinea pigs, chinchillas, and degus

These species present a significant obstacle to orotracheal intubation: The palatal ostium. This structure is a well-vascularized, thin, membrane of tissue from the soft palate, the palatoglossal arches laterally, and the base of the tongue. There is a very small laryngeal opening. The palatal ostium may very easily be traumatized, and significant edema and hemorrhage can occur with unsuccessful intubation attempts.[25,26] Guinea Pigs typically have ingesta in their oral cavity, further making visualization challenging (in awake and stable patients, syringe-feeding a 3- to 5-milliliter bolus of water can help clear the oral cavity; cotton-tipped swabs can otherwise be used).

The use of a dental rack or board with an endoscope (or potentially otoscope) is ideal to attempt intubation. Guinea pigs and chinchillas can accommodate a 2.0- to 2.5-millimeter diameter uncuffed ET tube, or an 8-Fr red rubber catheter (length cut to reduce dead space, and a Murphy eye should be added).[25]

Vascular Access: Intravenous

Vascular access should be obtained in any patient presenting in an urgent situation. Unless the patient is moribund, anxiolysis, sedation or anesthesia is typically required to safely place an intravenous or intraosseous catheter. The exception to this could be patients such as well-socialized rabbits or ferrets; however, the benefits of anxiolytics should not be neglected as even sociable animals are likely stressed when presenting for an emergency.

Local anesthesia is very helpful in obtaining vascular access. EMLA cream ® (Aspen Pharmacare Canada, Oakville, ON, Canada) is a lidocaine 2.5% and prilocaine 2.5% topical anesthetic that can be applied prior to IV catheter or even venipuncture attempts.

See Table 2 for a list of equipment needed for IV catheter placement. Placement sites vary between species.

Table 2. Equipment needed for intravenous catheter placement in mammals.

Equipment
Clippers with #40 blade
Rubber tourniquet, paper clip (marginal auricular vein in rabbits), optional
Aseptic preparation for skin: Chlorhexidine or iodophor, isopropyl alcohol or sterile saline
Gauze sponges
Intermittent infusion/PRN plug, or pediatric T-port (slip-tip/Luer-slip preferred for T-port)
Sterile saline ± heparin for flushing
Indwelling over-the-needle catheter (20-26 gauge, 0.75-1")
Porous tape, roll cling gauze 0.5-1", self-adhesive wrap 0.5-1"
Topical anesthetic (e.g., E,MLA cream®), optional

Ferrets

Cephalic and lateral saphenous vein sites are similar in location to canine patients, and 22- to 26-gauge indwelling catheter is typically effective. Jugular placement requires a surgical cutdown and can accommodate a 20- to 24-gauge catheter, but can be difficult to maintain in a patient that is not anesthetized or moribund.

Hedgehogs

Cephalic catheters can only be maintained if the patient cannot/does not curl/ball, and this vein can typically accommodate a 24- to 26-gauge catheter.

Sugar gliders, mice, and hamsters

There are no great sites for IV catheter placement in this species as while peripheral veins may be accessible, catheters are typically too long to feed appropriately. Opt for an intraosseous catheter in these patients, instead.

Rabbits

Cephalic and lateral saphenous sites are predictable; however, both of these sites may have significant positional occlusion in awake patients. Since rabbits vary in body size significantly compared to other companion mammals, appropriate catheter sizes can vary between 20-26 gauge. The marginal auricular vein is an excellent, accessible site in most rabbits except dwarf breeds with very small pinnae. The caudal marginal vein is usually more prominent, but the author has found that the cranial vein can be the better option in some individuals. This vein can typically accommodate 22- to 26-gauge catheters. Occluding the vein naturally makes it more easily visualized; the restrainer may gently occlude using the index finger and middle finger in a scissor fashion, or a paperclip may be placed at the proximal end of the vein temporarily.

Guinea pigs

Guinea pigs tend to have very tough skin, and a cut-down using a hypodermic needle can be helpful to avoid burring of the catheter at insertion. The cephalic, accessory cephalic (located more laterally on the forelimb), and lateral saphenous veins are effective sites. A 24- to 26-gauge catheter is typically suitable.

Chinchillas and degus

The cephalic and lateral saphenous sites are often accessible. The cephalic veins are very superficial. Because of their small size, 26-gauge catheters are typically suitable.

Rats

The lateral saphenous veins as well as the lateral tail vein are accessible for catheter placement, and typically accommodate a 24- to 26-gauge catheter. The jugular requires a surgical cutdown but can accommodate a larger catheter, 22- to 26-gauge.

Vascular Access: Intraosseous

Intraosseous access is advised in patients that are too small to place an IV catheter, if the patient is experiencing peripheral venous collapse and the typically accessible veins are too small, or if the accessible veins are no longer usable due to prior catheter placement or injury.

Contraindications to IO catheter placement include the presence of a skin infection at the insertion site, pre-existing fracture of the target bone, osteomalacia (typically due to nutritional secondary hyperparathyroidism, common in sugar gliders), or pre-existing osteomyelitis of the target bone.

Local anesthesia in the form of a lidocaine block in the subcutaneous tissue is advised, with the goal of blocking the periosteal tissue. Lidocaine has a quick onset of action, as fast as 2-5 minutes after injection.[27] Sedation is also necessary unless the patient is moribund. The placement of IO catheters is suspected to be mildly to moderately painful; however, there is evidence that infusion of drugs and/or fluids is quite painful. An excellent summary of the evidence is presented by Cummings (2023).[28]

See Table 3 for a list of equipment needed for IO catheter placement. Placement sites vary between species. Needle selection is based on the following measurements:

- The internal needle diameter should equal ⅓-⅔ the medullary cavity's thinnest diameter of the diaphysis.
- The ideal needle length should equal ⅓-½ the length of the medullary cavity.

Orthogonal view radiographs are needed to confirm placement. Aspirating bone marrow contents and smooth infusion of flush solution are not as sensitive as radiographs to confirm placement.

Table 3. Equipment needed for intraosseous catheter placement in mammals.

Equipment
Adequate sedation if patient is not moribund
Clippers with #40 blade
Rubber tourniquet, paper clip (marginal auricular vein in rabbits), optional
Aseptic preparation for skin: Chlorhexidine or iodophor, isopropyl alcohol or sterile saline
Gauze sponges
Intermittent infusion/PRN plug, or pediatric T-port (slip-tip/Luer-slip preferred for T-port)
Radiology suite to confirm placement
Sterile gloves
Sterile saline ± heparin for flushing
Hypodermic or spinal needle (18-25 gauge)
Local anesthetic and needle, syringe for infusion
Porous tape, non-absorbable sutures (size 3-0 to 4-0), needle drivers, tissue forceps, optional
Scalpel, #11 or #15, optional
Cerclage wire optional as a stylet in hypodermic needle, optional

The procedure for placement of an IO catheter is as follows (consider that the patient is either adequately sedated, or is moribund):

- Locate the anatomic reference points.
- Perform an aseptic preparation of the site (shave fur, scrub) and instill local anesthetic.
- Stabilize the limb with your non-dominant hand.
- A stab incision with a scalpel may be made over the site if desired.
- Using your dominant hand, a firm, rotating motion is used to insert the needle. A slight pop will be felt when the medullary cavity is breached.
- Flush with saline (heparinized or non-heparinized).
- Take orthogonal view radiographs to confirm.
- The catheter may be secured with a tape butterfly and simple interrupted or horizontal mattress sutures, if desired.

Intraosseous catheters can be maintained in place for up to 72 hours. The catheter insertion site should be inspected regularly for infection and extravasation of fluids.[29]

There are three commonly used IO catheter placement sites in mammals: The proximal femur, the proximal tibia, and the proximal humerus.

Proximal femur

The patient is placed in lateral recumbency. The non-recumbent limb is where the IO catheter will be placed. The target limb is slightly adducted, and the proximal limb is palpated to identify either the greater trochanter (rabbits) or trochanteric fossa (ferrets and rodents). The musculature here can be thick, making palpation a bit more challenging. The needle is inserted into the site as described above, in a proximodistal direction.

Proximal tibia

The patient is placed in lateral recumbency. The non-recumbent limb is where the IO catheter will be placed. The target limb is flexed at the stifle, and the tibial tuberosity/tibial crest is palpated (the insertion point of the patellar ligament), this is the insertion site. The needle is inserted into the site as described above, in a proximodistal direction.

Proximal humerus

The patient is placed in lateral recumbency. The non-recumbent limb is where the IO catheter will be placed. The shoulder of the target limb is flexed, and the greater tubercle of the humerus is palpated. This is the insertion point. The needle is inserted in a proximodistal direction.

Cardiopulmonary Cerebral Resuscitation

The goal of cardiopulmonary cerebral resuscitation (CPCR) in exotic mammals is the same as in other species: Return of spontaneous circulation (ROSC) and ventilation. The RECOVER Initiative (www.recoverinitiative.org) is a foundational resource for canine and feline emergency medicine, and the principles of modern CPCR in these companion species will be used as a basis for discussing exotic mammal resuscitation.

It is important to recognize that there is a significant knowledge gap for exotic species, as few studies have been performed to evaluate the effectiveness of CPCR protocols in these animals. There are no specific exotic animal guidelines from the RECOVER Initiative, at this time.

Basic life support versus advanced life support

Basic life support (BLS) is the recognition of cardiopulmonary arrest (CPA) and the immediate response to this event. It includes airway management and ventilation, as well as chest compressions.

Advanced life support (ALS) includes the basic interventions of BLS, as well as rescue drugs (e.g., anticholinergics, positive inotropes, vasopressors), correction of electrolyte imbalances, severe anemia, and/or volume deficits, and defibrillation where appropriate.

It is important to establish a hospital protocol to differentiate between BLS, ALS, and do not resuscitate (DNR) orders, to direct resuscitation efforts appropriately. In the author's experience, a client agreement for CPCR incorporates both BLS and ALS, versus a DNR order.

Some important statistics

Out-of-hospital CPA has a low success rate of CPCR for any species. CPA that occurs during general anesthesia has the highest success rate of CPCR for any species. Overall, CPA in exotic mammal species has a low success rate: Onuma and associates[30] found a survival rate of 0-5.6% and a discharge rate of 0-2.8% for exotic mammal species, including rabbits, hamsters, ferrets, and some other small mammal species.

There are many challenges with small exotic mammal species that could explain these poor survival and discharge rates. Some of those challenges are described below.

- Due to their relatively high metabolic rate, the time between the onset of CPA and the development of severe neurologic damage is likely shorter than in dogs or cats.
- If a facemask is used for ventilation, this can cause gastric bloat or tympany, which impedes diaphragmatic movements.
- Doses of rescue drugs are often extrapolated from other species, and efficacy has not been verified for all species.
- Animals with a small body size present various challenges, and simultaneous interventions are not always possible.

Ventilation

Always remember to ensure correct tube placement during ventilation, which can be reinforced by observing chest movements during intermittent positive pressure ventilation (IPPV), auscultating the lungs, and finally, capnography when ROSC is achieved.

Most patients can be ventilated at 10-20 breaths/minute, which equals to IPPV every 3-6 seconds (with a one second inspiratory time). If a ventilator is being used, a tidal volume of 10 ml/kg of body weight can be calculated. Respect a maximum airway pressure during IPPV of 10-20 mmHg to avoid barotrauma. High peak inspiratory pressure will also compromise venous return. If the $ETCO_2$ is greater than 45 mmHg (e.g., if spontaneous respirations have returned but are ineffective, or when ROSC is achieved), adjust the IPPV rate, inspiratory time, and tidal volume to better ventilate the patient.[31]

Chest compressions

Most exotic mammal patients encountered will be small enough to create a thoracic pump effect with one of two techniques: (1) place fingers and thumb on opposing sides of the thorax at the widest diameter; or (2) use the fingers of both hands on either side of the thorax, held straight, compressing using the full length of your fingers at the widest diameter of the thorax. The compression should make up half of the total compression and release cycle to allow for complete recoiling of the thorax.[31]

Electrocardiography

Electrocardiograms are helpful to identify arrhythmias, which helps direct treatment. The leads may be applied to the patient using atraumatic clips. Alternatively, hypodermic needles may be inserted intradermally in the lead placement locations, and then the leads attached to the needles.

In small mammals, there are several common rhythms that can be identified. They are described below.

- Sinus bradycardia: The sinus node signals appropriate cardiac muscle depolarization; however, the heart rate is slow.
- Asystole: Cessation of electrical and mechanical activity of the heart.
- Pulseless electrical activity (PEA): There is sufficient electrical activity in the heart; however, no pulse is detected (therefore, no perfusion occurring).
- Ventricular tachycardia: A fast heart rate that originates in the ventricles.

Doppler probe

A Doppler probe may be applied over the patient's heart, jugular/carotid, or limb arteries to attempt to detect a pulse in-between compression cycles.

Capnometry

The use of a capnograph is considered the best indicator of ROSC.[31,32] Side-stream capnography will produce a delay in $ETCO_2$ readings, and the sample volume of some units can surpass the tidal volume of very small patients.[33]

When monitoring the capnograph during a CPA event, consider the following:[31]

- $ETCO_2 < 10$ mmHg: If there is cardiac output, this can indicate hyperventilation (in response, the rate of IPPV should be reduced). This may also occur if the patient is not correctly intubated, or if CPCR is ineffective.

- $ETCO_2 > 45$ mmHg: A transient increase over 45 mmHg may occur at the onset of ROSC. This may also indicate hypoventilation if there is cardiac output.

- $ETCO_2$ 12-18 mmHg: These values indicate sustained ROSC.

CPCR algorithm

The following CPCR algorithms (Figs 1 and 2) have been adapted from Lichtenberger and Lennox,[34] and van Zeeland and Schoemaker.[31]

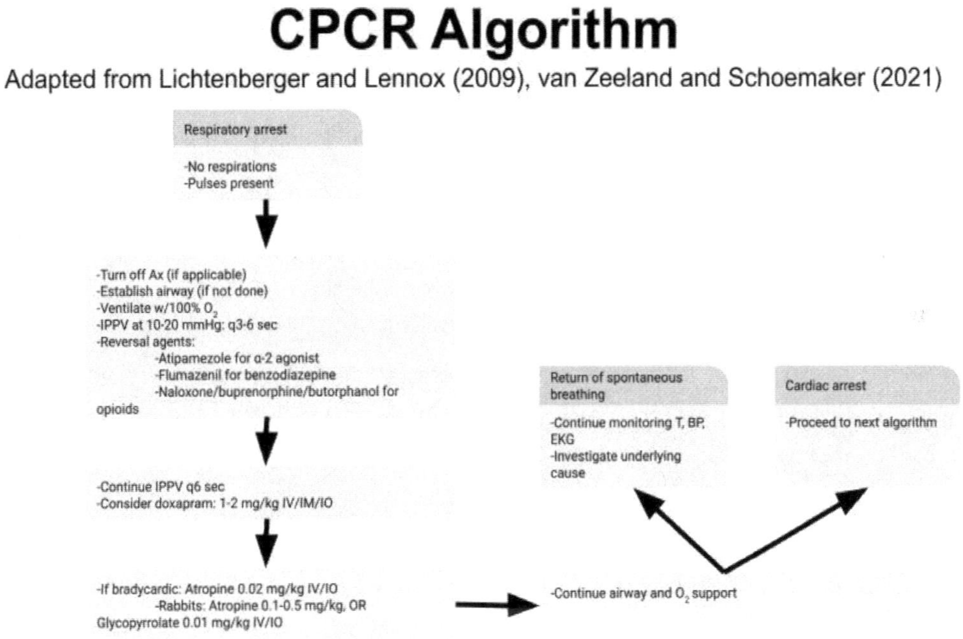

Figure 1. Respiratory arrest cardiopulmonary cerebral resuscitation (CPCR) algorithm.

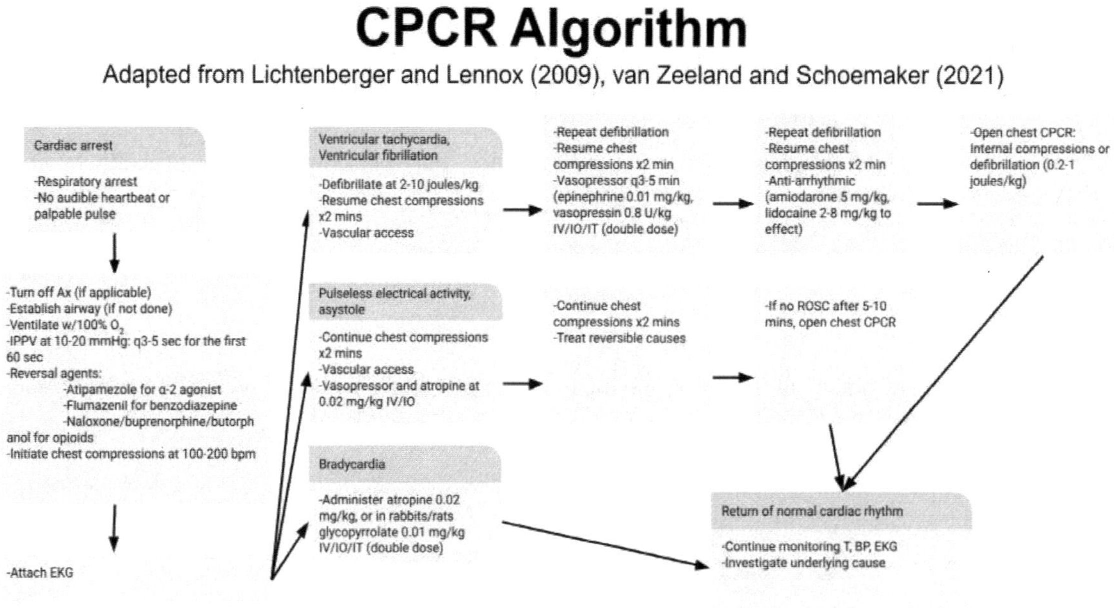

Figure 2. Cardiac arrest cardiopulmonary cerebral resuscitation (CPCR) algorithm.

Other preparations

Being prepared for emergencies will give your patients the best chance at survival. Aside from regular review of emergency protocols with staff and undertaking continuing education to be aware of current CPCR guidelines, what else can be done?

The crash cart should be kept stocked with adequate supplies, including drugs that are not out-of-date. See Table 4 for a list of supplies that are kept in the crash cart, which is similar to the canine and feline crash cart.

Anesthetic patients should have emergency drugs and fluid bolus volumes calculated in advance in the event of an anesthetic CPA. For patients with an American Society of Anesthesiologists (ASA) score over 3/5 (i.e., moderate to severe risk of death under anesthesia), emergency drugs can be pre-drawn to save vital seconds during a CPA event.

Generic drug charts with pre-calculated drug doses based on common body weight ranges can be prepared for commonly seen species. This can be kept with any CPCR guidelines or algorithm charts for quick access.

Table 4. Crash cart recommended supply list.

Supplies to be kept in the crash cart	Supplies to be kept near the crash cart
Emergency drugs, including epinephrine, atropine, glycopyrrolate, vasopressin, doxapram, lidocaine	CPCR protocols; quick drug reference charts
Various sizes of endotracheal tubes, supraglottic airway devices, oxygen masks	Defibrillator
1-ml, 3-ml syringes, ideally with needles attached	Syringe driver/pump
U-100 insulin syringes, ⅓ ml volume	Warming equipment for patient
Needles of various sizes (20-26 gauge)	Bag valve mask
Sterile physiologic saline flush	Crystalloids (e.g., Lactated Ringer's solution), colloids
IV catheters of various sizes (20-26 gauge)	Infusion sets and appropriate extensions
Spinal needles of various sizes (18-22 gauge)	Indirected blood pressure monitoring equipment
Fenestrated red rubber catheters (5- to 12-Fr)	Pulse oximeter

Conclusions

Many principles of exotic mammal emergency medicine are similar to canine and feline medicine; however, like with all exotic taxa there are many idiosyncrasies with these species that present unique challenges. Preparation of the treatment and surgical suites as well as staff members is critical for success.

References

1. Genovart M, Negre N, Tavecchia G, et al. The young, the weak and the sick: evidence of natural selection by predation. *PLoS ONE.* 2010;5(3):e9774. doi:10.1371/journal.pone.0009774.

2. Krumm CE, Conner MM, Hobbs NT, et al. Mountain lions prey selectively on prion-infected mule deer. *Biol Lett.* 2010;6:209-211. doi:10.1098/rsbl.2009.0742.

3. Mukherjee S, Heithaus M. Dangerous prey and daring predators: a review. *Biol Rev Camb Philos Soc.* 2013; 88:550-563. doi:10.1111/brv.12014.

4. McLaughlin A, Strunk A. Common emergencies in small rodents, hedgehogs, and sugar gliders. *Vet Clin North Am Exot Anim Pract.* 2016;19(2):465-99. doi:10.1016/j.cvex.2016.01.008. PMID: 27131160.

5. Covey E. How to triage. *Vet Nurse.* 2018;9(5). doi.org/10.12968/vetn.2018.9.5.262.

6. van Zeeland YRA, Schoemaker NK. Mammals: history and clinical rxamination. In: Graham JE, Doss GA, Beaufrère H, eds. *Exotic Animal Emergency and Critical Care Medicine.* Hoboken, NJ: Wiley-Blackwell; 2021:5-23.

7. Mazzaferro EM. Oxygen therapy. In: Silverstein DC, Hopper K, eds. *Small Animal Critical Care Medicine.* 2nd Ed. St. Louis, MO: Elsevier Saunders; 2015:77-80.

8. Boysen SR, Mathews KA. Respiratory emergencies. In: Mathews KA, ed. *Veterinary Emergency and Critical Care Manual.* 3rd Ed. Guelph, ON: Lifelearn, Inc; 2017:37-84.

9. Lee-Fowler T. Feline lower airway disease. In: Drobatz KJ, Hopper K, Rozanski E, Silverstein DC, eds. *Textbook of Small Animal Emergency Medicine.* Hoboken, NJ: Wiley; 2018:220-224.

10. Gardhouse S. Oxygen therapy. In: Graham JE, Doss GA, Beaufrère H, eds. *Exotic Animal Emergency and Critical Care Medicine.* Hoboken, NJ: Wiley-Blackwell; 2021:38-50.

11. Ferasin L, Ferasin H, Kilkenny E. Heart murmurs in apparently healthy cats caused by iatrogenic dynamic right ventricular outflow tract obstruction. *J Vet Int Med.* 2020;34(3):1102-1107. https://doi.org/10.1111/jvim.15774.

12. Warnefors E, Rueløkke ML, Gredal H. Results of a modified neurological examination in 26 healthy rabbits. *J Exot Pet Med.* 2019;30:54-59.

13. van der Woerdt A. Ophthalmologic diseases of small mammals. In: Quesenberry KE, Orcutt CJ, Mans C, Carpenter JW, eds. *Ferrets, Rabbits, and Rodents: Clinical Medicine and Surgery.* 4th Ed. St. Louis, MO: Elsevier; 2021:583-594.

14. Didier ES, Maddry JA, Brindley PJ, et al. Therapeutic strategies for human microsporidia infections. *Expert Rev Anti Infect Ther.* 2005;3(3):419-434.

15. Keeney CMH, Hung CS, Harrison TM. Comparison of body temperature using digital, infrared, and tympanic thermometry in healthy ferrets (*Mustela putorius furo*). *J Exot Pet Med.* 2021;36:16-21.

16. Di Girolamo N, Toth G, Selleri P. Prognostic value of rectal temperature at hospital admission in client-owned rabbits. *J Am Vet Med Assoc.* 2016;248(3):288-297. doi:10.2460/javma.248.3.288.

17. Levy IH, Di Girolamo N, Keller KA. Rectal temperature is a prognostic indicator in client-owned guinea pigs. *J Small Anim Pract.* 2021;62(10):861-865.

18. Heard D. Rodents. In: West G, Heard D, Caulkett N, eds. *Zoo Animal and Wildlife Immobilization and Anesthesia.* 2nd Ed. Ames, IA: Wiley-Blackwell; 2014:893-903.

19. Varga M. Airway management in the rabbit. *J Exot Pet Med.* 2017;26:29-35.

20. Thompson KL, Meier TR, Scholz JA. Endotracheal intubation of rabbits using a polypropylene guide catheter. *J Vis Exp.* 2017:1-8.

21. Docsinnovent. V-gel ADVANCED for rabbits. Available at: https://docsinnovent.com/products/v-gel-rabbit/. Accessed March 6, 2023.

22. Engbers S, Larkin A, Rousset A, et al. Comparison of a supraglottic airway device (V-gel ®) with blind orotracheal intubation in rabbits. *Front Vet Sci.* 2017;4:49.

23. Devalle JMS. Successful management of rabbit anesthesia through the use of nasotracheal intubation. *J Am Assoc Lab Anim Sci.* 2009;48:166-170.

24. Lafferty K. A guide to nasotracheal intubation in the rabbit. Available at: https://lafeber.com/vet/a-guide-to-nasotracheal-intubation-in-rabbits/. Accessed March 6, 2023.

25. Johnson DH. Endoscopic intubation of exotic companion mammals. *Vet Clin North Am Exot Anim Pract.* 2010;13:273-289.

26. Turner MA, Thomas P, Sheridan DJ. An improved method for direct laryngeal intubation in the guinea pig. *Lab Anim.* 1992;26:25-28.

27. Cousins MJ, Bridenbaugh PO. *Neural Blockade in Clinical Anesthesia and Management of Pain*. 4th Ed. Philadelphia, PA: Lippincott Williams & Wilkins; 2008.

28. Cummings CO. Are we providing adequate analgesia for intraosseous infusion in exotic animal practice? *J Exot Pet Med.* 2023;44:27.

29. van Zeeland YRA, Schoemaker NK. Mammals: Catheterization and Venipuncture. In: Graham JE, Doss GA, Beaufrère H, eds. *Exotic Animal Emergency and Critical Care Medicine.* Hoboken, NJ: Wiley-Blackwell; 2021b:51-69.

30. Onuma M, Kondo H, Ono S, et al. Retrospective investigation of cardiopulmonary resuscitation outcome in 146 exotic animals. *J Vet Med Sci.* 2017;79(9):1611-1614. doi:10.1292/jvms.16-0360.

31. van Zeeland YRA, Schoemaker NK. CPR and euthanasia. In: Graham JE, Doss GA, Beaufrère H, eds. *Exotic Animal Emergency and Critical Care Medicine.* Hoboken, NJ: Wiley-Blackwell; 2021:80-94.

32. Brainard BM, Boller M, Fletcher DJ. RECOVER evidence and knowledge gap analysis on veterinary CPR. Part 5: monitoring. *J Vet Emerg Crit Care.* 2012;22(1):S65-84.

33. Lafferty K, Pollock CG. Capnometry in exotic animal species. Available at: https://lafeber.com/vet/capnometry/. Accessed March 20, 2023.

34. Lichtenberger M, Lennox AM. Exotic companion mammal emergency medicine and critical care. *Proc Annu Conf Assoc Exot Mamm Vet.* 2009.

Exotic Animal Emergencies for Technicians, Part 2: Evaluation of the Emergent Reptile Patient

Christina Miller, RVT, BSc

Session #3508

Affiliation: From Clayton Park Veterinary Hospital, 278 Lacewood Dr, Halifax, NS, B3M 3N8, Canada.

Abstract: Reptiles commonly present to the veterinary hospital in a state of severe illness. This masterclass reviews emergency presentations in non-crocodilian and non-avian reptiles (e.g., squamates and chelonians). It covers the technician's role in the primary survey, obtaining vitals efficiently and safely, obtaining airway and vascular access, and proceeding with cardiopulmonary and cerebral resuscitation.

Introduction

The basic principles of triage and masking phenomenon are covered in "Exotic Animal Emergencies for Technicians, Part 1: Evaluation of the Emergent Exotic Mammal Patient."

Reptiles present unique challenges as emergent patients due to their anatomy and ectothermic nature. Ectotherms are animals which rely on external heat sources to maintain their body temperature within a preferred optimum temperature zone (POTZ), that is species-specific.[1,2] They are sometimes described as poikilotherms; however, this term relates to thermoconformers, which does not characterize the entire daily habits of many reptiles (i.e., some reptiles are always thermoconformers, whereas others only become thermoconformers when they are inactive and are thermoregulators when active). The ectotherm's mentation, metabolism, immune function, drug metabolism, among innumerable other physiological processes rely on the animals' access to an appropriate thermal gradient to reach their preferred body temperature (PBT) within the POTZ.[1,2]

A husbandry review and client education of how to correct deficiencies is critical for success once the patient is discharged from the hospital.[1]

The Reptile Primary Survey

At the initial assessment, the patient may not be within their POTZ, so their mentation and vitals may be altered. They may have dulled responsiveness, neurological deficits, bradycardia, and bradypnea.[1] Many lizard species adopt a darker overall body colour.[3] If the patient is cool to the touch on arrival, active warming should be initiated. Cloacal temperature is not a useful assessment except for detecting hyperthermia and monitoring warming progress with regards to environmental temperature. Most rectal thermometers intended for use in mammals will not read below 32-34 °C (89.6-93.2 °F), and many species will present with a body temperature below these values. A digital probe thermometer intended for use in reptile enclosures can instead be used; ensure that you are using a quality unit by verifying temperatures against another thermometer.

If the species' POTZ is unknown, provide thermal support at 28-29.4 °C (82-85 °F).[1,4] Ideally, the patient should be warmed slowly, over 4-6 hours.[1] Once the patient is warmed to their POTZ, the initial assessment should be performed. Latent clinical signs may be appreciated once the animal has been warmed within their POTZ.

Masking phenomenon may not be as prominent in these animals as other exotic taxa; however, many pet owners simply do not recognize signs of illness in reptiles. This may either be due to unfamiliarity with normal behaviours and anatomy, or because some illnesses develop very slowly in reptiles.

Reptiles are morphologically and behaviorally diverse, and activity cycle as well as posture varies with the species being presented. Generally, reptiles should be bright, alert, and responsive to your presence and/or touch. Squamates will generally exhibit tongue-flicking behavior. Most lizard species should be able to raise their body off and above the surface they are resting on, with the exception of some stocky, low-bodied taxa such as Scincidae, Gerrhosauridae, and a few others. Chameleons should perch when offered a branch or limb, with their body lifted. Snakes may coil into a defensive position, Ball Pythons (*Python regius*) most famously will curl into a ball with their head hidden in the center. Many snakes will instinctively wrap around a hand or arm when being handled, and their grip should be substantial. Chelonians should be bright and alert, monitoring their environment, but will recede into the shell if threatened.[5]

If the patient is dull, dyspneic, or non-responsive to touch, proceed to the ABCDE protocol. The ABCDE protocol is a systematic approach to patient assessment in order of the most acutely life-threatening problems to least life-threatening problems.

A: Airway

Airway and breathing assessment is effectively performed simultaneously; however, signs of airway obstruction should be considered immediately, including: Increased respiratory effort (especially on inspiration), cyanosis (rare in reptiles; some species have dark blue or purple oral mucous membranes naturally),[6] stertor (low-pitched sound, suggests nasal or pharyngeal obstruction) or stridor (high-pitched sound, suggests lower airway obstruction).[6,7] Upper airway obstructions can be easily identified as the glottis is relatively rostrally located in squamates and chelonians, at the base of the tongue. There is no epiglottis. Reptiles are nasal breathers, and open-mouth breathing (aside from threat displays, vocalizations, or thermoregulation, e.g., increasing evaporative water loss in lizards[2,6]) is considered abnormal.[7] Orthopneic posture may also be exhibited by reptiles: Snakes may elevate the cranial ¼-⅓ of the body against a wall or object in their enclosure, and lizards may adopt a similar posture with or without support of an object. Chelonians may extend limbs and the neck to increase the limited space for lung expansion due to the presence of the rigid, boney shell.[6]

Reptiles have a relatively low metabolic rate compared to endotherms, and during anaerobic metabolism they exhibit a substantial capacity to buffer increased concentrations of lactic acid and hydrogen ions.[8] These characteristics allow reptiles to compensate effectively for partial airway obstruction or chronic disease and may only present when they decompensate from chronic disease.[6]

Oxygen therapy:

Oxygen therapy can be counterproductive in some patients, as the reptile's respiratory drive is stimulated by hypoxia.[6,7] Oxygen therapy is indicated if there are signs of hypoxia: Dyspnea, tachycardia, and cyanosis (which is rare). A positive response to oxygen therapy may be considered confirmatory of hypoxia.[9] Exposure to airborne irritants and toxins, evidence of drowning, and decompensating heart disease are all indications for oxygen therapy in reptiles.[6,9]

Note that pulse oximetry and blood gas analysis are not validated in reptiles.[6,9] Pulse oximetry may be used to evaluate trends in a patient; however, the value obtained may not be representative.

Flow-by oxygen may not suppress the respiratory rate: The FiO_2 provided by flow-by oxygen is generally between 25-40% at a flow rate of 2-3 L/min,[10] and the oxygen should be humidified to prevent airway desiccation.[6] Intermittent positive pressure ventilation (IPPV) may be necessary in debilitated individuals if apnea occurs from a high FiO_2 of oxygen.[9]

If respiratory arrest occurs or the obstruction cannot be identified and relieved, airway access must be secured. If there is jaw tone present, use of sedation and/or induction agents may be necessary. Airway access will be discussed below. Apneic patients can be ventilated on room air using a bag valve mask unless there is an indication (as discussed above) for oxygen therapy.

B: Breathing

The quality and rate of the patient's respirations should be assessed. Rate correlates to various physiologic parameters: Hypoxia results in an increased respiratory rate, and while normoxia in theory results in eupnea, supplemental oxygen results in respiratory suppression. Tachypnea can occur with hyperthermia and bradypnea occurs when the patient is below their POTZ. Hypercapnia and acidosis result in increased tidal and minute volumes whereas hypocapnia and alkalosis result in a decrease in these volumes.[7]

Note that some species will exhibit buccal pumping; therefore, thoracic respirations should be counted in these animals.

Lung fields may be assessed with auscultation, but this is difficult due to the presence of scales or a shell. A moistened gauze placed between the patient's skin or shell and the stethoscope diaphragm can be helpful in dampening the interfering sounds.[6]

C: Circulation

Circulation is challenging to evaluate. Peripheral pulses are typically not palpable. Capillary refill time and mucous membrane color are not used to evaluate perfusion,[11] although mucous membrane pallor can be suggestive of anemia. Heart rate is currently the best indicator of cardiovascular function.[11]

Noninvasive and indirect blood pressure (NIBP) is considered inaccurate but may be used to observe trends in a patient. A Doppler with a sphygmomanometer or an oscillometric unit can be used.[12,13] Direct arterial blood pressure monitoring is not practical for clinical applications.[1]

The heart is located within the pectoral girdle in most lizard and chelonian species, although some lizard taxa (Chamaeleonidae, Varanidae, Teiidae, and a few others) have a more caudally positioned heart. In snakes, heart position varies with the species' ecological niche: Arboreal species' heart is located approximately 15% of the body length from the rostrum; in snakes that are not arboreal or aquatic, the heart is located approximately 25% of the body length from the rostrum; in marine and freshwater species, the heart is located at nearly 50% of the body length.[14-16] Heart rate can be evaluated using a Doppler either over the heart, or at major vessels such as the carotid and brachial arteries. Like with pulmonary auscultation, cardiac auscultation is possible with a dampened gauze but may not be rewarding, especially in species with the bony pectoral girdle overlying the heart. Murmurs may be auscultated in larger species where the heart is located more caudally using a pressure-sensitive acoustic stethoscope.[17]

D: Disability

As previously noted, patients should be within their POTZ in order to accurately evaluate mentation and neurological function.[1] When below the POTZ, ectotherms such as reptiles exhibit reduced nerve conduction. This results in apparent paresis, lethargy, abnormal behaviours, or even neurological deficits.[18]

A description of normal mentation and posture in some common species was described in the previous section, The Reptile Primary Survey.

In addition to observing mentation and posture, gait may be evaluated on a surface with reasonable traction, such as a towel or yoga mat (reptiles may have difficulty ambulating on smooth surfaces such as a metal exam table or tile floor). The presence of ataxia, paresis, and lameness may all be observed during a gait evaluation.[19] Aquatic and semiaquatic species benefit from evaluation in water deep enough for them to submerge;[19] the water should be warmed to be within the species' preferred temperature, and close attention must be paid to the patient in case of deficits to avoid aspiration of water or potential drowning.

Proprioception can be evaluated in some larger lizards but cannot be evaluated in snakes and is difficult to interpret in chelonians.[19] Righting reflex can be evaluated in snakes.[19]

The menace response is described in some texts;[19] however, it is observed to be unreliable or absent in some species.[20-22] Palpebral and corneal reflex cannot be assessed in species with a spectacle (fused, transparent eyelid).[19] Pupillary light response is unreliable in reptiles due to their ability to override the reflex.[19]

E: Exposure of environment

At this point in the triage evaluation, patient temperature has been evaluated already. Survey the body for skin lesions, fractures, bruises, and other injuries.

Airway Access

Airway access should be obtained in apneic patients. Unlike in mammals and birds, pre-oxygenation may not be beneficial, depending on the patient's presentation as discussed above. Commercially available uncuffed tubes can be used for many patients; however, smaller purpose-made endotracheal (ET) tubes made from items such as red rubber catheters and polypropylene IV catheters are commonplace. Many of these house-made ET tubes will attach to the anesthetic circuit using the smallest adapter for commercial 2-3 mm internal diameter tubes. These can be cut to length and a Murphy eye can be added using a scalpel blade; a brief pass over a flame can dull the edges.[6] Squamate reptiles have dorsally incomplete tracheal rings, and chelonians have complete tracheal rings;[2] therefore, cuffed tubes must be avoided with the latter, as inflating the cuff has a great risk of tracheal tears and/or necrosis.[6] Tubes should ideally be clear when possible to allow for observation of spontaneous breathing (fogging), since capnometry is often ineffective in small reptiles due to their small tidal volume.[23]

Snakes

The mouth is opened with a soft speculum, or gentle manual traction on the mandible. The glottis in snake species is mobile to accommodate ventilation during ingestion of large prey items.[2,24] It is located caudodorsal to the tongue sheath, and a muscular valve remains closed until inspiration. Genus *Pituophis* has an epiglottal keel across the transverse aspect of the glottis, thought to amplify their defensive hissing sound; care should be taken to avoid damaging this structure during intubation.[24,25] The ET tube is premeasured to the level of the heart. Most snakes have only a single lung on the right side; in boid snakes (boas and pythons) a smaller left lung exists.[24] The ET tube should be lubricated, and no resistance should be encountered when intubating.

Lizards

The mouth is opened using a soft speculum; the author avoids opening the mouth in lizards with manual traction on the mandible in case of osteomalacia, as iatrogenic fracture may occur. The glottis is located caudodorsal to the tongue sheath and has a muscular valve as in snakes. Agamidae and Chamaeleonidae possess a very fleshy tongue and expandable gular region, so the glottis may be more difficult to visualize in these taxa (using a finger in the gular region to gently prop the tissue dorsally can help the operator visualize and stabilize the glottis). The ET tube is measured to the level of tracheal bifurcation, in many species this coincides with the level of the heart at the pectoral girdle. In Chamaeleonidae, the trachea bifurcates cranial to the thoracic inlet.[6] The ET tube should be lubricated, and no resistance should be encountered when intubating.

Chelonians

The mouth is opened with a speculum. The glottis is located at the base of the tongue and may be obscured by the fleshy tongue. Gentle rostral traction of the tongue can be helpful in propping up the glottis for better visualization. Members of suborder Cryptodira have a variable trachea length that either bifurcates in the mid-cervical region,[6,26-28] or immediately caudal to the head in members of the genus *Testudo*.[29] In suborder Pleurodira, the trachea length may be 85% of the carapace length, or longer.[29] The ET tube should be measured appropriately as per the species, lubricated, and no resistance should be encountered during intubation.

Vascular Access: Intravenous

Vascular access should be obtained in any patient presenting in an urgent situation. Reptiles do not have obvious peripheral veins, and the presence of thick and often scaled skin or the presence of a shell makes palpation of vessels difficult. Small patient size also contributes to the difficulty (or impossibility) of placing intravenous (IV) catheters in some patients. Furthermore, reptiles have a relatively low blood pressure compared to similarly sized birds or mammals, and in patients with cardiovascular or perfusion deficits, the flash of blood that helps to confirm IV catheter placement may not occur. Sites for IV catheter placement will be discussed.

Jugular vein

The right side is typically larger than the left. In lizards and snakes, a cutdown is typically needed for this site. Note that in snakes this requires opening the coelomic cavity as the vein is located quite deep and runs parallel to the trachea.[30] Patients as small as leopard geckos (*Eublepharis macularius*) can accommodate a jugular catheter.[31] In chelonians, the jugular can be superficial and easily accessed.[31]

Cephalic vein

This site may be accessed in lizards via a surgical cutdown in a transverse direction across the antebrachium.[30] An ultrasound-guided technique has been described in chelonians under 15 kg in body weight.[32]

Ventral caudal coccygeal vein

This vein may be accessed via a lateral or ventral approach in squamate reptiles, at 25-50% of the tail length to avoid accidental trauma of the hemipenes, hemiclitori, or scent glands.[33] This site should be avoided in species that perform caudal autotomy.

Dorsal caudal coccygeal vein

This site can be accessed in chelonians; it is difficult to maintain in active animals.[30]

Cardiac, ventricular

This site may be used in snakes where the catheter can be introduced directly into the ventricle of the heart with the patient in dorsal recumbency; this catheter may only be maintained in moribund or anesthetized patients.[30,34] Long-term consequences of catheter placement do not appear to have been studied.

Vascular Access: Intraosseous

Venous access may not be possible in very small, dehydrated, or hypotensive patients, and intraosseous (IO) catheters are an acceptable alternative in many animals.

Contraindications to IO catheter placement include the presence of a skin infection at the insertion site, pre-existing fracture of the target bone, osteomalacia (typically due to nutritional or renal secondary hyperparathyroidism), or pre-existing osteomyelitis of the target bone.[30] Local anesthesia and sedation for IO catheter placement is preferred unless the patient is moribund. Ongoing pain management is advised for IO catheters used for fluid and/or drug infusion.[35]

Placement sites vary between species. Needle selection is based on the following measurements: the internal needle diameter should equal ⅓-⅔ the medullary cavity's thinnest diameter of the diaphysis. The ideal needle length should equal ⅓-½ the length of the medullary cavity.[30,31]

Proximal tibia

This site is accessible in lizards and may be easier to maintain than other sites. The stifle is flexed to expose the tibial crest, which is the insertion site.[30,31]

Distal femur

This site is accessible in lizards and chelonians, and it is typically easy to maintain in lizards. The stifle is flexed to expose the diaphysis of the distal femur, which is the insertion site.[30,31]

Distal humerus

This site is accessible in chelonians and lizards and is easy to maintain in lizards. The elbow is flexed to expose the great tubercle of the humerus, and the needle introduced at the craniolateral aspect.[30,31]

Gular plastron and plastrocarapacial junction

These sites are only present in chelonians and may be easier to maintain in awake patients compared to IO catheters placed in the limbs. A drill (e.g., EZ-IO® Intraosseous Vascular Access System, Teleflex, Morrisville, NC, USA) is generally needed for placement in larger specimens.[30,31] A 2012 study found that fluid uptake from these sites is less effective compared to IO catheters placed in limbs.[36]

Cardiopulmonary Cerebral Resuscitation

The goal of cardiopulmonary cerebral resuscitation (CPCR) in reptiles is the same as in other species: Return of spontaneous circulation (ROSC) and ventilation. The knowledge gap between reptiles and even other exotic taxa is significant, and reptiles present many challenges during CPCR. The following discussion is informed by the RECOVER Initiative (Reassessment Campaign on Veterinary Resuscitation), found at https://recoverinitiative.org.

As discussed in "Exotic Animal Emergencies for Technicians, Part 1: Evaluation of the Emergent Exotic Mammal Patient," it is important to operate with protocols for basic life support (BLS), advanced life support (ALS), and do not resuscitate (DNR) orders for individual patients. A basic algorithm for CPCR is found in Part 1, and deviations from this algorithm will be discussed below.

Ventilation

Ventilation with 100% oxygen is advised until ROSC is achieved. At this time, the patient may be transitioned to room air (or alternatively, 30-40% oxygen)[37] as hypoxemia is the primary stimulus for spontaneous respirations in reptiles.[6,7] Ventilating the patient with a maximum circuit pressure of 12 cmH_2O and 2-6 breaths per minute has been advised.[38]

Chest compressions

Chest compressions may be less effective in reptiles compared to other taxa. The heart is located within the pectoral girdle in many lizards, or protected by the plastron in chelonians: Both of these features limit the effectiveness of compressions; a Doppler probe may be placed over a major artery (heart, carotid, brachial; spectacle in applicable squamates) to determine the effectiveness of compressions.[38] Compressions in snakes can be very effective, recall that the location of the heart varies with the species.[14] There appear to be no published recommended rates for evidence-based CPCR in reptiles; the author typically aims for 60-100 bpm.

Electrocardiography

Electrocardiography can be challenging to interpret in reptiles,[39] and arrhythmias are not well described in the literature. Arrhythmias that can be identified are treated as in other species.

Capnometry

Capnometry may not provide accurate readings in reptiles. In small patients (< 100 g body weight), side-stream capnography may have a sampling volume greater than the patient's tidal volume, resulting in the sample becoming diluted with fresh gas.[40] End tidal CO_2 ($ETCO_2$) values may not be representative of $PaCO_2$ due to pulmonary shunting.[41] Values rarely surpass 25 mmHg, and a return to normal values observed during anesthesia (e.g., 11-25 mmHg) may indicate ROSC.[38]

Conclusions

Reptiles are challenging patients due to their unique anatomy and physiology. There is a lack of evidence-based recommendations for CPCR in these taxa.

References

1. Petritz OA, Son TT. Emergency and critical care. In: Divers SJ, Stahl SJ, eds. *Mader's Reptile and Amphibian Medicine and Surgery.* 3rd Ed. St. Louis, MO; Elsevier; 2019:63-74.

2. Vitt LJ, Caldwell JP. *Herpetology: An Introductory Biology of Reptiles and Amphibians.* 4th Ed. St. Louis, MO; Elsevier; 2014.

3. Barten S, Simpson S. Lizard taxonomy, anatomy, and physiology. In: Divers SJ, Stahl SJ, eds. *Mader's Reptile and Amphibian Medicine and Surgery.* 3rd Ed. St. Louis, MO; Elsevier; 2019:967-976.

4. Pollock, C. Lafebervet: Exotic ICU: Nursing Care for Reptiles. Available at: https://lafeber.com/vet/exotic-icu-nursing-care-for-reptiles/. Accessed March 5, 2023.

5. Durso AM, JC Maerz. Natural behavior. In: Divers SJ, Stahl SJ, eds. *Mader's Reptile and Amphibian Medicine and Surgery.* 3rd Ed. St. Louis, MO: Elsevier; 2019;90-99.

6. Kanda I, Brandão J. Herptiles: oxygen therapy. In: Graham JE, Doss GA, Beaufrère H, eds. *Exotic Animal Emergency and Critical Care Medicine.* Hoboken, NJ: Wiley-Blackwell; 2021:716-721.

7. Knotek Z, Divers SJ. Pulmonology. In: Divers SJ, Stahl SJ, eds. *Mader's Reptile and Amphibian Medicine and Surgery.* 3rd Ed. St. Louis, MO; Elsevier; 2019:786-804.

8. Bickler PE, Buck LT. Hypoxia tolerances in reptiles, amphibians, and fishes: life with variable oxygen availability. *Annu Rev Physiol.* 2007;69:145-170.

9. Schumacher J. Respiratory medicine of reptiles. *Vet Clin North Am Exot Anim Pract.* 2011;14:207-224.

10. Mazzaferro EM. Oxygen therapy. In: Silverstein DC, Hopper K, eds. *Small Animal Critical Care Medicine.* 2nd Ed. St. Louis, MO: Elsevier Saunders; 2015:77-80.

11. Schumacher J, Mans C. Anesthesia. In: Mader DR, Diver SJ, eds. *Current Therapy in Reptile Medicine and Surgery.* St. Louis: Elsevier; 2014:145-148.

12. Chinnadurai SK, Wrenn A, DeVoe RS. Evaluation of noninvasive oscillometric blood pressure monitoring in anesthetized boid snakes. *J Am Vet Med Assoc.* 2009;234:625-630.

13. Chinnadurai SK, DeVoe R, Koenig A, et al. Comparison of an implantable telemetry device and an oscillometric monitor for measurement of blood pressure in anaesthetized and unrestrained green iguanas (*Iguana iguana*). *Vet Anaesth Analg.* 2010;37:434-439.

14. Farrell AP, Gamperl AK, Francis ETB. Comparative aspects of heart morphology. In: Gans C, Gaunt A, eds. *Biology of the Reptilia.* Vol. 19. Morphology G. Visceral Organs. Ithaca, NY: Society for the Study of Amphibians and Reptiles; 1998:375-424.

15. Wyneken J. Normal reptile heart morphology and function. *Vet Clin North Am Exot Anim Pract.* 2009;12: 51-63, vi.

16. McCracken HE, Fowler ME, Miller RE. Organ location in snakes for diagnostic and surgical evaluation. In: Miller RE, Fowler ME, eds. *Zoo and Wild Animal Medicine: Current Therapy.* 4th Ed. Philadelphia: WB Saunders; 1999:243-248.

17. Schilliger L, Girling S. Cardiology. In: Divers SJ, Stahl SJ, eds. *Mader's Reptile and Amphibian Medicine and Surgery.* 3rd Ed. St. Louis, MO: Elsevier; 2019:669-698.

18. Long SY. Approach to reptile emergency medicine. *Vet Clin North Am Exot Anim Pract.* 2016;19:567-590.

19. Platt SR. Neurology. In: Divers SJ, Stahl SJ, eds. *Mader's Reptile and Amphibian Medicine and Surgery.* 3rd Ed. St. Louis, MO: Elsevier; 2019:805-826.

20. Tarbert DK, Murthy VD, Sanchez-Migallon Guzman D. Neurological examination in healthy adult inland bearded dragons (*Pogona vitticeps*). *J Am Vet Med Assoc.* 2022;260(9).

21. Hedley J, MacHale J, Rendle M, Crawford A. Neurological examinations in healthy juvenile bearded dragons (*Pogona vitticeps*) and adult leopard geckos (*Eublepharis macularius*). *J Herpetol Med Surg.* 2021;31(2):141-146.

22. Monchaux M. Neurology in exotic animals. *Proc Int Conf VIII FAUNA.* 2019.

23. Briscoe LA, Syring R. Techniques for emergency airway and vascular access in special species. *Semin Avian Exot Pet Med.* 2004; 13(3): 118-131.

24. Funk RS, Bogan JE, Jr. Snake taxonomy, anatomy, and physiology. In: Divers SJ, Stahl SJ, eds. *Mader's Reptile and Amphibian Medicine and Surgery.* 3rd ed. St. Louis, MO; Elsevier; 2019:50-62.

25. Young BA, Sheft S, Yost W. Sound production in *Pituophis melanoleucus* (Serpentes: Colubridae) with the first description of a vocal cord in snakes. *J Exp Zool.* 1995;273(6):472-481.

26. Boyer TH, Innis CJ. Chelonian taxonomy, anatomy, and physiology. In: Divers SJ, Stahl SJ, eds. *Mader's Reptile and Amphibian Medicine and Surgery.* 3rd Ed. St. Louis, MO; Elsevier; 2019:31-49.

27. Zehtabvar O, Tootian Z, Vajhi A, et al. Computed tomographic anatomy and topography of the lower respiratory system of the European pond turtle (*Emys orbicularis*). *Int J Vet Sci.* 2014;9(2):9-16.

28. Lambertz M, Bohme W, Perry SF. The anatomy of the respiratory system in *Platysternon megacephalum* Gray, 1831 (Testudines: Cryptodira) and related species, and its phylogenetic implications. *Comp Biochem Physiol A: Mol Integr Physiol.* 2010;156(3):330-336.

29. Habova M, Pyszko M, Horak O, et al. Differences in the anatomy of the lower respiratory tract in selected species of the order Testudines. *Vet Med-Czech.* 2022;67(2):78-86.

30. Pardo MA, Divers SJ. Catheter placement. In: Divers SJ, Stahl SJ, eds. *Mader's Reptile and Amphibian Medicine and Surgery.* 3rd Ed. St. Louis, MO: Elsevier; 2019:422-428.

31. Kanda I, Brandão J. Herptiles: catheterization and venipuncture. In: Graham JE, Doss GA, Beaufrère H, eds. *Exotic Animal Emergency and Critical Care Medicine.* Hoboken, NJ: Wiley-Blackwell; 2021:722-731.

32. Di Bello A, Valastro C, Freggi D, et al. Ultrasound-guided vascular catheterization in loggerhead sea turtles (*Caretta caretta*). *J Zoo Wildl Med.* 2010;41(3):516-518.

33. Olsson A, Simpson M. Analgesia and anaesthesia. In: Doneley B, Monks D, Johnson R, Carmel B, eds. *Reptile Medicine and Surgery in Clinical Practice.* Hoboken, NJ: Wiley-Blackwell; 2018:369-381.

34. Divers SJ. Merck Veterinary Manual: Clinical Procedures for Reptiles. Available at: https://www.merckvetmanual.com/exotic-and-laboratory-animals/reptiles/clinical-procedures-for-reptiles. Accessed March 5, 2023.

35. Cummings CO. Are we providing adequate analgesia for intraosseous infusion in exotic animal practice? *J Exot Pet Med.* 2023;44:27.

36. Young BD, Stegeman N, Norby B, Heatley JJ. Comparison of intraosseous and peripheral venous fluid dynamics in the desert tortoise (*Gopherus agassizii*). *J Zoo Wildl Med.* 2012;43(1):59-66. doi:10.1638/2011-0058.1.

37. Costello MF. Principles of cardiopulmonary cerebral resuscitation in special species. *Semin Avian Exot Pet Med.* 2004;13(3):132-141.

38. Doss GA, Sladky KK. Herptiles: CPR and euthanasia. In: Graham JE, Doss GA, Beaufrère H, eds. *Exotic Animal Emergency and Critical Care Medicine.* Hoboken, NJ: Wiley-Blackwell; 2021:738-745.

39. Doss GA, Mans C, Sladky KK. Herptiles: analgesia, anesthesia, and monitoring. In: Graham JE, Doss GA, Beaufrère H, eds. *Exotic Animal Emergency and Critical Care Medicine.* Hoboken, NJ: Wiley-Blackwell; 2021:746-757.

40. Ozeki L, Caulkett N. Monitoring. In: West G, Heard D, Caulkett N, eds. *Zoo Animal and Wildlife Immobilization and Anesthesia.* 2nd Ed. Ames, IA: Wiley-Blackwell; 2015:149-165.

41. Bertelsen MF. Squamates (snakes and lizards). In: West G, Heard D, Caulkett N, eds. *Zoo Animal and Wildlife Immobilization and Anesthesia.* 2nd Ed. Ames, IA: Wiley-Blackwell; 2015:657-659.

Exotic Animal Emergencies for Technicians, Part 3: Evaluation of the Emergent Avian Patient

Christina Miller, RVT, BSc

Session #3509

Affiliation: From Clayton Park Veterinary Hospital, 278 Lacewood Dr, Halifax, NS, B3M 3N8, Canada.

Abstract: Birds commonly present to the veterinary hospital in a state of severe illness. This masterclass reviews emergency presentations in avian patients (focusing on Psittaciformes, Galliformes, and Columbiformes), the technician's role in the primary survey, obtaining vitals efficiently and safely, obtaining airway and vascular access, and proceeding with cardiopulmonary and cerebral resuscitation.

Introduction

The basic principles of triage and masking phenomenon are covered in "Exotic Animal Emergencies for Technicians, Part 1: Evaluation of the Emergent Exotic Mammal Patient."

Birds are challenging patients due to their propensity to mask clinical signs from all but the most observant caretakers, only appearing obviously ill when the patient has become relatively unstable: What pet owners perceive as acute illness is often a chronic illness that has gone unrecognized.[1,2] Masking phenomenon and a failure to recognize early, subtle signs of illness on the part of the pet owner has resulted in birds having a reputation for being fragile patients.

The Avian Primary Survey

When initially triaging a bird, the patient should first be observed covertly. If the patient is dull, dyspneic, or non-responsive to touch, proceed to the ABCDE protocol. The ABCDE protocol is a systematic approach to patient assessment in order of the most acutely life-threatening problems to least life-threatening problems.

At presentation, birds should always be bright and alert, and able to perch. Psittacine birds will especially be exhibiting exploratory behavior. Birds should always notice the veterinary staff and respond to your presence. The author uses a brief mental algorithm to determine if an avian patient that does not qualify for immediate assessment with the ABCDE protocol is safe to handle, see Table 1.

Immediate supportive care in the form of a low-stress environment, heat support, and oxygen (if indicated), is paramount for birds that appear overtly ill.[1]

Table 1. "Is this bird stable?" algorithm.

Response when the patient is approached	Assessment and recommended support
It fluffs up, but still watches you. It reacts to your movements. It exhibits an increased respiratory rate, but the effort is normal.	This patient may be stable. Proceed with exam; however, monitor for signs of distress. If handling causes signs of distress, cease the exam and provide supportive care ± anxiolysis.
It is fluffed up and ignores you. It exhibits an increased respiratory rate and effort or is dyspneic. It appears weak: Unable to perch, abnormal splay-legged posture, resting at the bottom of the cage.	This patient is unstable and may deteriorate with handling. Supportive care ± anxiolysis is strongly recommended before proceeding with hands-on exam.

A: Airway

Airway and breathing assessment are effectively performed simultaneously; however, signs of airway obstruction should be considered immediately, including: Collapse or weakness, dyspnea (open-mouth breathing, exaggerated keel movements with respirations, tail bobbing with respirations).[1] Upper airway obstructions can be easily identified as the glottis is relatively rostrally located, at the base of the tongue. There is no epiglottis. The nares should also be surveyed for obstruction. Birds are nasal breathers, and open-mouth breathing (aside from threat displays, vocalizations, or thermoregulation) is considered abnormal.[1,2] Orthopneic posture is abnormal and may also be adopted by birds: The wings may be held away from the body, the neck outstretched.

Oxygen therapy

Birds have the highest metabolic rates among vertebrates, and therefore, the highest rates of oxygen consumption per unit of body mass. Any avian patient presenting for illness, especially those with an obviously increased respiratory effort or dyspnea, could benefit from oxygen supplementation.[3]

Oxygen provided by face mask may be very stressful for conscious bird patients due to the restraint necessary to maintain positioning, or they may find the face mask aversive.[3] FiO_2 of 35-60% can be achieved by using a face mask.[4] An oxygen chamber provides support that does not necessitate ongoing patient handling, resulting in less stress for the bird. The FiO_2 in an oxygen chamber can vary from 30-60%,[3] which is comparable to a face mask. Commercial incubators that allow for monitoring and control of O_2, CO_2, temperature, and humidity are ideal.

B: Breathing

The quality and rate of the patient's respirations should be assessed. Rate and effort correlates to various physiologic parameters, including $PaCO_2$, PaO_2, intrapulmonary PCO_2, and pH.[3] Rates correlate to patient body weight: Birds weighing < 300 g typically have a normal respiratory rate of 40-60 rpm, and birds weighing between 400-1000 g typically range between 13-30 rpm.[5] Respiratory rate and effort will increase will handling, so a hands-off assessment of these signs should always be performed first during the primary survey. After handling, respiratory rate and effort should normalize within 3-4 minutes in a healthy bird.[5]

Birds lack a diaphragm and rely on the bellows-like system of air sacs and some accessory muscles to ventilate the lungs. An increased respiratory effort may at first be subtle, involving accessory muscles (the tail depressor and suprapubic muscles) that create a "tail bob" movement.[6] Open-mouth breathing and an orthopneic posture suggest severe dyspnea.

While breathing should be assessed first as a hands-off observation, air sac and lung auscultation should be reserved for a more stable patient (or, moribund patients may not perceive this handling as stressful). The presence and location of abnormal sounds should be noted. Pediatric or infant stethoscopes are invaluable for evaluating small avian patients.

C: Circulation

Circulation is challenging to evaluate in birds presenting in distress, as the handling required to examine peripheral vessels, mucosal surfaces, or even performing a cardiac auscultation can be very stressful.

The heart may be auscultated at either side of the cranial portion of the keel bone, or immediately caudal to the scapulae on the patient's dorsum.[1] See Table 2 for a summary of normal heart rates in a variety of avian species.

Table 2. Normal heart rates in select avian species.

Species	Heart rate (bpm)
African grey parrot, Congo (*Psittacus erithacus*)	340-600
Amazon parrot (*Amazona* spp.)	340-600
Canary (*Serinus* spp.)	~274
Cockatiel (*Nymphicus hollandicus*)	~206
Eclectus parrot (*Eclectus roratus*)	340-600
Finch (Estrildidae and Fringillidae)	~274
Lovebird (*Agapornis* spp.)	206-274
Macaw, large (*Ara* spp. and others)	~275
Macaw, small (*Diospittaca* sp. and others)	~389
Rock dove (*Columba livia*)	160-300

Source: Lafferty K, Pollock CG. Monitoring vital signs in exotic animal species. May 17, 2018. LafeberVet Website. Available at: https://lafeber.com/vet/monitoring-vital-signs-in-exotic-animal-species/. Accessed March 6, 2023.

Few mucosal surfaces are easy to access on an alert bird. The cloacal (proctodeal) mucosa can be gently everted using a sterile, lubricated cotton-tipped applicator to assess mucous membrane color and capillary refill time. Likewise, the conjunctiva can be gently assessed for color. The choana can also be assessed for color but requires an oral exam,[1] which may not be tolerated by a distressed bird. Some species have dark pigmentation of the choanal mucosa, so pallor cannot be appreciated in this location.

Capillary refill time can likewise be assessed at the ulnar or basilic vein, or the dorsal metatarsal vein.[1]

D: Disability

As prey species, birds are excellent at exhibiting masking phenomenon and in the unfamiliar exam room will put maximum effort into appearing healthy. Thus, birds that are presenting with an abnormal mentation are likely severely compromised. Flighted birds prefer perching, so an inability to perch (e.g., the patient presents at the bottom of a cage or carrier where a perch is provided) is very concerning.

Typical reflexes should be present such as the corneal reflex and menace response. The pupillary light reflex is present in birds although it is non-consensual, and may be overridden by voluntary muscular control of the iris.[8]

E: Exposure of environment

The patient can be surveyed for injury such as obvious skin lesions, fractures, bruises, etc.

Cloacal temperature is not routinely taken in birds due to their propensity to exhibit elevated body temperature due to the stress of transport, illness, handling, among other things. A low cloacal temperature (hypothermia) is significant in that it suggests severe hemodynamic compromise.[9] The author will typically obtain a cloacal temperature on an avian patient that is moribund or during anesthetic monitoring.

Note that birds with feather loss due to either feather destructive behaviors or avian circovirus, and birds with a poor body condition score and/or muscle condition score, are at a higher risk of becoming hypothermic when compromised due to the lack of insulation or heat-generating muscle mass. These patients should be provided with heat support upon presentation unless there is a contraindication (e.g., presenting for heat stress).

Airway Access

Airway access should be obtained in apneic patients. Dyspneic birds with a possible lower airway obstruction may require air sac cannulation to secure an airway; however, this technique will not be discussed here.

Orotracheal intubation is relatively straightforward in birds. Birds lack an epiglottis, and the glottis is located at the base of the tongue. The fleshy tongue of psittacine birds may obscure visualization of the glottis, so the tongue may be gently retracted to better expose the tracheal opening. Birds have complete tracheal rings, so uncuffed tubes are preferred to avoid tracheal trauma. If a cuffed tube is used, the cuff should not be inflated. The trachea narrows a short distance from the glottis in many species,[2,10] so the width of the glottis is not a perfect predictor for the width of the lumen of the caudal trachea (i.e., the endotracheal tube selected should generally be narrower than the width of the glottis[10]). Cole tubes may reduce potential trauma.[2] Take note that the smaller the diameter of the tube, the greater the airway resistance encountered as per Poiseuille's law.[11] Because of this, it could be advantageous in very small patients (< 100 g body weight) to switch to a face mask once spontaneous respirations return.

Vascular Access: Intravenous

Anxiolysis, sedation, and/or anesthesia is typically required to obtain vascular access in birds, unless the patient is moribund.[12] Avian skin appears thin and translucent; however, it is also tough: A dermal cut down may be necessary immediately parallel to the vein to help prevent the intravenous (IV) catheter from burring.[13] Any site selected should ideally be plucked of feathers (if present) to allow for aseptic preparation.

It is important to note that birds can exsanguinate themselves if an IV catheter is displaced or removed by the patient, either due to excessive movement, or the patient's curious nature. Intraosseous (IO) catheters may be more suitable for some patients for these reasons. IV catheters should be appropriately secured and patients monitored closely if conscious.

Basilic or ulnar vein

This vein is located on the ventral surface of the wing, immediately distal to the elbow joint. The catheter is inserted in a proximal direction, and secured with either a transparent adhesive dressing or a tape butterfly around the hub of the catheter with simple dermal sutures. The wing can be stabilized with a figure-of-8 bandage, and distractor tabs (tape) added for conscious patients.[9,13]

Medial metatarsal vein

This vein is located on the medial surface of the tibiotarsus, immediately proximal to the tibiotarsus-tarsometatarsal joint, at the junction of scaled to scaleless skin. It is an excellent choice in birds with a body weight over 300 g. It can be secured with porous tape.[9,13]

Jugular vein

This vein is accessible in birds as small as 75 g in body weight. The right side is typically much larger than the left side and is generally preferred for catheterization (or venipuncture) attempts. The jugular can be found in a naturally featherless tract of skin (apteryla) in most species.[9,13] The jugular is somewhat mobile in the soft tissue, and the cervical air sac is sometimes precariously positioned immediately medial to the vein: A gentle digital manipulation of the area can improve its placement.

Vascular Access: Intraosseous

Venous access may not be possible in very small, dehydrated, or hypotensive patients, and intraosseous (IO) catheters are an acceptable alternative in many animals. Confirmation of placement requires orthogonal radiographic views. IO catheters can be maintained in place for up to 72 hours.[13]

Contraindications to IO catheter placement include the presence of a skin infection at the insertion site, pre-existing fracture of the target bone, osteomalacia (typically due to nutritional or renal secondary hyperparathyroidism), or pre-existing osteomyelitis of the target bone.[13] Local anesthesia and deep sedation or anesthesia for IO catheter placement is preferred unless the patient is moribund due to the stress and pain associated with placement. Ongoing pain management is advised for IO catheters used for fluid and/or drug infusion.[14]

Placement sites vary between species: The humerus, and in some species the femur, are pneumatic bones and should not be used for IO catheter placement. Infusing drugs or fluid into pneumatic bones can result in pneumonia, air sacculitis, or asphyxiation.[13] Needle selection is based on the following measurements: The internal needle diameter should equal ⅓-⅔ the medullary cavity's thinnest diameter of the diaphysis. The ideal needle length should equal ⅓-½ the length of the medullary cavity.[13]

Distal ulna

The dorsal tubercle of the distal ulna is localized, and the metacarpals slightly pronated to expose the insertion site. Once placed, flushing the catheter should result in blanching of the ulnar vein. A figure-of-eight bandage of the wing helps to stabilize this catheter site for prolonged use.[13]

Proximal tibiotarsus

The leg is flexed at the knee to expose the cnemial crest of the tibiotarsus. The needle is introduced in a disto-medial direction immediately posterior to the patellar tendon insertion site.[13] This site may be tolerated poorly in conscious birds, as standing and/or perching on the catheterized limb is likely uncomfortable.

Cardiopulmonary Cerebral Resuscitation

The goal of cardiopulmonary cerebral resuscitation (CPCR) in birds is the same as in other species: Return of spontaneous circulation (ROSC) and ventilation. The knowledge gap between birds and even other exotic taxa is significant, and these animals present many challenges during CPCR: Their metabolic rate and thus oxygen demands are higher than other taxa, drug doses are mostly extrapolated from other species, the small body size of many patients make interventions challenging, and the presence of the keel renders chest compressions less effective.[15] The following discussion is informed by the RECOVER Initiative (Reassessment Campaign on Veterinary Resuscitation), found at https://recoverinitiative.org.

As discussed in "Exotic Animal Emergencies for Technicians, Part 1: Evaluation of the Emergent Exotic Mammal Patient," it is important to operate with protocols for basic life support (BLS), advanced life support (ALS), and do not resuscitate (DNR) orders for individual patients. A basic algorithm for CPCR is found in Part 1, and deviations from this algorithm will be discussed below.

Ventilation

Ventilation with 100% oxygen is advised, at a rate of 10-20 breaths/min. Ventilation should be held for one second at a maximum airway pressure of 15-20 cmH_2O.[15]

Chest compressions

The large sternum shields the heart and is non-pliable. Chest compressions in birds function on the "thoracic pump" principle, where increasing intrathoracic pressure via compressions is presumed to move blood from the heart to emulate circulation.[16] Compressions over the lateral or dorsal body (approximating the anatomic location of the heart) may be more effective than ventral compressions over the keel.[15] Compressions at a rate of 120 compressions/min, or approximating a normal heart rate for the species, is advised.[15]

Electrocardiography

Electrocardiogram (ECG) leads may be placed at the wing and thigh skin adjacent to the body wall.[15] Avian taxa exhibit some differences in the normal electrocardiogram compared to mammals:

- The mean electrical axis is negative in most species.
- The S wave is very prominent, except in meat poultry (e.g., *Gallus gallus domesticus*).
- In smaller bird species with a relatively high heart rate, P on T phenomenon is common, and ST segment appears short (i.e., ST slurring).[15]

In birds, there are several common rhythms that can be identified as in mammals:

- Sinus bradycardia: The sinus node signals appropriate cardiac muscle depolarization; however, the heart rate is slow.
- Asystole: Cessation of electrical and mechanical activity of the heart.
- Pulseless electrical activity (PEA): There is sufficient electrical activity in the heart; however, no pulse is detected (therefore, no perfusion occurring).
- Ventricular tachycardia: A fast heart rate that originates in the ventricles.

Capnometry

In small patients (< 100 g body weight), side-stream capnography may have a sampling volume greater than the patient's tidal volume, resulting in the sample becoming diluted with fresh gas, although monitoring trends can still be helpful for patient assessment.[17] A return to normal $ETCO_2$ values observed during anesthesia may indicate ROSC (30-45 mmHg).[18]

Conclusions

Birds are challenging patients due to their unique anatomy and physiology. There is a lack of evidence-based recommendations for CPCR for avian species.

References

1. Doss GA, Mans C. Birds: history and clinical exam. In: Graham JE, Doss GA, Beaufrère H, eds. *Exotic Animal Emergency and Critical Care Medicine.* Hoboken, NJ: Wiley-Blackwell; 2021;435-444.

2. Doneley B. The clinical examination. In: Samour J, ed. *Avian Medicine.* 3rd Ed. St. Louis, MO: Elsevier; 2016:49-72.

3. Beaufrère H. Birds: oxygen therapy. In: Graham JE, Doss GA, Beaufrère H, eds. *Exotic Animal Emergency and Critical Care Medicine.* Hoboken, NJ: Wiley-Blackwell; 2021:450-456.

4. Manning AM. Oxygen therapy and toxicity. *Vet Clin Small Anim Pract.* 2002;32(5):1005-1020.

5. Hoppes SM. Merck Veterinary Manual: Management of Pet Birds - Exotic and Laboratory Animals. Available: at https://www.merckvetmanual.com/exotic-and-laboratory-animals/pet-birds/management-of-pet-birds. Accessed March 6, 2023.

6. Baumel J, Wilson J, Bergren D. The ventilatory movements of the avian pelvis and tail: function of the muscles of the tail region of the pigeon (*Columba livia*). *J Exp Biol.* 1990;151(1):263-277.

7. Lafferty K, Pollock CG. Monitoring vital signs in exotic animal species. May 17, 2018. LafeberVet Website. Available at: https://lafeber.com/vet/monitoring-vital-signs-in-exotic-animal-species/. Accessed March 6, 2023.

8. Williams D. Ophthalmology. In: Harrison GJ, Lightfoot TL, eds. *Clinical Avian Medicine.* Palm Springs, FL: Spix Publishing, Inc.; 2011:673-694.

9. Beaufrère H. Medical, nursing, and cosmetic procedures. In: Samour J, ed. *Avian Medicine.* 3rd Ed. St. Louis, MO: Elsevier; 2016:204-245.

10. Doneley B. *Avian Medicine and Surgery in Practice: Companion and Aviary Birds.* 2nd Ed. Boca Raton, FL: CRC Press; 2016.

11. Li Y, Lu Y, Tan X, Zhang Z, et al. Effect of endotracheal tube size on airway resistance and dynamic lung compliance. *Med (Baltimore).* 2022;101(43):e31410. doi:10.1097/MD.0000000000031410.

12. Dubé C, Dubois I, Struthers J. Intravenous and intraosseous fluid therapy in critically ill birds of prey. *J Exot Pet Med.* 2011;20(1):21-26.

13. Schnellbacher R, Beaufrère H. Birds: catheterization and venipuncture. In: Graham JE, Doss GA, Beaufrère H, eds. *Exotic Animal Emergency and Critical Care Medicine.* Hoboken, NJ: Wiley-Blackwell; 2021:457-467.

14. Cummings CO. Are we providing adequate analgesia for intraosseous infusion in exotic animal practice? *J Exot Pet Med.* 2023;44:27.

15. Kabakchiev C, Beaufrère H. Birds: catheterization and venipuncture. In: Graham JE, Doss GA, Beaufrère H, eds. *Exotic Animal Emergency and Critical Care Medicine.* Hoboken, NJ: Wiley-Blackwell; 2021; 477-487.

16. Costello MF. Principles of cardiopulmonary cerebral resuscitation in special species. *Semin Avian Exot Pet Med.* 2004;13(2):132-141.

17. Ozeki L, Caulkett N. Monitoring. In: West G, Heard D, Caulkett N, eds. *Zoo Animal and Wildlife Immobilization and Anesthesia.* 2nd Ed. Ames, IA: Wiley-Blackwell; 2015:149-165.

18. Sanchez Migallon-Guzman D, Beaufrère H. Birds: avian pain management and anesthesia. In: Graham JE, Doss GA, Beaufrère H, eds. *Exotic Animal Emergency and Critical Care Medicine.* Hoboken, NJ: Wiley-Blackwell; 2021:488-502.

Index

Symbols

3D
imaging 224
model 809
printing 809
3-way stop cock 65

A

abdominocentesis 2
ABV antibody 265
accident 254
acute pain 908
Adenium obesum 258
African gray parrot 261
after hours 735, 738
agency 88
AI 244, 245
air-sac cannula 63
airway access 928, 946, 956
alfaxalone 787
algal blooms 61
Amazon parrot 39, 261
analgesia 185
analyte 221
anesthesia 64, 799, 884
Animal and Plant Health Inspection Service 30
animal science 895
Animal Welfare Act 30
anorexia 42, 240
antibiotics 781
anticoagulant rodenticides 832
apex predator 254
APHIS 30
Apis mellifera 867
appetite 42, 240
stimulation 42, 237
Aquila chrysaetos 53
aspergillosis 233
atherosclerosis 200
atorvastatin 38, 39
augmentative alternative communication 45
average client transaction 750
avian 100, 185, 239, 908, 956
analgesia 899
anesthesia 899
behavior 271
bones 117
bornavirus 167, 265
ganglioneuritis 167
influenza 244, 247
nutrition 221
wing wrap 222
AWA 30

B

backyard chickens 247
bald eagle 53
ballistic 254
ball python 872
bandaging 100
barn owl 269
bat 785
beak 224
trimming 224
bearded dragon 245, 872
beekeeping 867
behavior 123, 256
behavioral
correspondence 45
cues 908
biodiversity 800
biosecurity 246, 820
bird 69, 94, 812, 956
bird of prey 133
bladder stones 839
bleomycin 853
blood
collection 2
lipids 38
boa constrictor 872
Board of Veterinary Medicine 776
box turtle 872
breeder 133
brevetoxicosis 61
butorphanol 66, 94

C

Cacatuidae 139
Candida 228
cannabinoids 185
cannabis 21
capromorelin 42, 237
car 254

cardiac
disease 200
glycoside 258
steroid 258
cardiopulmonary cerebral resuscitation 928, 946, 956
cardiovascular disease 53
Caribbean 785
catheter 2, 912
caudal vein 912
CBC 245
chameleon 872
chelonian 946
chicken 44, 66
chinchilla 918, 923, 928
Chlamydia sp. 69
chlamydiosis 69
C. hominis 215
chronic pain 908
ciclopirox 228
cisplatin 853
cladistics 895
climate changes 807
clinical signs 242
cloaca 139
coach 773
coaptation 222
cockatiel 215, 262
cockatoo 139, 261
coelomic masses 267
cohesive team 773
Columba livia domestica 235
Columbia livia 41
columbiform 956
Columbiformes 956
common eider 228
communication 765
board 45
complaint 776
compounding 742
computed tomography 63, 239
computer-aided design 809
conflict 765
conservation 106, 254, 800
conservative management 239
controlled drugs 755
copper 261
level 261
toxicity 261
corn snake 872
corticosterone 271
C. parvum 215
CPCR 928, 946, 956
crop
tubing 2
wash 2
Cryptosporidium 215
CT 63, 239, 809
culture 765
cyproheptadine 42

D

degu 928
desert rose 258
dexdomitor 66
dexmedetomidine 94
diabetes mellitus 43
diagnostic 245
imaging 181
diagnostics 133, 812
diazepam 240
diet 11, 221, 262
dinoflagellate 61
DNA next-generation sequencing 264
dogma 839
domestic pigeon 235, 237
drug 742
diversion 755
duck 228
dystocia 123

E

Eastern long-necked turtle 781
eco-tourism 785
ECT 853
egg 44
binding 154
eggshell 44
electrochemotherapy 853
electropermeabilization 853
electroporation 853
ELISA 233
emergency
cases 735, 738
medicine 928, 946, 956
practice 738
room 735

empirical sociologic veterinary studies 259
emu 244
encephalitis vaccine 244
end-tidal carbon dioxide 65
enrichment 23, 88, 256
Entyce® 237
epidemic 755
equipment 784
evolution 895
exotic
animal medicine 731
companion mammal 928
emergency medicine 812
mammal 918, 928
pets 784, 799
expenses 750

F
feather
destructive behavior 123, 265
plucking 839
fecal microbiome 262
feeding 923
fees 750
ferret 918, 928
figure-of-8 222
flamingo 785
fluconazole 228
FluDETECT® 242
fluoroscopy 181
food intake 42, 240
foraging 256
fracture 6, 117, 222, 254
fungus 228
furosemide 839

G
gabapentin 235
galah 52
Galliformes 956
gallinaceous 956
Gallus gallus 44
gastrointestinal stasis 839
gavage 2
gecko 872
gerbil 928
geriatrics 200
gliotoxin 233
glucocorticoids 41
glucose 43
monitor 43
Goffin's cockatoo 45
golden eagle 53
granuloma 267
grapiprant 185
great horned owl 62
grey parrot 256
guinea pig 918, 923, 928

H
Haliaeetus leucocephalus 53
hamster 928
harlequin duck 228
heart 44
hemangiolipoma 268
hematology 245
heterophil 245
highly pathogenic AI 244
avian influenza 242, 246
Hispaniolan Amazon parrot 64
honey bee 867
HPAI 242, 244, 246, 247
husbandry 106
hydromorphone 62
hydromorphone-3-glucuronide 62
hypothermia 64

I
ILE 21
image analysis 245
impaction 156
indexing 787
infectious disease 133, 820
intramuscular injections 2
intravenous 912
catheter 65
lipid emulsion 21
intubation 65, 799
invasive species 807
IV catheter placement 918

K
Klebsiella 264
pneumoniae 264

L

lead 832
leadership 765, 773
lipoma 268
liposarcoma 268
little corella 52
liver 44
lizard 872, 912, 946
local communities 797
long-tailed duck 228

M

macaw 63, 258, 261, 264
machine learning 245
Macrorhabdus 52
maggot 794
magnetic resonance imaging 809
malnutrition 221
mammal 812
management 773
medication 742
medicinal maggot 794
methodology 848
microsurgery 844
midazolam 94, 240
minor species 787
mirtazapine 42
MiSEQ next-generation sequencing 264
mouse 928
MRI 809
myelolipoma 268

N

nalbuphine 66
nasal flush 2
necropsy 53
neoplasia 156, 269
nesting 785
neuropathic pain 908
Newcastle disease 832
New World vulture 106
next-generation DNA sequencing 262
nontraditional companion animals 807
North American river otter parasites 832
NSAID 185
NTCAs impact 807
nursing 912
nutrition 11, 42, 240, 923
nutritional support 923
Nymphicus hollandicus 262

O

ocular 254
Old World vulture 106
opioid 185, 755
oral growths 264
orange-winged Amazon parrots 39
orthopedics 6
orthopedic surgery 117
osteoarthritis 200
outreach 797
owl 254

P

packed cell volume 60
pain
- assessment 908
- management 133

parasitism 254
parrot 23, 69, 88, 133, 139, 200, 221, 256, 261
pathogen 820
pet birds 259
pharmaceuticals 742
pharmacokinetics 39, 62, 781
phylogeny 895
physical exam 94
pigeon 41
Pionites leucogaster 21
Pionus parrot 261
plastronotomy 784
pleurodesis 63
poultry 247
practice management 731
pradofloxacin 781
prednisolone 267
preen gland 156
prioritization 812
prognosis 60
prolapse 123, 139
psittaciformes 256
Psittaciformes 200, 268, 956

psittacine 94, 956
  bird 11, 123, 133, 139, 256, 271
Psittacus erithacus 256
publication 848

Q

quarantine 246
queen 867

R

rabbit 781, 794, 844, 884, 918, 923, 928
  anesthesia 884
radiation therapy 269
radiography 44
Ramphastos toco 43
Ranavirus 832
raptor 133, 242, 246, 254, 269
rat 928
recession 750
red-eared slider 872
regulations 755
rehabilitation 60, 106
renal disease 200
reporting 848
reproduction 133
reproductive
  disease 123, 150, 200
reptile 245, 812, 872, 912, 946
  hospital care 872
research 848
revenue 750
rewilding 800
rhabdomyolysis 244
RNA next-generation sequencing 264
rodent 844
rosuvastatin 38
ruptured air-sac 63

S

salpingohysterectomy 150
salt 221
sea turtle 785
sedation 66, 94
skin grafts 100
small
  exotic mammal 799, 923
  psittacine birds 154
snake 912, 946
snuffles 839
socialization 23
Spironucleus 52
splint 222
squamate 912, 946
squamous cell carcinoma 269
squirrel 794
statin drugs 38
stereotypy 23, 88
stomatitis 264
stress 94, 271
stress-associated behaviors 235
subcutaneous injections 2
sugar glider 928
supervising 773
supervisor 773
surf scoter 228
surgery 63, 150, 844
synsacral fracture 239
synsacrum 239

T

taxonomy 895
team 773
  building 765
technique tutor 918
technology 23, 88
tetrahydrocannabinol 21
THC 21
therapeutics 41
toco toucan 43
tortoise 872, 946
total solids 60
toxic change 245
training 738
trauma 100, 254
triage 832, 928, 946, 956
trimethoprim sulfamethoxazole 61
turtle 946
Tyto alba 269

U

uropygial gland 156

V

vacuum-assisted closure 100
valvular pathology 53
vascular access 912, 918, 928, 946, 956
vehicular 254

ventral coccygeal vein 912
veterinarian patient owner relationship 259
veterinary
    care 807
    clinical practice 259
    technician 884
VFD 867
visceral pain 908
vulture 106

W

wasting syndrome 52
welfare 30
white bellied caique 21
white-winged scoter 228
wildlife 60, 794, 820, 832
    rehabilitation 797
worker 867
wound 794
    scare 100
writing 848

X

xanthogranulomatous disease 267

Z

zoogeography 895
zoology 895
zoonoses 69
zoonosis 69
zoonotic disease 820

Association of Avian Veterinarians
PO Box 9
Teaneck, NJ 07666
USA

ISBN 978-1-7138-9471-1

9 781713 894711

CURRAN ASSOCIATES INC.
proceedings.com